Food Engineering Handbook

Food Engineering Handbook

Edited by **Lisa Jordan**

Callisto Reference

New York

Published by Callisto Reference,
106 Park Avenue, Suite 200,
New York, NY 10016, USA
www.callistoreference.com

Food Engineering Handbook
Edited by Lisa Jordan

International Standard Book Number: 978-1-63239-336-4 (Hardback)

Printed in China.

Contents

Preface

This book consists of several important topics which will be of interest to food engineering and related professions. Some of the topics discussed within the book include the importance of working in multidisciplinary teams, and the advantages of developing the field of food engineering by assimilating specialists from the industry, academia and government. It also elucidates issues related to the relevance of developing food engineering based upon approved and accepted principles, and the unmatched advantage of working as globally as possible in the understanding and development of food engineering principles.

Various studies have approached the subject by analyzing it with a single perspective, but the present book provides diverse methodologies and techniques to address this field. This book contains theories and applications needed for understanding the subject from different perspectives. The aim is to keep the readers informed about the progresses in the field; therefore, the contributions were carefully examined to compile novel researches by specialists from across the globe.

Indeed, the job of the editor is the most crucial and challenging in compiling all chapters into a single book. In the end, I would extend my sincere thanks to the chapter authors for their profound work. I am also thankful for the support provided by my family and colleagues during the compilation of this book.

Editor

Section 1

Characteristics of Foods

Chapter 1

Oxygen Scavengers: An Approach on Food Preservation

Renato Souza Cruz, Geany Peruch Camilloto and Ana Clarissa dos Santos Pires

Additional information is available at the end of the chapter

1. Introduction

Many foods are very sensitive for oxygen, which is responsible for the deterioration of many products either directly or indirectly. In fact, in many cases food deterioration is caused by oxidation reactions or by the presence of spoilage aerobic microorganisms. Therefore, in order to preserve these products, oxygen is often excluded.

Oxygen (O_2) presence in food packages is mainly due to failures in the packaging process, such as mixture of gases containing oxygen residues, or inefficient vacuum. Vacuum packaging has been widely used to eliminate oxygen in the package prior to sealing. However, the oxygen that permeates from the outside environment into the package through the packaging material cannot be removed by this method (Byun et al., 2011).

Modified atmosphere packaging (MAP) is often used as an alternative to reduce the O_2 inside food packaging. However, for many foods, the levels of residual oxygen that can be achieved by regular (MAP) technologies are too high for maintaining the desired quality and for achieving the sought shelf-life (Damaj et al., 2009). The use of oxygen scavenging packaging materials means that oxygen dissolved in the food, or present initially in the headspace, can potentially be reduced to levels much lower than those achievable by modified atmosphere packaging (Zerdin et al., 2003).

In this context, research and developments in the food packaging area have been conducted, aiming to eliminate residual O_2. One of the most attractive subjects is the active packaging concept. Active packaging includes oxygen and ethylene scavengers, carbon dioxide scavengers and emitters, humidity controllers, flavor emitters or absorbers and films incorporated with antimicrobial and antioxidant agents (Santiago-Silva et al., 2009).

The most used active packaging technologies for food are those developed to scavenge oxygen and were first commercialized in the late 1970s by Japan's Mitsubishi Gas Chemical

Company (Ageless®). In the case of gas scavengers, reactive compounds are either contained in individual sachets or stickers associated to the packaging material or directly incorporated into the packaging material (Charles et al., 2006).

The first patent of an absorber was given in 1938 in Finland. This patent was developed to remove the residual oxygen in headspace of metallic packaging. The method of introduction of hydrogen gas in the packaging to react with oxygen in palladium presence was commercialized in 1960s however this method has never been popularized and well accepted because the hydrogen was unstable during manipulation and storage and, furthermore, it is expensive and unwholesome (Abe and Kondoh, 1989). Recently, more than 400 patents were recorded, mainly in EUA, Japan and Europe, due the great interest by absorbers use (Cruz et al., 2005).

Oxygen scavengers are becoming increasingly attractive to food manufacturers and retailers and the growth outlook for the global market is bullish. Pira International Ltd estimated the global oxygen scavenger market to be 12 billion units in Japan, 500 million in the USA and 300 million in Western Europe in 2001. This market was forecast to grow to 14.4 billion in Japan, 4.5 billion in the USA and 5.7 billion in Western Europe in 2007 (Anon., 2004). In addition, Pira International Ltd. estimated the global value of this market in 2005 to be worth $588 million and has forecast this market to be worth $924 million in 2010. The increasing popularity of oxygen scavenging polyethylene terephthalate (PET) bottles, bottle caps and crowns for beers and other beverages has greatly contributed to this impressive growth (Anon., 2005).

Overall, oxygen absorbing technology is based on oxidation or combination of one of the following components: iron powder, ascorbic acid, photosensitive polymers, enzymes, etc. These compounds are able to reduce the levels of oxygen to below 0.01%, which is lower than the levels typically found (0.3-3%) in the conventional systems of modified atmosphere, vacuum or substitution of internal atmosphere for inert gas (Cruz et al., 2007). A summary of the most important trademarks of oxygen scavenger systems and their manufacturers is shown in Table 1.

An appropriate oxygen scavenger is chosen depending on the O_2-level in the headspace, how much oxygen is trapped in the food initially and the amount of oxygen that will be transported from the surrounding air into the package during storage. The nature of the food (e.g. size, shape, weight), water activity and desired shelf-life are also important factors influencing the choice of oxygen absorbents (Vermeiren et al., 2003).

Oxygen scavengers must satisfy several requirements such as to be harmless to the human body, to absorb oxygen at an appropriate rate, to not produce toxic substances or unfavorable gas or odor, to be compact in size and are expected to show a constant quality and performance, to absorb a large amount of oxygen and to be economically priced (Nakamura and Hoshino, 1983; Abe, 1994; Rooney, 1995).

The most well known oxygen scavengers take the form of small sachets containing various iron based powders containing an assortment of catalysts. However, non-metallic oxygen scavengers have also been developed to alleviate the potential for metallic taints being imparted to food products and the detection of metal by in-line detectors. Non-metallic scavengers include those that use organic reducing agents such as ascorbic acid, ascorbate salts or catechol. They also include enzymatic oxygen scavenger systems using either glucose oxidase or ethanol oxidase (Day, 2003).

Company	**Trade Name**	**Type**	**Principle/Active substances**
Mitsubishi Gas Chemical Co., Ltd. (Japan)	Ageless	Sachets and Labels	Iron based
Toppan Printing Co., Ltd. (Japan)	Fresilizer	Sachets	Iron based
Toagosei Chem. Ind. Co. (Japan)	Vitalon	Sachets	Iron based
Nippon Soda Co., Ltd. (Japan)	Seaqul	Sachets	Iron based
Finetec Co., Ltd. (Japan)	Sanso-cut	Sachets	Iron based
Toyo Pulp Co. (Japan)	Tomatsu	Sachets	Catechol
Toyo Seikan Kaisha Ltd. (Japan)	Oxyguard	Plastic Trays	Iron based
Dessicare Ltd. (US)	O-Buster	Sachets	Iron based
Multisorb technologies Inc. (US)	FreshMax	Labels	Iron based
	FreshPax	Sachets	Iron based
Amoco Chemicals (US)	Amosorb	Plastic film	Unknown
Ciba Specialty chemicals (Switzerland)	Shelfplus O2	Plastic film	Iron based
W.R. Grace and Co. (US)	PureSeal	Bottle crowns	Ascorbate/metallic salts
	Darex	Bottle crowns, bottle	Ascorbate/sulphite
CSIRO/Southcorp Packaging (Australia)	Zero2	Plastic film	Photosensitive dye/organic compound
Cryovac Sealed Air Co. (US)	OS1000	Plastic film	Light activated scavenger
CMB Technologies (UK)	Oxbar	Plastic bottle	Cobalt catalyst/nylon polymer
Standa Industrie (France)	ATCO	Sachets	Iron based
	Oxycap	Bottle crowns	Iron based
	ATCO	Lables	Iron based
Bioka Ltd. (Finland)	Bioka	Sachets	Enzyme based

Table 1. Some manufacturers and trade names of oxygen scavengers (Ahvenainen and Hurme, 1997; Day, 1998; Vermeiren et al., 1999)

Structurally, the oxygen scavenging component of a package can take the form of a sachet, label, film (incorporation of scavenging agent into packaging film) (Figure 1), card, closure liner or concentrate (Suppakul et al., 2003).

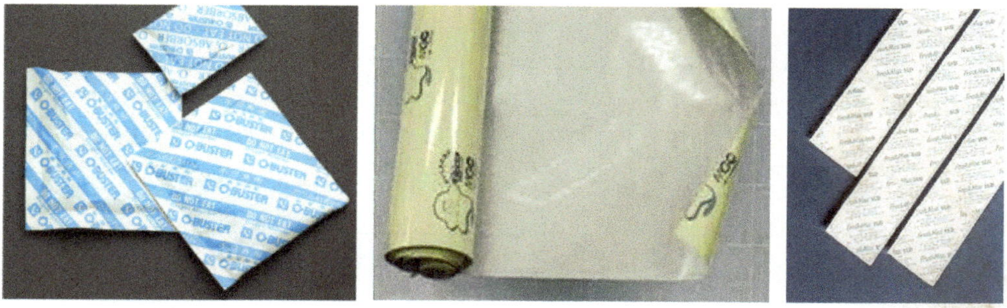

(a) (b) (c)

Figure 1. Oxygen scavengers: (A) O-Buster® sachet, (B) OMAC® film and (C) FreshMax™ SLD label.

Although the performance of oxygen-absorbing sachets was quite satisfactory for a wide range of food storage conditions, a number of limitations to their use in practice were recognized. The esthetics of inserts, coupled with a concern about possible ingestion or rupture, as well as their unsuitability for use with liquid foods, drove researchers to seek package-based solutions (Rooney, 2005). The incorporation of scavengers in packaging films is a better way of resolving sachet-related problems. Scavengers may either be imbedded into a solid, dispersed in the plastic, or introduced into various layers of the package, including adhesive, lacquer, or enamel layers (Ozdemir and Floros, 2004). In general, the speed and capacity of oxygen-scavenging systems incorporated in the packaging materials are considerably lower than those of (iron-based) oxygen scavenger sachets and labels (Kruijf et al., 2002).

For an oxygen scavenger sachet to be effective, some conditions have to be fulfilled (Nakamura and Hoshino, 1983; Abe, 1994; Smith, 1996). First of all, packaging containers or films with a high oxygen barrier must be used, otherwise the scavenger will rapidly become saturated and lose its ability to trap O_2. Films with an oxygen permeability not exceeding 20 ml/m^2.d.atm are recommended for packages in which an oxygen scavenger will be used. Secondly, for flexible packaging heat sealing should be complete so that no air invades the package through the sealed part. Finally, an oxygen scavenger of the appropriate type and size must be selected. The appropriate size of the scavenger can be calculated using the following formulae (Roussel, 1999; ATCO® technical information, 2002). The volume of oxygen present at the time of packaging (*A*) can be calculated using the formula:

$$A = \frac{(V - P) \text{ x } [O_2]}{100}$$

where *V* is the volume of the finished pack determined by submission in water and expressed in ml, *P* is the weight of the finished pack in g and $[O_2]$ is the initial O_2 concentration in package (= 21% if air).

In addition, it is necessary to calculate the volume of oxygen likely to permeate through the packaging during the shelf-life of the product (*B*). This quantity in ml may be calculated as follows:

$$B = S \text{ x } P \text{ x } D$$

where S is the surface area of the pack in m^2, P is the permeability of the packaging in ml/m^2/24h/atm and D is the shelf-life of the product in days.

The volume of oxygen to be absorbed is obtained by adding A and B. Based on these calculations, the size of the scavenger and the number of sachets can be determined.

According Cruz et al. (2005), the scavengers may be used alone or combined with modified atmosphere. This association requires the equipments to apply the modified atmosphere and decreases the filling velocity. However, this technique is generally used in the market to reduce the oxygen to desirable levels.

Oxygen scavengers have attracted interest of food researchers, and then in this chapter we will discuss the principles involved in scavenge of O_2, as well the main applications and researches in this field of active food packaging.

2. Oxygen scavengers systems

Nowadays, there are many systems of oxygen scavengers based on metallic and non-metallic coumpounds. The mechanism of each system is described below.

2.1. Iron powder oxidation

The commercially oxygen scavengers available are in form of small sachets containing metallic reducing agents, such as powder iron oxide, ferrous carbonate and metallic platinum. The majority of these scavengers are based on the principle of iron oxidation in water presence. A self-reacting type contains moisture in the sachet and as soon as the sachet is exposed to air, the reaction starts. In moisture-dependent types, oxygen scavenging takes place only after moisture has been taken up from the food. These sachets are stable in open air before use because they do not react immediately upon exposure to air therefore they are easy to handle if kept dry (Vermeiren et al., 1999; Cruz et al., 2005). The action mechanism of oxygen scavenger based on iron oxidation is very complicated and is described by the following reactions.

$$Fe \rightarrow Fe^{2+} + 2\,e^-$$

$$\frac{1}{2}\,O_2 + H_2O + 2\,e^- \rightarrow 2\,OH^-$$

$$Fe^{2+} + 2\,OH^- \rightarrow Fe(OH)_2$$

$$Fe(OH)_2 + \frac{1}{4}\,O_2 + \frac{1}{2}\,H_2O \rightarrow Fe(OH)_3$$

According Shorter (1982), if the oxidation rate of the food product and the oxygen permeability of the packaging were known, it is possible to calculate the required iron amount to maintain the desirable oxygen level during the storage time. A rule of thumb is that 1 g of iron will react with 300 ml of O_2 (Labuza, 1987; Nielsen, 1997; Vermeiren *et al.*, 1999). The LD_{50} (lethal dose that kills 50% of the population) for iron is 16 g/kg body weight. The largest commercially available sachet contains 7 grams of iron so this would amount to only 0.1 g/kg for a person of 70 kg, or 160 times less than the lethal dose (Labuza and Breene, 1989).

Cruz et al. (2007) evaluate the efficiency of O-Buster® oxygen - absorbing sachets at relative humidity of 75%, 80% and 85% and different temperatures, 10 ± 2 ºC and 25 ± 2 ºC. They observed that oxygen absorption by the sachet increased as the relative humidity increased for both temperature. Therefore the oxygen - absorbing sachets were most active under 25 ± 2 ºC and 85 % relative humidity. At ambient condition (25 ± 2 ºC/75 % RH) the rate of oxygen absorbed was 50 ml/day and 18.5 ml/day for 10 ± 2ºC.

Some important iron-based O_2 absorbent sachets are Ageless® (Mitsubishi Gas Chemical Co., Japan), ATCO® O2 scavenger (Standa Industrie, France), Freshilizer® Series (Toppan Printing Co., Japan), Vitalon (Toagosei Chem. Industry Co., Japan), Sanso-cut (Finetec Co., Japan), Seaqul (Nippon Soda Co., Japan), FreshPax® (Multisorb technologies Inc., USA) and O-Buster® (Dessicare Ltd., USA).

2.2. Ascorbic acid oxidation

The ascorbic acid is another oxygen scavenger component which action based on ascorbate oxidation to dehydroascorbic acid. Most of these reactions is slow and can be accelerated by light or a transition metal which will work as catalyst, e.g., the copper (Cruz et al, 2005).

The ascorbic acid reduce the Cu^{2+} to Cu^{+} to form the dehydroascorbic acid (Equation I). The cuprous ions (Cu^{+}) form a complex with the O_2 originating the cupric ion (Cu^{2+}) and the superoxide anionic radical (Equation II). In copper presence, the radical leads to formation of O_2 and H_2O_2 (Equation III). The copper-ascorbate complex quickly reduces the H_2O_2 to H_2O (Equation IV) without the OH^- formation, a highly reactive oxidant. The following reactions show the process of oxygen absorber by ascorbic acid.

$$\mathrm{AA} + 2\,\mathrm{Cu}^{2+} \rightarrow \mathrm{DHAA} + 2\,\mathrm{Cu}^{+} + 2\,\mathrm{H}^{+} \tag{1}$$

$$2\,\mathrm{Cu}^{+} + 2\,\mathrm{O}_2 \rightarrow 2\,\mathrm{Cu}^{2+} + 2\,\mathrm{O}_2^{-} \tag{2}$$

$$2\,\mathrm{O}_2^{-} + 2\,\mathrm{H}^{+} + \mathrm{Cu}^{2+} \rightarrow \mathrm{O}_2 + \mathrm{H}_2\mathrm{O}_2 + \mathrm{Cu}^{2+} \tag{3}$$

$$\mathrm{H}_2\mathrm{O}_2 + \mathrm{Cu}^{2+} + \mathrm{AA} \rightarrow \mathrm{Cu}^{2+} + \mathrm{DHAA} + 2\,\mathrm{H}_2\mathrm{O} \tag{4}$$

These equations can be summarized as described below:

$$\mathrm{AA} + \frac{1}{2}\mathrm{O}_2 \rightarrow \mathrm{DHAA} + \mathrm{H}_2\mathrm{O},$$

where AA is the ascorbic acid and DHAA is the dehydroascorbic acid.

The total capacity of the O_2 absorption is determined by the amount of ascorbic acid. The complete reducing of 1 mol of O_2 requires 2 moles of ascorbic acid (Cruz et al., 2005).

Ascorbic acid and ascorbate salts are being used in the design of scavengers in both sachet and film technologies. A patent from Pillsbury describes the oxygen-reducing properties of these substances. The active film may contain a catalyst, commonly a transition metal (Cu, Co), and it is activated by water; therefore, this technology is specially indicated for aqueous food products, or when the packaged product is sterilized because the water vapor inside the autoclave is capable of triggering the scavenging process (Brody et al., 2001a).

2.3. Enzymatic oxidation (e.g., glucose oxidase and alcohol oxidase)

Some O_2-scavengers use a combination of two enzymes, glucose oxidase and catalase, that would react with some substrate to scavenge incoming O_2. The glucose oxidase transfers two hydrogens from the -CHOH group of glucose, that can be originally present or added to the product, to O_2 with the formation of glucono-delta-lactone and H_2O_2. The lactone then spontaneously reacts with water to form gluconic acid (Labuza and Breene, 1989; Nielsen, 1997). A negative factor of this process is the catalase presence, a natural contaminant found in the glucose oxidase preparation, since the catalase reacts with the H_2O_2 forming H_2O and O_2 and, therefore, decreasing the system efficiency. However, the glucose oxidase production without catalase is so expensive. The reactions can be expressed as follows:

$$2\text{ glucose} + 2\ O_2 + 2\ H_2O \rightarrow 2\text{ gluconic acid} + 2\ H_2O_2$$

where glucose is the substrate.

Since H_2O_2 is an objectionable end product, catalase is introduced to break down the peroxide (Brody and Budny, 1995):

$$2\ H_2O_2 + \text{catalase} \rightarrow 2\ H_2O + O_2$$

According the reaction, 1 mol of glucose oxidade reacts with 1 mol of O_2. So, in an impermeable packaging with 500 ml of headspace only 0.0043 mol of glucose (0.78 g) is necessary to obtain 0 % of O_2. The enzymatic efficiency depends on the enzymatic reaction velocity, the substrate amount and the oxygen permeability of the packaging.

Coupled enzyme systems are very sensitive to changes in pH, a_w, salt content, temperature and various other factors. Additionally, they require the addition of water and, therefore, cannot be effectively used for low-water foodstuffs (Graff, 1994). One application for glucose oxidase is the elimination of O_2 from bottled beer or wine. The enzymes can either be part of the packaging structure or put in an independent sachet. The immobilization occurs by different process, such as, adsorption and encapsulation. Both polypropylene (PP) and polyethylene (PE) are good substrates for immobilizing enzymes (Labuza and Breene, 1989). A commercially available O_2-removing sachet based on reactions catalyzed by food-grade enzymes is the Bioka O_2-absorber (Bioka, Finland). It is claimed that all components of the reactive powder and the generated reaction products are food-grade substances safe for both the user and the environment (Bioka technical information, 1999). The oxygen scavenger eliminates the oxygen in the headspace of a package and in the actual product in 12–48 hours at 20 ºC and in 24–96 hours at 2–6 ºC. With certain restrictions, the scavenger can also be used in various frozen products. When introducing the sachet into a package, temperature may not exceed 60ºC because of the heat sensitivity of the enzymes (Bioka technical information, 1999). An advantage is that it contains no iron powder, so it presents no problems for microwave applications and for metal detectors in the production line.

Besides glucose oxidase, other enzymes have potential for O_2-scavenging, including ethanol oxidase which oxidises ethanol to acetaldehyde. It could be used for food products in a wide aw range since it does not require water to operate. If a lot of oxygen has to be absorbed from the package, a great amount of ethanol would be required, which could cause an off-odour in the package. In addition, considerable aldehyde would be produced which could give the food a yoghurt-like odour (Labuza and Breene, 1989).

2.4. Unsaturated hydrocarbon oxidation

The oxidation of polyunsaturated fatty acids (PUFAs) is another technique to scavenge oxygen. It is an excellent oxygen scavenger for dry foods. Most known oxygen scavengers have a serious disadvantage: when water is absent, their oxygen scavenging reaction does not progress. In the presence of an oxygen scavenging system, the quality of the dry food products may decline rapidly because of the migration of water from the oxygen scavenger into the food. Mitsubishi Gas Chemical Co. holds a patent that uses PUFAs as a reactive agent. The PUFAs, preferably oleic, linoleic or linolenic, are contained in carrier oil such as soybean, sesame or cottonseed oil. The oil and/or PUFA are compounded with a transition metal catalyst and a carrier substance (for example calcium carbonate) to solidify the oxygen scavenger composition. In this way the scavenger can be made into a granule or powder and can be packaged in sachets (Floros *et al.*, 1997).

In many patent applications (Ackerley et al., 1998; Akkapeddi and Tsai, 2002; Barski et al., 2002; Cahill and Chen, 2000; Goodrich et al., 2003; Kulzick et al., 2000; Mize et al., 1996; Morgan et al., 1992; Roberts et al., 1996; Speer and Roberts, 1994; Speer et al., 2002), it was disclosed that ethylenic-unsaturated hydrocarbons, such as squalene, fatty acids, or polybutadiene, had sufficient commercial oxygen scavenging capacity to extend the shelf-life of oxygen-sensitive products. These unsaturated hydrocarbons, after being functionally terminated with a chemical group to make them compatible with the packaging materials, can be added during conventional mixing processes to thermoplastics such as polyesters, polyethylene, polypropylene, or polystyrene, and the films can be obtained using most conventional techniques for the plastic processing such as coinjection or coextrusion. 1,2-Polybutadiene is specially preferred because it exhibits transparency, mechanical properties, and processing characteristics similar to those of polyethylene. In addition, this polymer is found to retain its transparency and mechanical integrity, and exhibits a high oxygen-scavenging capacity (Roberts et al., 1996). Transition metal catalysts, such as cobalt II neodecanoate or octoate (Barski et al., 2002; Mize et al., 1996; Speer et al., 2002), are also included in the oxygen scavenger layer in order to accelerate the scavenging rate. Photoinitiators can also be added to further facilitate and control the initiation of the scavenging process. Adding a photoinitiator or a blend of photoinitiators to the oxygen-scavenging composition is a common practice, especially where antioxidants were added to prevent premature oxidation of the composition during processing and storage.

The main problem of this technology is that during the reaction between these polyunsaturated molecules and oxygen, by-products such as organic acids, aldehydes, or ketones can be generated that affect the sensory quality of the food or raise food regulatory issues (Brody et al., 2001a). Indeed, some of these compounds are used to determine the quality and shelf-life of fatty foodstuffs because they are intrinsically related to rancidity (Jo et al., 2002; Van Ruth et al., 2001). This problem can be minimized by the use of functional barriers that impede migration of undesirable oxidation products. This functional layer must provide a high barrier to organic compounds, but allow oxygen to migrate, and it has to be inserted between the food product and the scavenger layer. Another solution comes from the use of adsorber materials. Some polymers present inherent organic compound-scavenging properties. Others incorporate adsorbers within the polymer structure (i.e., silica gel, zeolites, etc). It has also been found that when the ethylenic unsaturation is contained within a cyclic group, substantially fewer by-products are produced upon oxidation as compared with analogous noncyclic materials. The Oxygen Scavenging Polymer developed by Chevron Chemical is an example of this kind of technology. This system is reported to scavenge oxygen without degrading into smaller, undesirable compounds. Ten percent of the polymer is a concentrate that contains a photoinitiator plus a transition metal catalyst that maintains the polymer in a nonscavenging state until triggered by ultraviolet (UV) radiation (Rooney, 1995).

Oxbar™ is a system developed by Carnaud-Metal Box (now Crown Cork and Seal) that involves cobalt-catalyzed oxidation of a MXD6 nylon that is blended into another polymer. This system is used especially in the manufacturing of rigid PET bottles for packaging of wine, beer, flavored alcoholic beverages, and malt-based drinks (Brody et al., 2001b).

Another O_2 scavenging technology involves using directly the closure lining. Darex® Container Products (now a unit of Grace Performance Chemicals) has announced an ethylene vinyl alcohol with a proprietary oxygen scavenger developed in conjunction with Kararay Co. Ltd. In dry forms, pellets containing unsaturated hydrocarbon polymers with a cobalt catalyst are used as oxygen scavengers in mechanical closures, plastic and metal caps, and steel crowns (both PVC and non-PVC lined). They reportedly can prolong the shelf life of beer by 25% (Brody et al., 2001b).

2.5. Immobilization of microorganisms in solid holders

At least two patents from the 1980s and 1990s describe the use of yeast to remove oxygen from the headspace of hermetically sealed packages. One patent, from enzyme manufacturer Gist Brocades, focused on the incorporation of immobilized yeast into the liner of a bottle closure (Edens et al., 1992). The other patent used the yeast in a pouch within the package (Nezat, 1985). The concept of the patents was that, when moistened, the yeast is activated and respires, consuming oxygen and producing carbon dioxide plus alcohol. In the bottleclosure application, any carbon dioxide and alcohol produced would enter the contents, in this case beer, without causing measurable changes in the product.

Other researchers proposed an alternative approach: the use of entrapped aerobic microorganisms, capable of consuming oxygen (Tramper et al., 1983; Doran and Bailey, 1986; Gosmann and Rehem, 1986 and Gosmann & Rehem 1988). Natural and biological oxygen scavengers, based on the use of microorganisms entrapped in a polymeric matrix, effective in preserving foods, safe to use, agreeable to consumer, inexpensive, environment friendly, could be a very interesting concept to modern food technology. In fact, the possibility to create a new package, having many desirable characteristics, is very promising, also taking into account the new consumers' demand for mildly preserved convenience foods, having fresh-like qualities and being environmental friendly. In the field of biotechnology, immobilization of whole cells is gaining increasing importance (Gosmann and Rehem, 1988). Alginate, agar, and gelatin (Tramper et al., 1983; Doran and Bailey, 1986; Gosmann and Rehem, 1986 and Gosmann and Rehem, 1988) have been successfully used. Unfortunately, the above study cannot be used for the development of a biological oxygenscavenger. In fact, the cycle life of a biological oxygen-scavenger film includes the entrapment of the microorganisms in an appropriate polymeric matrix (film manufacturing), the maintenance of the desiccated film till its use (film storage and distribution), and the re-hydration (film usage, obtained by putting the film in contact with the food).

Altiere et al. (2004) develop an environmental friendly oxygen-scavenger film using microorganisms as the active component. In particular, hydroxyethyl cellulose (HEC) and polyvinyl alcohol (PVOH) were used to entrap two different kinds of microorganisms: *Kocuria varians* and *Pichia subpelliculosa*. In this work a new method is proposed to produce oxygen-scavenger films using aerobic microorganisms as the "active compound". The manufacturing cycle of the investigated oxygen-scavenger film was optimized both to prolong the microorganisms viability during storage and to improve the efficiency of the film to remove oxygen from the package headspace. It was found that it is possible to store the desiccated film over a period of 20 days without monitoring any appreciable decrease of microorganism viability. It was also pointed out that the highest respiratory efficiency of the proposed active film is obtained by entrapping the microorganisms into polyvinyl alcohol, and by using the active film as a coating for a high humidity food.

2.6. Photosensitive dye oxidation

Another technique of oxygen absorption is a photosensitive dye impregnated onto a polymeric film. When the film is irradiated by ultraviolet (UV) light, the dye activates the O_2 to its singlet state, making the oxygen-removing reaction much faster (Ohlsson and Bengtsson, 2002).

Australian researchers have reported that reaction of iron with ground state O_2 is too slow for shelf-life extension. The singletexcited state of oxygen, which is obtained by dye sensitisation of ground state oxygen using near infra-red, visible or ultraviolet radiation, is highly reactive and so its chemical reaction with scavengers is rapid. The technique involves sealing of a small coil of ethyl cellulose film, containing a dissolved photosensitising dye and a singlet oxygen acceptor, in the headspace of a transparent package. When the film is illuminated with light of the appropriate wavelength, excited dye molecules sensitise oxygen molecules, which have diffused into the polymer, to the singlet state. These singlet oxygen molecules react with acceptor molecules and are thereby consumed. The photochemical reaction can be presented as follows (Vermeiren et al., 2003, Cruz et al., 2005).

$$\text{photon} + \text{dye} \rightarrow \text{dye}^*$$

$$\text{dye}^* + O_2 \rightarrow \text{dye} + O_2^*$$

$$O_2^* + \text{acceptor} \rightarrow \text{acceptor oxide}$$

This scavenging technique does not require water as an activator, so it is effective for wet and dry products. Its scavenging action is initiated on the processor's packaging line by an illumination-triggering process (Vermeiren et al., 2003).

Cryovac® 0S2000™ polymer based oxygen scavenging film has been developed by Cryovac Div., Sealed Air Corporation, USA. This UV light-activated oxygen scavenging film (Figure 2), composed of an oxygen scavenger layer extruded into a multilayer film, can reduce headspace oxygen levels from 1% to ppm levels in 4–10 days and is comparable in effectiveness with oxygen scavenging sachets. The OS2000™ scavenging films have applications in a variety of food products including dried or smoked meat products and processed meats (Butler, 2002).

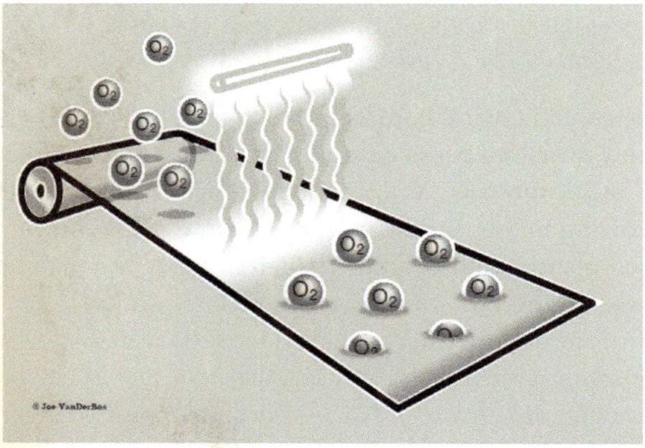

Figure 2. Light-activated oxygen scavenging films Cryovac® OS Films (Cryovac Food Packaging, Sealed Air Corporation, USA).

A similar UV light-activated oxygen scavenging polymer ZERO$_2$®, developed by CSIRO, Division of Food Science Australia in collaboration with Visy Pak Food Packaging, Visy Industries, Australia, forms a layer in a multilayer package structure and can be used to reduce discoloration of sliced meats. The active ingredient of the ZERO$_2$® is integrated into the polymer backbones of such common packaging materials as PET, polyethylene, polypropylene and EVA. The active ingredient is nonmetallic and is activated by UV light once it is incorporated into packaging material (Graff, 1998).

Another successful commercial example for use with meat is the OSP™ system (Chevron Philips Chemical Company, USA). The active substances of OSP™ systems are ethylene methacrylate and cyclohexene methacrylate, which need to be blended with a catalyst or photoinitiator in order to activate the oxygen scavenging mechanism.

2.7. Others

Sulphites have also been proposed as active substances for use, not only in sachets, but also in plastic gasket liners of bottle closures, as liquid trapped between sheets of flexible packaging material, or directly incorporated into plastic film structures to pack products such as wine or ketchup. For example, potassium sulfite is cited as an O_2 scavenger that can be readily triggered by the moist high temperature of the retorting process, and it also has enough thermal stability to pass unchanged through thermoplastic processes. However, any oxygen scavenger producing an end-product compound such as sulfur dioxide is viewed with concern because these by-products can exert a sensory change, or even an allergic effect on a susceptible consumer (Brody et al., 2001a).

Antioxidants, incorporated into flexible and thermoformable plastic packaging materials, are intended to reduce oxygen passage through the plastic structure or to remove oxygen from packages containing dry food products such as breakfast cereals (Floros et al., 1997). Butylated hydroxytoluene (BHT), a commonly used plastic antioxidant, has been proven to prolong the shelf-life of packed oat flakes (Miltz et al., 1989), but there is some concern related to the physiological effects of consuming it because it seems that BHT tends to accumulate in the adipose tissue (Wessling et al., 1998).

Nowadays, tendencies lead toward natural products and, therefore, natural antioxidants are being explored. There are a number of naturally occurring compounds that have antioxidant properties, including tocopherols, lecithin, organic acids and rosemary extracts. Among them, there is a growing interest in the use of vitamin E (also known as a-tocopherol) and vitamin C to be incorporated into polymers. Vitamin E has been marketed as a food-grade odor remover in packaging materials. For example, Laermer et al. (1996) showed that addition of vitamin E to high density polyethylene (HDPE)-ethylene/vinyl acetate (EVA)-HDPE flexible packaging system could reduce the "plastic" taste and preserve the fresh taste of breakfast cereals. Ho et al. (1994) showed that vitamin E was effective in reducing off-flavor compounds released from HDPE bottles. Vitamin E has somehow superior antioxidant behavior than BHT related to the off-flavor generation, stability and solubility, in polyolefins. The incorporation of vitamins E or C into the plastic material presents another advantage when compared with the addition of synthetic antioxidants because the possible migration of these compounds into the food not only does not produce adverse effects, but also improves the nutritional characteristics of the food product. However, being a bigger molecule than BHT, it is less mobile (Wessling et al., 1998). The amount of antioxidant added to the polymer must be controlled, as high levels of antioxidant incorporated into films can alter the polymer properties. Oxygen permeability of the film would increase and somemechanical properties of the film would change (Wessling, 2000).

Many patents have been issued for UV light activated oxygen scavengers, however, these UV activation steps reduces packaging line speeds and resulting in reduced profitability. In addition, there is a significant cost increases for oxygen scavenging films due to the high cost of photoinitiators and the operation and maintenance costs of the UV machine. Therefore the development of new oxygen scavenging systems that don' t require a UV activation step should be valuable to the food packaging industry.

Oxygen scavenging systems that utilise natural compounds as the basis for the oxygen scavenger may provide added benefit. One such potential compound is α-tocopherol which is a natural free radical scavenger with a positive consumer perception (Hamilton et al., 1997). It has been incorporated into the polymer materials as a stabilizer (Al-Malaika et al., 1999) and as an antioxidant in controlled release packaging to reduce the oxidation in food products (Byun et al., 2010, Lacoste et al., 2005, Siro et al., 2006 and Wessling et al., 2000).

The oxygen scavenging principle for the use of α-tocopherol was that oxygen free radicals can be produced by a transition metal. Oxygen free radicals are derived from the non-enzymatic reactions of oxygen along with transition metals (Bagchi and Puri, 1998). The transition metal activates oxygen to the singlet electron state oxygen. Then, this activated oxygen undergoes subsequent reduction to reactive oxygen species (ROS), which is an oxygen free radical. α-Tocopherol is a strong free radical scavenger which can also react irreversibly with singlet oxygen and produce tocopherol hydroperoxydienone, tocopherylquinone, and quinine epoxide (Choe and Min, 2006). α-Tocopherol can donate its electrons to scavenge the oxygen free radical. When the free radical gains the electron from α-tocopherol, it returns to its ground state and the free radical is eliminated.

There are two chemical reaction steps in this oxygen scavenging reaction as follow. In first step, oxygen free radicals are produced in the presence of a transition metal. In the second step, the oxygen free radicals are eliminated by receiving electrons from α-tocopherol (Smirnoff, 2005). Therefore, the presence of both the transition metal and α-tocopherol are essential conditions for the oxygen scavenging system. Furthermore, thermal processing can accelerate oxygen scavenging reaction.

$$\text{Initiation step: Oxygen} + \text{transition metal} \xrightarrow{\Delta} \text{oxygen free radical}$$

$$\text{Scavenging step: } \alpha - \text{tocopherol} + \text{oxygen free radical} \rightarrow \text{dimer or tocopherylquinone}$$

3. Practical application and researches

Oxygen scavengers have been studied for many researchers. There are many different types of oxygen scavengers that have been successfully applied to reducing food spoilage. In this section, we will discuss about the main and recent studies involving this technology.

Acid ascorbic is degraded to dehidroascorbic acid in the presence of oxygen, and the rate at which dehydroascorbic acid is formed is approximately first order with respect to the concentrations of ascorbic acid, oxygen, and metal catalysts. To evaluate the ascorbic acid loss in orange juice due to oxygen presence, the product was packed in oxygen scavenging film and oxygen barrier film. The initial concentration of ascorbic acid in the orange juice was 374 mg/l and this decreased by 74 and 104 mg/l after 3 days of storage at 25 ºC in the O_2 scavenger film and O_2 barrier film, respectively. The rapid loss in ascorbic acid was related to the high oxygen content initially present in the headspace and that dissolved in the juice. This content of oxygen could not be eliminated by O_2 barrier film. The authors concluded that the rapid removal of oxygen is an important factor to maintain the ascorbic acid content in orange juice over long storage times (Zerdin et al., 2003).

Altieri et al. (2004) purposed a new method to produce oxygen-scavenger film based on aerobic microorganisms (*Kocuria varians* and *Pichia subpelliculosa*). These microorganisms were entrapped into hydroxyethyl cellulose and polyvinyl alcohol and maintained their viability over 20 days. Both films were able to reduce oxygen content present into vials, however the highest respiratory efficient was obtained by entrapping the microorganism into polyvinyl alcohol.

Mohan et al. (2009) studied the effect of commercial oxygen scavenger in reducing the formation of biogenic amines during chilled storage of fish. It was observed that the O_2 scavenger was able to reducing the oxygen content of the pack up to 99.95% within 24 h and it extended the fish shelf-life up to 20 days compared to only 12 days for air packs. The biogenic amine content was significantly higher in air packs compared to the O_2 scavenger packs. Inhibition of enzymatic activity of food or bacterial decarboxylase activity and prevention of bacterial growth are essential to control the production of biogenic amine. The authors verified that the use of oxygen scavengers associated to chilled storage temperature helps in reducing the formation of biogenic amines in fish. In conclusion, the authors believe that by using O_2 scavengers, use of vacuum packing machine can be avoided.

The health benefits of the Mediterranean diet are often related to the consumption of olive oil. The container material has been related to influence the oi quality and sensorial characteristics. Glass is the most used material, however the use of polyethylene terephthalate (PET) bottle have increased, since it is transparent, recyclable, unbreakable, inexpensive and it has demonstrated the ability to preserve the characteristics of olive oil during its shelf-life. In the other hand, the permeability of the PET bottle to gases and vapour, such as oxygen limits the use of these containers to olive oil, since rancidity is the main cause of oil spoilage. In this context, Cecchi et al. (2010) evaluated the quality of extra-virgin olive oil packed into PET bottles containing or not commercial oxygen scavenger. Results of the 13-months experimental study indicate that the presence of the O_2 scavenger in the plastic matrix was able to better maintain the quality and authenticity attributes of the oil. A reduced flux of oxygen through the PET bottle keeps the level of primary and secondary oxidation products lower than that obtained in simple PET bottles stored under the same conditions. The active barrier reduces the olive oil antioxidant activity decline during storage. The chlorophylls content decay can only be prevented via the storage of the sample in the dark, while the active barrier is able to diminish the carotenes loss at the end of the shelf-life. On the whole, the performance of the tested innovative packageing proved to better preserve the extra virgin characteristics of the oil during its shelf-life.

A variety of oxygen scavengers have been commercialised for use in the food packaging industry. These oxygen scavenging system are used in various forms such as; sachets, plastic films, labels, plastic trays, and bottle crowns. The most used O_2 scavengers are based on the principle of iron oxidation.

Cruz et al. (2006) evaluated an O_2 absorbent system on the inhibition of microorganisms growth in fresh lasagna pasta during storage at 10 ± 2ºC. Fresh lasagna pasta was produced with and without potassium sorbate and acondicionated in high O_2 barrier bags containing an O_2-absorber sachet in the headspace. Three treatments were obtained: pasta with potassium sorbate, pasta without potassium sorbate packed with sachet and pasta without potassium sorbate packed without sachet (Figure 3). Oxygen absorbers were efficient in controlling the growth of filamentous fungi and yeasts, *Staphylococus* spp, total coliforms and *E. coli* in lasagna type fresh pasta without the addition of potassium sorbate, vacuum-packed in O_2-absorbent sachets, stored at 10 ± 2 °C. Therefore, the O_2-absorber sachet can be used as a hurdle technology, associated with vacuum packaging and applying the good manufacturing practices, to preserve lasagna pasta without additives.

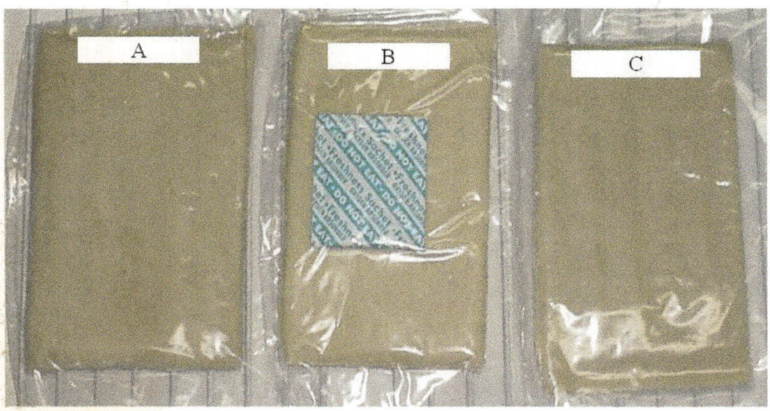

Figure 3. Fresh lasagna pasta vacuum-packed with potassium sorbate (A), without potassium sorbate and with oxygen scavenger (B) and without potassium sorbate.

However, nowadays, many consumers have a negative view of the term "iron-based." Therefore, Byun et al. (2011) studied the development of na oxygen scavenger using a natural compound: α-tocopherol. A natural free radical scavenger, α-tocopherol, and a transition metal in an oxygen scavenging system were evaluated as a possible oxygen scavenger. An initial, cup headspace oxygen content (%) of 20.9% was decreased to 18.0% after thermal processing and 60 days of storage at room temperature when the oxygen scavenging system containing α-tocopherol (500 mg) and transition metal (100 mg) was utilised. The oxygen content (%) decreased further to 17.1% when the amount of transition metal increased from 100 to 150 mg. The authors concluded that α-tocopherol (500 mg) and transition metal (150 mg) had an oxygen scavenging capacity of 6.72 ml O_2/g and an oxygen scavenging rate of 0.11 ml O2/g•day.

Others authors also research alternative systems able to scavenging oxygem. Anthierens et al. (2011) developed an O_2 scavenger using an endospore-forming bacteria genus *Bacillus amyloliquefaciens* as the "active ingredient". Spores were incorporated in poly(ethylene terephthalate, 1,4-cyclohexane dimethanol) (PETG), na amorphous PET copolymer having a considerable lower processing temperature and higher moisture absorption compared to PET (Figure 4). The work showed that endospores were able to survive incorporation in PETG at 210 ºC, and the spores could consume oxygen for minimum 15 days, after na activation period of 1-2 days at 30 ºC under high moisture conditions. According to the authors, the usse of viable spores as oxygen scavengers could have advantages towards consumer perception, recyclability, safety, material compatibility and production costs compared to currently available chemical oxygen scavengers.

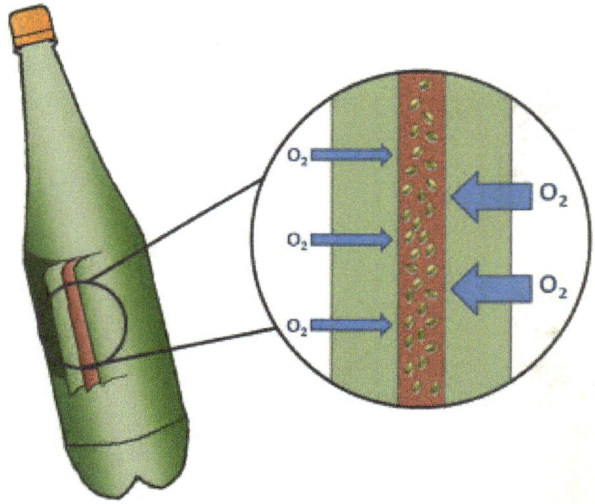

Figure 4. Schematic representation of a multilayer PET bottle consisting of a PETG middle layer containing bacterial spores surrounded by two outer PET layers. The inside of the bottle is in contact with the product, allowing moisture uptake of the bottle needed for spore germination. The system allows scavenging of residual oxygen from the in-bottle environment and scavenging from atmospheric oxygen permeating through the bottle wall (Anthierens et al., 2011)

An ascorbyl palmitate-β-cyclodextrin inclusion complex was produced and used as oxygen scavenger by Byun and Whiteside (2012). Cyclodextrin inclusion complex is one microencapsulation technique that has a significant potential for oxygen scavenging technology. Cyclodextrins (CDs) are cyclic oligosaccharides with a hydrophilic exterior and a hydrophobic central cavity. Its molecular dimensions allow total or partial inclusion of guest compounds. Among conventional microencapsulation methods, β–cyclodextrin inclusion is the most effective for protecting flavors. Production of off-flavors is a common problem of conventional oxygen scavenging sachets and films. Therefore, eliminating or reducing these potential off-flavors is a major concern for developing new oxygen scavenger. Cyclodextrin has other advantages, such as its thermal and chemical stability. The new O_2 scavenger based on ascorbyl palmitate-β-cyclodextrin inclusion complex was able to reduce oxygen content under 4 and 23 ºC more than iron powder based sachet. In addition, the effect of thermal processing on oxygen scavenging capability was also investigated, and the O_2 scavenger developed maintained good oxygen scavenging capability after thermal processing. The results indicated that ascorbyl palmitate-β-cyclodextrin inclusion complex is an effective O_2 scavenger.

Gibis and Rieblinger (2011) incorporate the oxygen scavenger into the packaging material aiming to achieve better quality preservation and longer shelf-life of the chilled food. First investigations concentrated on defining the influence of temperature to the oxygen consumption of an oxygen scavenger film. Reducing the temperature from 23 °C to 5 °C caused a decrease (factor 3.0) in the oxygen consumption rate of the oxygen scavenger multilayer film PE /AL(SP2400; PE) within the first four days (RH 100 %; 0.5 % initial headspace-oxygen). Moreover the influence of using a polymer with a higher oxygen permeation rate than PP (commonly used) to the oxygen consumption of the scavenger film was investigated. Thus the masterbatch SP2500 was mixed with EVA that shows higher oxygen permeability than PP (by factor 2.3). Consequently the oxygen scavenger multilayer film PE/AL(SP2500; EVA) showed a faster oxygen consumption than the film PE/AL(SP2500; PP) (by factor 2.3). Finally the oxygen concentration in measuring cells with scavenger film PE/AL(SP2500; EVA) and with sausage were compared at 5 °C (initial oxygen concentration in headspace: 0.5 %). The combination (calculated) with oxygen scavenger film showed a faster oxygen decrease in the headspace of the measuring cell than the sausage alone. This leads to the assumption of a certain protection of the sausage against oxygen deterioration. Better protection of the sausages might be achieved by storing the food sample in combination with the scavenger film in darkness for the first few days. This would allow the scavenger to absorb the oxygen much faster than the sausage because the fast photo-oxidation processes in the food do not appear without light-exposure.

Absorption kinetics of two commercial O_2 and CO_2 scavengers (ATCO® LH-100 and ATCO® CO-450, respectively) commonly used in active modified atmosphere packaging (MAP), were studied. Individual scavenger sachets were placed in polyvinylidene chloride pouches filled with air or modified atmosphere at 0% or 100% relative humidity and at 5, 20 and 35 °C. The headspace gas composition was measured as a function of time. Absorption kinetics were described by a first-order reaction with an Arrhenius type behaviour. The absorption capacity, absorption rate constant, energy of activation, Arrhenius constant and variation of all these parameters were evaluated This study illustrated the importance to take into account the temperature effect and the variation of the scavenger absorption kinetics to understand gas kinetics inside pouches, as well as to predict the product quality in modified atmosphere packaging (Charles et al., 2006).

Rodrigues et al. (2012) evaluated the antioxidant capacities of gum arabic and maltodextrin microcapsules containing antioxidant molecules (trolox, α-tocopherol, β-carotene, apo-8′-carotenal and apo-12′-carotenal) against reactive oxygen and nitrogen species. The scavenging capacities were influenced by the wall material, the reactive species, namely ROO•,H2O2,HO•, HOCl and $ONOO^-$, and the antioxidant molecule. In general, a more pronounced enhancement of the antioxidant capacity due to incorporation of antioxidant molecules was observed in gum arabic microcapsules. The empty microcapsules showed capacity to scavenge reactive oxygen species (ROS) and reactive nitrogen species (RNS), being gum arabic a more potent antioxidant than maltodextrin. Apo-8'-carotenal incorporation promoted the highest increase in the scavenging capacities among the evaluated antioxidants, varying from 50% to 132% and from 39% to 85% for gum arabic and maltodextrin microcapsules, respectively, suggesting that this carotenoid presented the best balance between the molecule localization inside the microcapsules and the reactivity against the specific reactive species. These results contribute to the development of multi-functional microcapsules that are able to scavenge a broad range of reactive species of biological relevance, serving as a dietary supplement or as antioxidants for food products, and can also be used as colourants in hydrophilic matrices, such as foods and drugs, without raising the fat content.

Zeolites (mostly faujasites) with adsorbed terpenes ((R)-(+)-limonene or D-pinene) or phenol derivatives (thymol, resorcin, pyrocatechol) have been applied as effective oxygen scavengers of oxygen in packing bags. Their efficiency depends on type of zeolite and on cation modification. Na- and Cu-forms of zeolites X and Y accelerate the oxidation of terpenes greatly, whereas the H-forms retard the reaction with oxygen. The reactivity of phenol derivatives with oxygen is also affected by the zeolite support markedly. Although the reactivity of phenols does not increase after adsorption on zeolites, the oxidation products remain adsorbed and do not affect the packing system (Frydrych et al., 2007).

An oxygen scavenging system (OSS), composed of oxygen scavenging nanoparticles α-tocopherol and iron chloride (II), was incorporated into warm-water fish gelatin film and their oxygenvscavenging capability was investigated. The initial oxygen content (%) in the cup headspace, 20.90%, was decreased to 4.56% after 50 days of storage. The oxygen scavenging fish gelatin (OSFG) film had good oxygen scavenging capacity,1969.08 cc O_2/m^2/mil, and moisture was used as the activator to trigger the oxygen scavenging reaction (Byun et al., 2012).

The researches briefly presented above show that there is an increasing interest in the oxygen scavengers field, and that the role of packaging in food preservation is more active, contributing for extending food shelf-life.

Author details

Renato Souza Cruz,
Technology Department, State University of de Feira de Santana, Feira de Santana, BA, Brazil

Geany Peruch Camilloto and Ana Clarissa dos Santos Pires*
Food Tecnhology Department, Federal University of Viçosa, Viçosa, MG, Brazil

4. References

Abe Y, Kondoh Y (1989) CA/MA vacuum packaging of foods. In: Oxygen absorbers. Trumbull, Westport, CT, pp. 149-158.

* Corresponding Author

Abe Y (1994) Active packaging with oxygen absorbers. In Ahvenainen R, Nattila-Sandholm T, Ohlsson T. Minimal processing of foods. VTT symposium 142, Espoo, pp. 209–233.

Ackerley DF, Barski RFJr, Cashhill PJ, Chen SY, Johnson DC, Nyderek WM, Rotter GE (1998) Zero Oxygen Permeation Plastic Bottle for Beer and Other Applications. WO Patent 9,812,127.

Ahvenainen R, Hurme E (1997) Active and smart packaging for meeting consumer demands for quality and safety. Food addit contam. 14:753–763.

Akkapeddi MK, Tsai M (2002) Ultra High Oxygen Barrier Films and Articles Made Therefrom. US Patent 6,479,160.

Al-Malaika S, Goodwin C, Issenhuth S, Burdick D (1999) The antioxidant role of α-tocopherol in polymers II. Melt stabilizing effect in polypropylene. Polym Degrad Stabil. 64: 145–156.

Altieri C, Sinigaglia M, Corbo MR, Buonocore GG, Falcone P, Del Nobile MA (2004) Use of entrapped microorganisms as biological oxygen scavengers in food packaging applications. LWT 37: 9-15.

Anon. (2004) US invasion – it dominates the US, but Multisorb is pushing its oxygen scavengers in Europe. Active & intelligent pack news. 2: 5.

Anon. (2005) Up and active – Pira's latest market report plots a healthy future for active packaging. Active & intelligent pack news. 3: 5.

Anthierens T, Ragaert P, Verbrugghe S, Ouchchen A, De Geest BG, Noseda B, Mertens J, Beladjal L, De Cuyper D, Dierickx W, Du Prez F, Devlieghere F (2011) Use of endospore-forming bacteria as an active oxygen scavenger in plastic packaging materials. Innov. Food Sci. Emerg. Tech. 12:594-599.

ATCO® Technical Information (2002) ATCO® oxygen absorbers, Standa Industrie, France.

Bagchi K, Puri S (1998) Free radicals and antioxidants in health and disease. East Mediterr Health J. 4: 350–360.

Barski RFJr, Richardson JA, Cahill PJ, Rotter GE, Wass RV, Nyderek WM, Smyser GL (2002) Active Oxygen Scavenger Compositions and Their Use in Packaging Articles. US Patent 6,346,308.

Bioka Technical Information (1999) Bioka oxygen absorber, Bioka Oy, Finland.

Brody AL, Budny JA (1995) Enzymes as active packaging agents. In: Rooney ML (Ed) Active Food Packaging. London, Blackie Academic Professional. pp.174-192.

Brody AL, Strupinsky ER, Kline LR (2001a) Oxygen scavenger systems. In: Active Packaging for Food Applications. Pennsylvania, USA, Technomic Publishing Company.

Brody AL, Strupinsky ER, Kline, LR (2001b) Active packaging for food applications. Lancaster: Technomic Publishing Co., Inc. 218 p.

Butler, B.L. (2002). Cryovac® OS2000TM Polymeric oxygen scavenging systems. Presented at Worldpak 2002. http://www.sealedair.com/library/articles/article-os2000.html

Byun Y, Bae HJ, Whiteside S (2012) Active warm-water fish gelatin film containing oxygen scavenging system. Food Hydrocol 27: 250-255.

Byun Y, Darby D, Cooksey K, Dawson P, Whiteside S (2011) Development of oxygen scavenging system containing a natural free radical scavenger and a transition metal. Food Chem. 124: 615-619.

Byun Y, Kim Y, Whiteside S (2010) Characterization of an antioxidant polylactic acid (PLA) film prepared with α-tocopherol, BHT and polyethylene glycol using film cast extruder. J Food Eng. 100: 239–244.

Byun Y, Whiteside S (2012) Ascorbyl palmitate-β-cyclodextrin inclusion complex as an oxygen scavenging microparticle. Carbohydrate Pol. 87:2114-2119.

Cahill PJ, Chen SY (2000) Oxygen Scavenging Condensation Copolymers for Bottles and Packaging Articles. US Patent 6,083,585.

Cecchi T, Passamonti P, Cecchi P. (2010) Study of the quality of extra virgin olive oil stored in PET bottles with or without an oxygen scavenger. Food Chem. 120: 730-735.

Charles F, Sanchez J, Gontard N (2006) Absorption kinetics of oxygen and carbon dioxide scavengers as part of active modified atmosphere packaging. J. Food Eng. 72: 1-7.

Choe E, Min DB (2006) Chemistry and reactions of reactive oxygen species in foods. Crit Rev Food Sci Nutr. 46: 1–22.

Cruz RS, Soares NFF, Andrade NJ (2005) Absorvedores de oxigênio na conservação de alimentos: Uma revisão. Rev Ceres. 52: 191-206.

Cruz RS, Soares NFF, Andrade NJ (2006) Evaluation of oxygen absorber on antimicrobial preservation of lasagna -type fresh pasta under vacuum packed. Cienc. Agrotec. 30: 1135-1138.

Cruz RS, Soares NFF, Andrade NJ (2007) Efficiency of oxygen - absorbing sachets in different relative humidities and temperatures. Cienc. Agrotec. 31: 1800-1804.

Damaj Z, Naveau A, Dupont L, Hénon E, Rogez G, Guillon E (2009) Co(II)(L-proline)$_2$(H_2O)$_2$ solid complex: Characterization, magnetic properties, and DFT computations. Preliminary studies of its use as oxygen scavenger in packaging films. Inorg. Chem. Commun. 12: 17-20.

Day BPF (1998) Active packaging of foods. CCFRA New technologies bulletin. 17: 1-23.

Day BPF (2003) Active packaging. In: Coles R, McDowell D, Kirwan M. (Eds) Food Packaging Technologies. Boca Raton, FL, USA: CRC Press, pp. 282–302.

Doran PM, Bailey JE (1986) Effects of immobilization on growth, fermentation properties, and macromolecular composition of *Saccharomyces cerevisiae* attached to gelatin. Biot Bioeng. 28: 73–87.

Edens L, Farin F, Ligtvoet AF, Van Der Platt JB (1992) Dry yeast immobilized in wax or paraffin for scavenging oxygen. US Patent 5106633.

Floros JD, Dock LL, Han JH (1997) Active packaging technologies and applications. Food Cosmet Drug Packag. 20:10–17.

Frydrycha E, Foltynowicz Z, Kowalakb S, Janiszewskab E (2007) Oxygen scavengers for packing system based on zeolite adsorbed organic compounds. In: Xu R, Gao Z, Chen J, Yan W (Eds) Zeolites to Porous MOF Materials – the 40th Anniversary of International Zeolite Conference. Elsevier B.V., pp: 1597-1604.

Gibis D, Rieblinger K (2011) Oxygen scavenging films for food application. Proc Food Sci 1:229–234.

Goodrich JL, Schmidt RP, Jerdee GD, Ching TY, Leonard JP, Rodgers BD (2003) Oxygen Scavenging Packaging. US Patent 6569506.

Gosmann B, Rehem HJ (1986) Oxygen uptake of microrganisms entrapped in Ca-alginate. Appl Microbiol Biot. 23: 163–167.

Gosmann B, Rehem HJ (1988) Influence of growth behaviour and physiology of alginate-entrapped microorganisms on the oxygen consumption. Appl Microbiol Biot. 29: 554–559.

Graff E. (1994) Oxygen removal. US patent 5284871

Graff G (1998) O_2 scavengers give "smart" packaging a new lease on shelflife. Mod Plast. 75: 69–72.

Hamilton RJ, Kalu C, Prisk E, Padley FB, Pierce H (1997) Chemistry of freeradicals in lipids. Food Chem. 60: 193–199.

Ho YC, Yam KL, Young SS, Zambetti PF (1994) Comparison of vitamin E, IRGANOX 1010 and BHT as antioxidants on release of off-flavors from HDPE bottles. J. Plast Film Sheet. 10:194–212.

Jo C, Ahn DU, Byun MW (2002) Irradiation-induced oxidative changes and production of volatile compounds in sausages prepared with vitamin E-enriched commercial soybean oil. Food Chem. 76: 299–305.

Kulzick MA, Cahill PJ, Barnes TJ, Bauer CW, Johnson DC, Lynch T-Y, Rotter GE, Nyderek WM (2000) Improved Active Oxygen Scavenger Packaging. WO Patent 0037321.

Kruijf N, van Beest M, Rijk R, Sipiläinen-Malm T, Paseiro Losada P, De Meulenaer B (2002) Active and intelligent packaging: applications and regulatory aspects. Food Addit Contam. 19: 144-162.

Labuza TP (1897) Oxygen absorber sachets. Food Res. 32: 276-277.

Labuza TP, Breene WM (1989) Applications of active packaging for improvement of shelf-life and nutritional quality of fresh and extended shelf-life foods. J Food Process Pres. 13: 1-69.

Lacoste A, Schaich KM, Zumbrunnen D, Yam KL (2005) Advancing controlled release packaging through smart blending. Packag Tech Sci. 18: 77–87.

Laermer SF, Young SS, Zambetti PF (1996) Could your packaging use a dose of vitamin E? Converting Magazine. 80–82.

Miltz J, Hoojjat P, Han J, Giacin JR, Harte BR, Gray IJ (1989) Loss of antioxidants from high density polyethylene—its effect on oatmeal cereal oxidation. In: Food and Packaging Interactions. Washington D.C. ACS Symposium 365.

Mize JAJr, Stockley HWIII, Logan RH, Miranda NR (1996) Peelable Package with Oxygen Scavenging Layer. EP Patent 0698563.

Mohan CO, Ravishankar CN, Srinivasa Gopal TK, Ashok Kumar K, Lalitha KV (2009) Biogenic amines formation in seer fish (*Scomberomorus commerson*) steaks packed with O_2 scavenger during chilled storage. Food Res. Int. 42: 411-416.

Morgan CR, Speer DV, Roberts WP (1992) Methods and Compositions for Oxygen Scavenging. EP Patent 0520257 A3 B1.

Nakamura H, Hoshino J (1983) Techniques for the preservation of food by employment of an oxygen absorber. Mitsubishi Gas Chemical Co., Tokyo, Ageless® Division, 1–45.

Nezat JW (1985) Composition for absorbing oxygen and carrier thereof. US Patent 4510162.

Nielsen T (1997) Active packaging-a literature review, The Swedish Institute for Food and Biotechnology, 631, Sweden.

Ohlsson T, Bengtsson N (2002) Active and intelligent packaging: oxygen absorbing packaging materials. In: Minimal Processing Technologies in the Industry. Cambridge, Woodhead Publishing Limited and CRC Press LLC.

Ozdemir M, Floros JD (2004) Active Food Packaging Technologies. Crit Rev Food Sci Nut. 44:185–193.

Roberts WP, Vanputte AW, Speer DV, Morgan CR (1996) Multilayer Structure for A Package for Scavenging Oxygen. US Patent 5350622.

Rodrigues E, Mariutti LBR, Faria AF, Mercadante AZ (2012) Microcapsules containing antioxidant molecules as scavengers of reactive oxygen and nitrogen species. Food Chem (article in press).

Rooney ML (1995) Active packaging in polymer films. In: Active Food Packaging. London, Blackie Academic Professional.

Rooney ML (2005) Introduction to active food packaging technologies. In: Han JH (Ed) Innovations in Food Packaging. London, UK: Elsevier Ltd., pp. 63–69.

Roussel (1999) Les emballages absorbeurs d'oxygène. In: Gontard N (Ed) Les emballages actifs. Paris, Tec and Doc, pp. 31–37.

Santiago-Silva P, Soares NFF, Nóbrega JE, Júnior MAW, Barbosa KBF, Volp ACP, Zerdas ERMA, Würlitzer NJ (2009) Antimicrobial efficiency of film incorporated with pediocin (ALTA® 2351) on preservation of sliced ham. Food Control 20: 85-89.

Shorter AJ (1982) Evaluation of rapid methods for scavenging oxygen in flexible pouches. LWT. 5: 380-381.

Siro I, Fenyvesi E, Szente L, Meulenaer BD, Devlieghere F, Orgovanyi J, Sényi J, Barta, J (2006) Release of alpha-tocopherol from antioxidative low density polyethylene film into fatty food simulant: influence of complexation in beta-cyclodextrin. Food Addit Contam 23: 845–853.

Smith JP (1996) Improving shelf life of packaged baked goods by oxygen absorbents. AIB research department technical bulletin. 18: 2–7.

Smirnoff N (2005) Ascorbate, tocopherol and carotenoids: metabolism, pathway engineering and functions. In: Antioxidants and reactive oxygen species in plants. New York: Blackwell Publishing, pp. 53–86.

Speer DV, Cotterman RL, Kennedy TD (2002) Process for Pasteurizing an Oxygen Sensitive Product and Triggering an Oxygen Scavenger and the Resulting Package. US Patent 2002142168.

Speer DV, Roberts WP (1994) Oxygen Scavenging Compositions for Low Temperature Use. US Patent 5310497.

Suppakul P, Miltz J, Sonneveld K, Bigger SW (2003) Active packaging technologies with an emphasis on antimicrobial packaging and its applications. J Food Sci. 68: 408-420.

Tramper J, Luyben KCAM, van der Tweel WJJ (1983) Kinetic aspects of glucose oxydation by *Gluconobacter oxydans* cells immobilized in Ca-alginate. Eur J Appl Microbiol. 17: 13–18

van Ruth SM, Shaker ES, Morrissey PA (2001) Influence of methanolic extracts of soybean seeds and soybean oil on lipid oxidation in linseed oil. Food Chem. 75:177–184.

Vermeiren L, Devlieghere F, van Beest M, de Kruijf N, Debevere J (1999) Developments in the active packaging of foods. Trends Food Sci. Tech. 10: 77–86.

Vermeiren L, Heirlings L, Devlieghere F, Debevere J (2003) Oxygen, ethylene and other scavengers. In: Ahvenainen R (Ed.) Novel food packaging techniques. Boca Raton, USA, CRC Press.

Wessling C (2000) Antioxidant-Impregnated Polymers Antioxidant Ability and Interactions with Food. Ph.d Thesis Cornell University Ithaca NY USA.

Wessling C, Nielsen T, Leufven A (2000) The influence of α-tocopherol concentration on the stability of linoleic acid and the properties of low density polyethylene. Packag Tech Sci. 13: 19–28.

Wessling C, Nielsen T, Leufven A, Jagerstad M (1998) Mobility of tocopherol and BHT in LDPE in contact with fatty food simulants. Food Addit Contam. 15: 709–715.

Zerdin K, Rooney ML, Vermuë J (2003) The vitamin C content of orange juice packed in an oxygen scavenger material. Food Chem. 82: 387-395.

Chapter 2

Protein-Based Edible Films: Characteristics and Improvement of Properties

Thawien Wittaya

Additional information is available at the end of the chapter

1. Introduction

Protein-based edible films have received consideration attention in recent years because of their advantages, including their use as edible packaging materials, over the synthetic films. In addition, protein-based edible films can also be used for the individual packaging of small portions of food, particularly products that are not currently individually packaged for practical reasons, such as beans, nuts and cashew nuts. In addition, protein-based edible films can be applied inside heterogeneous foods at the interfaces between different layers of components. They can be tailored to prevent the deterioration of inter-component moisture and solute migration in foods such as pizzas, pies and candies. Moreover, protein-based edible films can function as carriers for antimicrobial and antioxidant agents. Through a similar application they also can be used at the surface of food to control the diffusion rate of preservative substances from the surface to the interior of the food. Another possible application for protein-based edible films could be their use in multilayer food packaging materials together with non edible films. In this case, the protein-based edible films would be the internal layers in direct contact with food materials. Through functions relating to mechanical and barrier properties, protein-based edible films may be able to substitute synthetic polymer films.

In their natural states, proteins generally exist as either fibrous proteins or globular proteins. The first type are water insoluble and serve as the main structural materials of animal tissues., Globular proteins are soluble in water or aqueous solutions of acids bases or salts and function widely in living systems. The fibrous proteins are fully extended and associated closely with each other in parallel structures, generally through hydrogen bonding, to form fibers. The globular proteins fold into complicated spherical structures held together by a combination of hydrogen, ionic, hydrophobic and covalent (disulfide) bonds. The chemical and physical properties of these proteins depend on the relative amounts of component amino acid residues and their placement along the protein polymer chain. Regarding the fibrous proteins, collagen has received the most attention in the production of edible films. Several globular proteins, including wheat gluten, corn zein, soy protein, whey protein and mung bean protein, have been investigated for their film properties.

Protein-based edible films are generally formed from solutions or dispersions of the protein as the solvent/carrier evaporates. The solvent/carrier is generally limited to water, ethanol or ethanol-water mixtures. Generally, proteins must be denatured by heat, acid, bases, and/or solvents in order to form the more extended structures that are required for film formation. Once extended, protein chains can associate through hydrogen, ionic, hydrophobic and covalent bonding. The chain-to-chain interaction that produces cohesive films is affected by the degree of chain extension and the nature and sequence of amino acid residues. The uniform distribution of polar, hydrophobic, and/or thiol groups along the polymer chain increase the likelihood of the respective interactions. The promotion of polymer chain-to-chain interaction results in films that are stronger but less flexible and less permeable to gases, vapors and liquids. Polymers containing groups that can associate through hydrogen or ionic bonding result in films that are excellent oxygen barriers but susceptible to moisture. Thus, protein films are expected to be good oxygen barriers at low relative humidity.

Polymers containing a preponderance of hydrophobic groups are poor oxygen barriers but excellent moisture barriers. However, the fact that proteins are not totally hydrophobic and contain predominantly hydrophillic amino acid residues limits their moisture-barrier properties. The creation of protein to edible films with low water vapor permeability requires the addition of hydrophobic components. This is analogous to the situation with synthetic polymers. Here moisture-sensitive oxygen-barrier polymers must be either co polymerized with a hydrophobic polymer or sandwiched between hydrophobic polymer layers to limit the ability of water to reduce barrier properties. Because of the poor water vapor resistance of protein-based edible films and their lower mechanical strength in comparison with synthetic polymers their application is limited in food packaging. The improvement of protein-based edible films properties has been investigated to seek suitable applications. This chapter provides details of protein-based edible films and their properties, the production of protein-based edible films, the methods used for the formation of protein-based edible films, the improvement of properties of protein-based edible films and their potential applications.

2. Type of protein-based edible films

2.1. Collagen films

Collagens are the major structural proteins of connective tissue such as bone, hide, tendons cartilage, and ligaments. They constitute about one-third of the total body protein in mammals. There are different types of collagen. Each type has its own amino acid sequence, but all contain a significant amount of triple helical structure. Because of its biological properties and ready availability, type I collagen is widely used as a biomaterial. It is used in a variety of physical forms such as sponges, films and membranes (Sisken et al., 1993). Collagen possesses characteristics as a biomaterial that is distinct from those of synthetic polymers. Collagen is used to make the most commercially successful edible protein films. Collagen films offer several advantages: it is biocompatible and non-toxic to most tissues; it has well-documented structural, physical, chemical, and immunological properties; it can be processed into a variety of forms; and it is readily isolated and purified in large quantities. The production of collagen films from animal hides can be accomplished using a dry or wet process with some similarities. These include: (a) alkaline treatment to de-hair and remove collagen from carbohydrates and other proteins; (b) acid swelling and homogenization to form a ~ 4.5% moisture gel (wet process) or ~ 10% moisture gel dough (dry process); (c) extrusion into a tube; and (d) neutralization of the extruded tube, washing the tube of salts, treating the tube with plasticizer and cross-linkers and drying to 12-14% moisture (the order depends on whether the wet or dry process is used) (Hood, 1988).

2.2. Gelatin films

Gelatin is unique among hydrocolloids in forming a thermo-reversible substance with a melting point close to body temperature, which is particularly significant in edible and pharmaceutical applications. Basically, gelatin is obtained by controlled hydrolysis from the fibrous insoluble protein, collagen, which is widely found in nature as the major constituent of skin, bones and connective tissue. Gelatin is composed of a unique sequence of amino acids. The characteristic features of gelatin are the high content of the amino acids glycine, proline and hydroxyproline. Gelatin also has a mixture of single and double unfolded chains of a hydrophilic character (Ross-Murphy, 1992). At approximately 40 oC, gelatin aqueous solutions are in the sol state and form physical, thermoreversible gels on cooling. During gelation, the chains undergo a conformational disorder–order transition and tend to recover the collagen triple-helix structure (Ross-Murphy, 1992).

Gelatin is used to encapsulate low moisture or oil phase food ingredients and pharmaceuticals. Such encapsulation provides protection against oxygen and light, as well as defining the amount of ingredient or drug dosage. In addition, gelatin films have been formed as coatings on meats to reduce oxygen, moisture and transport of oil (Gennadios et al., 1994). Addition, gelatin is able to form clear and strong films and is used for microencapsulation and capsule coatings in food and pharmaceutical manufacturing. Gelatin films could be formed from 20-30% gelatin, 10-30% plasticizer (glycerin or sorbitol) and 40-70% water followed by drying the gelatin gel (Guilbert, 1986). However, gelatin films, as with most protein films, do not have an ideal water vapor barrier, which limits its application as edible film and biomaterial. However, modification of the polymer network through cross-linking of the polymer chains can be applied to improve the functionality of protein film.

2.3. Corn zein films

Zein is the most important protein in corn. It is a prolamin protein and therefore dissolves in 70–80% ethanol (Dickey & Parris, 2002). Zein is a relatively hydrophobic and thermoplastic material. The hydrophobic nature of zein is related to its high content of non-polar amino acids (Shukla & Cheryan, 2001). Technically, the films made from an alcohol soluble protein like zein, have relatively high barrier properties compared to other proteins. Zein has excellent film forming properties and can be used for the fabrication of biodegradable films. Zein film is formed through the development of hydrophobic, hydrogen and limited disulfide bonds between zein chains (Guilbert, 1986). The formation of corn zein films is believed to involve the development of hydrophobic, hydrogen and limited disulfide bonds between zein chains in the film matrix (Gennadios et al., 1994). The resulting films are brittle and therefore require the addition of plasticizer for increasing the flexibility. Zein films are relatively good water vapor barriers compared to other edible films (Guilbert, 1986). Zein coating have also shown an ability to reduce moisture and loss of firmness and delay color change (the reduction of oxygen and carbon dioxide transmission) in fresh fruit. In addition, zein may also take part in the coating of conventional packaging plastics. Although, zein is definitely not water soluble at a neutral pH, it has high water vapor permeability compared with typical synthetic polymers. However, the water vapor barrier properties can be improved by adding fatty acids or by using a cross-linking reagent. However, when cross-linking agents are used, the edibility of the resulting films needed to be considered.

2.4. Wheat gluten films

Wheat gluten is a water insoluble protein of wheat flour which is comprised of a mixture of polypeptide molecules, and considered to be globular proteins. The cohesiveness and elasticity of the gluten gives integrity to wheat dough and facilitates film formation. Wheat gluten is composed of two main groups of water insoluble proteins: gliadins, consisting of low molecular weight proteins; and glutenins containing high molecular weight proteins. Gliadins are single monomeric proteins in which disulfide bonds make up intra-chains or are absent, while glutenins form high molecular weight polymers maintained by inter-chain disulfide bonds. Films from glutenins were stronger and had better barrier properties than films from gliadins or whole gluten. Gliadin films presented better optical properties but were not water resistant. The properties of these films make them sometimes sensitive to thermal treatments, which could result in improved film properties. Addition, gliadin is soluble in 70% ethanol; glutenin is not (Gennadios & Weller, 1990). Although insoluble in natural water, wheat gluten dissolves in aqueous solutions of high or low pH at low ionic strength (Krull & Inglett, 1971). Wheat gluten films can be fully biodegraded after 36 days in aerobic fermentation and within 50 days in farmland soil without releasing toxic products (Domenek et al., 2004).

Technically, wheat gluten films can be formed by drying aqueous ethanol solutions. Cleavage of native disulfide bonds during the heating of film-forming solutions. The formation of new disulfide bonds during film drying is then believed to be important to the formation of wheat gluten films structure, along with hydrogen and hydrophobic bonds (Gennadios & Weller, 1990). The addition of plasticizer such as glycerin in gluten films is necessary to improve the flexibility of wheat gluten films. However, increasing film flexibility by increasing plasticizer content may reduce the strength, elasticity and water vapor barrier properties of the resulting film. In addition, the purity of wheat gluten also affects the film's appearance and mechanical properties; greater purity gluten results in stronger and clearer films. Wheat gluten films are effective oxygen barriers, but poor water vapor barriers. The poor resistance of wheat gluten films to water vapor is due to the hydrophilic nature of the protein and to the substantial amount of hydrophilic plasticizer added to impart adequate film flexibility. The properties of wheat gluten films can be improved by using a cross-linking agent such as glutaraldehyde, or heat curing.

2.5. Soy protein films

Soy protein from soybeans has been extensively used as a food ingredient in nearly every food product available to the consumer, since it contains high nutrition and excellent functional properties. The protein content of soybeans (38-44%) is much higher than the protein content of cereal grain (8-15%). The major soybean proteins have molecular weights ranging from 200 to 600 kDa. Most soy proteins (90%) are globulins, which can be fractionated into 2S, 7S, 11S and 15S according to their sedimentation coefficients. 7S and 11S are the main fractions, being about 37% and 31% of the total extractable protein and have a capacity for polymerization (Cho & Rhee, 2004). Soy proteins consist of both polar and non-polar side chains. There are strong intra- and inter-molecular interactions, such as hydrogen bonding, dipole–dipole, charge–charge, and hydrophobic interactions. The strong charge and polar interactions between side chains of soy protein molecules restricts segment rotation and molecular mobility, which increase the stiffness, yield point, and tensile strength of soy protein films (Zhang et al., 2001).

Because soy proteins are abundant, inexpensive, biodegradable, and nutritional, they show the potential to be developed as edible and biodegradable films. The formation of films from soy proteins has been described as a two-step process involving: (a) the heating of film solutions to disrupt the protein structure, cleave native disulfide bonds and expose sulfhydryl groups and hydrophobic groups; and (b) the formation of new disulfide, hydrophobic and hydrogen bonds. The unfolded proteins link through intermolecular interactions, such as the disulfide bonds and hydrophobic interactions, leading to the formation of a network which occur during drying processes.

Technically, protein-based edible films can form bonds at different positions and offer high potential for forming several linkages. However, soy protein films still have poor moisture barrier properties due to their hydrophilic properties and the substantial amount of hydrophilic plasticizer used in film preparation. One extensively used method to enhance the water vapor barrier of films has been the incorporation of hydrophobic compounds such as lipids into the film forming solution. In addition, the other way to improve the properties of soy protein film is to modify the protein network through cross-linking of the protein chains. The presence of reactive functional groups in the amino acid side chain of protein makes this cross-linking process possible through chemical, enzymatic or physical treatments.

2.6. Casein films

Milk proteins can be classified into two types: casein and whey protein. Casein comprises of three principal components,α, ß, and κ, which together form colloidal micelles in milk containing large numbers of casein molecules and are stabilized by a calcium-phosphate bridge (Kinsella, 1984). The casein molecules possess little defined secondary structure, exhibiting instead an open random-coil structure. Casein, which comprises 80% of milk protein, precipitates when skim milk is acidified to the casein isoelectric point of approximately 4.6 (Dalgleish, 1989). Acidification solubilizes the calcium phosphate, thus releasing individual casein molecules, which associate to form insoluble acid casein. The acid casein can be converted to functional soluble caseinates by neutralization through the addition of alkali. Edible protein films based on various caseinates can be obtained by solubilization in water followed by casting and drying.

Caseinates films are made from aqueous solutions without heat treatment due to their random coil nature. Interactions in the film matrix are likely to include hydrophobic, ionic, and hydrogen bonding (Avena-Bustillos & Krochta, 1993). Caseinate films are transparent and flexible, but have poor water barrier properties. At comparable test conditions, caseinate films appear to have similar moisture barriers to wheat gluten films and soy protein films but poorer moisture barriers than corn zein films. Casein has been investigated for the formation of free standing films and coatings on food products. Laminated films that included casein did protect dried fruit and vegetables from moisture absorption and oxidation. Caseinate-lipid emulsion coatings were successful in reducing moisture loss from peeled carrots and zucchini (Avena-Bustillos et al., 1993).

2.7. Mung bean protein films

Mung beans are of interest as a potential component of biopolymeric films because of their high protein content. The whole seeds of mung beans contain approximately 25-30% protein. The proteins from mung bean are large and much of the protein has a molecular weight (MW) between 24 and 55 kDa with some traces having less than 24 kDa. However, there were small amounts of proteins of a MW between 24 and 14.2 kDa. The amino acid composition of mung bean means they are rich in essential amino acids such as leucine, isoleucine, lysine, and phenylalanine and are also rich in acidic amino acids such as glutamic acid and aspartic acid. However, the sulfur containing amino acid, such as methionine and cysteine, were also detected in mung bean protein (2.75 and 3.62%) (Keereekasetsuk et al., 2009).

Bourtoom (2008) prepared and analyzed the films from mung bean protein. It was found that the mechanical properties (tensile strength and elongation at break) of mung bean protein films had superior mechanical properties and water vapor barrier properties. These were better than other protein sources such as casein, soy protein isolate, wheat gluten, peanut proteins and water-soluble fish proteins films. However, the mung bean protein films still showed substantially lower mechanical and water vapor barrier properties compared with some synthetic polymers (high density polyethylene, polyvinyl chloride, cellulose acetate and polyester). The resistance of protein mung bean protein films to water vapor permeability is limited due to the inherent hydrophilicity of proteins. The transmission of water vapor through protein-based edible film is also facilitated by the presence of, a hydrophilic plasticizer, which favors adsorption of water molecules. However, the properties of mung bean protein films can be improved by addition of hydrophobic materials and also using chemical and enzymatic cross linking.

3. Formation of protein-based edible films

Protein-based edible films may be formed by two different methods: surface film formation and the deposition method.

3.1. Surface films formation

Films are obtained by the prolonged heating of film solutions and films are periodically harvested from the surface, drained and dried. The use of this method has been described as a two-step process involving the heat denaturation of the proteins followed by surface dehydration. Heating changes the three-dimensional structure of proteins and the existing functional groups. These, such as CO and NH of peptidic bonds, side chain amine groups and hydrophobic groups, engage in intra-molecular hydrogen bonding and electrostatic interaction (Wang & Damodaran, 1991). During drying, the unfolded proteins approach each other and become linked through intermolecular interactions (disulfide and hydrophobic interaction). This yields the formation of a protein network that acts as the matrix for entrapping film components such as plasticizing agents (Gennadios & Weller, 991). When the formation of film occurs in denaturated conditions, it is assumed that the protein remains in the fully denaturated state in the film.

However, it is quite possible that the denatured protein may undergo partial refolding, thus regaining some secondary structure during the film process. It is conceivable that the extent of such refolding affects the number of functional groups available for intermolecular interactions and thus the formation and stability of the film network (Subirade et al, 1998). Wu & Bates (1973) have prepared films from peanut milk. They reported that during the heating of the peanut milk, the high molecular weight peanut protein is broken down into lower molecular weight moiety. Heat first dissociates the conarachin and then arachin fractions to form small subunits of insoluble complexes at the surface. In addition, the interfacial forces may initiate the formation of protein matrixes capable of trapping oil droplets and water released from the surface facilitating the formation of protein matrixes (Farnum et al., 1976).

3.2. Deposition method

Films obtained from this method generally are made by casting and drying film forming solution on a non-stick surface. Technically, the casting process consists of drying a film solution or a gel for producing films with controlled thickness. This technique is useful to mimic some industrial processes for forming free standing starch films as is the case for dip-molding. In this method, which is used for food coatings as well as for non-food applications, the gelled state is usually preferred to set hot solutions on a surface upon cooling. Jaynes & Chou (1975) used this method to produce soy protein-lipid films. They used a protein isolate solution at natural pH 6.6 casting on Teflon coated baking pan and drying at 100 °C.

Films made from deposition techniques are more uniform films compared to the surface forming method. The film thickness can be controlled by the amount of total solid in the film solutions which is not the case when the surface formation method is employed. Most researchers have been using the deposition technique in recent years to produce edible films. However, casting material and casting temperature may vary depending upon the state and type of substrate. The deposition technique has been used to make protein films from wheat gluten, corn zein, casein, whey protein isolate, soy protein isolate and rice protein concentrate.

4. Factors affecting protein-based edible films

4.1. Type of raw material

Raw materials used in film solutions are classified, according to their solubility characteristics, into two categories-hydrophilic and hydrophobic. Hydrophilic materials such as soy protein isolate, whey protein isolate, water soluble fish protein and mung bean proteins are water soluble. Hydrophobic materials such as corn zein, wax are water-insoluble but they dissolve in non-polar liquids such as alcohol. The difference in soluble properties of these raw materials influence the amount of energy needed to obtain dried films and their use on foods. Carbohydrates such as alginate, carregeenan, pectin, starch, cellulose and cellulose derivatives provide a strong matrix free standing film, but these films are poor water barrier properties because of the hydrophilic nature of raw materials used (Kester and Fennema, 1986). Proteins provide good gas barrier but poor water vapor barrier properties. However, some protein films such as corn zein films exhibit better water resistance than other protein films because zein contains high amount of hydrophobic side chain amino acid. Lipid films, made from hydrophobic materials such as wax, fatty acid, show excellent water vapor barriers but poor mechanical properties.

4.2. Polymer chemistry

The regular structure molecule is more diffusible than the irregular stereochemical structure whereas branched molecules may provide a greater cohesive strength than non-branched molecules. A lower molecular weight fraction shows a greater cohesion and a greater change in cohesion with temperature change. In highly polar polymers such as protein and cellulosic, self-adhesion by diffusion is not significant due to the minimal flexibility and fixed order of the macromolecule. This is caused by the internal molecular forces holding the polymer chains. Cellulosics have a back bone with a rigid ring structure chain whereas proteins tend to form helical chain structure (Banker, 1966).

Kinsellar & Phillips (1989) summarized the desired molecular characteristic for formation of protein films: 1) high soluble molecules promote rapid diffusion; 2) the large molecules allow more interactions at the interface resulting in strong film; 3) amphiphatic molecules provide an unbalanced distribution of charged and apolar residuals for improved interfacial interaction; 4) flexible domains facilitate phase behavior and unfolding at interface; 5) the dispersion of charged groups affect protein-protein interaction in the films and charge repulsion between neighboring bubbles; 6) polar residue can provide hydratable or charged residues to keep bubbles apart, and binding and retaining water; 7) the retention of structure could be enhance overlap and segmental interaction in film; and 8) interactive regions can affect the deposition of different functional segments and facilitate secondary interactions in the air and aqueous phases.

4.3. pH

pH plays an important role in protein films made from water-soluble materials, such as soy protein isolate and whey protein isolate, as the solubility of these proteins depend on their isoelectric point (pI). During the dissolution of macromolecular substances, the cohesive forces between the solute macromolecules are neutralized by unions with the solvent molecules (Banker, 1966). The functionality of the polymer is related to the solution's properties which further influences film characteristics. The charge groups repel each other and produce a stretching of the polymer chain when the functional groups on a linear polymer become ionized during dissolution. The greater the degree of dissolution and the more extensively the chain is charged, the greater is the uncoiling of the chain.

The interaction between the charged polymer molecules and the molecules of the polar solvent increases with the increasing charge on the chain. The maximum protein solubility is obtained at pH away from its isoelectric point (pI). But to produce an edible film at extreme pH, the sensory property must also be considered along with other film properties. Gennadios et al. (1993) studied the effect of pH on soy protein isolate film and found that highly acidic ($pH < 1$) or alkaline conditions ($pH > 12$) inhibit the formation of soy protein isolate film. Kinsella & Phillip (1989) reported that films formed near the isoelectric point of major proteins are more condensed and stronger.

4.4. Drying temperature

Protein-based edible films are usually obtained by the casting method. This technique involves the drying of a complex colloidal solution made up of the protein, a solvent and, usually, a plasticizer previously poured on an appropriate support. The effect of a specific drying temperature depends on the various characteristics of the raw material, such as the occurrence of a preexisting gel phase or the occurrence of thermal gelation during drying. Furthermore, various phenomena may occur such as the transition from a rubbery to a vitreous phase, a phase separation (thermodynamic incompatibility) or crystallization. The interaction between the physicochemical nature of biopolymers and the drying conditions is very important (Devani et al., 2009).

The interaction forces in protein structures are affected by temperature. The temperature is a strong denaturing factor for proteins, and even the thermal stability and conformation of protein depend on the amino acid composition. During the drying period, when water is progressively eliminated, the conformation of the proteins change. Furthermore, the degree of protein unfolding determines the type and proportion of covalent (S–S bonds) or non-covalent (hydrophobic interactions, ionic and hydrogen bonds) interactions that can be established between protein chains. It is known that chains can interact more strongly and easily, especially through disulfide bonds, when proteins are denatured (Mauri & Anon, 2006). The cohesion of the final network would then be a function of these bonds and determines the properties of the films obtained.

In addition, hydrophillic interactions increase, hydrogen bonds and electrostatic interaction decreases when temperature increases. This results in the facilitation of adhesion between polymer films and the substrate (Banker, 1966). High temperature (70-100 °C) affects the forming of rigid structures in protein solutions because of protein denaturation (Chefel et al., 1986). Excessive heat or an excessive solvent evaporation rate during processes may produce non-cohesive films (Guilbert et al., 1986). Water soluble proteins such as soy protein and whey protein need a higher temperature and longer time for film formation than films from alcohol-soluble protein such as corn zein or wheat gluten. The higher drying temperature of water-soluble based-films may limit a film's use. However, low relative humidity can also be employed for film formation at low temperature

4.5. Concentration

Protein films make up complex structures. Protein–protein interactions within aggregates can be of a different nature (such as electrostatic or hydrophobic, etc.), which can lead to different kinds of cohesiveness. This could affect the mobility of proteins and their ability to form films. The concentration of film solutions affect the self adhesion of high polymers and the rate of matrix forming in film preparations. Besides, the protein concentration in film solution can also influence the formation of the protein matrix. At a lower protein concentration there is probably less protein-protein interaction, while at the higher protein concentration self-diffusion is promoted resulting in inferior properties. At the optimum concentration of film solutions, an intermediate viscosity could be obtained which result in the highest cohesive strength. However, the optimum concentration of each protein films requires various concentrations. The production of films with whey protein isolate requires a relatively high protein concentration (>8%) in the film forming solution so that the formation of S–S bridges occur (Sothornvit & Krochta, 2001). Whereas the films produced from muscle fish protein prepared with 1.5-2% showed stronger films than other concentrations.

4.6. Relative humidity

Water interaction with protein-based edible films are considered a priority in order to explain the physical property modifications induced by the presence of the moisture content. The adsorption of water vapor by dried materials is generally assumed to involve the binding of water molecules to specific hydrophilic sites, such as carboxylic, amino and hydroxy residues, in addition to backbone peptide groups. At high relative humidity, multimolecular adsorption occurs through swelling and conformational changes in the macromolecular structure. Basically, the property of protein-based edible films varies to the same degree as storage time, especially if compared with those of synthetic films, due to the intrinsic instability of their raw materials. These variations could affect their properties as a result of changing characteristics. Thus the oxidation of the protein sulfhydryl groups could cause degradation of the polymeric chains (Micard et al., 2000). Physical changes include polymer rearrangement and may be due to the migration of the low molecular weight components used in film formulation, such as plasticizers (Anker et al., 2001). The migration of additives could be considered the most important cause of physical instability of protein films.

The relationship between equilibrium relative humidity and film water content has been assessed by measuring water sorption isotherms. In addition, an understanding of water sorption properties is necessary in order to tailor film applications. The influence of relative humidity on the mechanical properties and permeability of protein film has been tentatively explained by the decrease of glass transition temperature induced by a plastification phenomena created by the water on protein films.

Cuq et al. (1996) determined the effect of relative humidity on the mechanical and water vapor barrier properties of myofibrillar protein-based films. The plasticizing effect of water related to rapid changes in the functional properties of the resulting films was explained by disruptive water-polymer hydrogen bonding. Relatively sharp decreases in force at break, the elastic modulus and water vapor barrier properties, and increases in deformation at break were observed when relative humidity increased. Pochat-Bohatier et al. (2006) studied the influence of relative humidity on carbon dioxide permeability in wheat gluten films. They reported that the increasing of the permeability of the gas was observed at 96% relative humidity. This was attributed to the swelling of the polymer matrix with water, allowing chemical interactions to take place between amino acids and the gas. The increase in the water content of the films resulted in promoting the affinity between the gas and the protein matrix, leading to outstanding sorption values at high relative humidity.

4.7. Film additive

Various materials can be incorporated into protein films to influence the mechanical, protective, sensory, or nutritional properties. Plasticizers are additives that are an important class of low molecular weight non-volatile compounds that are widely used in polymer industries. The primary role of such substances is to improve the flexibility and capacity for processing of polymers by lowering the second order transition temperature, the glass transition temperature (T_g). The council of the IUPAC (the International Union of Pure and Applied Chemistry) defined a plasticizer as "a substance or material incorporated in a material (usually a plastic or elastomer) to increase its flexibility, workability, or distensibility". These substances reduce the tension of the deformation, hardness, density, viscosity and electrostatic charge of a polymer. At the same time they increase the polymer chain flexibility, resistance to fracture and dielectric constant. Among the other properties also affected are the degree of crystallinity, optical clarity, electric conductivity, fire behavior and resistance to biological degradation (Vieira et al., 2011). The compatibility between polymer and plasticizer is a major effective part of plasticization and various parameters can indicate this feature, including polarity, hydrogen bonding, dielectric constant and solubility parameters (Choi et al., 2004). In addition, another impact factor is solvation, as plasticizers with solubility parameters close to those of the polymer require less energy to fuse or solvate the polymer. The temperature of fusion or gelation is related to the solvation strength of the plasticizer and to the size of its molecule (Rahman & Brazel, 2004).

Generally, two types of plasticizers can be distinguished. Internal plasticization is a result of modifications to the chemical structure of the polymer; external plasticization is obtained by adding an agent which modifies the structure and energy within the three dimensional arrangement of the film polymer (Banker, 1966). In practice the addition of a plasticizer to protein films produces films which are less likely to break and are more flexible and stronger. The reduction of the intermolecular bonds between the polymer chains, and thus the overall cohesion, facilitates elongation of the films and reduces its glass transition temperature. This is manifested by a reduction in the barrier properties to gases, vapors, and film solutes (Banker, 1966). The plasticizers that are most usually used in protein films are mono-, di-, and oligosaccharides (generally glucose syrups or glucose fructose honey), polyols (principally glycerol and its derivatives, polyethylene glycols, and sorbitol) and lipids and its derivatives (fatty acids, monoglycerides and their esters, acetoglycerides, phospholipids, and other emulsifiers). The molecular size, configuration and total number of functional groups of the plasticizer, as well as its compatibility with the polymer, could affect the interactions between the plasticizer and the polymer (Yang & Paulson, 2000).

Jangchud & Chinnan (1999) reported that glycerin was found to be the most suitable plasticizer to incorporate into peanut protein concentrate when compared with sorbitol, propylene glycol and polyethylene propylene glycol or polyethylene glycol. These showed very poor mechanical properties, resulting in unsuccessful permeability tests. The concentration of glycerin between 0.67 to 1.67% of protein did not affect the water vapor permeability and oxygen permeability, but affected the tensile strength and percent elongation properties. Glycerin levels of greater than 0.67% of protein were needed to obtain free-standing film.

Vanin et al. (2005) studied the influence of type and content of plasticizers and the properties of gelatin-based films. Four polyols (glycerol-GLY, propylene glycol-PPG, di-DTG and ethylene glycol-ETG) were tested in different concentrations. The results showed that plasticizer type and content had significant effects on the properties of the resulting films. A higher plasticizing effect on thermal properties was observed with the DTG, followed by PPG, GLY and ETG, principally with low plasticizer content. In relation to the mechanical properties, the GLY showed the greater plasticizing effect and efficiency, but the plasticizing efficiency of DTG on the puncture deformation was also considerable. The mechanical resistance could be related to the glass transition temperature of films. It was not possible to observe the plasticizer effect on the water vapor permeability. However the DTG had shown greater plasticizer efficiency, followed by GLY and ETG, while a counter effect was observed with the addition of PPG.

Andreuccetti et al. (2009) determined the effect of hydrophobic plasticizers derived from citric acid (tributyl citrate, acetyltributyl citrate, triethyl citrate, acetyltriethyl citrate) on the functional properties of gelatin-based films. They found that the addition of ester citrate derivatives in gelatin-based film formulations proved feasible, generating flexible materials. The increasing addition of hydrophobic plasticizers had significantly diminished the tensile strength. Such behavior is similar to that described for films prepared with hydrophilic plasticizers. Regarding water vapor permeability, it was determined that the use of hydrophobic plasticizers caused a slight reduction as observed for gelatin films prepared with glycerol and sorbitol.

5. Improvement of protein-based edible films

5.1. Modification of protein-based edible films by chemical treatment

Proteins are promising biomaterials since films made with them are gas barriers. However, the main limitations of protein films, similar to other biopolymers, are their lack of mechanical strength and poor water vapor barrier because of their hydrophilic nature. A very powerful method to improve water resistance, cohesion, rigidity, and mechanical strength and the barrier properties of films to water is cross-linking. To do this different functional groups of proteins can be used. Protein networks have the ability to interact with a wide range of active compounds. This is done via functional groups acting on their reactive side groups. This has the potential to modify via chemical, physical or enzymatic cross-linking to enhance the functional properties of the films.

Chemical treatments with acid, alkali or cross linking agents have been extensively used to improve film properties. Theoretically, the more protein interaction from chemical treatment occurs then chain structure extends and less permeability and greater tensile strength should be obtained. However, Brandenburg et al. (1993) found that alkaline treatment on soy protein isolate did not affect water vapor permeability, oxygen permeability and tensile strength but alkaline treatment improved the film's appearance (it was clearer, more uniform, and with less air bubble) and elongation at break. Chemical agents used for the covalent cross-linking of protein have included aldehydes such as glutaraldehyde, glyoxal or formaldehyde, and others natural cross linking agents.

Because of the cross-linking used in the protein films, formaldehyde is the simplest of cross linking agents, and has the widest reaction specificity. In addition to the amine group of lysine, it reacts with the side chains of cysteine, tyrosine, histidine, tryptophan, and arginine. Although formaldehyde contains a single functional group, it can react bi-functionally and therefore crosslink. Glutaraldehyde is more specific than formaldehyde; it can react with lysine, cysteine, histidine and tyrosine (Tae, 1983). Protein cross-linking by glyoxal involves lysine and arginine side chain groups (Marquie, 2001) at alkaline pH.

Basically, the reaction between formaldehyde and protein contains two step processes: the first step corresponds to the formation of the methylol compound; and the second one corresponds to the formation of methylene bridges that are cross-links between protein chains. The expected reaction scheme was according to Figure 1.

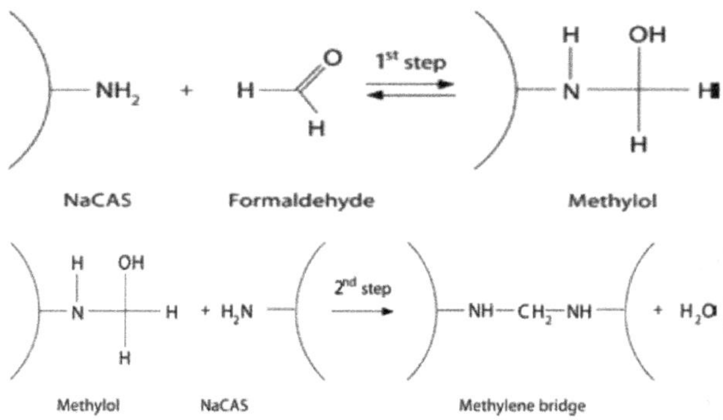

Figure 1. Scheme of cross-linking between formaldehyde and ε-amino groups of protein Source: Gueguen et al. (1998)

Hernandez-Munoz et al. (2004) studied the effect of cross-linking using aldehydes on properties of glutenin-rich films. They reported that the water vapor permeability values of glutenin rich films decreased by around 30% when cross-linking agents such as formaldehyde, glutaraldehyde and glyoxal were incorporated. The highest tensile strength values were obtained using formaldehyde, followed by glutaraldehyde and glyoxal. In addition, the glass transition temperature of cross-linked films shifts to slightly higher values when cross-linking agents were used. Because of this formaldehyde was more efficient than glutaraldehyde and gossypol in cross-linking. The better properties of protein films treated with formaldehyde may be due to the lack of specificity of this chemical with respect to the different amino acid side chain groups. In addition to amines, formaldehyde reacts with sulphydryl, phenolic, imidazolyl, indolyl and guanidinyl groups. Addition, some researchers have reported that the formation of methylene bridges occurred between lysine and tyrosine in formaldehyde-treated of protein (Hernandez-Munoz et al., 2004). However, although highly reactive to aldehyde, they also have a major disadvantage - their toxicity. This must be taken into account when synthesizing biodegradable materials. All of the aldehyde used must be cross-linked in a permanent protein network; the fate of the aldehyde in the environment at the end of the material's life must be considered.

Because of the toxicity of aldehydes, many operators have been trying to use the natural cross linking agents to improve the protein film properties. Orliac et al. (2002) determined the effects of natural cross linking agents (tannins and gallic acid) on the properties of thermo-moulded films produced from sunflower protein isolate. The results showed that the incorporation of tannins and gallic acid resulted in films with higher mechanical properties than for control films, but lower than the films obtained with aldehydes. This was probably because they act through weak interactions rather than covalent bonds in the case of aldehydes. Furthermore, their extremely low volatility at the processing temperature prevents the elimination of the non-bound tannin and gallic acid parts.

Later, Cao et al. (2007) improved the mechanical properties of gelatin films by using ferulic acid and tannic acid. The results showed that the ferulic acid and tannic acid act as natural cross linking agents and had cross-linking effects on gelatin film. The maximal mechanical strength of gelatin film could be obtained when the pH value of the film-forming solution was 7 for ferulic acid as the cross-linked agent, or when the pH value was 9 as for tannic acid. In addition, the properties of gelatin films treated by tannic acid could become better after being stored for more than 90 days, while the storage time had little effect on ferulic acid-modified films. The reason may be because tannic acid could cross link with gelatin later step-by-step with time during drying and storage.

5.2. Modification of protein-based edible films by enzymatic treatment

Protein- based edible films have good barrier characteristics against gas, organic vapor and oil as compared to synthetic films. However, the weak mechanical characteristics and the high water permeability of protein films limit their application as a packaging material. Many studies have been carried out in an attempt to improve the performance of protein-based edible films. The technique for improving the functionality of protein-based edible films is to modify the polymer network through the cross linking of the polymer chains. One effective technique for improving the barrier properties and mechanical strength of protein-based edible films is a cross linking technique using enzymatic methods. Some enzymes that have been used for cross linking proteins include transglutaminase (TGase; EC.2.3.2.13), lipoxygenase, lysyl oxidase, polyphenol oxidase and peroxidase. However, transglutaminase is a kind of enzyme which can catalyze the covalent cross linking reactions between proteins to form high molecular weight (MW) biopolymers. De Jong & Koppelman (2002) reported that transglutaminase catalyzes acyltransfer reactions between λ-carboxyamide groups of glutamine residues (acyl donor) and ε-amino groups of lysine residues (acyl acceptor), resulting in the formation of ε-(λ-glutaminyl) lysine intra and intermolecular cross-linked proteins. The reaction catalyzed of glutamyltransferases is shown in Figure 2 (Yee et al., 1994)

Polymerization using transglutaminase has been investigated with various protein sources including casein, soy proteins and gelatin, where different responses in gel strength were dependent on the reaction conditions and on the different protein sources (Sakamoto et al., 1994). The increase in gel strength of proteins submitted to the action of transglutaminase depended on the order and intensity by which the enzyme produced cross links, and the extent to which these new covalent linkages could impede the 'physical' cross-linkages occurring during renaturation and formation of the triple helix during gel formation (Babin & Dickinson, 2001). Larre et al. (2000) reported that transglutaminase was effective in introducing covalent bonds into films obtained from slightly deamidated gluten. The establishment of these covalent bonds induced the formation of polymers of high molecular weight that were responsible for the greater insolubility of the treated films but a reduced surface hydrophobicity. Mechanical properties showed that the addition of covalent bonds by the use of transglutaminase increased the film's integrity and heavy-duty capacity as well as its capacity to stretch.

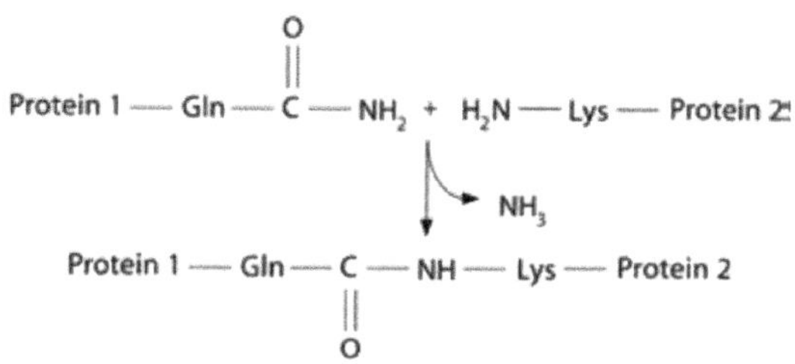

Figure 2. The reaction catalyzed of glutamyltransferases Source: Yee et al. (1994)

Generally, the cross-linkage by transglutaminase improves the tensile strength of protein films, while it decreases the elongation at break and solubility properties. In some cases, such as with isolated soy protein and deamidated gluten films, the transglutaminase treatment also significantly increased the surface hydrophobicity of films (Tang et al., 2005).

However, there are many problems involved in this cross linking technique that have to be investigated before it is commercially applied to films. For example, it is uncertain that this cross-linking by enzyme can improve the properties of films cast from various food proteins. Furthermore, for a certain proteins there are optimal processing parameters to obtain the best effects in the improvement. The improvement in the properties of protein-based edible films by using enzyme seems to be dependent upon the types of substrate protein and some processing parameters, such as the amount of enzyme applied. Thus it is expected that the improvement in the properties of protein-based edible films by enzyme treatment is also affected by the enzyme concentration, since coagulation or aggregation decreases the mechanical properties of protein-based edible films. Jiang et al. (2007) reported that the properties of soy protein films, especially the tensile strength and the hydrophobicity, could be modified by the transglutaminase. However, their work indicated that the modification of properties of soy protein films by transglutaminase was, to a various extent, dependent upon many processing parameters. These included the concentration of enzyme, the condition of the film-forming solutions and the air drying temperature. The influence of these processing parameters could account for the aggregation of soy protein films induced by transglutaminase. Thus, the improvement of the tensile strength of soy protein films by transglutaminase could be achieved by inhibiting or delaying the occurrence of this kind of aggregation.

5.3. Modification of protein-based edible films by irradiation treatment

Although proteins are known for showing good film forming abilities, protein films have rather moderate barrier properties. A need exists therefore to search for new substances and processes in order to obtain better products. Inducing cross linking through using irradiation was found to be an effective method for the improvement of both barrier and mechanical properties of protein-based edible films. Generally, irradiation affects proteins by causing conformational changes, oxidation of amino acids, rupture of covalent bonds, and the formation of protein free radicals (Cheftel et al., 1985). Proteins can be converted to higher molecular weight aggregates through the generation of inter-protein cross-linking reactions, hydrophobic and electrostatic interactions, and the formation of disulfide bonds (Davies & Delsignore, 1987). For example, the hydroxyl and superoxide anion radicals that are generated by the radiation of film-forming solutions can modify the molecular properties of proteins. This can result in the alteration of the protein films by covalent cross-linkages formed in the protein solution after irradiation (Garrison, 1987).

The formation of high molecular weight aggregates was negligible at the low-dose range, but increased significantly with higher doses. Irradiation treatment for the improvement protein films, such as gamma irradiation, has been extensively used to modify protein. Ouattara et al. (2002) used gamma irradiation cross-linking to improve the water vapor permeability and the chemical stability of milk protein films. The results showed that gamma irradiation significantly (p< 0:05) reduced water vapor permeability and increased resistance to microbial and enzymatic biodegradation. An increase in the concentration of high molecular weight proteins in the film forming solution was also observed. They pointed out that two hypotheses may explain the effect on gamma irradiation: (i) The participation of more molecular residues in intermolecular interactions when used in proteins with different physicochemical properties. (ii) The formation of inter- and/or intra-molecular covalent cross-links in the film-forming solutions.

Ciesla et al. (2004) also investigated the effect of gamma irradiation on the physical properties of milk proteins. They found that the viscosity of the irradiated proteins film solutions was increased as compared to the control films. This was because the gamma irradiation developed a "fine-stranded" structure of the protein gel. The creation of the better ordered gels after irradiation corresponds well to the rearrangement of the cross linked β-phase (accompanied by the reorganization of a periodic phase). Using gamma irradiation therefore causes more improvement in well organized β-conformation than non irradiated milk protein films. In addition, the presence of the better ordered protein conformations in gels obtained from irradiated solutions leads to production of more "crystalline" films. These films are characterized by improved barrier properties and mechanical resistance and higher rigidity than those prepared from the non irradiated solutions.

Lee et al. (2005) reported that the gamma irradiation treatment of the gluten film solutions caused the disruption of the ordered structures of the protein molecules. It changed tensile strength, elongation at break and water vapor permeability. They observed that the increased tensile strength of gluten films suggests that cross-linking occurred as a result of gamma irradiation treatment. The increase in tensile strength was possibly caused by the increase of the aggregation of polypeptide chains under experimental condition in this study. It can be assumed that the formation of high molecular weight proteins, aggregated from cleaved polypeptide chains by gamma-irradiation, may be responsible for the reduction of water vapor permeability by reducing the rate of diffusion through the film.

Soliman et al. (2009) studied the influence of gamma irradiation on the mechanical and water barrier properties of corn protein-based films. This study showed that gamma irradiation treatment has potential for modifying the physicochemical properties of zein based films, particularly the water barrier properties. Through the formation of high molecular weight proteins from disaggregated protein particles and cleaved polypeptide, chains can be generated by gamma irradiation. The linkages formed can reduce the absorption of water molecules into the film and the diffusion through the film.

5.4. Modification of protein-based edible films by combination with hydrophobic materials

Generally protein films had good mechanical properties. However, the hydrophilic nature of protein films causes them to be less effective moisture barriers. Conversely, lipid films are good moisture barriers, but they are usually opaque, relatively inflexible, unstable (they tend to be rancid) and taste like waxy. Improved film performances are obtained with a multi-component system where proteins form a continuous and cohesive network, and the lipids provide moisture barrier properties. Lipids can form a layer over the hydrocolloid matrix (bilayer films) or can be dispersed into the matrix (emulsified films). In practice emulsified films have received more interest than bilayer films. Two models have been proposed to describe transfer through emulsified films. Ukai et al. (1976) proposed the 'microvoid model' and suggested that the mass transfer of gases and vapors occurs through microvoids. These are formed between the micro particles of the hydrophobic material and the hydrocolloid matrix during emulsion drying. Krochta et al. (1990) proposed an alternative model, by the 'micro pathway model'. This attributes mass transfer through the high polymer matrix itself. This can occur because proteins are often quite compatible with moisture and gases and can offer little resistance to their transmission.

The addition of lipids to protein films may interfere with polymer chain-to-chain interactions and/or provide flexible domains within the film. Because of their lack of cohesive structural integrity they could also affect the mechanical properties of the protein film. Pérez-Gago & Krochta (2000) reported that the type and content of the lipids were important in controlling the water vapor permeability of protein emulsion films but also had a negative effect on their mechanical properties. Consequently, the components of protein films should be carefully selected and in accord with the final application of the protein films. A protein film should be resistant in order to withstand manipulation during its application and to maintain its integrity and also its barrier properties.

McHugh & Krochta (1994) produced whey protein and lipid emulsion films and found that the water vapor permeability of the films was reduced through the incorporation of lipids. Fatty acid and beeswax emulsion films exhibited very low water vapor permeability. Gontard et al. (1994) also reported that beeswax was the most effective lipid to improve the moisture barrier of films prepared from wheat gluten. Combining wheat gluten protein with diacetyl tartaric ester monoglycerides reduced water vapor permeability, increased tensile strength and maintained transparency. Anker et al. (2002) produced composite whey protein isolated lipid films (laminate and emulsion films) to improve barriers against water vapor. The laminated whey protein lipid film decreased the water vapor permeability 70 times compared with the whey protein film. The water vapor permeability of the emulsion films was half the value of the whey protein isolated film. Regarding the mechanical properties, the results showed that the lipid functioned as an apparent plasticizer by enhancing the fracture properties of the emulsion films. Bertan et al. (2005) incorporated Brazilian elemi (highly hydrophobic resinous oil) into a gelatin film using a blend of palmitic and stearic acids. They evaluated the physicochemical characteristics of the resulting films, all of which contained triacetin as the plasticizer. For films with added acids, the blend and the elemi presented better water vapor barrier properties as compared to the gelatin/triacetin film. However, the mechanical resistance decreased with the addition of the lipids and the opacity and soluble matter increased.

5.5. Modification of protein-based edible films by combination with synthetic polymers

Generally protein-based edible films show an excellent oxygen barrier property at low to intermediate relative humidity as well as fairly good mechanical properties. However, their barrier against water vapor is poor due to their hydrophilic nature (Avena-Bustillos & Krochta, 1993). Many researchers have focused on improving the film properties of protein-based edible films, particularly their mechanical and water vapor barrier properties, by the modification of the films. However, despite the improvement in protein film properties, their physical, thermal, and mechanical properties are still not satisfactory and these present difficulties in many applications. Existing composite films containing layers of different film materials may be needed. Therefore, there is an increasing interest in the development of protein-based edible films for packaging materials. These must have suitable properties for application and can be disposed of after use in an economically and ecologically acceptable way.

Oxygen-barrier layers in food packaging materials typically consist of expensive synthetic barrier polymers including low density polyethylene (LDPE) and polypropylene (PP). These are commonly used in the form of coextruded or laminated films and coatings. It is important to choose the proper polymer matrix as the mechanical and barrier properties can be improved to a great extent by forming the composite plastic films on conventional protein films. To obtain fundamental data on these composite structures further investigations are needed. These should focus on both the optimization of coating formulations used to prepare biopolymer-coated plastic films, and the evaluation of their physical properties.

Hong & Krochta (2006) reported that smooth and transparent oxygen-barrier coatings based on whey proteins can be formed on common plastic films such as polyethylene and polypropylene. The resulting whey protein-coated plastic films with glycerol as a plasticizer have excellent oxygen-barrier properties at low to intermediate RH, and are comparable to synthetic oxygen barriers. These coated films also have good visual quality and adhesion between the coating and the substrate. Because of this, it is suggested that whey protein coatings could be a new biopolymer oxygen barrier. These have great potential for partly replacing existing expensive synthetic barrier polymers in various composite packaging structures.

Lee et al. (2008) characterized the protein-coated polypropylene films as a new composite structure for application in food packaging. The composite structure of polypropylene film coated with three kinds of proteins (soy protein isolate: SPI, whey protein isolate: WPI, corn zein: CZ) could be obtained by a simple casting method. High glossy surfaces were observed on the films coated with WPI and CZ. Proteins exerted a noticeable effect on color of the coated films. WPI coated films also showed greater transparency and tensile strength than the other coated films. These results suggested that WPI coatings with a proper plasticizer possess excellent visual and mechanical characteristics and have great potential for application in food packaging systems.

Tihminlioglu et al. (2010) determined the water vapor and oxygen-barrier performance of corn–zein coated polypropylene films. They found that significant improvements in water vapor and oxygen barrier properties of uncoated polypropylene films were obtained with corn–zein coating. The water vapor permeability of the coated films decreased significantly with increasing corn–zein concentration. The application of plasticized corn–zein coating on polypropylene films showed nearly a three times reduction in oxygen permeability. The high water vapor and oxygen-barriers were obtained for films coated with coating formulation consisting of higher amounts of corn–zein.

6. Application of protein-based edible films

Protein-based edible films offer alternative packaging without adversely affecting the environmental costs. However, edible films are not meant to totally replace synthetic packaging or to limit moisture, aroma and lipid migration between food components where traditional packaging cannot function. For instance, protein-based edible films can be used for versatile food products to reduce loss of moisture, to restrict absorption of oxygen, to lessen migration of lipids, to improve mechanical handling properties, to provide physical protection, or to offer an alternative to commercial packaging materials (Kester & Fennema, 1986).

Protein- based edible films have impressive gas barrier properties compared with those prepared from lipids and polysaccharides. When they are not moist, the O_2 permeability of soy protein-based film was 500, 260, 540 and 670 times lower than that of low-density polyethylene, methylcellulose, starch and pectin respectively (Cuq et al., 1998). In addition, the mechanical properties of protein-based edible films are also better than those of polysaccharide and fat-based films. This is because proteins have a unique structure which confers a wider range of functional properties, especially a high intermolecular binding potential. Protein- based edible films may be able to partially replace some of the conventional synthetic packaging materials used to preserve and protect foods. These protein-based edible films should not be used alone, since contamination during food handling could occur, but would be used to wrap foods inside a secondary synthetic package during food distribution and storage. The wraps could also be used in the home to cover leftovers in the refrigerator, peeled fruit mixtures or as a sandwich bag for lunch. Because such wraps are biodegradable and may even be eaten, they are not harmful to the environment. This attribute could also reduce waste disposal costs (McHugh et al., 1996).

Several researchers have studied the application of protein-based edible films in food use and an excellent review of these is given by Gennadios et al. (1994). They reviewed the applications of several protein-based edible films, such as corn zein on nut and fruit products, casein emulsion film on fruit, and whey protein films on fruit products. However, the application of edible films from water-soluble fish protein in surimi wash-water was not mentioned in their review. One of the potential uses of protein films is in reducing lipid oxidation due to the film's excellent barrier properties. Stuchell & Krochta (1995) used whey protein isolate and acetylated monoglyceride to maintain the quality of frozen king salmon and found a delay in the lipid oxidation onset and a reduction in moisture loss rate. Herald et al. (1996) used corn zein with an antioxidant and emulsifier to maintain the quality of cooked turkey. They found that dipping it in corn zein resulted in very dry products in terms of sensory evaluation. However, corn zein with an antioxidant and emulsifier reduced hexanol after 3 days when compared with PVDC films. Schou et al. (2005) used the sodium caseinate films for wrapping bread. They found that bread samples wrapped with single or double layered sodium caseinate film had the lower value in the compression test among the unwrapped samples. These films prevented the hardening of the bread to the same extent as the polyvinyl chloride films after 3 h storage. After 6 h storage, the sodium caseinate films were not as effective as synthetic film but still reduced the hardening of the bread relative to the unwrapped samples.

7. Conclusion

The protein-based edible films have advantages, including their use as edible packaging materials, over synthetic films. In addition, these films have impressive gas barrier properties compared with those prepared from lipids and polysaccharides. The mechanical properties of protein-based edible films are also better than those of polysaccharide and fat-based films because proteins have a unique structure. Technically, the properties of protein-based edible films depend on the type of proteins, polymer chemistry, processing conditions and the additives used. However, protein-based edible films still show poor both water vapor resistance and mechanical strength in comparison with synthetic polymers and this limits their application in food packaging. Their properties could be improved by chemical and enzymatic methods, by combining them with hydrophobic material or some synthetic polymers or by using a physical method. The resulting film properties depend on the methods of modification and conditions. The chemical and enzyme modifications were efficient in increasing mechanical and water barrier properties. However, modification with chemical compounds, especially aldehydes, would cause toxicity. This must be a matter for concern in the application of chemical modification.

Edible protein-based edible films combination with hydrophobic materials and synthetic polymer can result in better functionality than films produced with only proteins, especially with respect to their mechanical and barrier properties. Irradiation was found to be an effective method for the improvement of both barrier and mechanical properties of protein-based edible films. Protein-based edible films show impressive gas barriers; hence protein films would be a suitable food packaging material for preventing the growth of aerobic microbial and lipid oxidation in lipid enriched foods.

Author details

Thawien Wittaya
Prince of Songkla University, Department of Material Product Technology, Thailand

8. References

Andreuccetti, C.; Carvalho, R.A. & Grosso, C.R.F. (2009). Effect of hydrophobic plasticizers on functional properties of gelatin-based films. *Food Research International,* Vol.42, No.8, (October 2009), pp. 1113–1121. ISSN: 0963-9969.

Anker, M.; Berntsen, J.; Hermansson, A.-M. & Stading, M. (2002). *Improved water vapor barrier of whey protein films by addition of an acetylated monoglyceride.* Innovative Food Science and Emerging Technologies, *Vol.3, No.1, (April 2002), pp. 81–92. ISSN: 1466-8564.*

Anker. M.; Stading, M. & Hermasson, A. (2001). Aging of whey protein films and the effect on mechanical and barrier properties. *Journal of Agricultural and Food Chemistry,* Vol.49, No.2, (February 2001), pp. 989-995. ISSN: 1520-5118.

Avena-Bustillos, R. J. & Krochta, J. M. (1993). Water vapor permeability of caseinate based edible films as affected by pH, calcium cross-linking and lipid content. *Journal of Food Science,* Vol.58, No. 4 (July 1993), pp. 904-907. ISSN: 1750-3841.

Avena-Bustillos, R. J.; Cisneros-Zevallos, L. A.; Krochta, J. M. & Saltveit, M. E. (1993). Optimization of edible coatings on minimally processed carrots using response surface methodology. *Transactions of the ASAE (American Society of Agricultural Engineers),* Vol. 36, No. 3 (May- June, 1993), pp. 801-805. ISSN: 0001-2351.

Babin, H. & Dickinson, E. (2001). Influence of transglutaminase treatment on the thermoreversible gelation of gelatin. *Food Hydocolloids,* Vol.15, No. 3, (May 2001), pp. 271–276. ISSN: 0268-005X.

Banker, G. S. 1966. Film coating theory and practice. *Pharmaceutical Sciences,* Vol. 55, No.1, (January 1966), pp. 81-89. ISSN:1520-6017.

Bertan, L. C.; Tanada- Palmu, P. S.; Siani, A. C. & Grosso, C. R. F. (2005). Effect of fatty acids and 'Brazilian elemi' on composite films based on gelatin. *Food Hydrocolloids,* Vol.19, No.1 (January 2005), pp. 73–82. ISSN: 0268-005X.

Bradenburg, A. H.; Weller, C. L. & Testin, R. F. (1993). Edible films and coatings from soy protein. *Journal of Food Science,* Vol.58, No. 5, (September 1993), pp. 1086-1089. ISSN: 1750-3841.

Bourtoom, T. (2008). Factors Affecting the Properties of edible film prepared from mung bean proteins. *International Food Research Journal,* Vol. 15, No. 2, (May 2008), pp. 167-180. ISSN: 2231-7546.

Cao, N.; Fu, Y.; & He, J. (2007). Mechanical properties of gelatin films cross-linked, respectively, by ferulic acid and tannin acid. *Food Hydrocolloids,* Vol.21, No.4, (June2007), pp. 575-584. ISSN: 0268-005X.

Cheftel, J. Cuq, C, J. L. & Lorient, D. (1985). Amino acids, peptides, and proteins. In: *Food Chemistry,* O. R. Fennema, 245-370. Marcel Dekker Inc., ISBN: 0-8342-1307-9, New York.

Cho, S. Y. & Rhee, C. (2004). Mechanical properties and water vapor permeability of edible films made from fractionated soy proteins with ultrafiltration. *Lebensmittel-Wissenschaft und-Technologie,* Vol.37, No.8, (December 2004), pp. 833–839. ISSN: 0023-6438.

Choi, J. S.; Lim, S. T.; Choi, H. J.; Mohanty, A. K.; DrzaL, L. T.& Misra, M. (2004). Preparation and characterization of plasticized cellulose acetate biocompostite with natural fiber. *Journal of Materials Science,* Vol.39, No.21, (November 2004), pp. 6631–6633. ISSN: 1573-4803.

Ciesla, K.; Salmieri, S.; Lacroix, M.& Le Tienb, C. (2004). Gamma irradiation influence on physical properties of milk proteins. *Radiation Physics and Chemistry,* Vol. 71, No.1-2, (September–October 2004), pp. 93–97. ISSN: 0969-806X.

Cuq, B.; Gontard, N. & Guilbert, S. (1998). Proteins as agricultural polymers for packaging production.*Cereal Chemistry,* Vol.75, No.1, (January/February 1998), pp.1-9. ISSN: 0009-0352.

Cuq, B.; Gontard, N.; Cuq, J. L. & Guilbert, S. (1996a). Stability of myofibrillar protein-based biopackagings during storage. *Lebensmittel-Wissenschaft und-Technologie,* Vol.29, No.4, (1996), pp.344-348. ISSN: 0023-6438.

Davies, K. J. A. & Delsignore, M. E. (1987). Protein damage and degradation by oxygen radicals. III. Modification of secondary and tertiary structure. *Journal of Biological Chemistry,* Vol. 262, No. 20, pp. 9908–9913. ISSN: 0021-9258.

De Jong, G. A. H. & Koppelman, S. J. 2002. Transglutaminase catalyzed reactions: impact on food applications. *Journal of Food Science,* Vol. 67, No. 8 (October 2002), pp. 2798– 2806. ISSN: 750-3841

Devani, G.; Tapia- Blancido, D. R.; Anon, M. C.; Sorbral, P. J. A.; Mauri, A. N. & Menegalli, F. C. (2009). Effects of drying conditions on some physical properties of soy protein films. *Journal of Food Engineering,* Vol. 90, No. 3, (February, 2009), pp. 341-349. ISSN: 0260-8774

Dickey, L. C. & Parris, N. (2002). Serial batch extraction of zein milled maize. *Industrial Crops and Products an International Journal,* Vol. 15, No. 1, (January 2002), pp. 33-42. ISSN: 0926-6690.

Domenek, S.; Feuilloley, P.; Gratraud, J.; Morel, M-H. & Guilbert, S. (2004). Biodegradability of wheat gluten based bioplastics. *Chemosphere,* Vol. 54, No. 4, (January 2004), pp. 551–559. ISSN: 0045-6535.

Farnum, C.; Stanley, D. W.; & Gray, J. I. (1976). Protein–lipid interactions in soy films. *Canadian Institute of Food Science and Technology Journal,* Vol.9. pp. 201–206. ISSN: 0315-5463.

Garrison, W. M. (1987). Reaction mechanism in the radiolysis of peptides, polypeptides, and proteins. *Chemical Review,* Vol. 87, No.2 (April, 1987), pp.381-398. ISSN:1520-6890.

Gennadios, A.; Brandenburg, A. H.; Weller, C. L. & Testin, R. F. (1993). Effect of pH on properties of wheat gluten and soy protein isolate films. Journal of Agricultural and Food Chemistry, Vol. 41, No. 11, (November, 1993), pp. 1835-1839. ISSN: 0021-8561

Gennadios, A.; McHugh, T. H.; Weller, C. L. & Krochta, J. M. (1994). Edible coating and films based on protein. In: *Edible Coatings and Films to Improve Food Quality,* Krochta, J. M., Balwin, E. A. & Niperos Carriedo, M. O., pp. 201-277, Technomic Publishing, ISBN: 978-1-4200-3198-0 , Basel.

Gennadios, A. & Weller, C. L. (1991). Edible films and coatings from soymilk and soy protein. *Cereal Foods World,* Vol. 36, No. 6 (November, 1991), pp.1004-1009. ISSN: 0146-6283.

Gennadios, A. & Weller, C. L. (1990). Edible films and coating from wheat and corn proteins. *Food Technology,* Vol.44, No. 10, (November 1990), pp. 63-69. ISSN: 1365-2621.

Gontard, N.; Duchez, C. L Cuq, J.-L. & Guilbert, S. 1994. Edible composite films of wheat and lipids: Water Vapour permeability and other physical properties. *International Journal of Food Science and Technology,* Vol.29, No.1, (February 1994), pp.39–50. ISSN: 1365-2621.

Gueguen, J, Viroben, G. ; Noireaux, P. and Subirade, M. 1998. Influence of plasticizers and treatments on the properties of films from pea protein. *Industrial Crops and Products an International Journal,* Vol. 7, No. 2 (January 1998), pp. 149–57. ISSN: 0926-6690

Guilbert, S. 1986. Technology and application of edible protective films. In: Food packaging and preservation, Theory and Practice, Matholouthi, M., pp. 371-394. Elsevier Applied Science Publishing, ISBN: 0853343772. London:

Herald, T. J.; Hachmeister, K. A.; Huang,S. & Bowers,J. R. (1996). Corn zein packaging materials for cooked turkey. *Journal of Food Science,* Vol.61, No.2, (March 1996), pp. 415-418. ISSN:1750-3841.

Hernandez-Mun, P.; Villalobos, R. & Chiralt, A. (2004). Effect of cross-linking using aldehydes on properties of glutenin-rich films. *Food Hydrocolloids,* Vol.18, No.3, (May 2004), pp. 403–411.ISSN: 0268-005X.

Hong,Seok-In.& Krochta.J.M. (2006). Oxygen barrier performance of whey-protein-coated plastic films as affected by temperature, relative humidity, base film and protein type. *Journal of Food Engineering,* Vol.77, No.3, (December 2006), pp. 739–745. ISSN: 0260-8774.

Hood, E. E.; Shen, Q. X. & Varner, J. E. (1988). A developmentally regulated hydroxyproline-rich glycoprotein in maize pericarp cell walls. *Plant Physiology,* Vol. 87 ,No.1, (May1988), pp.138–142. ISSN: 1532-2548.

Jangchud, A. & Chinnan, M. S. (1999). Peanut protein film as affected by drying temperature and pH of film forming solution. *Journal of Food Science,* Vol.64, No.1, (January 1999), pp. 153-157. ISSN: 1750-3841.

Jaynes, H.O. & Chou, W.N. (1975). New method to produce soy protein-lipid films. *Food Product Development,* Vol.9, No. 4 (August, 1976), pp. 86-90. ISSN: 2223-9944.

Jiang, Y.; Tang, C. H.; Wen, Q. B.; Li, L. & Yang, X. Q. (2007). Effect of processing parameters on the properties of transglutaminase-treatedsoyprotein isolate films. *Innovative Food Science & Emerging Technologies* ,Vol.8, No.2, (June 2007), pp. 218–225. ISSN: 1466-8564.

Keereekasetsuk, S. & Bourtoom, T. (2008). Influence of plasticizers on the properties of edible film from mung bean protein, *14th World Congress of Food Science and Technology,* Shanghai, China, October 2008.

Kester, J. J. & Fennema, O. R. (1986). Edible films and coatings: a review. *Food Technology,* Vol.40, No.12, (December 1986), pp. 47-59, ISSN: 0015-6639.

Kinsella, J. E. & Phillips, L. G. (1989). Film properties of modified protein. In: *Food Protein,* Kinsella, J. E. & Soucie, W. G., pp. 78-99. Champaign, The American Oil Chemists's Society, ISBN: 9780935315264, Illinois, USA.

Krochta, J. M.; Pavlath, A. E. & Goodman N. (1990). Edible films from casein-lipid emulsions for lightly processed fruit and vegetables. *Postharvest Biology and Technology,* Vol. 4, No. 4, (September 1994), pp. 319-329, ISSN: 0925-5214.

Krull, L. H. & Inglett, G. E. (1971). Industrial uses of gluten. *Cereal Science Today,* Vo.16, No.8, (August 1971), pp. 232-261. ISSN: 0009-0360.

Kinsella, J. (1984). Milk proteins physicochemical and functional properties. *Critical Reviews in Food Science and Nutrition,* Vol. 21, No.3, (September 1984), pp. 197–204. ISSN: 1549-7852.

Larre, C.; Desserme, J.; Barbot, J. & Gueguen, J. (2000). Properties of deamidated gluten films Enzymatically cross-linked. *Journal of Agricultural and Food Chemistry,* Vol.48, No.11, (November 2000), pp. 5444–5449. ISSN: 1520-5118.

Lee, J. W.; Son, S. M. & Hong, S. I. (2008). Characterization of protein-coated polypropylene films as a novel composite structure for active food packaging application. *Journal of Food Engineering,* Vol. 86, No. 4 (June 2008), pp. 484-493, ISSN: 0260-8774.

Lee, S. L. ; Lee, M. S. & Song, K. B. (2005). Effect of gamma-irradiation on the physicochemical properties of gluten films. *Food Chemistry,* Vol. 92, No.4, (October 2005), pp. 621-625. ISSN: 0308-8146.

Marquie, C. (2001). Chemical reactions in cottonseed protein cross-linking by formaldehyde, glutaraldehyde, and glyoxal for the formation of protein films with enhanced mechanical properties. *Journal of Agricultural and Food Chemistry,* Vol.49, No.10, (October 2001), pp. 4676–4681. ISSN: 1520-5118.

Mauri, A. N. & Anon, M. C. (2006). Effect of solution pH on solubility and some structural properties of soybean protein isolate films. *Journal of the Science of Food and Agriculture,* Vol.86, No.6, (May 2006), pp. 1064–1072.ISSN:1097-0010.

McHugh, T. H.; Huxsoll, C. C.& & Krochta, J. M. (1996). Permeability properties of fruit puree edible films.*Journal of Food Science,* Vol.61, No.1, (January 1996), pp. 88–91. ISSN: 1750-3841.

McHugh , T. H. & Krochta, J. M. (1994). Sorbitol-vs glycerol-plasticized whey protein edible films: Integrated oxygen permeability and tensile property evaluation. *Journal of Agricultural and Food Chemistry,* Vol. 42, No. 4, (April 1994), pp. 841-845. ISSN: 1520-5118.

Micard, V.; Belamri, R.; Morel, M. & Guilbert, S., (2000). Properties of chemically and physically treated wheat gluten. *Journal of Agricultural and Food Chemistry,* Vol.48, No.7, (July 2000), pp. 2948–2953. ISSN: 1520-5118.

Orliac, O.; Rouilly, A.; Silvestre, F. & Rigal, L. (2002). Effect of additives on the mechanical properties,hydrophobicity and water uptake of thermo-moulded films produced from sunflower protein isolate. *Polymer,* Vol.43, No.20, (September 2002), pp. 5417–5425. ISSN: 0032-3861.

Ouattara, B.; Canh, L. T.; Vachon, C.; Mateescu, M.A. & Lacroix, M. (2002). Use of γ-irradiation cross-linking to improve the water vapor permeability and the chemical stability of milk protein films. *Radiation Physics and Chemistry,* Vol.63, No.3, (March 2002), pp.821-825. ISSN 0969-806X.

Perez-Gago, M. B.& Krochta, J. M. (2000). Drying temperature effect on water vapor permeability and mechanical properties of whey protein-lipid emulsion films. *Journal of Agricultural and Food Chemistry,* Vol.48, No.7, (July 2000), pp. 2687–2692. ISSN: 1520-5118.

Pochat- Bohatier,C.; Sanchez, J.& Gontard, N. (2006). Influence of relative humidity on carbon dioxide sorption in wheat gluten films. *Journal of Food Engineering,* Vol.77, No.4, (December 2006), pp. 983–991. ISSN: 0260-8774.

Rahman, M. & Brazel, C. S. (2004). The plasticizer market: an assessment of traditional plasticizers and research trends to meet new challenges. *Progress in Polymer Science,* Vol.29, No.12, (December 2004), pp. 1223–1248. ISSN: 0079-6700.

Ross-Murphy, S. B. 1992. Structure and rheology of gelatin gels-Recent progress. *Polymer,* Vol. 33, No. 12. (September 1991), pp. 2622–2627. ISSN: 0032-3861.

Sakamoto, H.; Kumazama, Y. & Motoki, M. (1994). Strength of protein gels prepared with microbial transglutaminase as related to reaction conditions. *Journal of Food Science,*Vol.59, No.4, (July1994), pp.866–871. ISSN: 1750-3841.

Schou, M.; Longares, A.; Montesinos-Herrero, C.F.; Monahan, D. J.;O'Riordan, M.& O'Sullivan.(2005). Properties of edible sodiumcaseinatefilms and their application as food wrapping. *Lebensmittel-Wissenschaft und-Technologie,* Vol.38, No.6, (September 2005), pp. 605–610. ISSN: 0023-6438.

Sisken, B.F.; Zwick, M.; Hyde, J.F.& Cotrill, C.M. (1993). Maturation of the central nervous system: comparison of equine and other species. Equine Veterinary, Vol. 25, No.14, (April 1993), pp.31–34. ISSN: 2042-3306.

Shukla, R. & Cheryan, M. (2001). Zein: The industrial protein from corn. *Industrial Crops and Products anInternational Journal,* Vol. 13, No. 3, (May 2001), pp. 171–192. ISSN: 0926-6690.

Soliman, E. A.; Tawfik, M. S.; Hosni, E.-S. & Moharram, Y. G. (2007). Preparation and characterization of soy protein based edible/biodegradable films. *American Journal of Food Technology,* Vol.2, No.6, (November 2007), pp. 462–476. ISSN: 1557-4571.

Sothornvit, R. & Krochta, J. M. (2001). Plasticizer effect on mechanical properties of b-lactoblobulin films. *Journal of Food Engineering,* Vol.50, No.3, (November 2001), pp. 149–155. ISSN: 0260-8774.

Stuchell Y. M.& Krochta J. M., (1995). Edible Coating on Frozen King Salmon : effect of whey protein isolateand acetylated monoglycerides on moisture loss and lipid oxidation. *Journal of Food Science,* Vol.60, No.1, (January 1995), pp. 28-31. ISSN: 1750-3841.

Subirade, M. ; Mungara, P.; Kelly, I.; Gueguen, J. & Pezolet, M. (1998). Molecular basis of film formation from soybean protein: comparison between the conformation of glycinin in aqueous solution and in films. *Biological Macromolecules,* Vol.23, No. 4, (November 1998), pp. 241-249. ISSN: 0141-8130.

Tae, H. J. (1983). Bifunctional reagents. *Methods in Enzymology,* Vol.91, pp.580–609. ISBN: 0-12-181991-4.

Tang, Y.; Finlay, J. A.; Kowalke, G. L.; Meyer, A. E.; Bright, F. V.; Callow, M. E.; Wendt, D. E. & Detty, M. R., (2005). Hybrid xerogel films as novel coatings for antifouling and fouling release. *Biofouling,* Vol.21, No.1, pp.59 – 71. ISSN: 0892-7014.

Tihminlioglu,F.; Dogan, A. I. & Banu, O. (2010). Water vapor and oxygen-barrier performance of corn–zein coated polypropylene films. *Journal of Food Engineering,* Vol.96, No.3, (February2010), pp. 342–347. ISSN: 0260-8774.

Ukai, N. Y.; Shingo, I.; Kurume, T. T. & Fukuoka, K. M. (1976). Preservation of agricultural products, U.S.Patent, 3,997,674. December, 14.

Vanin, F. M.; Sobral, P. J. A.; Menegalli, F. C.; Carvalho, R. A., & Habitante, A. M. Q. B. (2005). Effects of plasticizers and their concentrations on thermal and functional properties of gelatin-based films. *Food Hydrocolloids,* Vol.19, No.5, (September 2005), pp. 899-907. ISSN: 0268-005X.

Vieira, M. G. A.; Altenhofen, D.; Silva, M.; Oliveira, D. S. L. & Beppu, M. M. (2011). Natural-based plasticizers and biopolymer films: A review. *European Polymer Journal,* Vol.47, No.3, (March 2011), pp. 254-263. ISSN: 0014-3057.

Wang, D.H. & Damodaran, S. (1991). Thermal gelation of globular proteins: influence of protein conformation on gel strength. *Journal of Agricultural and Food Chemistry,* Vol.39, No. 3, (March 1991), pp. 433-438. ISSN: 1520-5118.

Wu, L. C. & Bates, R. P. (1973). Influence of ingredients upon edible protein–lipid film characteristics. *Journal of Food Science,* Vol.38, No. 5, (July 1973), pp. 783-787. ISSN: 1750-3841.

Yang, L. & Paulson, A. T. (2000). Effects of lipids on mechanical and moisture barriers properties of edible gellan flim. *Food Research International,* Vol.33, No.7, (August 2000), pp. 571-578. ISSN: 0963-9969.

Yee, V. C. ; Pedersen, L. C. ; Trong, I. L. ; Bishop, P. D. ; Stenkamp, R. E. & Teller, C. C. 1994. Threedimensional structure of a transglutaminase: Human blood coagulation factor XIII. *Procedure National Academy Science,* Vol. 91, No. 5 (July 1994), pp. 7296-7300. ISSN: 1091-6490.

Zhang, J. & Jane, J. (2001). *Mechanical and thermal properties of extruded soy protein sheets.* Polymer, *Vol. 42, No. 6, (March 2001),* pp. 2569–2578. ISSN: 0032-3861.

Chapter 3

Antioxidant, Anticancer Activity, and Other Health Effects of a Nutritional Supplement (Galaxy®)

Gary M. Booth, Tory L. Parker and Christopher M. Lee

Additional information is available at the end of the chapter

1. Introduction

Approximately one in four prescription drugs from pharmacies in the U.S., Canada, and Western Europe have active ingredients that are plant derived (Balick and Cox, 1997). Edible and even non-edible plants have long been considered sources of anticancer drugs. Indeed, our laboratory (Dr. Booth) published a paper two years ago (Capua et al., 2010) showing that even non-edible and non-tropical desert plants have bioactivity against a wide variety of human cancer cells. From the literature, it is clear that diets rich in grains, fruits, and vegetables are known to reduce cancer risk (Ferguson, et al., 2004; Guthrie and Carrol, 1998) especially those rich in antioxidant activity.

Over the last 20 years, a number of juices and nutritional liquid supplements have appeared on the market purporting high amounts of antioxidant activity and suggesting a number of benefits to human health. In the corridor of Utah Valley, for example, there are at least a dozen manufacturing centers for plant-derived supplements within 50 miles of each other. While these supplements have caught the interest of health enthusiasts throughout the world, it is of interest that very few toxicology studies have been published on these supplements especially with reference to anticancer dose-response curves and antioxidant activity. Three years ago, our laboratory was asked to evaluate the potential health effects of a new nutritional supplement referred to as Galaxy®. This product was of interest to us because it contained a variety of bioactive ingredients including several superfruits (a marketing term for fruits with unusual nutrient and antioxidant properties).

We agreed to investigate this product on five conditions:

1. We develop our own research protocols.

2. Principal investigators accept no personal compensation for the studies.
3. We were allowed to present our data at professional conferences and symposia.
4. We are allowed to publish our results (regardless of the outcome) in the peer-reviewed literature.
5. They provide the product to our lab free of charge.

The company agreed without hesitation which was a fresh departure from the traditional position of the industry that generally requires non-disclosure documents, no-publication policies, and restricted professional presentations of the data sets. Thus, with that agreement, we pressed forward with the following objectives:

1. Investigate the antioxidant activity of the freeze-dried product.
2. Develop dose-response curves using Galaxy® on a variety of human cancer cell lines including calculation of EC50s.
3. Determine if a correlation exists between anticancer activity and antioxidant activity using Galaxy® and several selected superfruits.
4. Compare the toxicity of an approved FDA drug (paclitaxel) with the particulate (most active) fraction of Galaxy®.
5. Develop Bioactivity Indices for Galaxy® and selected superfruits.
6. Develop Selectivity Indices which compares the cytotoxicity between cancer cells and normal cells.
7. Show the effect of Galaxy® on blood glucose levels of senior athletes (n = 308).
8. Demonstrate the effect of Galaxy® on the white blood cell (WBC) count of an acute lymphocytic leukemia (ALL) patient over a number of months without chemotherapy or any other treatment intervention.
9. Determine the simple carbohydrate and amino acid content of a freeze-dried sample of Galaxy®.

2. Materials and methods

Galaxy, a nutritional supplement, was provided by JoyLife International. This blend contains 32 bioactive ingredients. Bioassay data from this product were collected using the straight sample (from the container), as a freeze-dried sample (Fig. 1), supernatant, or particulate fraction. Bioassay procedures for all these matrices were completed using methods developed previously from Brigham Young University (BYU) laboratories (Capua et al., 2010).

All cell lines were grown in the laboratories of BYU or Reaction Biology Corporation. The purity of the cell lines was checked periodically using an inverted microscope (Fig. 2).

The ORAC assay was performed according to published protocols (Parker et al., 2007; Fig. 3). Preparation of the freeze-dried Galaxy sample for sugar extraction was completed by extracting 5 g of the sample with 50 mL of 70 % (w/v) methanol solution in a 100 mL Ehrlenmeyer Flask for 24 hours using a stirring bar. The extract was then analyzed for sugars and amino acids standardized GC/MS procedures from protocols developed at the BYU College of Life Sciences Chromatography Facility.

Figure 1. Dr. Gary M. Booth (left) and Kyle Lorenzen preparing freeze-dried samples of Galaxy®.

Figure 2. Dr. Gary M. Booth (standing) and Matt Dungan viewing the effect of Galaxy® on cancer cells in an inverted microscope.

Figure 3. Dr. Tory L. Parker preparing Galaxy® samples for determination of oxygen radical absorption capacity (ORAC).

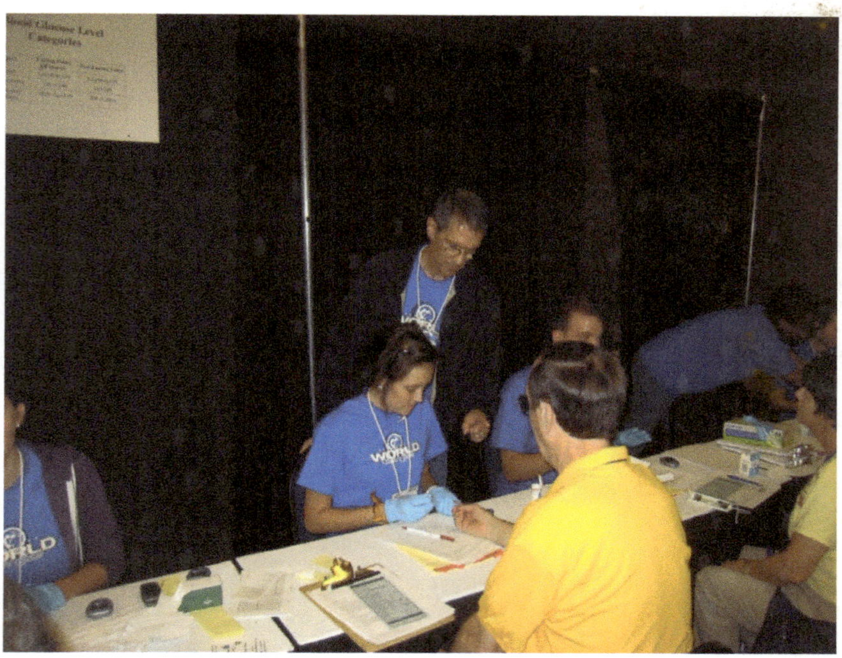

Figure 4. Collection of blood glucose samples from the athletes (n = 308) of the 2011 Huntsman Senior World Games for investigation of the effect of Galaxy® on blood glucose.

Glucose measurements were determined using a Bayer Contour Glucosometer on 308 senior athletes during the 2011 Huntsman Senior Games held in St. George, Utah. The average age of the participants was 64.5 ± 4 years (Fig 4).

Onxol® (paclitaxel; also called Taxol®) an FDA-approved drug for breast cancer was purchased from Cancer Care Northwest in Spokane, WA. This drug was used to compare with the most active fraction (particulate) of Galaxy®.

3. Results and discussion

It is now well documented through epidemiological studies that diets rich in fruits and vegetables can reduce cancer and other chronic diseases (Boivin et al., 2007; Block et al., 1992; Steinmetz et al., 1991). It is believed that the reduction in these chronic diseases is related to the diversity and high concentrations of antioxidants, known collectively as phytochemicals. Furthermore, it has been generally documented that the complex mixtures of phytochemicals in fruits and vegetables are more effective than their individual components in preventing cancer through a variety of mechanisms that include both additive and synergistic effects (Liu et al., 2004; Seeram et al., 2005; Mertens-Talcon et al., 2003; Zhou et al., 2003).

In the U.S., a recent survey (Yang et al., 2011) showed that the total antioxidant capacity (TAC) was positively correlated with daily consumption of fruits and fruit juices, vegetables, and antioxidant-containing beverages. The major sources of the dietary TAC in the U.S. were teas, dietary supplements, and fruits and fruit juices which accounted for 28%, 25%, and 17% of the TAC respectively. Unfortunately, vegetables only contributed 2% of the dietary TAC. Recognizing that their diets may not be optimum, many people supplement their diet with powdered or liquid supplements, such as Galaxy® to get their "daily dose" of dietary TAC.

The nutritional supplement/juice industry has shown consistent growth over the last several years, mostly through consumption by the older generation. However, even though this economic growth spurt has stimulated the economy and possibly even contributed to the health of our population, the supporting scientific data from these products has often been lacking. The experimental data from this product now follows:

A freeze dried sample of Galaxy® had an ORAC value of 178.10 ± 15.02 μmoles TE/g (n=4), while the fresh product was 35.6 ± 3.1 μmoles TE/g wet weight. The USDA recommends (USDA, 1999) that each person consumes approximately 3000-5000 ORAC units per day. As of this writing, the average person in the U.S. consumes less than 1000 ORAC units/day. Based on our data (200 mg particulates/one mL Galaxy®), if a person consumes 30 mL of Galaxy® per day, they would consume about 1068 ORAC units per day or 21-36% of the recommended units per day. And if one consumes it 2X/day, the values double, or 42-72% of the recommended units. Thus, one bottle of Galaxy® contains about 26,715 ORAC units using the freeze-dried data or 26,700 ORAC units if the fresh weight data are used. Thus, Galaxy® juice from the bottle has a higher ORAC value than 21 of the 22 fruit juices listed by the USDA (Haytowitz and Bhagwat, 2010). Only black raspberry juice (ORAC = 10,460 μmoles TE/100 g wet weight) had a higher antioxidant rating.

The freeze-dried Galaxy® ORAC value of 178.1 μmoles TE/g is approximately the same antioxidant capacity as that reported for cranberry (Sun et al., 2002) which is considered the reference standard for calculation of Biological Indices (BI). Thus, Galaxy® has a BI of 3.40 which of course is 3.40X higher than the cranberry standard (BI = 1). Table 1 shows a summary of the ORAC values, EC50s, and BI for Galaxy® and several superfruits of which three (mangosteen, GAC, and Acai) are contained in this product. The BI is a useful parameter to compare the combined effects of anticancer activity and antioxidant activity. Any BI above 1 is considered relatively good. Thus, Galaxy has a good BI compared against the individual fruits. Except for mangosteen and GAC, all of these superfruits were in the same order of magnitude ranging from 1.73 to 8.46. There was no correlation ($r^2 = 0.07$; $p = 0.473$) between the anticancer (EC50) parameter and the ORAC value for several superfruits suggesting that antioxidative activity may not be totally related to in vitro anticancer activity (Fig. 5). Still, a careful look at Fig. 5 shows that five of the nine samples (including Galaxy® particulates) had ORAC values above 150 μmoles TE/g and an EC50 less than 1mg/mL. Thus, ORAC and EC50s values for these superfruits appear to be related, just not in a clear inverse linear fashion.

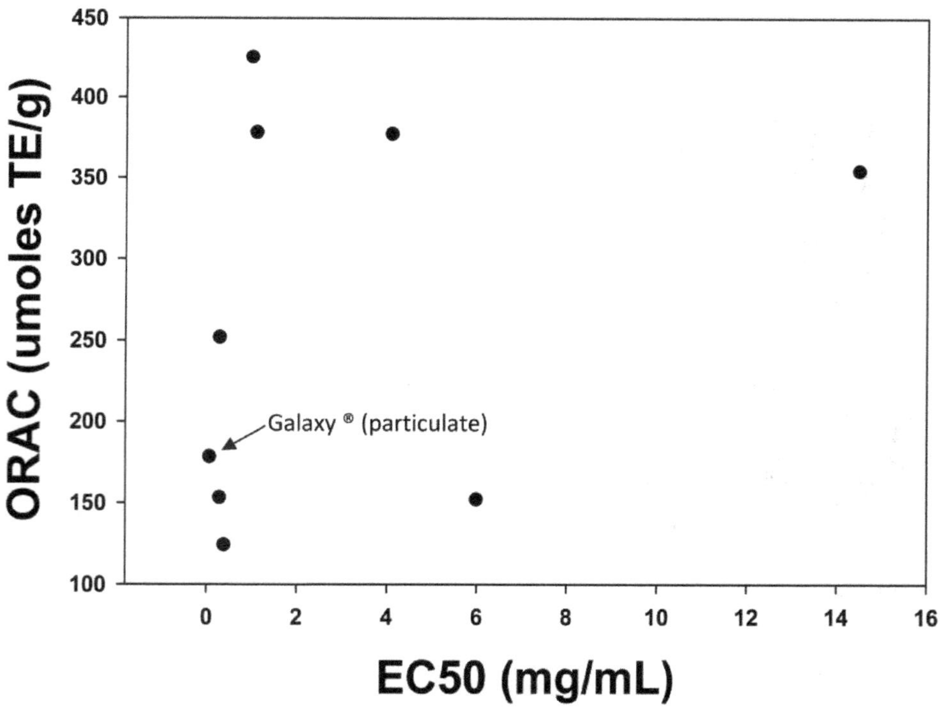

Figure 5. Plot of the relationship between ORAC values and EC50 values for Galaxy® and eight superfoods

Correlation studies on these parameters with other fruits have also been equivocal; some are positively correlated (Wang and Lewers, 2007; Faria et al., 2006) while others are not (Boivin et al., 2007). A recent review (Wang et al., 2011) demonstrated that there is no conclusive proof that high antioxidant activity is a good indicator of high anticancer activity but left the reader with the challenge to continue to test this hypothesis until it is unequivocally resolved one way or the other. Similar studies with total phenolics have also shown mixed results. Some studies of total phenolics contained in traditional fruit, have shown to be highly correlated with anticancer activity (Silva et al., 2006; Atmani et al., 2011) and they (phenolics) are also highly correlated with ORAC values (Atmani et al., 2011; Kalt et al., 2003). However, some researchers have shown that there is no correlation between total phenolics and anticancer activity (Thompson et al., 2009; Sun et al., 2002; Weber et al., 2001). Thus, the expected link between ORAC values and total phenolics and anticancer activity continues to be a subject of debate in the literature.

Fruit	ORAC[a]	Score[b]	EC50 (mg/mL)	Score[c]	BI[d]
Mangosteen	251.60	1.42	0.30	48.33	24.88
GAC	153.00	0.86	0.30	48.33	24.60
Acai	425.10	2.42	1.00	14.50	8.46
Goji	378.10	2.14	1.10	13.20	7.67
Galaxy®	178.10	1.00	2.30	6.30	3.65
Sea Buckthorn	377.10	2.13	4.10	3.54	2.84
Noni	151.80	0.86	5.60	2.59	1.73
Cranberry	354.27	1.00	14.50	1.00	1.00

[a] Oxygen Radical Absorption Capacity (ORAC) = umoles TE/g freeze-dried product
[b] Score for total antioxidant activity = sample ORAC value/cranberry total antioxidant activity
[c] Score of antiproliferative activity = cranberry EC50 value/sample EC50 value
[d] BI = score of total antioxidant activity + score of antiproliferative activity/2

Table 1. Summary of the total antioxidant activity (ORAC), antiproliferative (EC50), and the ranked Bioactivity Index (BI) for Galaxy® and seven superfruits; mangosteen, acai, goji, and cranberry are found in Galaxy found in Galaxy® compared against the cranberry standard.

Fig. 6 shows the in vitro effect of Galaxy® straight (raw product) on MDA-MB-231 breast cancer cells. Galaxy® has an EC50 = 2.3 which is an order of magnitude lower (more cytotoxic) than most traditional fruits (Weber et al., 2001) but higher (less cytotoxic) than most chemotherapy drugs (Frankfurt and Krishan, 2003).

Figs. 7 and 8 show the in vitro dose-response curve for the effect of the Galaxy® supernatant and particulate fractions respectively on breast cancer cell line MDA-MB-231. It is clear that the supernatant (Fig. 7: EC50=3.4) and the particulate fraction (Fig. 8: EC50=0.075 mg/mL) were both cytotoxic to this cell line, but the particulate fraction was 47X more toxic than the supernatant fraction. This indicates that most of the phytochemicals (98%) in Galaxy® contributing to the anticancer activity are membrane-bound. These data are in contrast to those reported by Wang et al. (1996) who found less than 10% of the ORAC activity in a wide variety of different fruit pulp or particulates; they found most of the bioactivity in the extracted juice. However, the USDA ORAC database (Haytowitz and Bhagwhat, 2010) apparently also showed that most of the antioxidant activity in a number of fruits was associated with the pulp, and not the juice.

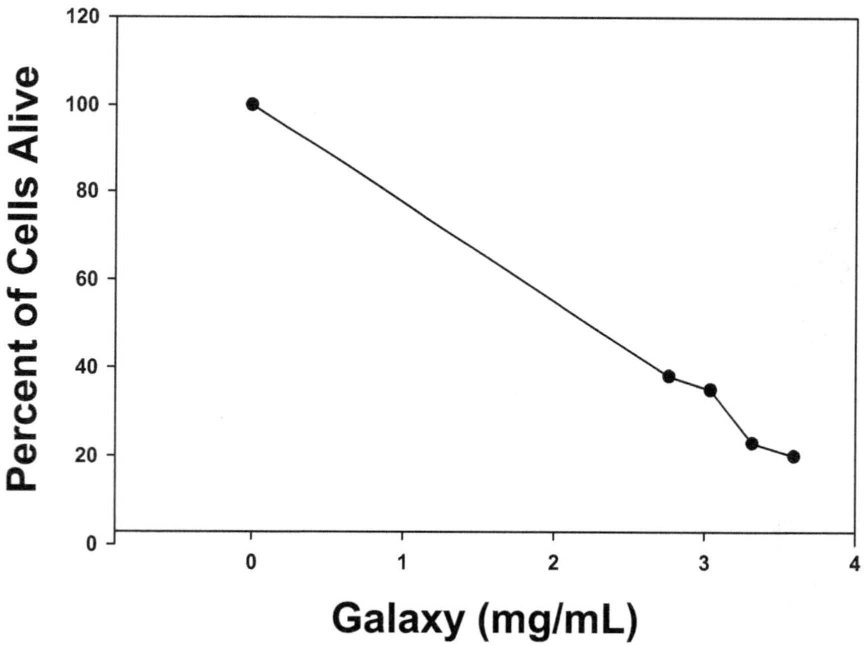

Figure 6. Galaxy® (straight from container) effect on MDA-MB-231 breast cancer Cells (EC50= 2.3 mg/mL).

When the particulate dose-response curve was compared to the FDA-approved breast cancer drug Onxol® (paclitaxel), the two curves tended to track each other with both EC50s being in the same order of magnitude (Fig. 8; EC50=0.03 mg/mL for paclitaxel; EC50=0.075 mg/mL for the Galaxy® particulates). The EC50 for paclitaxel in our study was in the same order of magnitude as that reported by Danhier et al. (2009) for HeLa cells, but higher than that reported by Yu et al. (2004) for MCF-7 and MDA-MB-231 cell lines.

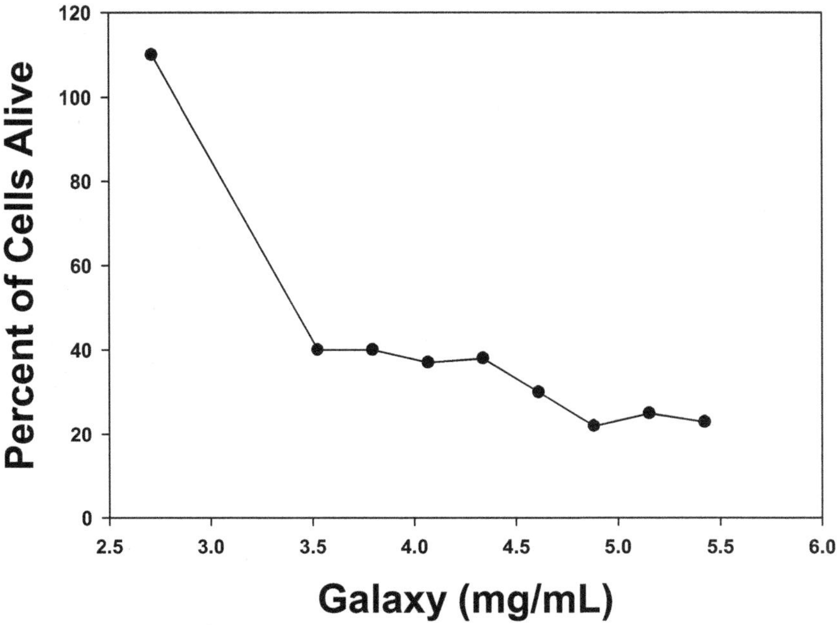

Figure 7. Galaxy® (supernatant) toxicity to MDA-MB-231 Human Breast Cancer Cells (EC50 = 3.4).

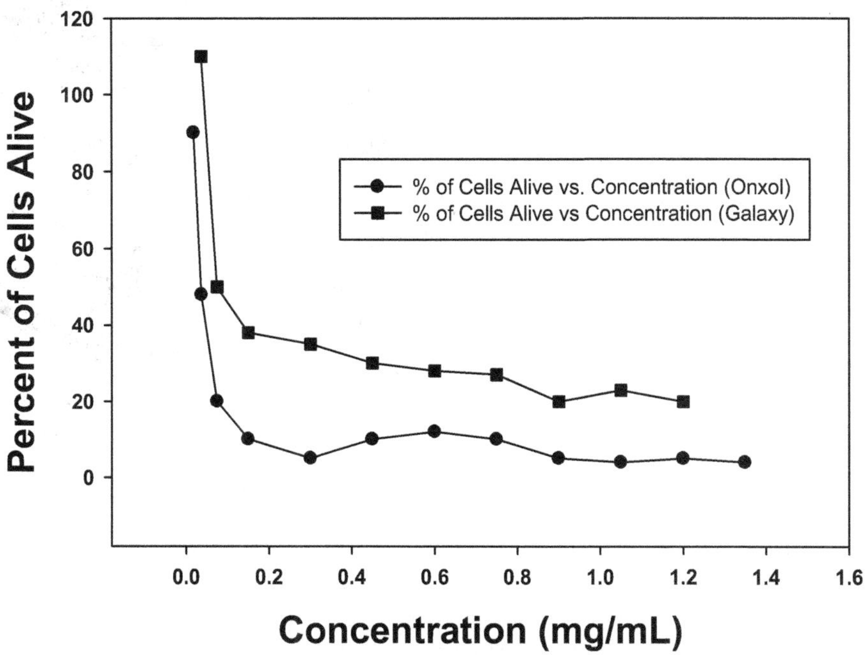

Figure 8. Comparison of Galaxy® Particulates and Onxol® (FDA approved drug) against Human Breast Cancer Cell Line MDA-MB-231 (Galaxy® EC50 = 0.075 mg/mL; Onxol® EC50 = 0.03 mg/mL).

We have also tested Galaxy® against several other human cancer cell lines and calculated a Selectivity Index (SI) which compares EC50s of cancer cells and normal cells. Table 2 summarizes the EC50s and ranked (SI) for various samples of Galaxy® on breast cancer, lung cancer, and liver cancer compared to its effect on normal tissues. The higher the SI values, the higher the selectivity of Galaxy® against cancer cells compared to the normal tissues. Any SI above 2 is considered a reasonably selective SI (Basida et al., 2009). Note that the SI values for Galaxy® ranged from 3.2 (for lung cancer) to over 370 for the particulate fraction. Badisa et al. (2009) reported that pure anticancer compounds (piperidinyl-DES, Pyrrolidinyl-DES, and 4-hyroxy tamoxifen) had SI values that ranged from 1.29 to >2.84. They further indicated that an SI value less than 2 probably indicates general toxicity of pure compounds (Koch et al., 2005). Thus, one of their three compounds, pyrrolidinyl-ES showed a high degree of cytotoxic selectivity, while 4-hydroxy tamoxifen (an anti-breast-cancer drug) showed an SI of less than 2 suggesting general toxicity to cells according to the recommendation of Koch et al.(2005).

Cell Line	Cell Type	Galaxy Sample	EC50 (mg/mL)	SI*
MDA-MB-231	Breast cancer	Particulates	0.075	371
	Breast cancer	Unfractionated	2.30	12
	Breast cancer	Supernatant	3.5	8
Hs578Bst	Normal breast cells	Freeze-dried	27.80	-
HepG2	Liver cancer	Freeze-dried	3.44	3.5
THLE-3	Normal hepatocytes	Freeze-dried	11.99	-
A-549	Lung cancer	Freeze-dried	6.20	3.20
MRC-5	Normal lung cells	Freeze-dried	19.68	-

*SI = EC50 on normal tissue/EC50 on cancer cells

Table 2. Summary of the EC50s and ranked Selectivity Indices (SI) for Galaxy® samples on human breast cancer, lung cancer, and liver cancer compared to normal human tissues.

In addition, Al-Qubaisi et al. (2011) reported an SI of 7.6 for liver cell comparisons using Goniothalamin isolated from a plant compared to Galaxy®'s SI of 3.5; however, Galaxy® was over 1,700X less toxic to normal liver cells when the EC50s of Galaxy® and Goniolthalamin were compared. Apparently, phytotherapy (using mixtures of dozens of fruit-based and vegetable-based polyphenols and other antioxidants such as those found in Galaxy®) are often less toxic than the purified compounds used in mainstream cancer therapy. Indeed, Krishna et al. (2009) has suggested that no anti-neoplastic drug is devoid of side effects which includes the widely used anti-cancer drug, paclitaxel. Our in vitro data suggests that the phytochemicals found in Galaxy® are quite selective for cancer cells. While these preliminary data on Galaxy® are encouraging, the authors caution all readers to not extrapolate conclusions from the in vitro data to what might happen in large randomized double-blind in vivo investigations.

Figs. 9-12 summarizes the dose-responses curves of freeze-dried Galaxy® on colon cancer, prostate cancer, lung cancer, and liver cancer. The EC50s ranged from about 4.0-11 mg/mL which is an order of magnitude higher in cytotoxicity (lower EC50) than that reported for four raspberry varieties and one apple variety (Weber et al., 2001) and for a variety of other fruits (Sun et al., 2002). Thus, Galaxy® seems to have a broad-spectrum of in vitro antiproliferative and antioxidant activities. However, it is still not known if or how the individual components in the product interact to produce these toxicities.

Three hundred and eight senior athletes had their baseline blood glucose levels taken, then drank 30 mL of Galaxy®, and then waited about one hour and had their blood glucose taken again (post-treatment). The data show that Galaxy® apparently does not spike blood glucose (Table 3) which is probably a result of the balance of simple sugars found in the product (Fig. 13) which was 16.5% xylitol; 42.2% fructose; and 41.3% glucose. Galaxy® also contains nine essential amino acids.

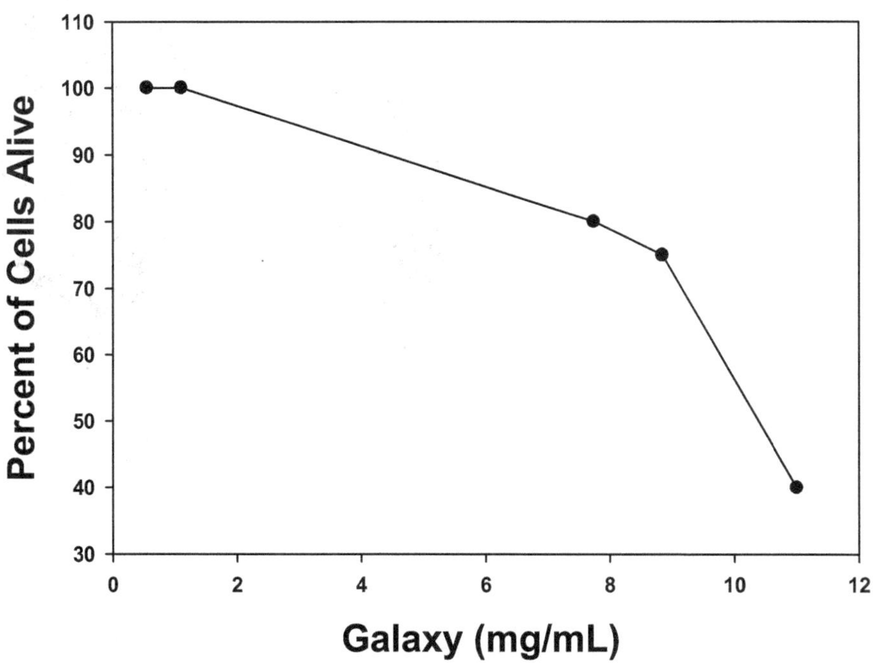

Figure 9. Effect of Freeze-dried Galaxy® on Human Colon Cancer Cells (EC50 = 10.5 mg/mL).

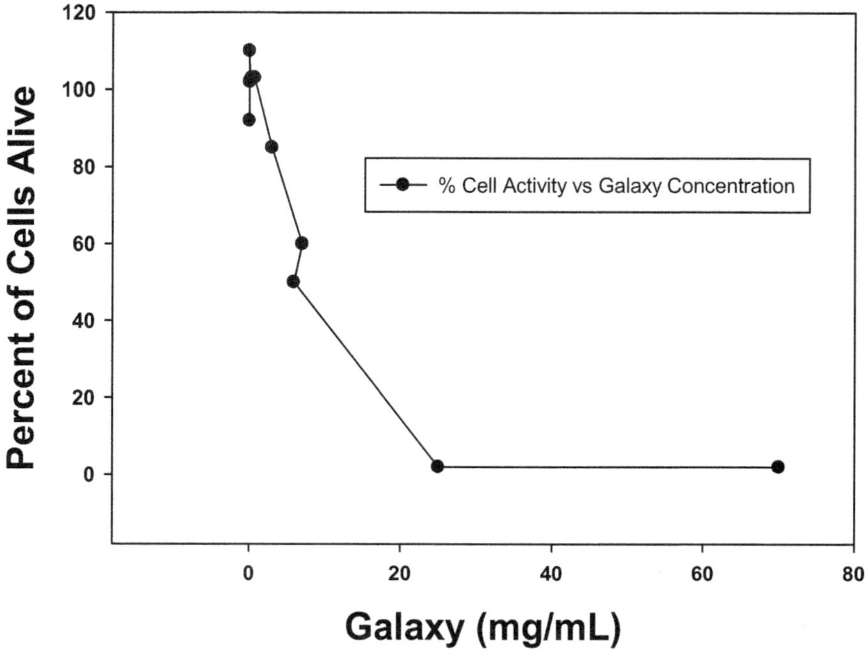

Figure 10. Effect of Freeze-dried Galaxy® on DU 145 Human Prostate Cancer Cells (EC50 = 5.3 mg/mL).

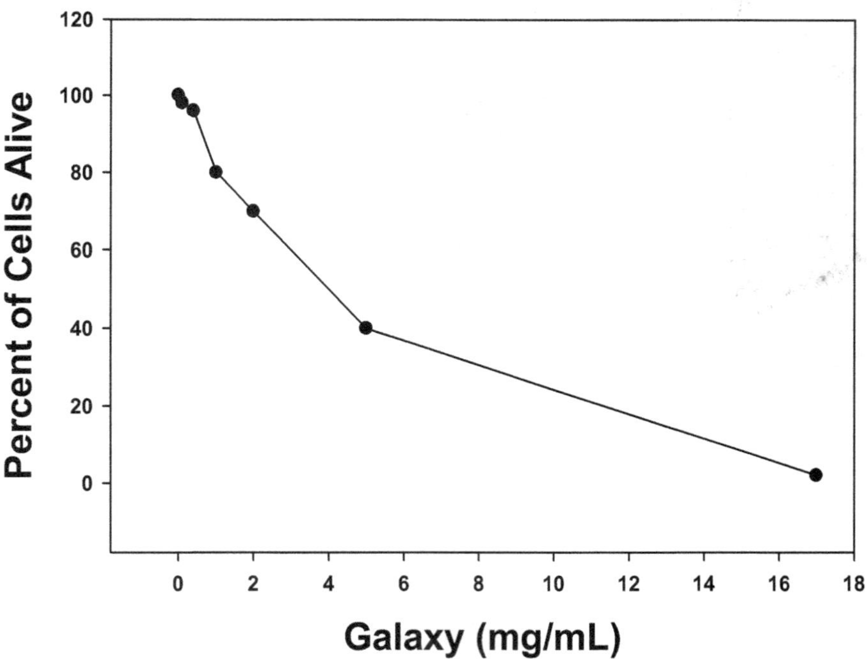

Figure 11. Effect of Freeze-dried Galaxy® on A-549 Human Lung Cancer Cells (EC50 = 5.0 mg/mL).

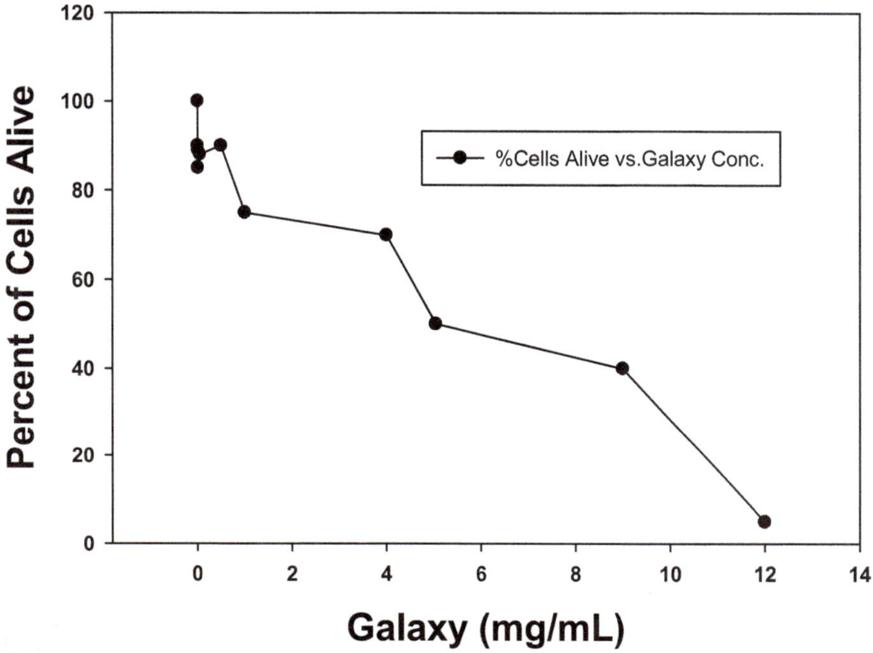

Figure 12. Effect of Freeze-dried Galaxy® on HepG2 Human Liver Cancer Cells (EC50 = 5.0 mg/mL).

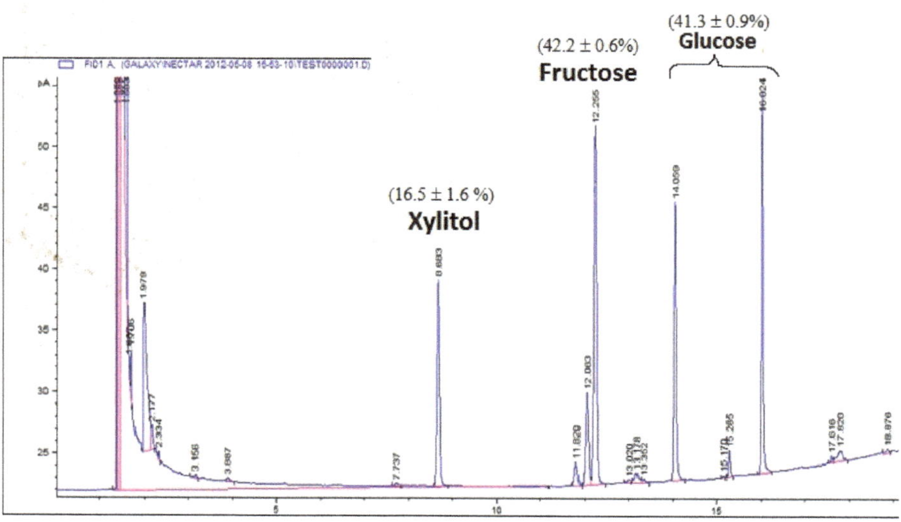

Figure 13. Chromatogram of the types of natural sugar in Galaxy®.

JoyLife International does not make health claims for their product because it is marketed as a nutritional supplement, not as a medicine. However, we do have one data set on a 33 year old Caucasian female who suffers from Acute Lymphocytic Leukemia (ALL) who agreed to take the product for seven months without any other medical intervention, e.g. no chemotherapy. This was done under the strict supervision of her health-care providers. Fig. 14 shows that her white blood cell count dropped from 68,000 to 21,700 cells/μL, coupled with a substantial improvement of her secondary health parameters associated with the disease (e.g., lethargy, bruising, muscular weakness, etc). This resulted in a 68% drop in the WBC over the seven month treatment period. The regression ($r^2 = 0.77$) relationship between the seven-month Galaxy® treatment and changes in WBC suggests that 77% of the variability in the changes in the slope of the graph was likely due to the Galaxy® treatment. The authors are not drawing any conclusions from this single data set, nor should the reader, but it is provided since it was done under the close scrutiny and supervision of her oncologists and without any other intervention. This study is on-going.

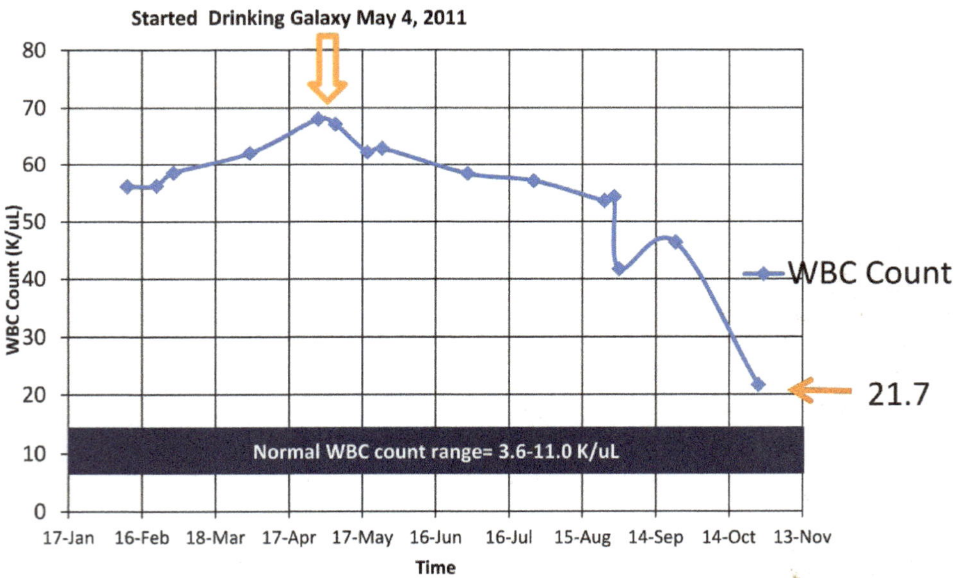

Figure 14. Effect of Galaxy® on WBC Count in an Acute lymphocytic Leukemia Patient

4. Summary and conclusions

Galaxy® is a nutritional food blend that contains 32 bioactive components including thirteen high antioxidant fruits. A freeze-dried sample of this blend has an ORAC score of 178.10 ± 3.1 μmoles TE/ g of freeze-dried product or 35.6 μmoles TE/g fresh product. Based on a dose of 30 mL/day, a person would consume 1068 ORAC units/dose. This value represents 21-36 % of the daily recommended ORAC units (3000-5000 ORAC units/day) suggested by the USDA. Thus, the entire bottle of Galaxy® contains about 26,700 ORAC units. It is likely that the antioxidants contained in this product contributed to the in vitro anticancer activity. These anticancer EC50s ranged from 0.075 to 11 mg/mL on breast, colon, prostate, lung, and liver cancers. The raw product, the supernatant, and particulate fractions of Galaxy® were all bioactive, although the particulate fraction was over 40X more active than either the raw product or supernatant suggesting that about 98% of the bioactivity is contained in the particulate fraction. In fact, the particulate fraction dose-response curve tracked the dose-response curve of Onxol® (an FDA-approved drug for breast cancer) with EC50s (EC50 = 0.075 mg/mL for Galaxy® particulates and EC50 = 0.03 mg/mL for Onxol®) that were in the same order of magnitude of each other.

There was a no correlation between the anticancer activity (EC50s) and the ORAC values of the 9 superfruits investigated (four of which are found in the product) ($r^2 = 0.0758$; $p = 0.473$), although four of the nine superfruits (including Galaxy®) had an EC50 < 1.0 and ORAC values greater than 150 umoles TE/g. From these data, it appears that ORAC units may somehow still be connected, though it is apparently not a clear inverse relationship. Galaxy® was also shown to have a Bioactivity Index (BI) of 3.40, which was 3.40X more active than the cranberry BI reported in the literature.

Galaxy® additionally did not spike blood glucose levels of 308 participants during the 1 ½ hours following the normal dose of 30 mL (n=308). This was likely due to the balance of simple sugars found in the product: 16.5% xylitol; 42.2% fructose; and 41.3% glucose (Fig. 13). In addition, Galaxy® contains nine essential amino acids.

JoyLife International makes no health claims for this product because it is marketed as a nutritional supplement, not a medicine; however, in cooperation with her physician, one patient with ALL showed a 68% drop in WBC count (from 68,000/uL to 21,700/uL) after seven months post-treatment with Galaxy® (30 mL/day) without any other medical interventions including chemotherapy. In addition, there were significant qualitative improvements in the patient's secondary health parameters. This study is on-going.

Research has clearly shown that people in the U.S. are not eating enough vegetables, since only 2 % of TAC consumed in any form comes from vegetables.

These data have suggested other experiments that need to be considered. For example, investigations are needed to determine how these cancer cells are dying, i.e., by apoptosis or necrosis as well as the toxicological contribution of each of the individual components of the blend. The issue of possible synergistic activity among the ingredients also needs further study. The effect of Galaxy® on rodent models with induced inflammatory diseases need to also be considered. And finally, larger randomized double-blind experiments are needed to validate the data from the ALL patient and from other people suffering from chronic diseases.

Author details

Gary M. Booth*
Brigham Young University, Department of Plant and Wildlife Sciences, Provo, Utah, USA

* Corresponding Author

Tory L. Parker
Brigham Young University, Department of Nutrition, Dietetics, and Food Science, Provo, Utah, USA

Christopher M. Lee
Research for Cancer Care Northwest and Gamma Knife of Spokane,
Department of Radiation Oncology at the University of Washington, Spokane, WA, USA

Acknowledgement

The authors are not associated with the sale, marketing, or distribution of Galaxy®. All of the data were generated from protocols developed by the principal investigators without contact with the company. All of the data contained in this manuscript has not been submitted to any other publisher.

5. References

Al-Qubaisi, M., R. Rozita, S-K. Yeap, A-R Omar, A-M Ali, and N.B. Alitheen. 2011. Selective cytotoxicity of Goniothalamin against hepatoblastoma HepG2 cells. Molecules 16:2944-2959.

Atmani, Dina., M. Begna Ruiz-Larrea, J.I. Ruiz-Sanz, L.J. Lizcano, F. Bakkali, and Djebbar Atmani. 2011. Antioxidant potential, cytotoxic activity and phenolic content of Clematis flammula leaf extracts. J. Med. Pl. Res. 5(4):589-598.

Badisa, R.B., S.F. Darling-Reed, P. Joseph, J.S. Cooperwood, L.M. Latinwo, and C.B. Goodman. 2009. Selective cytotoxic activities of two novel synthetic drugs on human breast carcinoma MCF-7 cells. Anticancer Res. 29:2993-2996.

Balick, Michael J. and Paul Alan Cox. 1996. Plants, People, and Culture. Scientific American Library: New York.

Block, G., B. Patterson, and A. Subar. 1992. Fruit, vegetables, fruit, and cancer prevention: a review of the epidemiological evidence. Nutr. Cancer 18:1-29.

Boivin, D., M. Blanchette, S. Barrette, A. Moghrabi, and R. Beliveau. 2007. Inhibition of cancer cell proliferation and suppression of TNF-induced activiation of NFkB by edible berry juice. Anticancer Res. 27:937-948.

Capua, C., N.P. Hopson, M.M. Stewart, G. Robert Johnston, K.L. O'Neil, G. B. Schaalje, C.M. Lee, and G.M. Booth. 2010. Cytotoxicity of Atriplex confertifolia. J. Toxicol. 2010:Article ID 976548, 7 pages. Doi:10.1155/2010/976548.

Danhier, F., N. Lecouturier, B. Vroman, C. Jerome, J. Marchannd-Brynaert, O. Feron, and V. Preat. 2009. Paclitaxel-loaded PEGylated PLGA-based nanoparticles: In vitro and in vivo evaluation. J. Control. Rel. 133:11-17.

Faria, A., C. Calhau, V. Freitas, and N. Mateus. 2006. Procyanidins as antioxidants and tumor cell growth modulators. J. Agric. Fd Chem. 54:2392-2397.

Ferguson, P. J., E. Kurowska, D. J. Freeman, A. F. Chambers, and D. J. Koropatnick. 2004. A flavonoid fraction from cranberry extract inhibits proliferation of human tumor cells. J. Nut. 134(6):1529-1535.

Frankfurt, O.S. and A. Krishan. 2003. Apoptosis-based drug screening and detection of selective toxicity to cancer cells. Anti-Cancer Drugs. 14(7):555-561.

Haytowitz, D.B. and S. Bhagwat. 2010. USDA database for the oxygen radical absorbance capacity (ORAC) of Selected foods, Release 2. Nutrient Data Laboratory, Beltsville Human Nutrition Research Center (BHNRC), Beltsville, MD.

Guthrie, N. and K.K . Carrol. 1998. Inhibition of human breast cancer cell growth and metastasis in nude mice by citrus juices and their constituent flavonoids. Packer, L. and A.S.H. Ong. Eds. Biological Oxidants and Antioxidants. Molecular Mechanisms and Health Effects. 1998:310-316. AOCS Press. Champaign, IL.

Kalt, W., C. Lawand, D.A.J. Ryan, J.E. McDonald, H. Donner, and C.F. Forney. 2003. Oxygen radical absorbing capacity, anthocynanin and phenolic content of highbush blueberries (*Vaccininium corymbosum L.*) during ripening and storage. J. Amer. Hort. Sci. 128(6):917-923.

Koch, A., P. Tamez, J. Pezzuto, and D. Soejarto. 2005. Evaluation of plant used for antimalarial treatment by the Massai of Kenya. J. Ethnopharmacol. 101:95-99.

Krishna, I.V., G.R. Vnaja, N.S. Kumar, and G. Suman. 2009. Cytotoxic studies of anti-neoplastic drugs on human lymphocytes—in vitro studies. Cancer Biomark. 5(6):261-262.

Liu, R.H. 2004. Potential synergy of phytochemicals in cancer prevention: mechanism of action. J. Nutr. 134:3479S-3485S.

Mertens-Talcott, S.U., S.T. Talcott, and S.S. Percival. 2003. Low concentrations of quercetin and ellagic acid synergistically influence proliferation, cytotoxicity and apoptosis in MOLT-4 human leukemia cells. J. Nutr. 133:2669-2674.

Parker, T.L., X-H. Wang, J. Pazimino, and N.J. Engeseth. 2007. Antioxidant capacity and phenolic content of grapes, sun-dried raisins, and golden raisins and their effect on ex vivo serum antioxidant capacity. J. Agric. Fd. Chem. 55:8472-8477.

Seeram, N.P., L.S. Adams, S.M. Henning, Y. Niu, Y. Zhang, M.G. Nair, and D. Heber. 2005. In vitro antiproliferative, apoptotic, and antioxidant activities of punicalagin, ellagic acid and a total pomegranate tannin extract are enhanced in combination with other polyphenols as found in pomegranate juice. J. Nutr Biochem. 16:360-367.

Silva, E.M., J.N.S. Souza, H. Rogez, J.F. Rees, and Y. Larondelle. 2006. Antioxidant activities and polyphenolic contents of fifteen selected plant species from the Amazonian region. Food Chem. 101: 1012-1018.

Steinmetz, K.A. and J.D. Potter. 1991. Vegetables, fruit, and cancer. I. Epidemiology. Cancer Causes Control 2:325-357.

Sun, Jie, Yi-Fang Chu, Xianzhong Wu, and Rui Hai Liu. 2002. Antioxidant and antiproliferative activities of common fruits. J. Agric. Fd Chem. 50:7449-7454.

Thompson, M.D. 2009. Chemical composition and mammary cancer inhibitory activity of dry bean. Crop Sci. 49(1):179-186.

USDA Human Research Center on Aging. Feb. 1999. "Can foods forestall aging?" Agri. Res. Feb. 1999.

Wang, Hong, Guohua Cao, and Ronald L. Prior. 1996. Total antioxidant capacity of fruits. J. Agric. Fd Chem. 44(3):701-705.

Wang, S., K.A. Meckling, M.F. Marcone, Y. Kakuda, and R. Tsao. 2011. Can phytochemical antioxidant rich foods act as anti-cancer agents? Food Res. Intern. 44:2545-2554.

Wang, S.Y. and K.S. Lewers. 2007. Antioxidant activities and anticancer cell proliferation properties of wild strawberries. J. Amer. Soc. Hort. Sci. 132:647-658.

Weber, C., M. Liu, X.Q. Li, and R. H. Liu. 2001. Antioxidant capacity of anticancer properties of red raspberry. NY Fruit Quart. 9(3): 13-15.

Yang, Meng, Sang-Jin Chung, Chin Eun Chung, Dae-Ok Kim, Won O. Song, Sung I. Koo, and Ock K. Chun. 2011. Estimation of total antioxidant capacity from diet and supplements in US adults. Brit. J. Nut. 106:254-263.

Zhou, J.R., L. Yu, Y. Zhong, and G.L. Blackburn. 2003. Soy phytochemicals and tea bioactivity components synergistically inhibit androgen-sensitive human prostate tumors in mice. J. Nutr 133:516-521.

Chapter 4

Staphylococcus aureus: Characterisation and Quantitative Growth Description in Milk and Artisanal Raw Milk Cheese Production

Alžbeta Medveďová and Ľubomír Valík

Additional information is available at the end of the chapter

1. Introduction

The safety and quality of fermented raw foods are generally determined by the presence of pathogenic and spoilage microorganisms, their interaction with lactic acid bacteria, intrinsic, extrinsic and technological factors [1]. This fact concerns also the short ripened ewes' lump cheese traditionally produced immediately after milking in Slovakian upland cottages. The cheese is curdled with rennet, fermented by native lactic acid bacteria and briefly ripened for 7 to 10 d. Then it is usually sent to a cheese factory for production of the soft Slovakian „Bryndza" cheese [2].

This chapter deals with the behaviour of coagulase-positive staphylococci as their populations belong to the ubiquitous microflora of ewes' milk. *S. aureus* is able to multiply rapidly, especially during the initial phase of preparation when natural lactic acid bacteria are in lag phase and a sufficient amount of lactic acid has not been produced. The initial period to reach pH 5.3 lasted on average up to 30 h in upland artisanal ewes' cheese production stations [3]. However, *S. aureus* is competitive in milk and dairy environments; it is quite sensitive to higher lactic acid concentration. The growth of *Staphylococcus aureus* and potential production of heat-stable enterotoxins with respect to the food matrices and conditions of food preparation represent a potential, even actual threat of a public health menace residing in food poisoning outbreaks. That is why the control of *S. aureus* growth during the fermentation of young raw milk cheese means prevention against staphylococcal enterotoxin production.

2. *Staphylococcus aureus* – general description

Staphylococcus aureus subsp. *aureus* (*S. aureus*) belongs to the genus *Staphylococcus* and to the family *Staphylococcaceae* [4]. It was firstly described by Sir Alexander Ogston in 1882 and 2

years later Rosenbach isolated it in a pure culture and introduced the name *Staphylococcus aureus*. The name of the organism is derived from Greek words *staphyle* (a bunch of grapes) and *coccus* (grain or berry) [5,6].

S. aureus is a Gram-positive, facultative anaerobic, catalase-positive, oxidase-negative, non-motile microorganism that does not form spores. It creates smooth, convex, lustrous, circular colonies reaching a size of 0.5-1.5 μm in diameter and growing in an irregular three-dimensional bunch of grapes-like clusters of cells. In dependence on growth conditions, the colony pigmentation varies from grey, grey-white with yellowish to orange shades with typical β-haemolysis on the blood agar [6-9].

For growth it requires B vitamins (thiamine and nicotic acid), inorganic salts and amino acids as a nitrogen source, especially arginine, cystein, proline and valine. Glutamic acid, leucine and tyrosine are not required for growth, but they are essential for enterotoxin production. Deprivation of any amino acid is much less responsive in SEA production than for SEB or SEC production. Arginine seems to be essential for enterotoxin B production [5,7,10].

S. aureus belongs among chemo-organotrophs with a respiratory and fermentative metabolism. Under aerobic conditions, acids are produced from glucose, lactose, maltose and mannitol, under anaerobic conditions acids are produced from many other sugars and alcoholic sugars [6,7].

Most strains hydrolyse native animal proteins (casein, gelatine, fibrin), lipids, phospholipoproteins and Tween. They also coagulate animal plasma with the assistance of a coagulase and the clumping factor. Besides that, the typical enzymatic activity of *S. aureus* includes production of coagulase, alkaline phosphatase, proteases, lipases, and esterases and some strains also produce lecithinase [5-7].

3. Production of enterotoxins and other virulent factors

S. aureus produces a wide range of virulence factors which can be divided into different groups. Due to the production of surface-associated factors like microbial surface components recognizing adhesive matrix molecules (MSCRAMM), protein A, polysaccharide A, peptidoglycan and a clumping factor, *S. aureus* is responsible for resistance to opsonophagocytosis, the formation biofilm and adhesion to the host cell matrix [11,12]. Following colonization, *S. aureus* secretes various toxins and enzymes which are responsible for the lesions during the development of the infection. Once *S. aureus* penetrates the subcutaneous tissues and reaches the blood stream, it can infect almost any organ, most notably bone tissue and cardiac valves [12].

The role of enzymes like coagulase, catalase, hyaluronidase, lipase, heat-resistant nuclease, staphylokinase and β-galactosidase is to disrupt cell structure, degrade cell lipids and hyaluronic acid, and to convert fibrinogen to fibrin. All those mechanisms promote *S. aureus* *i)* to affect leukocytes, sebaceous glands and subcutaneous tissues; *ii)* to increase propagation of infection and *iii)* to inactivate the effect of β-lactam antibiotics [9,11,13].

Toxins (leukocidins, haemolysins and epidermolytic toxin) possess haemolytic, cytotoxic, dermonecrotic and lethal activity. They are able to paralyse smooth and skeletal muscles, damage blood vessels, cause extensive lesions on the skin and reveal a moist glistering surface (called also Ritter's disease) and finally have a toxic effect on the central nervous system [5,11,14].

In addition to surface factors, enzymes and cytotoxins, strains of *S. aureus* are also equipped with superantigenic toxins, including shock syndrome toxin-1 (TSST-1) and enterotoxins. They not only modulate host immune response but are also able to cause food poisoning in human [11]. The release of TSST-1 into the bloodstream may give rise to a variety of severe clinical difficulties, such as toxic shock syndrome, sudden infant death syndrome and Kawasaki syndrome [15].

From the food point of view, the production of one or more staphylococcal enterotoxins (SEs) is crucial, because they are causative agents of staphylococcal food poisoning (SFP) outbreaks in human.

Staphylococcal enterotoxins are heat-stable exoproteins consisting from 236 to 296 aminoacids with a molecular mass of 25-35 kDa. Upon hydrolysis, 18 amino acids are present, mostly aspartic acid, glutamic acid, lysine and tyrosine. For the majority of these, an isolectric point of pH 5.7-8.6 is considered. There are five different types of classical enterotoxins (SEA-SEE) which are distinct in antigen reaction. Recently, new types of enterotoxins and enterotoxin-like types (SEG-SEV) have been described in *S. aureus*. Classical enterotoxins are encoded by phage (SEA), chromosome (SEB and SEC) or by plasmid genes (SED). They are produced during all phases of growth (SEA and SED) or only as secondary metabolites in late exponentially or in stationary phase (SEB and SEC). Most strains are capable of producing one or more enterotoxins. Enterotoxins are resistant to proteolytic enzymes, such as trypsin, chymotrypsin, rennin and papain, but at pH of about 2, they are sensitive to pepsin [5,6,9,10,16,17].

The SFP is characterized as a relatively mild intoxication which occurs after ingestion of at least 20 ng of staphylococcal enterotoxins presented in the food. Although the numbers of outbreaks caused by bacterial toxins are generally underestimated, official EU data [18] reported 558 outbreaks in 2009 from which almost 53% were caused by *Staphylococcus* spp. Two cases were from a verified outbreak and one from a possible outbreak was fatal.

3.1. Resistance of *S. aureus* and its enterotoxins to environmental factors

3.1.1. Heat resistance (D-values)

S. aureus is a mesophilic organism with optimum growth temperature in the range from 37 °C to 40 °C [7-9,17]. The minimal temperature for growth is about 7.0 °C [5,8,10], but some strains do not even show growth at 8 °C [19]. *S. aureus* survives freezing, in meat at -18 °C it will survive for at least 6 months with no change in counts [6]. On the other hand, a temperature higher than 46 °C is not acceptable for the majority of strains, with some exceptions that do grow up to 50 °C [6,7,9,10]. Heating causes damage to the cell. A $D_{60°C}$

value of 1-6 minutes in foods with high water activity or $D_{60°C}$ of 1-2.5 minutes in phosphate buffer is expected. Cells heated in oil, fat or in low water activity environments showed higher D-values, e.g. $D_{60°C}$ of 5.3 minutes in milk and $D_{60°C}$ of 42.3 minutes in milk with 57% sucrose, $D_{60°C}$ of 6 minutes in meat containing 3-4% of NaCl and $D_{60°C}$ of 25 minutes at a salt content of 8%. Contrary to this, changes of pH value out of optimal values decrease heat resistance [6].

Enterotoxins are produced in a narrower range of temperature than the growth is noticed. In general, enterotoxins production is expected in a temperature range of 10-46 °C, with the optimum temperature for production in the range 40-45 °C [6,8,10,20]. Enterotoxins are heat-stable in milk. Their resistance to heating is represented by D-values at 121°C and 100°C ranging from 9.9-11.4 to 70.0 minutes, respectively [21,22]. Their heat-resistance decreases following SEC>SEB>SEA and is also significantly reduced in acidic conditions [10]. It should also be noticed that 99.6% of cells are destroyed by the pasteurisation of milk at 72 °C for 15s, and at 72 °C for 35s all cells are killed. Enterotoxins can resist both the process of milk pasteurisation or sterilisation of canned foods [6,7].

3.1.2. Acid tolerance

Regarding pH, *S. aureus* is able to grow in a range of pH 4.0-9.8, with an optimum of 6-7 pH [6,8-10]. The minimal values of pH for growth are influenced by other environmental factors. The growth of *S. aureus* is inhibited by 0.1% of acetic acid and also by the presence of lower (C_1-C_4) fatty acids [17]. Moreover, *S. aureus* is more sensitive to acidification when salt concentration is increased, although it is a halotolerant microorganism.

Fast acidification down to values unacceptable for growth is the most efficient way of *S. aureus* inhibition. Acids do not have the same inhibition capacity and for a given pH value, the impact on *S. aureus* physiology will vary with the nature of the acid used. Organic acids at pH values equivalent to those obtained by using inorganic acids are more effective against *S. aureus*. The effectiveness of organic acids generally depends on the concentration of their undissociated form, which is determined by the dissociation constant of organic acids. Thus, acetic acid and propionic acid with pK_a of 4.8 and 4.9 (pK_a is pH at which the ratio of dissociated to undissociated forms is 50:50) are more inhibitive than lactic acid whose pK_a is 3.9 [6,23].

In general, a tolerance of *S. aureus* to pH values higher than 5.5 is caused due to maintaining of the intracellular pH by the sequestering or releasing protons from cytoplasm and also by the expression of genes responsible for cytoplasm buffering. These genes include genes encoding intracellular chaperones, urease operon and genes involved in the metabolism and transport of amino acids (histidine, lysine, arginine), carbohydrates and phosphoric acid [23,24].

Complete inhibition of *S. aureus* is achieved at pH lower than 5.0. An acidic stress and the drop of intracellular pH alter the membrane structure and lead to a decrease in the activity of several enzymes which are pH-sensitive. Non-dissociated form of acid acts as uncouplers of the respiratory chain. The protonated form diffuses into the cell at low pH and is followed by a dissociation of the proton. Bacterial growth is then strongly altered because most of the energy available in the cell is used for the de-acidification of the cytoplasm by generating a proton gradient across the cytoplasm membrane [24].

Similarly to temperature effect on enterotoxins production, the pH range allowing production of enterotoxins is also more limited than those for growth. The practical limit in acidic foods is pH 5.0, with an optimum of 7.0. The SEA is produced under a wider range of pH than SEB or SEC [6,20].

3.1.3. Salt resistance

A characteristic feature that distinguishes *S. aureus* from other pathogenic bacteria is its high tolerance to low water activity values and NaCl concentrations up to 20%. Generally it is reported that the minimal water activity for the *S. aureus* growth is in the range of a_w from 0.83 to 0.86 [7,8,10,21]. Those values are dependent on the specific strain, the actual values of pH, temperature, humectants and atmospheric conditions. No growth of a mixture of *S. aureus* strains in BHI broth containing NaCl and sucrose was observed at 8 °C, pH 4.3 and a_w 0.85 (19% of NaCl) or at 12 °C, pH < 5.5 and a_w 0.9 (14% of NaCl) or at 12 °C, pH < 4.9 and a_w 0.96 (8% of NaCl) [10]. A single strain of *S. aureus* in PCA or BHI broth containing NaCl could not withstand concentrations of NaCl of 5% (a_w 0.97) at 12 °C, 13% (a_w 0.91) at 15 °C, 15% (a_w 0.89) at 18-21 °C and 18% (a_w 0.86) at temperatures in the range 25-30 °C. At optimal growth temperatures in the range from 35 °C to 37 °C it could multiply up to concentration of 20% of NaCl (a_w 0.84) [25].

The ability of *S. aureus* to grow at such high concentrations is related to its adaptive response to osmotic stress. It is due to the intracellular accumulation of compatible solutes including proline, betaine, choline, taurine which can occur by *de novo* synthesis or by transport from the growth medium. The transport systems appear to be constitutively synthesised and to be activated in a very specific way by osmotic stress. There are multiple transport systems for betaine and proline. There is probably a single specific system for each one and a less specific system which is strongly activated by osmotic stress and results in the accumulation of both proline and betaine. Compared to other pathogenic organisms, *S. aureus* does not accumulate sugars as compatible solutes and free peptides serve as a source of proline [26]. Besides the accumulation of compatible solutes to maintain turgor caused by the increased NaCl concentration, *S. aureus* also undergoes an extensive program of gene and protein expression in response to NaCl stress. One of them is probably an *ars* encoding the resistance to arsenate, arsenite and antimonite. However, mutation in the *ars* operon significantly decreases the ability of *S. aureus* to grow in the presence of NaCl, since the low expression of *ars* impedes the ability of *S. aureus* to rid itself of cytoplasmic Na^+ in NaCl-stressed cells [27].

With respect to enterotoxins production requirements, values of water activity for their production are mostly in the same range as for the growth of the producer. In food with decreased water activity and at aerobic conditions, the enterotoxins can be produced even if the value is from 0.86 to 0.89 a_w. The production of SEB appears to be more sensitive to reduced water activity than SEA production, whereas SEA is produced up to a_w 0.87-0.89, SEB is produced only in the narrow range of water activity values 0.99-0.97 [10,28].

3.1.4. Tolerance of S. aureus against sanitizing agents and antibiotic resistance

In generally, *S. aureus* is sensitive to sorbic acid, peracetic acid and hydrogen peroxide. Unsaturated fatty acids and alkaline dyes also affect inhibitory. On the other hand, it is resistant to phenol, compounds of mercury, cadmium and arsenates. The ionization radiation kills cells with a D-value of 0.2-0.4 kGy in meat and fish products, but the enterotoxins are not affected even by a sterilization dosage of radiation [6]. The effect of ethanol is also not unique. Concentrations up to 7% may have an inhibitive effect, but concentrations higher than 9% act bactericidal.

The majority of disinfectants routinely used in the food industry (halogens, quarternary ammonium salts) will be effective when applied correctly. After inappropriate sanitation however, the cells can recover and become more resistant [8]. *S. aureus* has also a high degree of tolerance to compounds such as tellurite, mercuric chloride, neomycin, polymyxin and sodium azide, all of which have been used as selective agents in culture media [10].

Pathogenic *S. aureus* is regarded as a "superbug", due to its amazing capacity to be resistant to a wide range of antibiotics. *S. aureus* strains resistant to methicillin (MRSA), vancomycin (VISA/VRSA), and to many other antibiotics represent an urgent problem in both community- and hospital-acquired infections. According to Girish et al. [29], the resistance results from *i)* surface protein modifications which promote colonization of host tissues, *ii)* biochemical variations which enhance survival in phagocytes and evasion of the host immune system, *iii)* enhanced release of toxins which lyse eukaryotic cell membranes and active efflux of antibiotics coupled with mutation events in target molecules.

The perspective targets for drugs in *S. aureus* may be the enzymes involved in lysine biosynthesis or genes encoding the activities essential for the life of the cell that have not been used for therapeutic intervention. In this context, the following antibiotics are used: *i)* linezolid by blocking the formation of the ribosomal initiation complex, *ii)* clarithromycin by the inhibition of the proteosynthesis, *iii)* phosphomycin by inhibition of the cell wall synthesis, *iv)* daptomycin by the insertion into the cell membrane, causing rapid depolarisation and the release of potassium ions, resulting in the inhibition of DNA, RNA and protein synthesis, *v)* tigecykline, erythromycins, tetracyclins, oxazolidinones and aminoglycosides by inhibition of the protein synthesis, *vi)* fluoroquinolones by inhibition of the DNA replication and repair [29-31].

3.2. Determination and identification of *S. aureus*

Staphylococci compete poorly with indigenous bacteria and are inhibited by the antagonistic activities of other organisms. Therefore the presence of *S. aureus* in foods must be considered in relation to the amount and types of the accompanying flora. Numerous methods to isolate and identify *S. aureus* have been described and standardized by international and national organizations. The principal approach is to isolate it on solid agar media and subsequently identify it by the use of microbiological, biochemical and molecular methods.

3.2.1. Determination of S. aureus counts

Media for isolation and determination of *S. aureus* can be divided into three groups [6,7].

- In the first group are media such as tryptone soya broth, brain heart infusion broth, mannitol-salt agar, salt meat broth. They use sodium chloride as the selective agent and metabolizable substrates such as mannitol, blood or milk as diagnostic agents are incorporated. However, higher concentrations of salt and the lack of resuscitators in the media may inhibit injured or stressed cells (false negative results). Moreover, other microorganisms are salt-tolerant or can metabolize substrates, so the media are not specific enough.
- In the second group are media which contain combinations of selective and diagnostic agents. The list of selective agents which includes sodium azide, sodium chloride, lithium chloride, potassium tellurite, glycine and antibiotics (polymyxin or sulphametathine) is not large but provides many combinations. Media like tellurite-polymyxin agar, KRANEP agar, Giolitti-Cantoni broth, Baird-Parker agar and its modifications, and some other media are found in this group. The mode of diagnostic action is fermentation of mannitol, egg yolk reaction – clear zones around colonies, black colonies (reduction of tellurite to tellurium) and pigment production [5]. The problems of this media are that some animal strains of *S. aureus* do not use lipovitellenin from egg yolk, and competing microorganisms (spp. *Enterococcus, Proteus, Micrococcus*) are also able to reduce tellurite. In spite of this, some of them are widely used and are also recommended by the ISO, IDF or AOAC organisations.
- To correct the discrepancies of media in the previous groups, the addition of plasma (from rabbit, pig or rat) with bovine fibrinogen instead of egg yolk is used. These media allow the detection of coagulase directly on the plate due to the formation of fibrin zones around the colonies. Such media include Baird-Parker agar with plasma and Rabbit-plasma fibrinogen (RPF) medium. Because of the cost and variable performance of commercially available plasmas, they are not used in routine examinations.

Nowadays, there is also the possibility to use chromogenic media for detection of *S. aureus* [32]. To minimize false negative or positive false results further confirmatory tests are necessary.

3.2.2. Identification of S. aureus

The first step in the identification of suspected colonies is the Gram-staining, microscopic examination of the morphology, catalase test and also β-haemolysis surrounding colonies on the sheep-blood agar [16,17,33-37].

One of the preferred examinations is the coagulase test, either as a tube format for the presence of unbounded extracellular coagulase or as a slide coagulase test for the presence of a clumping factor - cell wall associated enzyme. There are commercially available rapid and convenient tests, and also laboratory procedures are permissible to detect the presence of coagulase. It should also be noted that the production of coagulase is not a property of only *S. aureus*, but also of some Gram-negative bacteria and other staphylococci. In addition to this, coagulase is not exclusively produced by *S. aureus* and coagulase-negative strains may be also enterotoxigenic. Also the test to detect nucleases (deoxyribonuclease - DNAse and heat-stable endonuclease - thermonuclease) is useful by either the spectrophotometric method or by microbiological methods. The effect of lysostaphin on the cell wall destruction distinguishes staphylococci from micrococci, since staphylococci but not micrococci are lysed by an extracellular enzyme produced by *S. staphylolyticus* [13].

From among biochemical tests, the API-Staph system and the VITEK Gram-positive Identification Card are widely used. They are based on the reaction of microorganism with a set of specific substrates. There is also the possibility of fluorescence microscopy detection without previous growth of culture on selective media by the use of the VIT-Staphylococcus system. This is based on the penetration of a specific gene probe into the bacteria cell, marking the individual signature of the gene sequence with the dye and illuminating them. Subsequently, the samples are examined under fluorescence microscopy. Bacteria belonging to the genus *Staphylococcus* light up in green, bacteria belonging to the species *S. aureus* additionally light up in red [38].

However, the most reliable way to identify a suspicious colony as *S. aureus* is to investigate the presence of highly specific genes by the use of PCR technology. So from among the most employed genes, there is the possibility to detect the presence of 16S or 23S rRNA sequence, *tst* gene (encoding toxic shock syndrome), *coa* gene (encoding coagulase), *eta, etb* genes (encoding exfoliative toxin A and B), *clfA* and *clfB* genes (encoding clumping factors), *femA* gene (encoding resistance to methicilin), *cat* gene (encoding production of catalase), and *nuc* gene (encoding thermostable nuclease) [11,15,37,39-42]. Since the *nuc* gene is present in all *S. aureus* strains and is well conserved in this species at the nucleotide level but is either absent from or distinct in other bacterial species including coagulase-negative staphylococci it has been reliably used for *S. aureus* identification [43].

From the human health point of view, methods for the detection of staphylococcal enterotoxins are required. Firstly, the presence of genes encoding enterotoxins (*sea-sev*) are searched for by the use of PCR assays. Subsequently, the expression of the enterotoxin under the current conditions is investigated. One of the options is the use of immunological test system for routine use established in the ELISA procedure based on the monoclonal or polyclonal antibodies against enterotoxins detection. By using the reversed passive latex agglutination test (RPLA), enterotoxins antibodies are bounded to particles of latex, but the nonspecific agglutination is also possible. The immunoflourescence methodology has also been used to detect cell-associated enterotoxins, but this method has not been used to any great extent. An alternative to the fluorescence method, radioimmunoassay can be employed by the radioactive iodine as a marker, but also it is not widely used. For scientific, not for routine examinations, other procedures including the electrophoresis, the electroimmuno-diffusion reversed immunoosmophoresis and the affinity chromatography methods may also be used [13,20].

4. *S. aureus* in milk and dairy products

S. aureus is a ubiquitous organism frequently isolated from raw milk manually draw from individual animals, bulk raw milk and naturally, from milk of dairy cattle suffering from mastitis. In proper drawn milk, the typical counts of *S. aureus* are 100-200 CFU/ml. In the case of a contaminated udder, the counts may increase up to 10^4 CFU/ml [7].

4.1. Source of contamination and occurrence in the environment

The natural ecological niches of *S. aureus* are the nasal cavity and the skin of warm-blooded animals. The skin, mucosa membranes, teats and udders of milking animals are the most important reservoir of this contaminant. In the case of an infected udder, *S. aureus* can contaminate milk during milking in a density ranging from 10^1-10^8 CFU/ml, mostly about 10^4 CFU/ml [6,7,34]. It is responsible for approximately 30-40% of all mastitis cases in the world [35].

In primary production and the dairy environment, except for milk producing animals, human beings and operational environment belong among the main sources of product contamination. One third of people are the asymptomatic carriers of *S. aureus*. It is frequently found on the skin, in nose, axilla, umbilicus, gastrointestinal and urogenital tracts of humans. The frequency of enterotoxigenic strains isolated from humans is high, varying between 40% and 60%. The organisms find their way into food through hands (infected wounds, skin lesions) or by coughing and sneezing [6,7,12,34,44].

According to references [34,45,46], the frequency of *S. aureus* occurrence varied from 6% to 28% in samples of raw milk. However, Rall et al. [37] found that *S. aureus* was present in 70.4% of raw milk samples. Although the density of *S. aureus* was not analysed, the prevalence of enterotoxigenic strains in these isolates ranged from 25.5% to more than 72%, with SEA and SEC as the predominant enterotoxins. It is assumed, that SEA together with SED were the most frequent agents in SFP outbreaks [6,17,47,48]. Furthermore, SEA is predominantly produced by human strains, so the contamination of food samples during manufacture is possible [33,48]. On the other hand, SEC is the most important cause of SFP associated with the consumption of dairy products [17].

In Slovakia, a similar incidence (4-9%) of *S. aureus* in raw cows' or ewes' milk was reported [49,50]. In our investigations of raw milk, we found that *S. aureus* was present in 20% of samples, with a density of 2.2 log CFU/ml. And, 33% of those isolates were enterotoxigenic, with *sea* as the only enterotoxin encoding gene found.

The lack of proper hygienic measures during food processing would also increase the counts of *S. aureus*, especially in manually prepared foods. Therefore, *S. aureus* can contaminate also heat-treated milk and can subsequently be present in cheeses prepared from both raw and pasteurized milk. In this connection, the presence of *S. aureus* in 46% of Slovakian cheeses (fresh lump cheese, "Bryndza" cheese) and even in 83% of whey after lump cheese manufacture was not surprising. Densities of 0.5, 1.6 and 4.5 log CFU/g or ml in "Bryndza" cheeses, whey and in lump cheeses were determined, respectively. 14% of those isolates possessed the gene for only one SE and the other 14% possessed the genes for two SEs. In the majority of the isolates, the gene for SEA was detected, in 11% of isolates the combination of *sea* and *sec* genes was found and *see* gene or *sea/see* genes combination occurred in one of the isolates. Neither *seb* nor *sed* genes were found throughout the collection of isolates.

In the study performed by Kousta et al. [51], 96% of both unpasteurized and pasteurized milk cheeses met the EU regulations for *S. aureus* either absent or present in very low numbers. The rest of them consistently had a density higher than 4 log counts but none of these tested positive for enterotoxin. By investigation of mostly dairy products including cheeses, whey, butter, but also some samples from meat, meat products, sausages and eggs, *S. aureus* was detected in 13-20% of samples [16,17,20,52,53] or 35-45% [33,46] and even in 70-80% [42,54]. The prevalence of enterotoxigenic strains was higher than 30% in all mentioned studies. The *sea* and *sec* genes were again the most frequent. But there were also found some strains with a presence of *seb, sed, see* genes or combinations of all of them.

The correlation between the presence of a respective gene and real enterotoxin production is about 70-80%, which might be explained by the incomplete expression of the enterotoxigenic genes. This is influenced by environmental conditions, such as temperature, pH and water activity which are important both for the growth and production of enterotoxins [20,42,52,54]. For this reason, it is necessary to know cardinal values of intrinsic and extrinsic factors preventing the growth of *S. aureus* in specific raw milk cheese production.

4.2. *S. aureus* in milk: Quantitative assessment of growth

S. aureus requires a complex organic source of energy. The main substrates used by this organism are sugars (glucose, fructose, galactose, mannose, ribose, maltose, sucrose, trehalose), alcohols (mannitol), organic acids (acetate), and in some conditions amino acids (glutamine, arginine). Genome sequence analysis revealed the presence of lactose phosphotransferase systems that enabled the growth of *S. aureus* in milk [55].

4.2.1. Effect of incubation temperature on the S. aureus growth in milk

The growth of various *S. aureus* strains is now well documented in databases of predictive microbiology tools such as Combase or Predictive Modelling Program [56]. As an example, the growth of two strains in relation to temperature is demonstrated in Fig. 1. The range in which the SED was detected is also shown (Fig. 1a) as well as the average growth parameters in Table 1 [57]. According to references [17,58,59], SED is the second most common serotype of enterotoxins among staphylococcal strains isolated from dairy products associated with food poisoning. Fig. 1a indicates the fact that SED was already detected at the level of *S. aureus* of $10^{6.5}$ CFU/ml at the lower temperature of 12 °C. At the higher temperatures of 18 and 21 °C, the detectable amount of SED toxin was determined when *S. aureus* reached the density of 10^7 CFU/ml. Based on the previous literature data [7,47,58,60,61], the minimal concentration of *S. aureus* of 10^6 CFU/ml needed for enterotoxin production in food was confirmed.

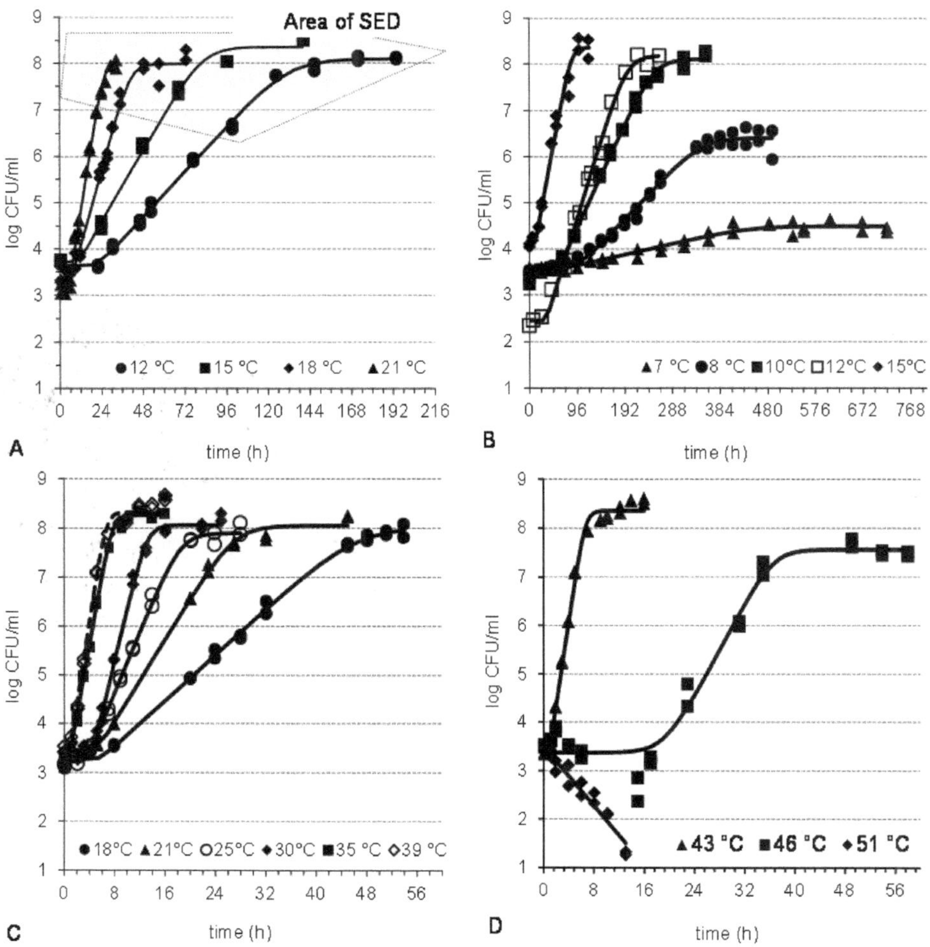

Figure 1. Growth of *S. aureus* strains D1 (a) and 2064 (b, c, d) in milk at temperatures from 7 °C to 51 °C. The growth data were fitted using DMFit tool [64]

T [°C]	D1		2064	
	μ [h^{-1}]	t_d [h]	μ [h^{-1}]	t_d [h]
7			0.006	120.3
8			0.026	27.0
10			0.055	12.7
12	0.103	6.8	0.082	8.5
15	0.148	4.7	0.145	4.8
18	0.313	2.2	0.264	2.6
21	0.545	1.3	0.484	1.4
25			0.711	1.0
30			1.215	0.6
35			1.664	0.4
39			1.931	0.4
43			1.903	0.4
46			0.562	1.2

Table 1. Specific growth rates and t_d of *S. aureus* strain D1 and 2064 in milk
μ – specific growth rate, t_d – time to double

Despite the slow growth of the 2064 strain, the temperature of 7 °C can be considered as the minimal temperature for growth of *Staphylococcus aureus* 2064 as proposed by Tatini [62]. However, some authors [19,63] did not observe *S. aureus* growth at 8 °C even after 1 week of incubation. On the other hand, other literature sources mentioned the lowest *S. aureus* growth temperature of T_{min} 6.5-7.0 °C [6-10].

In order to know the variability of growth rates as calculated from the growth curves, we performed static cultivations of the 28 confirmed *S. aureus* isolates in duplicate at the same temperature (15 °C). The results of the descriptive statistics are summarised in Table 2. The highest variability among the growth parameters was associated with the lag phase duration, as the most variable parameter. It reflects the previous history of the inoculum, the physiological state of the cells, the time necessary for production of the biological components needed for replication and the period of adjustment to the new environment.

Comparing the determined parameters with values generated by the Combase Predictor (μ = 0.170 h^{-1}, lag = 14.3 h) or in the Pathogen Modeling Program ver. 7.0 (μ = 0.177 h^{-1}, lag = 8.9 h) [56,65], it can be concluded that all values are very close. The average values of growth rates of isolates were slightly lower than those predicted by world programmes and also, the lag phase duration of our isolates was longer. This difference may be attributed to the fact that both software programs processed data from growth experiments carried out in broth media, not in milk.

Taking into account that 12-37% of the bound of reliability during cultivation experiments is tolerable; these findings demonstrate that the duration of the lag phase and the growth rate of *S. aureus* in milk can be predicted with a defined degree of reproducibility. Prediction of growth dynamic and effects of environmental factors on growth parameters, described further, resulting from analyzing the growth of the model *S. aureus* 2064 isolate in milk can be effectively and reliably used in food practice to reduce the risk of staphylococcal food poisoning outbreaks.

Parameter	μ	lag	N_0	N_{max}	t_d
aver	0.163	13.8	2.93	8.17	4.3
s_d	0.025	3.2	0.59	0.29	0.6
%v_c	15.3	23.0	20.0	3.6	14.0
min	0.104	4.4	0.78	7.19	2.2
max	0.318	21.4	3.72	9.02	6.6
med	0.159	14.3	3.13	8.19	4.4

Table 2. Growth parameters of *S. aureus* isolates in UHT milk at 15 °C (n = 28)
μ [h^{-1}] – specific growth rate in exponential phase, lag [h] - duration of lag phase, N_0 [log CFU/ml] - initial concentration of *S. aureus*, N_{max} [log CFU/ml] - maximal concentration of *S. aureus* in stationary phase, t_d [h] - time to double, aver - average value, s_d - standard deviation, v_c - coefficient of variation, min - minimal value, max - maximal value, med - median of the value

4.2.2. *Effect of temperature on S. aureus growth parameters*

Within quantitative predictive microbiology the secondary models are used to characterise the influence of intrinsic or extrinsic food factors on specific growth parameters. Among the temperature models, the Ratkowsky-type and cardinal temperature models are appreciated by users despite the basically empirical nature of the relationships [66].

The specific growth rates of three *S. aureus* strain determined in the suboptimal temperature range 7-39 °C were analyzed with Ratkowsky square root model and graphically compared with the Combase Predictor data [65]. The results presented in Fig. 2 showed high linearity with correlation coefficients R^2 from 0.962 to 0.995 when modelled with a square root model [64]. The following equations resulted from fitting the growth rates with the square root model in the temperature range from 7 to 39 °C for the strains 2064, D1 and B1, respectively:

$$\sqrt{\mu_{2064}} = -0.0883 + 0.042(T - T_{min}) \qquad R^2 = 0.9948;\ \%V = 99.46$$

$$\sqrt{\mu_{D1}} = -0.0804 + 0.0455(T - T_{min}) \qquad R^2 = 0.9784;\ \%V = 97.73$$

$$\sqrt{\mu_{B1}} = -0.0008 + 0.039(T - T_{min}) \qquad R^2 = 0.9623;\ \%V = 95.85$$

Based on the testing of goodness of fit, the per cent of variance (%V) confirmed high correlation coefficients R^2 (above) for strains 2064, D1 and B1, respectively. Their model coefficients b ($\sqrt{\mu} = a + b(T - T_{\min})$), except for the B1 strain, were very close not only to each other but also to the coefficient of Combase line b_{Comb} = 0.048 or b = 0.0442 found by Fujikawa and Morozumi [61].

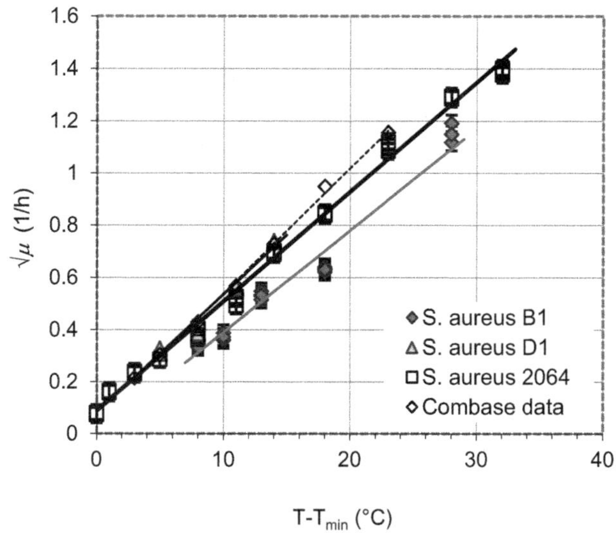

Figure 2. Comparison of the Ratkowsky model applied to the strains of *S. aureus* and selected data from the Combase Predictor within the sub-optimal growth temperatures

The effect of temperature in the whole range from 7 °C to 51 °C on the ability of *S. aureus* to grow in milk is depicted in Fig. 3a. *S. aureus* growth in milk was positively determined with the increasing of the incubation temperature, resulting in a shortening of the lag phase duration and more intensive growth in the exponential phase. Within an empirical approach, the extended model introduced by Ratkowsky [68] which includes data beyond the growth optimum, could be used for describing the impact of temperature on growth rate. The accuracy of the model was validated by comparison with accessible data for other *S. aureus* isolates. Since the data are very similar to each other, the prediction of *S. aureus* 2064 growth in milk can be reliably used for *S. aureus* generally.

According to the recommendation of Ratkowsky [68], maximal temperature for *S. aureus* 2064 of 47 °C was derived from data points in the high-temperature region. By use of this model, the optimal temperature for growth of *S. aureus* in milk of 38.5 °C was also calculated and validated by the use of the Gibson model. From the survival line, with the rate of -0.35 h^{-1} a D-value of 6.7 h at 51 °C (Fig. 1d) was calculated.

From the food practice point of view, the model of Gibson et al. [69] is useful for the prediction of the time (t_3) to increase counts of *S. aureus* by 3 log, if the parameter of specific growth rate is replaced by the t_3 function. In the original equation, a useful $b_w = \sqrt{(1-a_w)}$ transformation appears, in which the value of 1 represents maximal water activity. Analogically, in the case of temperature $T_w = \sqrt{(T_w - T)}$.

In the case of initial *S. aureus* counts in milk meant for cheese production of 10^3 CFU/ml, *S. aureus* will increase its counts during fermentation at 18 °C in 10 h to the level ordered by European Commission Regulation 1441/2007 [70] and the enterotoxin production will occur in 30 h at the same conditions. As is shown in Fig. 3b, in the case of optimal temperature, an increase of about 2 log or 4 log counts will occur in 2 h or 4 h, respectively.

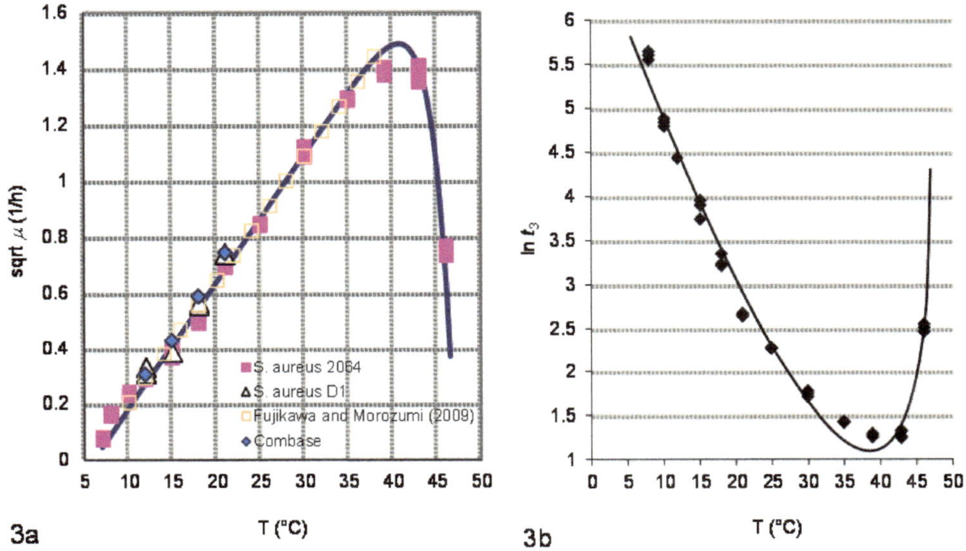

Figure 3. Plots of the square root of specific growth rates (*sqrt μ*) of *S. aureus* 2064 versus incubation temperature. Symbols indicate *sqrt μ* calculated from growth curves at each incubation temperature. The continuous line indicates the fitted of *sqrt μ* vs. *T* function, where $\sqrt{\mu} = 0.0456(T - T_{min})\,[1 - e^{0.447(T-T_{max})}]$
b. Plots of the natural logarithm of time (t_3) necessary for an increase of *S. aureus* 2064 counts about 3 log counts against its initial numbers versus incubation temperature. Symbols indicate the t_3 calculated from growth curves at each incubation temperature. The continuous line indicates the fitted $ln\ t_3 = 3/\mu$, where $\mu\mu = \exp^{(-0.378T_w^2 + 2.202T_w - 2.371)}$ and $T_w = \sqrt{(T_{\max} - T)}$

As temperature was the only modifying environmental factor, lag phase was described by means of the model developed by Davey et al. [71] according to the following equation ($R^2 = 0.962$) in the range from 8 to 43 °C: $\ln\ln\left(1/_{lag}\right) = 1.973 - 87.92/_{T} + 285.09/_{T^2}$.

4.3.2. Effect of pH value and LAB presence on S. aureus growth in milk

In dairy practice, the initial numbers of *S. aureus* play an important role especially at the beginning of the milk fermentation within the first 6 h or in 24h-old cheese. As described above, one of the most effective tools to inhibit the growth of *S. aureus* is to acidify the environment as soon as possible. This is performed by adding a sufficient amount of dairy starters, which are able to ferment lactose and to produce lactic acid very rapidly. As is obvious from Fig. 4, pH 6.0 and 5.5 influenced neither the growth dynamics nor the *S. aureus* counts in the stationary phase. However, pH 5.0 resulted in a decrease of growth rate for about 3.5-time and also in a reduction of total growth. If the pH of growth media is adjusted to pH 4.5, a total diminution of *S. aureus* counts is observed. The same effect is achieved at pH 4.0 if inorganic acids are used.

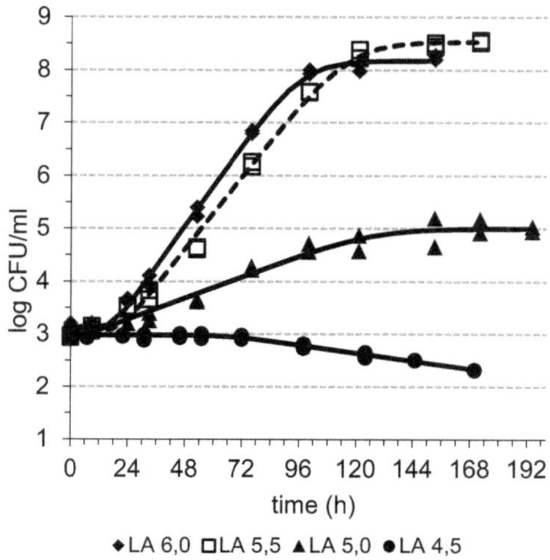

Figure 4. Growth dynamics of *S. aureus* 2064 in nutrient broth at 15 °C in dependence to initial pH value as adjusted with lactic acid to the values 6.0, 5.5, 5.0 and 4.5

Growth and fermentative metabolism of lactic acid bacteria, as a permanent component of raw milk microflora, are offered by a wide variety of fermented dairy products. Besides the most effective inhibitive activity against pathogen and spoilage microorganisms, which includes production of organic acids and subsequent pH decrease, they produce bacteriocins, H_2O_2, and aromatic compounds and act as a strong competitor for nutritional factors (nicotineamide, biotine or niacine) [23,72,73].

If there is slow and insufficient acid production in the growth environment, no inhibitive effect is observed. This was the case of *Lactobacillus rhamnosus* GG and VT1 which did not produce the required amount of lactic acid in milk under aerobic conditions. Despite their inhibitive effect against *Candida maltosa* and *Geotrichum candidum* [74], no inhibition was achieved during the co-cultivation with *S. aureus* in milk. It was also found that the inhibition level of 7% of *Lactococcus lactis* strains was variable and ranged from bacteriostatic to no inhibitory effect on *S. aureus* growth, mainly due to low acidification ability [60]. As *S. aureus* is catalase-positive, we may also expect its resistance against hydrogen peroxide, approx. up to 8% [75].

Thus, it is interesting to select an appropriate starter culture of LAB which is able to efficiently inhibit *S. aureus* growth together with improving the sensorial quality of the final products. However, the strong acidification may limit the activities of other bacterial populations involved in the development of the sensorial properties of ewes' lump cheese [76].

The requirements assigned to a starter culture of LAB include fast growth and survival in dairy environment, rapid production of lactic acid resulting in pH value diminution and no production of toxic or other technologically and sensorially unacceptable metabolites.

The effectiveness of a starter culture of LAB is related to the rate at which it can produce sufficient amounts of lactic acid, predominantly in the first six hours of fermentation. It is connected with the phenomena of the pH lag phase. It is obvious in Fig. 5 that the higher the incubation temperature, the more intensive the metabolism of LAB, and the sooner a pH decrease will occur. The intensity of pH drop is determined by the initial counts of the starter culture, as was confirmed in our co-culture experiments with *S. aureus* 2064 and culture Fresco DVS 1010 (Christian Hansen, Hørsholm, Denmark). The following relation between the duration of pH lag phase resulted from the linear regression analysis shown in Fig. 5: $\ln lag_{pH} = 6.494 - 0.129 \times T - 0.230 \times N_{0\ Fr}$ ($R^2 = 0.970$) where T is temperature in °C, $N_{0\ Fr}$ are initial counts of lactic acid bacteria of the culture Fresco.

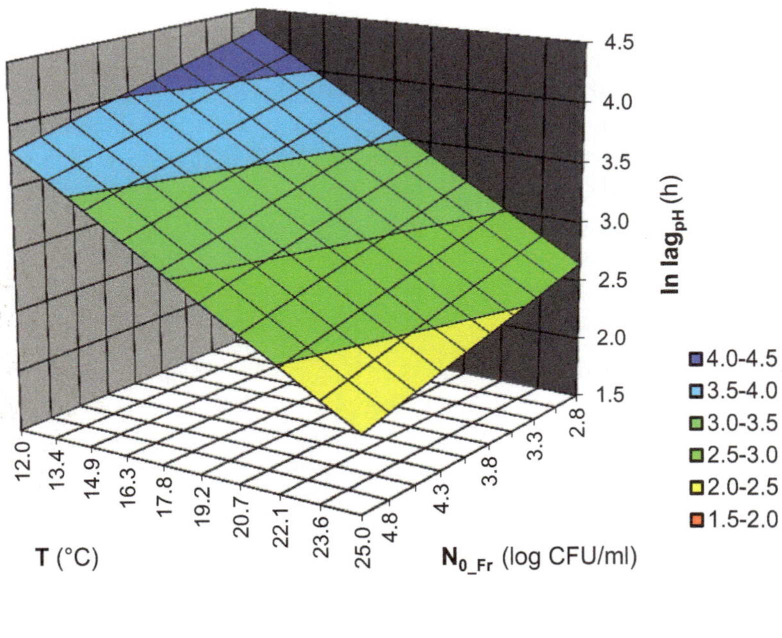

Figure 5. Period without any pH change (lag_{pH}) found during static cultivation of the starters Fresco 1010 (Christian Hansen, Hørsholm, Denmark) and *S. aureus* 2064 in co-culture, as dependent on temperature and initial LAB count ($N_{0\ Fr}$)

It was also observed that during co-cultivation of Fresco culture with *S. aureus* 2064 in milk, *S. aureus* was able to grow only during the pH lag phase. When the pH started to decrease, the growth of the pathogen stagnated or declined. This period was influenced by an appropriate amount of starter culture at a specific incubation temperature.

The ratio between the initial inoculum of S. *aureus* and LAB in a culture determines the efficiency of the inhibition as well. It was observed that when the S. *aureus* population was higher than that of *L. lactis*, S. *aureus* reached the counts as in a pure culture. On the other hand, for ratios of 1:1 or 1:10 for *L. lactis*, maximal S. *aureus* population reached counts about 4 to 5 log CFU/ml lower [23].

However, our data did not showed a direct relation between the inhibition of S. *aureus* counts in the stationary phase and the ratio of mesophilic culture Fresco DVS 1010 and S. *aureus* 2064. However, linear regression analysis (Fig. 6, 7) revealed strong relations between independent variables (as temperature, initial number of the starter) and specific growth rate of S. *aureus*, even between increases in the numbers of S. *aureus* during its growth (Nmax Sa).

Besides initial concentration of LAB, the applied temperature also has a strong effect on the microbial growth dynamics. With an increasing of the incubation temperature, the duration of pH and microbial lag phase is shortened. On the other hand, the higher the temperature, the higher the growth rates.

The combined effect of temperature and the initial Fresco culture is depicted in Fig. 7. From it, one is able to calculate the necessary addition of Fresco culture and thermal mode during milk or young cheese fermentation to ensure a minimal increase in the numbers of S. *aureus* ($N_{max\ Sa}$-$N_{0\ Sa}$). According to the EU regulation, the total S. *aureus* amounts in raw milk cheeses should not exceed 4 log CFU/g. Assuming properly drawn milk with 100 CFU/ml of S. *aureus*, to keep its increase in number at a level lower than 2.0 log CFU/ml, the initial Fresco density should be at least 4.0 or 2.5 log CFU/ml at 21 °C or 18 °C, respectively. Similarly, also the culture A, which contains *Lactobacillus acidophilus*, was able to inhibit growth of S. *aureus* 2064 or S. *aureus* D1 in milk co-cultures [74].

Alomar et al. [76] also found that S. *aureus* SA15 did not grow when *Lactococcus garvieae* was at a concentration of 7.8 log CFU/ml at temperature 22 °C or at 20 °C and initial concentration higher than 6.5 log CFU/ml. At an initial concentration of *L. garvieae* higher than 7.4 log CFU/ml and at temperatures between 22 °C and 34 °C, S. *aureus* growth was not negatively influenced. Higher temperatures favoured the growth of S. *aureus* and *L. garvieae* had no inhibitory effect regardless of concentration.

Although acidification plays an important role in S. *aureus* inhibition, other mechanisms of LAB inhibitive potential should not be excluded. If pH and LAB play only a minor role in the inhibition, it can still be hypothesized that the cessation of the growth is due to the accumulation of antistaphylococcal substances produced by the LAB [77]. Results from literature suggest that S. *aureus* is able to grow under stringent acid conditions (pH 5.25 at 15 °C and 4.48 at 30 °C). On the other hand, inhibition of S. *aureus* by some starter culture was observed at pH 6.8, which cannot be attributed to a drop in pH value. Indirect inhibitory effect may also be involved. The availability of nutrients may trigger other mechanisms, leading for instance to the secretion of metabolites, peptides or signalling molecules, which would in turn be responsible for the inhibitory effect of LAB [60].

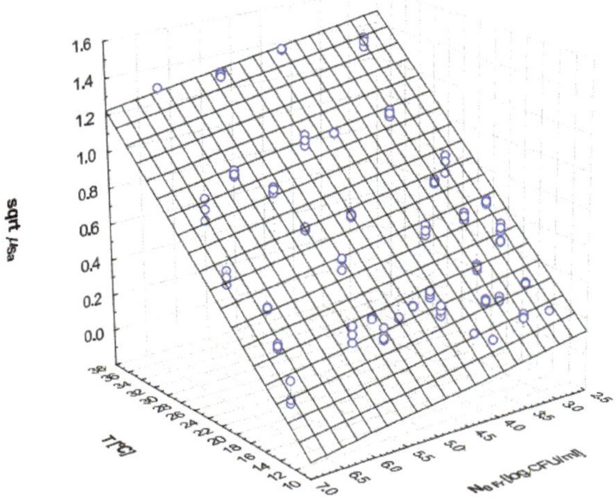

sqrt μ_{Sa} = -0.2111+0.0478T-0.0541$N_{0\ Fr}$ (R^2 = 0.948)

Figure 6. Dependence of specific growth rate (μ) of *S. aureus* 2064 on temperature and initial concentration of Fresco culture in milk

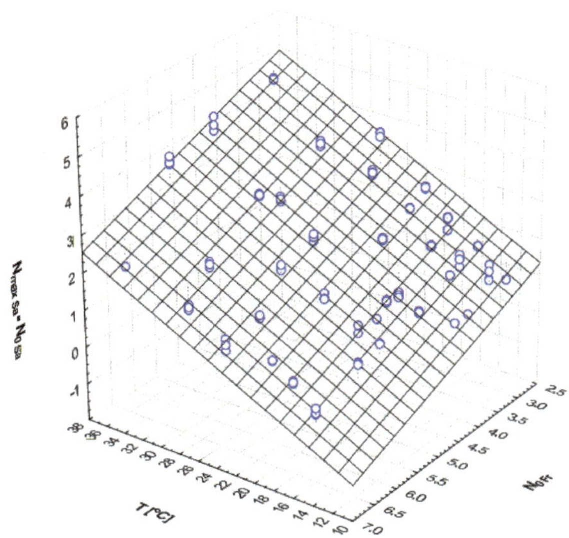

$N_{max\ Sa}$ - $N_{0\ Sa}$ = 1.3989+0.1283*T-0.5345*$N_{0\ Fr}$ (R^2=0.913)

Figure 7. Dependence of increase in *S. aureus* counts ($N_{max\ Sa}$-$N_{0\ Sa}$) on temperature and initial concentration of Fresco culture (Fresco 1010, Christian Hansen, Hørsholm, Denmark) in milk in co-culture with lactic acid bacteria

5. Artisanal raw milk cheese production in Slovakia

5.1. Technology and microbiology: Description based on flow diagram

Original ewes' lump cheese is an artisanal full-fat, soft rennet cheese from raw ewes' milk manufactured on the farm level in Slovakian mountain areas according to the technological steps described and pictured in Fig. 8 and 9, respectively. After two weeks of ripening at temperatures from 18 °C to 21 °C, it is used for industrial production of the popular Slovakian "Bryndza" cheese [2]. Fermentation of the lump cheese relies on native mesophilic lactic acid bacteria (LAB) such as *Lactococcus lactis, Enterococcus faecalis, Lactobacillus casei, Lb. lactis* and *Lb. plantarum*. During ripening, the essential role is played by the milk mould *Geotrichum candidum* and oxidative yeasts of the genera *Torulopsis, Candida* and *Kluyveromyces* [1,3].

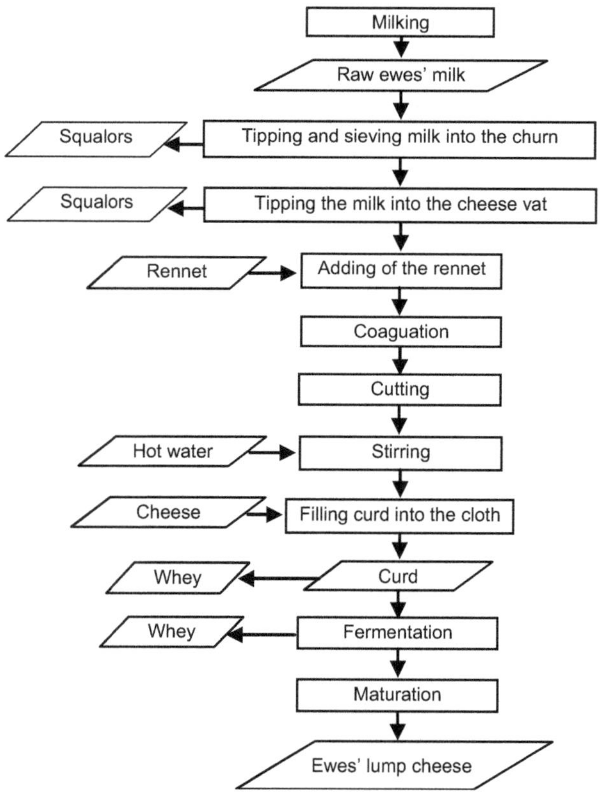

Figure 8. Flow diagram of artisanal production of ewes' lump cheese

Generally, cheeses are considered as one of the safest foods currently consumed. However, pathogenic bacteria which can be transmitted by dairy products cannot be underestimated. Historically, there have been several outbreaks related to the consumption of cheeses. The predominantly responsible organisms *Listeria monocytogenes, Escherichia coli, Salmonella* spp.

and *Staphylococcus aureus* have been reported. The sources of their contaminations were raw milk, inadequately pasteurized milk, or post-pasteurization contaminated milk [47,51,78,79]. In this context, microbiological specifications related to the finished cheeses made from raw milk defined by the Commission Regulation No. 1441/2007 [70] concern with food safety and process hygiene criteria. They comprise of absence *Salmonella* spp., *Listeria monocytogenes* and staphylococcal enterotoxins in 25 g and number of coagulase-positive staphylococci not exceeding $m = 10^4$ and $M = 10^5$ CFU/g at $c = 2$.

Figure 9. Photo-documentation illustrating the artisanal manufacture of ewes' lump cheese on the farm level in Slovakian mountain areas (Author: Ľ. Valík)

Despite the raw milk origin and substantial proportion of raw milk cheeses containing enterotoxigenic *S. aureus*, ewes' lump cheese is also consumed as a fresh cheese at a regional level or it is used as a raw material for the production of original "Bryndza" cheese in Slovakia. The safety and quality of fermented original cheeses manufactured from raw milk at a primary level is generally determined by various specific hygienic, technological, and intrinsic and extrinsic environmental factors. The factors which contribute to the safety of cheeses with respect to pathogenic bacteria include milk quality, native lactic acid bacterial growth during cheese manufacture, pH, salt, environmental conditions and chemical changes during ripening. However, the most important role during fermentation is played by metabolism of the LAB participating in effective competition with pathogenic and spoilage microorganisms and subsequently in inhibition of undesirable microorganisms.

According to our investigations of eight products manufactured under upland farm conditions, the acidification of the curd started after a 10-20 h period and went on intensively for 20 h. Thus, a level of acidity equivalent to pH of 5.2-4.9 was usually reached in young cheese after 30-40 h. Such a fairly long time permits to the growth not only LAB but also of undesirable bacteria, including *S. aureus*. Within these field trials, the initial numbers of *S. aureus* in ewes' milk were about 2.2 log CFU/ml, but in cheeses after 3 days of fermentation they reached 6.2 log CFU/g.

The first 24 h of the process of making raw milk cheese appeared to be critical for *S. aureus* growth, with the most troublesome period taking place within the first 6 h, during which the exponential growth of *S. aureus* mainly occurs. In cheeses with relatively slow acidification during the first 6 h, pH has no effect on the initial growth phase of *S. aureus* before 6 h but may have a modulating effect on subsequent growth of up to 24 h. High pH value in the fresh cheese suggests a weak lactose fermentation ability by the non-starter LAB [7,58,76].

In order to prevent *S. aureus* from reaching the density of 10^6 CFU/g, it is necessary to shorten the pH lag phase by making the fermentative metabolism of LAB more effective under conditions related to lump cheese manufacture. As our previous experiments in model milk media confirmed, the Fresco culture is effective in the *S. aureus* inhibition. Verification of the microbial populations' behaviour under predicted safety conditions was analyzed in laboratory conditions during raw milk ewes' or cows' lump cheese fermentation at 18 °C with or without addition of 1% Fresco culture prior to coagulation.

As seen in Fig. 10a and 10b, a pH of 5.0, unacceptable for growth of *S. aureus*, was achieved after five hours of fermentation if Fresco culture was added. Such a short pH lag phase is crucial in pathogen growth inhibition during cheese manufacture, as has already been mentioned. Higher diminution of pH during the first 6 hours of fermentation means lower *S. aureus* increase in number. The *S. aureus* increase in ewes' lump cheese with the addition of Fresco culture was only 0.96 log counts, despite the growth rate in the exponential phase was more than twice as high as in milk mono-culture at 18 °C or in cheese prepared without Fresco. An increase in microbial counts in the first 24 h is a normal process in cheese making. This is partly due to the physical retention of microorganisms in the coagulum and also due to the microbial multiplication during coagulation and whey drainage [7,80,81]. In contrast to cheese without the addition Fresco starter, the *S. aureus* increase in number was about 2.9 log counts and its maximal counts exceeded 6 log CFU/g. Consumption of such a cheese might represent a potential threat of food poisoning outbreak if the enterotoxigenic strains are present.

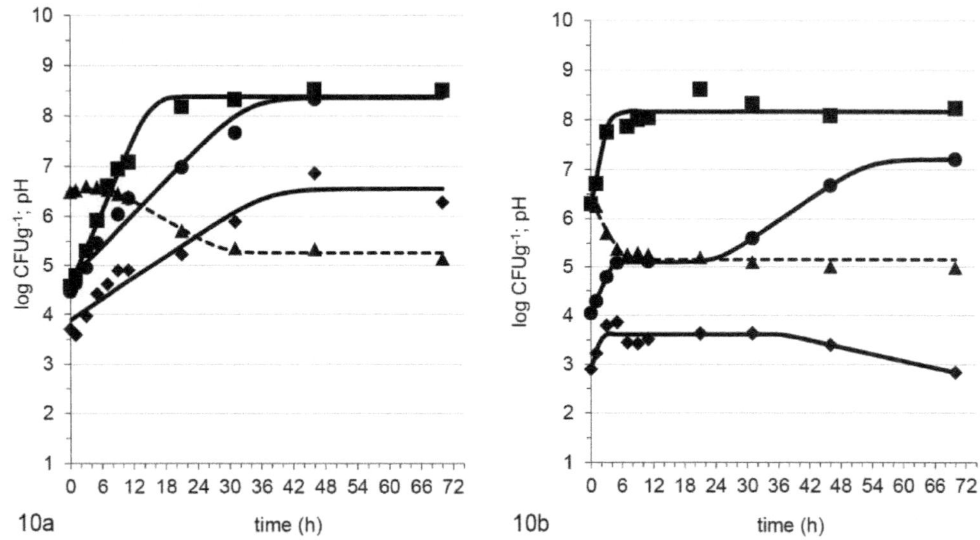

Figure 10. Growth dynamics and pH value changes during fermentation of ewes' lump cheese at 18 °C without starter culture (10a) and with addition of 1% Fresco culture (10b); ■ presumptive LAB on M17 agar, • presumptive LAB on MRS agar, ♦ counts of *S. aureus*, ▲ pH value

In order to keep the numbers of *S. aureus* under the limit defined by the EU regulation No. 1441/2007 [70], the initial addition of Fresco culture into the raw milk should be higher than 10^5 CFU/ml. These initial counts would be accompanied with the suitable timing of pH decrease down to pH 5.0. The addition of an appropriate amount of mixed mesophilic LAB culture, which produces inhibitory substances, provides opportunities to add additional barriers to the growth of bacterial pathogens. Moreover, it could be essential for the improvement of both the fermentation process and the quality of ewes' lump cheese.

This assumption was also confirmed by some other authors. Olarte et al. [82] observed differences in *S. aureus* growth in dependence on the addition of starter culture. In cheese without added starter culture, *S. aureus* exceeded concentrations higher than 5 log counts in 5 days. This was in contrast to the cheese prepared with starter culture, where decreases from counts higher than 4 log CFU/g to 2 log were observed within the fermentation and ripening process. The addition of starter culture of LAB during the manufacture of goats' milk cheese affected the pH value dynamics which after 5 days of fermentation decreased to pH 5.1 compared to 6.61 pH of cheese made without starter culture.

A rapid decrease in pH values from 6.7 to an average value of 5.24 was observed in 24 h old raw cows' milk cheeses [58]. From an initial average density of 1.89 log *S. aureus* grew rapidly during the first 6 h up to 5 log in average and then slowly up to 24 h, when the population reached a peak. In those cheeses *S. aureus* never reached 7 log CFU/g, but in 2 samples where SE production occurred, not only did they exceed 5 logs, but pH values of the cheeses at 6 h also exceeded 6.3. In Tenerife goats' raw milk cheese after 48 h pH reached value of 4.93, which led to a decrease in *S. aureus* counts from 3.14 log CFU/g in 2 days old cheeses to 1.62 log CFU/g in 30 day old cheeses [83]. Also in raw cows' milk cheeses counts of *S. aureus* in the end of 2-3 weeks fermentation were lower than 2 log CFU/g, mostly due to the rapid pH value decrease down to pH 5.09 [84]. During the ripening of Turkish White cheese made from raw cows' milk at 6 °C, pH was not changed and fluctuated from 4.63 to 5.06. Such acidic conditions contributed to the decrease in *S. aureus* counts from 5.03 log to 2.36 log CFU/g over 4 months [85].

In raw milk cheeses collected by Jakobsen et al. [80], a significant decline in pH values was observed after 5-6 h of fermentation and pH lower than 5.5 was achieved in all samples after 24 h. The highest *S. aureus* contamination was reached in 5-6 h old samples, in some samples higher than 4 log counts. Nevertheless, none of the sample exceeded counts higher than 5 logs and so it was concluded that *S. aureus* did not produce enterotoxins. A correlation between the contamination level of the milk and contamination level of 5-6 h old cheeses was noticed. The initial *S. aureus* level in raw milk greatly influences the level of staphylococci during this first period of cheese-making.

In raw goats' milk cheeses, the initial log counts of *S. aureus* were 4.86; 6.23 and 5.88 in winter, spring and summer cheeses. They were covered with brine (12%) for 10 days at 15 °C and then stored at 4 °C for 3 months. During the ripening, the counts of *S. aureus* decreased to 2.04-2.30 log in winter and spring cheeses or to 1.02 log counts in summer cheeses, respectively. During ripening of all 3 types of cheeses, pH was practically stable, reaching values in the range 5.23-6.06 [86]. In cheeses made from raw ewes' milk, *S. aureus* was not detected in either fresh or mature cheese. The pH of fresh or ripened cheese was 6.31 and 5.79 pH, respectively [87].

On the other hand, in raw milk Mexican cheese Fresco, pH decline from pH 6.7 to value 5.6 was achieved only after 10 days of ripening at 4 °C. Counts of *S. aureus* were close to the 10^7 CFU/g level and did not undergo any noticeable change during cheeses storage. It may be due to the *S. aureus* capacity to withstand a wide range of temperatures, pH and water activity [81].

Based on these results and observations from literature, it is strongly recommended, to use the starter culture at least in artisanal cheese production. Rapid fermentation process prevents against the growth of *S. aureus* and other pathogenic and undesirable microorganisms. Even in mountain areas, this can be performed by the inoculation of LAB, e.g. in the form of fresh fermented milk. Moreover, the addition of adjunct starter culture can improve flavour, reduced bitterness and increase the concentration of peptides, which impart desirable flavour, and of precursors of flavour volatiles [88]. It was also confirmed, that the Fresco culture addition had no negative effect on the sensorial descriptors of ewes' lump cheese and compared to the cheese made without starter culture it even achieved better sensorial acceptance. But it has been suggested that positive results for flavour and texture development are strongly strain-dependent [85], so the selection of appropriate starter culture is necessary.

The variation in the responses of *S. aureus* to pH value and its dynamic during cheese making process may be attributed to the variations in each dairy farm management of hygiene practices, environmental and personal diversities, process of manufacturing, herd characteristics, multiple sources of *S. aureus* contamination and geographical distribution of *S. aureus* strains [20,48,80]. Similarly, many factors are known to influence the SEs production, e.g. NaCl content, water activity value, pH, temperature, atmosphere, amino acid composition and competing microflora. For that reason, it is crucial to understand which factors control enterotoxin production in raw milk cheese, in order to be able to assess their safety and to prevent staphylococcal food poisoning.

Even if pasteurization kills *S. aureus* cells, a previous population reaching higher than 5 log counts may lead to enterotoxins production and once enterotoxins are produced they retain their activity. Besides this, pasteurization eliminates also enzymes and indigenous microflora, which are partly responsible for the development of the typical raw milk cheese flavour and texture [80]. Moreover, the raw milk contains a heat-labile lactoperoxidase system which has inhibitory effect to the growth of some pathogens [79]. The post-pasteurization addition of a starter culture may lead to losses in the unique organoleptic properties of the raw milk cheeses and to end products with uniform sensorial features [81,85,87]. Hence, there is today a renewed interest in traditionally produced raw milk cheeses due to consumer demands for increased varieties of cheese flavours and textures [80]. Consequently, as regards the safety of raw milk cheeses, potential pathogens associated with milk or milk products including *Staphylococcus aureus*, should still be of interest.

6. Conclusions

The inhibitory potential of LAB on *S. aureus* growth results from different factors described here and in the scientific sources used in this chapter. Taking into account the growth data obtained with a few *S. aureus* strains isolated from artisanal cheeses it was shown that they grew well in milk alone, in co-culture with LAB as well as in raw milk cheeses prepared in laboratory. As the specific strains isolated from raw ewes' milk and cheese were used in the study, the growth data may provide useful information for artisanal cheese practice. Taking into account that the initial *S. aureus* numbers in raw milk fluctuate about 3 log CFU/ml, artisanal ewes' lump cheese producers may apply our prediction of the time (t_3) directly as these t_3 times are in coincidence with the time to reach a critical density of 10^6 CFU/ml for possible enterotoxin presence.

Artisanal raw milk cheese production poses a few critical factors limiting its safety. With reference to the growth of *S. aureus*, many factors should be taken into consideration, such as its natural contamination in milk, quantitative growth data, cheese type, nature, activity and type of the starter culture and mutual relation between *S. aureus* and LAB populations. Factors that may prevent the reaching of *S. aureus* counts higher than 10^5 CFU/ml and production of enterotoxin are: low initial contamination of milk (less than 10^2 CFU/ml), high initial number of active LAB (higher than 10^5 CFU/ml), the capacity to pH during the first 6 h of fermentation and cease the growth of *S. aureus* as fast as possible within 24 h (in the best case within the first 6 h of young cheese fermentation). Inhibitory starters producing bacteriocins may also be used. Thus, the adding of a starter culture in artisanal cheese production is strongly recommended.

Author details

Alžbeta Medveďová

Department of Nutrition and Food Quality Assessment, Faculty of Chemical and Food Technology, Slovak University of Technology, Bratislava, Slovak Republic

Ľubomír Valík

Corresponding Author

Department of Nutrition and Food Quality Assessment, Faculty of Chemical and Food Technology, Slovak University of Technology, Bratislava, Slovak Republic

Acknowledgement

The authors would like to thank Kevin Slavin for reading the manuscript and for the language corrections. The results of research used in this chapter were supported by the Slovak Research and Development Agency, projects no. APVV-0590-10 and APVV-20-005605.

7. References

[1] Görner F, Valík Ľ (2004) Applied Microbiology of Foodstuffs. Bratislava: Malé centrum. 528 p.

[2] Palo V, Katáb M (1984) Slovak sheep cheeses. Milchwissenshaft 39: 518-521.

[3] Valík Ľ, Görner F, Polka P, Sonneveld, K (2004) Fermentation of ewes' lump cheese in sheep-farm manufacture. Sheep goats 24: 23-24.

[4] Euzéby JP (2012) List of Prokaryotic names with Standing in Nomenclature - Family *Staphylococcaceae* e Stand. Available http://www.bacterio.cict.fr/s/staphylococcaceae.html. Accessed 2012 Jan. 2.

[5] Baird-Parker A (1990) The staphylococci: an introduction. J. appl. bacteriol. symp. suppl. 1S-8S. [30]

[6] Baird-Parker T (2000) *Staphylococcus aureus*. In: Lund B, Baird-Parker T, Gould G, editors. The Microbiological Safety and Quality of Food. Gaithersburg: Aspen Publishers. pp. 1317-1330.

[7] Asperger H, Zangerl P (2003) *Staphylococcus aureus*. In: Roginski H, Fuquay J, Fox P, editors. Encyclopedia of Dairy Science. San Diego: Academic Press pp. 2563-2569.

[8] Bremer P, Fletcher G, Osborne C (2004) *Staphylococcus aureus*. Christchurch: New Zealand Institute for Crop and Food Research Limited. 6 p.

[9] Halpin-Dohnalek M, Marth E (1989) *Staphylococcus aureus*: Production of extracellular compounds and behaviour in foods - a review. J. food protect. 52: 267-282.

[10] Jay J (2000) Staphylococcal gastroenteritis. In: Jay J, editor. Modern Food Microbiology. Gaithersburg: Aspen Publishers. pp. 441-459.

[11] Ote I, Taminiau B, Duprez J-N, et al. (2011) Genotypic characterization by polymerase chain reaction of *Staphylococcus aureus* isolates with bovine mastitis. Vet. microbiol. 153: 285-292.

[12] Garzoni C, Kelley W (2009) *Staphylococcus aureus*: new evidence for intracellular persistance. Trends microbiol. 17: 59-65.

[13] Post D (1999). *S. aureus*. Food-borne pathogens. Basingstoke: Oxoid. 6 p.

[14] Arbuthnott J, Coleman D, DeAzavedo J (1990) Staphylococcal toxins in human disease. J. appl. bacteriol. symp. suppl. 101S-107S.

[15] Deurenberg R, Nieuwenhuis R, Driessen C, et al. (2005) The prevalence of the *Staphylococcus aureus tst* gene among community- and hospital-acquired strains and isolates from Wegener's Granulomatosis patients. FEMS microbiol. lett. 245: 185-189.

[16] Kérouanton A, Hennekinne J, Letertre C, et al. (2007) Characterization of *Staphylococcus aureus* strains associated with food poisoning outbreaks in France. Int. j. food microbiol. 115: 369-375.

[17] Normanno G, LaSalandra G, Dambrosio A, et al. (2007) Occurence, characterization and antimicrobial resistance of enterotoxigenic *Staphylococcus aureus* isolated from meat and dairy products. Int. j. food microbiol. 115: 290-296.

[18] The European Union Summary Report on Trends and Sources of Zoonoses, Zoonotic Agents and Foodborne Outbreaks in 2009. (2011). EFSA j. 9: 378.

[19] Notermans S, Heuvelman CJ (1983) Combined effect of water activity, pH and sub-optimal temperature on growth and enterotoxin production of *Staphylococcus aureus*. J. food sci. 48:1832-1840.

[20] Ertas N, Gonulalan Z, Yildirim Y, Kum E (2010) Detection of *Staphylococcus aureus* enterotoxins in sheep cheese and dairy desserts by multiplex PCR technique. Int. j. food microbiol. 142: 74-77.

[21] Sutherland JP, Bayliss AJ, Roberts TA (1994) Predictive modelling of growth of *Staphylococcus aureus*: the effects of temperature, pH and sodium chloride. Int. j. food microbiol. 21: 217-236.

[22] Notermans S, van Hoeij K (2008) The food safety file: *Staphylococcus aureus*. Woerden: Food Doctors, 39 p.

[23] Charlier C, Cretenet M, Even S, LeLoir Y (2009) Interactions between *Staphylococcus aureus* and lactic acid bacteria: An old story with new perspectives. Int. j. food microbiol. 131: 30-39.

[24] Baker-Austin C, Dopson M (2007) Life in acid: pH homeostasis in acidophiles. Trends microbiol. 15: 165-171.

[25] Medveďová A, Valík Ľ, Studeničová A (2009a) The effect of temperature and water activity on the growth of *Staphylococcus aureus*. Czech j. foods sci. 27: S2-28-S2-35.

[26] O'Byrne C, Booth IR (2002) Osmoregulation and its importance to food-borne microorganisms. Int. j. food microbiol. 74: 203-216.

[27] Scybert S, Pechous R, Sitthisak S, et al. (2003) NaCl-sensitive mutant of *Staphylococcus aureus* has a *Tn917-lacZ* insertion in its *ars* operon. FEMS microbiol. lett. 222: 171-176.

[28] Ewald S, Notermans S (1988) Effects of water activity on growth and enterotoxin D production of *Staphylococcus aureus*. Int. j. food microbiol. 6:25-30.

[29] Girish T, Sharma E, Gopal B (2008) Structural and functional characterization of *Staphylococcus aureus* dihidrodipicolinate synthase. FEBS lett. 582: 2923-2930.

[30] Casey A, Lambert P, Elliott T (2007) Staphylococci. Int. j. antimicrob. ag. 3: S23-S32.

[31] Götz F (2004) Staphylococci in colonization and disease: prospective targets for drugs and vaccines. Curr. op. microbiol. 7: 477-487.

[32] Perry JD, Rennison C, Butterworth LA, et al. (2003). Evaluation of *S. aureus* ID, a New Chromogenic Agar Medium for detection of *Staphylococcus aureus*. J. clin. microbiol. 41: 5695-5698.

[33] Akineden O, Hassan A, Scheider E, Usleber E (2008) Enterotoxigenic properties of *Staphylococcus aureus* isolated goats' milk cheese. Int. j. food microbiol. 124: 211-216.

[34] Boynukara B, Gulhan T, Alisarli M, et al. (2008) Classical enterotoxigenic characteristics of *Staphylococcus aureus* strains isolated from bovine subliclinical mastitis in Van, Turkey. Int. j. food microbiol. 125: 209-211.

[35] Peles F, Wagner M, Varga L, et al. (2007) Characterization of *Staphylococcus aureus* strains isolated from bovine milk in Hungary. Int. j. food microbiol. 118: 186-193.

[36] Sridhar Rao PN (2011) Coagulase test. available at http://www.microrao.com. Accessed 2011 March 28.

[37] Rall V, Vierira F, Rall R, et al. (2008) PCR detection of staphylococcal enterotoxin genes in *Staphylococcus aureus* strains isolated from raw and pasteurized milk. Vet. microbiol. 132: 408-413.

[38] Vermicon (2004) User manual VIT-Staphylococcus: detection of *Staphylococcus* spp. and *Staphylococcus aureus*. Munich: Vermicon. 17p.

[39] Pinto B, Chenoll E, Aznar R (2005) Identification and typing of food-borne *Staphylococcus aureus* by PCR-based techniques. Syst. appl. microbiol. 28: 340-352.

[40] Riyaz-Ul-Hassan S, Verma V, Qazi G (2008) Evaluation of three different molecular markers for the detection of *Staphylococcus aureus* by polymerase chain reaction. Food microbiol. 25: 452-459.

[41] Brakstad O, Aasbakk K, Maeland J (1992) Detection of *S. aureus* by polymerase chain reaction amplification of the *nuc* gene. J. clin. microbiol. 30: 1654-1660.

[42] Morandi S, Brasca M, Lodi R, et al. (2007) Detection of classical enterotoxins and identification of enterotoxin genes in *Staphylococcus aureus* from milk and dairy products. Vet. microbiol. 124: 66-72.

[43] Palomares C, Torres M, Torres A, et al. (2003) Rapid detection and identification of *Staphylococcus aureus* from blood culture specimens using real-time fluorescence PCR. Diagn. micr. infec. dis. 45: 183-189.

[44] Lawrynowicz-Paciorek M, Kochman M, Piekarska K, et al. (2007) The distribution of enterotoxin and enterotoxin-like genes in *Staphylococcus aurues* strains isolated from nasal carriers and food samples. Int. j. food microbiol. 117: 319-323.

[45] Mørk T, Kvitle B, Mathisen T, Jorgensen H (2010) Bacteriological and molecular investigations of *Staphylococcus aureus* in diary goats. Vet. microbiol. 141: 134-141.

[46] Pelisser M, Klein C, Ascoli K, et al. (2008) Occurence of *Staphylococcus aureus* and multiplex PCR detection of classic enterotoxin genes in chceese and meat products. Braz. j. microbiol. 40: 145-148.

[47] Lindqvist R, Sylvén S, Vågsholm I (2002) Quantitative microbial risk assessment exemplified by *Staphylococcus aureus* in unripened cheese made from raw milk. Int. j. food microbiol. 78: 155-170.

[48] Rosengren A, Fabricius A, Guss B, et al. (2010) Occurence of food-borne pathogens and characterizations of *Staphylococcus aureus* in cheese produced in farm-dairies. Int. j. food microbiol. 144: 263-269.

[49] Zigo F, Vasiľ M, Kadáši M, et al. (2011) Bacteria *Staphylococcus* spp. isolated from mastitis of sheep and their enterotoxigenic properties. Potr. 5: 70-72.

[50] Dudríková E, Poľaková L, Pukáčová J (2010) Health and hygienic conditions of ewes' milk processeing from the aspect of food safety. Potr. 4: 14-18.

[51] Kousta M, Mataragas M, Skandamis P, Drosinos E (2010) Prevalence and sources of cheese contamination with pathogens at farm and processing level. Food control 21: 805-815.

[52] Ayidin A, Sudagidan M, Muratoglu K (2011) Prevalence of staphylococcal enterotoxins, toxin genes and genetic-relatedness of foodborne *Staphylococcus aureus* strains isolated in the Marmara Region of Turkey. Int. j. food microbiol. 148: 99-106.

[53] Trnčíková T, Piskernik S, Kaclíková E, et al. (2010). Characterization of *Staphylococcus aureus* strains isolated from food produced in Slovakia and Slovenia with regard to the presence of genes encoding for enterotoxins. J. food nutr. res. 49: 215-220.

[54] Pereira V, Lopes C, Castro A, et al. (2009) Characterization for enterotoxin production, virulence factors, and antibiotic susceptibility of *Staphylococcus aureus* isolates from various foods in Portugal. Food microbiol. 26: 278-282.

[55] Kuroda M, Ohta T, Uchiyama I, et al. (2001) Whole genome sequencing of methicilin-resistant *Staphylococcus aureus*. Lancet 357: 1225-1240.

[56] Pathogen Modelling Programm available at http://pmp.arserrc.gov/.

[57] Medveďová A, Valík Ľ, Sirotná Z, Liptáková D (2009b) Growth characterization of *Staphylococcus aureus* in milk: a quantitative approach. Czech j. food sci. 27: 443-453.

[58] Delbes C, Alomar J, Chougui N, et al. (2006) *Staphylococcus aureus* growth and enterotoxin production during the manufacture of uncooked, semihard cheese from cows' milk. J. food protect. 69: 2161-2167.

[59] Balaban N, Rasooly A (2000) Staphylococcal enterotoxins. Int. j. food microbiol. 61: 1-10.

[60] Charlier C, Even S, Gautier M, LeLoir Y (2008) Acidification is not involved in the early inhibition of *S. aureus* growth by *Lactococcus lactis* in milk. Int. dairy j. 18: 197-203.

[61] Fujikawa H, Morozumi S (2006) Modeling *Staphylococcus aureus* growth and enterotoxin production in milk. Food microbiol. 23: 260-267.

[62] Tatini SR (1973) Influence of food environments on growth of *Staphylococcus aureus* and production of various enterotoxins. J. milk food technol. 36: 559-563.

[63] Dengremont E, Membré JM (1995) Statistical approach for comparison of the growth rates of five strains of *Staphylococcus aureus*. App. environ. microbiol. 61: 4839-4395.

[64] Baranyi J, Roberts TA, McClure P (1993) A non-autonomous differential equation to model bacterial growth. Food microbiol. 10: 43-59.

[65] Combase Database available at http://modelling.combase.cc

[66] Geeraerd AH, Valdramidis VP, et al. (2004) Development of a novel approach for secondary modelling in predictive microbiology: incorporation of microbiological knowledge in black box polynomial modelling. Int. j. food microbiol. 91: 229-244.

[67] Ratkowsky DA, Olley J, McMeekin TA, Ball A (1982) Relationship between temperature and growth rate of bacterial culture. J. food bacteriol. 149: 1-5.

[68] Ratkowsky D, Lowry R, McMeekin T, et al. (1983) Model for bacterial culture growth rate throughout the entire biokinetic temperature range. J. bacteriol. 154: 1222-1226.

[69] Gibson A, Baranyi J, Pitt J, et al. (1994) Predicting fungal growth: the effect of water activity on *Aspergillus flavus* and related species. Int. j. food microbiol. 23: 419-431.

[70] Commission Regulation (EC) No. 1441/2007 amending Regulation (EC) No. 2073/2005 on microbiological criteria for foodstuffs. Official Journal of the European Union, L322/12: 18 p.

[71] Daughtry B, Davey K, King K (1997) Tempteraure dependence of growth kinetics of food bacteria. Food microbiol. 14: 21-30.

[72] DeVuyst L, Leroy F (2007) Bacteriocins from lactic acid bacteria: production, purification and food applications. J. molecul. microbiol. biotechnol. 13:194-199.

[73] Makras L, Triantafyllou V, Fayol-Meddaoudi D, et al. (2006) Kincetic analysis of the antibacterial activity of probiotic lactobacilli towards *Salmonella enterica* serovar Typhimurium reveals a role for lactic acid and other inhibitory compounds. Res. microbiol. 157:241-247.

[74] Liptáková D, Hudecová A, Valík Ľ, Medveďová A (2009) The effect of *Lb. rhamnosus* GG on growth of *Geotrichum candidum* and *Candida maltosa* in milk. In: Flynn K, Garriga M, Hofstra H, Monfort JM, editors. The 2nd SAFE Consortium International Congress on Food Safety 2009. Girona: The SAFE Consortium. pp. 156-157.

[75] Elkins CHA, Munoz ME, Mullis LB, et al. (2008) *Lactobacillus*-mediated inhibition of clinical toxic shock syndrome *Staphylococcus aureus* strains and its relation to acid and peroxide production. Anaerobe 14: 261-267.

[76] Alomar J, Lebert A, Montel M (2008a) Effect of temperature and pH on growth of *S. aureus* in co-culture with *Lactococcus garvieae*. Curr. microbiol. 56: 408-412.

[77] LeMarc Y, Valík Ľ, Medveďová A (2009) Modelling the effect of the starter culture on the growth of *Staphylococcus aureus* in milk. Int. j. food microbiol. 129: 306-311.

[78] Little C, Rhoades J, et al. (2008) Microbiological quality of retail cheeses made from raw, thermized and pasteurized milk in the UK. Food microbiol. 25: 304-312.

[79] Donnelly C (2004) Growth and survival of microbial pathogens in cheese. In: Fox P, McSweeney P, Cogan T, Guince T, editors. Cheese: chemistry, physics and microbiology. Hardbound: Academic Press. pp. 541-559.

[80] Jakobsen R, Heggebo R, Sunde E, Skjervheim M (2011) *Staphylococcus aureus* and *Listeria monocytogenes* in Norwegian raw milk cheese production. Food microbiol. 28: 492-496.

[81] Ttores-Llanez M, Vallejo-Cordoba B, Díaz-Cinco M, et al. (2006) Characterization of the natural microflora of artisanal Mexican Fresco cheese. Food control. 17: 683-690.

[82] Olarte C, Sanz S, Gonzalez-Fandos E, Torre P (2000) The effect of a commercial starter cultures addition on the ripening of an artisanal goat´s cheese (Cameros cheese). J. appl. microbiol. 88: 421-429.

[83] Zárate V, Belda F, Pérez C, Cardell E (1997) Changes in the microbial flora of Tenerife Goats' milk cheese during ripening. Int. dairy j. 7: 635-641.

[84] Menéndez S, Godínez R, Centeno J, Rodríguez-Otero J (2001) Microbiological, chemical and biochemical characteristics of "Tetilla" raw cows' milk cheese. Food microbiol. 18: 151-158.

[85] Öner Z, Karahan A, Aloğlu H (2006) Changes in the microbiological and chemical characteristics of an artisanal Turkish white cheese during ripening. LWT. 39: 449-454.

[86] Psoni L, Tzanetakis N, Litopoulou-Tzanetaki E (2003) Microbiological characteristics of Batzos, a traditional Greek cheese from raw goats' milk. Food microbiol. 20: 575-582.

[87] Prodromou K, Thasitou P, Haritonidou E, et al. (2001) Microbiology of "Orinotyri", a ewes' milk cheese from the Greek mountains. Food microbiol. 18: 319-328.

[88] Katsiari M, Voutsinas L, Kondyli E (2002) Improvement of sensory quality of low-fat Kefalograviera-type cheese with commercial adjuct cultures. Int. dairy j. 12: 757-764.

Chapter 5

Rice Starch-Based Biodegradable Films: Properties Enhancement

Thawien Wittaya

Additional information is available at the end of the chapter

1. Introduction

The claims of environmental protection and the lack of petroleum resources provide a new opportunity for developing plastic materials derived from biopolymer resources. Starch is one of the most studied and promising raw materials for the production of biodegradable plastics, because starch is quite cheap, abundant, biodegradable and edible. Starch consists of two major types of molecules, primarily linear amylose and highly branched amylopectin. Normal starch consists of about 75% amylopectin and 25% amylose; waxy starches consist of mainly amylopectin and 0–8% amylose; and high-amylose starches consist of 40–70% amylose. Amylose is composing of D-glucose molecules, which are linked in a α-1, 4 conformation. The glucose monomers therefore form a linear straight chain polymer. Amylose is the key component involved in water absorption, swelling and gelation of starch in food and processing of material (Hoseney, 1986).

Amylopectin is the major component of most starches, and consists of a large number of shorter chains that are bound together at their reducing end side by a α-1, 6 linkage (Hoseney, 1986). Amylopectin is therefore highly branched as the α-1, 4 linear chains are punctuated with the α-1, 6 linkages. The α-1, 6 constitute about 5% of the structure of amylopectin and gives rise to branching. The amylopectin molecule is much larger than the amylose molecule. Minor components, such as lipids, phospholipids, and phosphate monoester derivatives, are found in starch and have profound effects on the properties of starch. Lipids and phospholipids are found in cereal starches. Normal cereal starches contain up to 1% lipids, and the level of lipid content is proportional to the amylose content of the normal starch (Morrison, 1995). Starches of different botanical origins consist of different species of lipids. For example, normal maize starch consists of mainly free fatty acids, glycerides, and little phospholipids; normal rice starch contains substantial amount of phospholipids and some free fatty acids; and wheat, barley, rye, and triticale starches consist exclusively of phospholipids (Kasemsuwan & Jane, 1996; Morrison, 1995). Cereal waxy starches contain few lipids, whereas high-amylose starches contain substantially more lipids. Root and tuber starches contain very little lipids and no detectable phospholipids (McPherson & Jane, 1999).

Rice is the most widely consumed basic food in the world. Each year over 500 million tons of rice are harvested, providing sustenance to many countries and people throughout the world. The unique properties of rice starches are found in its many varieties. Due to different climates, soil characteristics and cultures, over 240,000 registered varieties of rice exist in the world. These varieties lead to a wide range of rice starches with many different characteristics including: different onset gelatinization temperatures; textures; processing stability; and viscosities. Rice starch with high amylose is an attractive raw materials for use as barriers in packaging materials. They have been used to produce biodegradable films to partially or entirely replace plastic polymers because of the low cost and renewability, as well as possessing good mechanical properties. However, compared to the common thermoplastics, biodegradable rice starch products still reveal many disadvantages. These include low mechanical properties and lack of efficient barrier against high polarity compounds. The disadvantages are mainly attributed to the highly hydrophilic character of rice starch polymers. To cope with these problems while preserving the biodegradability of the materials, the improvement of rice starch film properties has been investigated to meet suitable applications. This chapter provides details of rice starch films, and the formation and factors affecting rice starch film properties. It also deals with improving rice starch film properties with various methods such as using chemical treatments, physical treatments, combination with other biodegradable materials and fiber reinforcement.

2. Formation of rice starch-based biodegradable films

The formation of rice starch films requires the use of at least one constituent capable of forming a matrix with adequate continuity and cohesion. Generally, this is composed of polymers or compounds which, under the preparation conditions, are used to form continuous crystalline or amorphous products. In the case of coatings for which the films' system can be applied directly to the product, two forces are relevant: that between the molecules of the coating material (cohesion) and that between the coating and the support structure (adhesion). The degree of cohesion produces the barrier and mechanical properties of the film. High structural cohesion is manifested by a reduction in flexibility, porosity, and permeability to gases and solutes (Banker, 1966). The degree of cohesion depends on the chemical structure of the film material, the presence of plasticizing and cross-linking agents, the nature of the solvent used and its dilution, the method of application, the procedure used for removal of the solvent, and the final thickness of the film.

The highest cohesion is generally obtained for ordered polar polymers with long chains which were precipitated in crystalline form. The preparation for average dilution which constitutes a compromise between the salvation and extension of polymer molecules and good initial viscosity is preferable (Baker, 1966). The cohesion of films generally increases in proportion to their thickness up to a threshold beyond which it remains constant. The speed of evaporation of the solvent and/or excessive temperature may be manifested by inadequate cohesion on account of the premature immobilization of the polymer molecular. Regarding the adhesion of the coatings to the foodstuff, this is generally facilitated by hot application.

The formulation of films base on starch may require one of the following methods. (1) Casting, which is a process consisting of drying a solution or a gel is a simple method for producing films with controlled thickness. This technique is useful to mimic some industrial processes for forming free-standing starch films as is the case for dip-molding. In this method, used for food coatings as well as for non-food applications, the gelled state is usually preferred to set hot solutions on a surface upon cooling. (2) Extrusion or the thermo pressing process is a process used to create objects of a fixed cross sectional profile. A material is pushed or drawn through a die of the desired cross section. The two main advantages of this process over other manufacturing processes are its ability to create very complex cross sections and work materials that are brittle, because the material only encounters compressive and shear stresses. It also gives finished parts an excellent surface finish. Next is (3) Electrohydrodynamic atomization (EHDA), referred to as electrostatic atomization or electro-spraying. This is a process in which a liquid is forced through a capillary and a potential difference of the order of kilo volts is applied between the capillary and the collection electrode (Pereta & Edirisinghe, 2006). EHDA can conduct in various modes but the stable cone-jet mode is the most desirable as it provides near-monodispersed droplets of a few micrometers in size. The droplets size can be controlled by the flow rate and applied as appropriate (Clopeau & Prunet-Foch, 1990).

3. Factor affecting rice starch-based biodegradable films

3.1. Amylose and amylopectin content

Starch consists of two polysaccharides, the essentially linear amylose, and the branched amylopectin (Manners, 1989). The pure amylose structure is very stable, with strong molecular orientation, forming films denser and stronger than amylopectin films (Lourdin et al., 1995). The ability of amylose to produce self supporting films has been known for a long time and this is attributed to the ability of its linear chains to interact by hydrogen bonds to a higher extent than the branched amylopectin chains. Amylopectin films, on the other hand, are rather frail due to the higher degree of entanglement caused by the extensive branching and the short average chain length (Rindlav-Westling et al., 1998). Rindlav-Westling et al. (2002) prepared films from potato starch, amylose, and amylopectin and blends by solution casting. Results showed that amylose films had a relative crystallinity of about 30% whereas amylopectin films were entirely amorphous. The blending of amylose and amylopectin resulted in films with a considerably higher degree of crystallinity than could be predicted. This is explained by co-crystallization between amylose and amylopectin and possibly by crystallization of amylopectin. The crystallized material gave rise to an endothermic detected with differential scanning calorimetry. The enthalpy and peak temperature of the transition also increased as the amylose content decreased. When the amylose proportion in the blends was low, separate phases of amylose and amylopectin were observed by light microscopy. At higher amylose proportions, however, the phase separation was apparently prevented by amylose gelation and the formation of continuous amylose network. Addition, the amylose network in the films, observed with transmission electron microscopy, consisted of stiff strands and open pores and became less visible as the amylose proportion decreased.

Alves et al. (2007) studied the effect of amylose enrichment on cassava starch films properties. This study showed the mechanical and barrier properties of cassava films were influenced by the amylose contents. The amylose enrichment originated stronger films and this could be explained because during drying of film-forming solutions, water evaporates, allowing the formation of starch network. During this stage the proximity of starch chains induced by higher amylose contents could facilitate the formation of matrix with more polymer content per area.

Ming et al. (2011) characterized the biodegradable films from corn starch with different amylose content. They concluded that amylose content had significantly affected the mechanical and thermal properties of the biodegradable starch-based films. The high amylose starch films exhibited better mechanical properties, such as higher modulus and tensile strength, and very high impact strength. The reasons for this include not only the easy entanglement of long linear amylose chains, but also the retained granular structure in high amylose films, which may act as self reinforcement.

Muscat et al. (2012) studied the effect of low and high amylose starches on film forming behavior. They found that, films with high amylose content showed higher glass transition temperature, tensile strength and modulus of elasticity values and lower elongation values than low amylose starch films. There was an increase in thermal and mechanical properties of high amylose starch films. This could be because of what happens when the drying of film-forming solutions, water evaporates, and allowing the formation of starch network takes place. During this stage, the proximity of starch chains induced by higher amylose contents could facilitate the formation of a matrix with more polymer content per area as well (Alves, et al., 2007).

3.2. Type and content of plasticizers

Native starch films are brittle compared with synthetic polymers such as polyethylene, and technically need to be plasticized. A plasticizer is substance that is incorporated into rigid materials to increase its flexibility, workability, and dispensability. By reducing the grass transition temperature and increasing chain lubricity, plasticizers could also improve processing and extrusion characteristics. They could also reduce the minimum required processing temperature, reduce the plastic's hardness and improve low temperature flexibility.

Generally, two types of plasticizers are distinguished. Internal plasticization is a result of modifications to the chemical structure of polymer. External plasticization is obtained by adding an agent which modifies the structure and energy within the tree-dimensional arrangement of the film polymer (Banker, 1966). It is the second method which, on the basis of the type of materials and the technology used, is mainly used for biodegradable packaging. The addition of a plasticizer to a film produces a film which is less likely to break and is more flexible and stronger.

Basically, the plasticizers should be generally compatible to the structure of the polymer that they plasticize and the permeability be present within the solvent-polymer system and under the conditions used. To be compatible, it must be compatible with the polymer, which results in the inter-molecular reactions. It is important to note that the formulation of the whole film system (polymer, solvent, plasticizer, and other additives) has a direct effect on the nature and characteristics of the film produced. As a result, the polymer and the plasticizer must not only be compatible, but must also have similar solubility in the solvent used. A soluble plasticizer will generally be sought for the development of soluble coating and an insoluble plasticizer (or a dispersible one) for an insoluble coating or for a slow solubilization.

The permanence of a plasticizer is also of prime importance since this influences the physical and mechanical stability of the film. The plasticizer should not be volatile (or not only very slightly volatile) and its degree of retention by the film should be high. Other properties, such as its chemical stability, hygroscopicity, color, flavor, and so on, are also more or less important depending on the type of film under consideration. In addition, the content of plasticizer necessarily varies from 10-60% (dry basis) according to the nature and type of film and the method of application. The plasticizers that are most usually used in the field of rice starch films are mono-, di-, and oligosaccharides, polyols and lipids and its derivatives. The molecular size, configuration and total number of functional groups of the plasticizer as well as its compatibility with the polymer, could affect the interactions between the plasticizer and the polymer (Yang & Paulson, 2000).

Bourtoom & Chinnan (2008) determined plasticizer effect on the properties of biodegradable blend film from rice starch-chitosan. The results of these studies demonstrated that sorbitol plasticized films provided the films with highest mechanical resistance, but the poorest film flexibility. In contrast, glycerol and polyethylene glycol plasticized films exhibited flexible structure; however, the mechanical resistance was low, while inversely affecting the water vapor permeability.

The effectiveness of glycerol in biodegradable blend films from rice starch-chitosan is most likely due to its small size which allows it to be more readily inserted between the polymer chains. It consequently exerts more influence on the mechanical properties than the larger polyethylene glycol molecule. In addition, at an equal percentage of concentration, the total number of glycerol molecules in the film solution is greater than that of the higher molecular weight polyethylene glycol. Therefore glycerol has more functional groups (-OH) than polyethylene glycol which should promote the plasticizer-polymers interactions in the films. As a result of the glycerol, plasticized films provided the films with higher water vapor permeability than polyethylene glycol, and sorbitol should be the result of the high hydrophillicity of the glycerol molecule, which is favorable to the adsorption of water molecules and could also contribute to the increase in the film water vapor permeability. In addition, at high glycerol concentration, glycerol could cluster with itself to open the polymer structure, enhancing the permeability of the film to moisture (Lieberman & Gilbert, 1973). An increase in inter chain spacing due to the inclusion of glycerol molecules between the polymer chain may promote water vapor diffusivity through the film and hence accelerate the water vapor transmission (Yang & Paulson, 2000).

Dai et al. (2010) reported that type and content of plasticizer affected the properties of corn starch films. Increasing the plasticizer content resulted in increasing water vapor permeability of the resulting film. These results would be related to structural modifications of the starch network brought about by the plasticizer concomitant with the hydrophilic character of plasticizer, which favored the absorption and desorption of water molecules. Plasticizers reduced intra- and inter-molecular forces in starch. In addition, plasticizers could extend, dilute and soften the structure effectively; then the starch chain mobility would be increased.

3.3. Type and content of lipids

Biodegradable starch films generally provide a good barrier against oxygen at low and intermediate relative humidity, and have good mechanical properties, but their barrier against water vapor is poor due to their hydrophilic nature (Kester & Fennema, 1986). In contrast, films prepared with lipid materials have good water vapor barrier properties, but are usually opaque and relatively inflexible. Lipid compounds commonly used for the preparation of lipid-based biodegradable films include neutral lipids, fatty acids, waxes, and resins (Kester & Fennema, 1986; Hernandez, 1994). The way to obtain a better water vapor barrier in starch films is to produce a composite film by adding hydrophobic components such as lipid and wax materials. A composite starch-lipid film is particularly desirable, since it has acceptable structural integrity imparted by the starch materials and good water vapor barrier properties contributed by the lipid materials (Greener & Fennema, 1989). The efficiency of the lipid materials in composite films depends on content and the nature of the lipid used such as structure, chemical arrangement, crystal type, shape, size, distribution of lipids, nature of barrier components, the film structure (including homogeneity, emulsion, multilayer.), and thermodynamics such as temperature, vapor pressure, or the physical state of water in contact with the films (Rhim & Shellhammer, 2005).

Haggenmaier & Shaw (1990) tested the effect of stearic acid concentration on the water vapor permeability of hydroxypropyl methylcellulose composite films. It was found that the water vapor permeability of the composite films decreased about 300 times with the addition of 40-50% of stearic acid. However, excessive levels of lipid materials result in the film becoming brittle. Yang & Paulson (2000) investigated gellan/lipid composite films through emulsification and determining the effect of lipid (beeswax and 1:1 blend of stearic-palmitic acids) on the moisture barrier, and mechanical and optical properties of the films. The results depicted that the addition of the lipids to gellan films significantly improved the water vapor permeability ($p < 0.05$), but lowered the mechanical properties and caused the films become opaque. Beeswax was more effective than stearic-palmitic acids in reducing the water vapor permeability and films with beeswax showed better mechanical properties overall than those with stearic-palmitic acids. Srinivasa et al. (2007) studied the effect of fatty acids (stearic and palmitic acids) on the mechanical and permeability characteristics of chitosan films. No considerable differences in water vapor permeability were observed in fatty acid blend films.

Bourtoom & Chinnan (2009) investigated the effect of lipid types (oleic acid, palm oil, and margarine) and content of lipids on water vapor permeability, tensile strength, percentage of elongation at break, and structure of rice starch-chitosan composite film. Tensile strength and water vapor permeability of rice starch-chitosan composite film decreased with the addition of lipids, whereas elongation at break increased in these films. In general, lipid films lack the structural integrity of polysaccharide films. Therefore, incorporation of lipids into hydrophilic polysaccharide films in an effort to decrease their water vapor permeability can negatively affect film strength as expressed by tensile strength measurements. The increase in the lipid content causes a partial replacement of lipids in the film matrix. The interactions between non-polar lipid molecules and between the polar polymer and non-polar lipid molecules are believed to be much lower than those between the polar polymer molecules. In addition, rice starch-chitosan films added with oleic acid provided the films with smoother surface and higher values of tensile strength and elongation at break but lower water vapor permeability than with margarine and palm oil, respectively. The differences in mechanical and barrier properties between these films could be related to their physical state, structure, and chemical nature of the lipids.

3.4. Relative humidity

Relative humidity is a term used to describe the amount of water vapor in a mixture of air and water vapor. It is defined as the ratio of the partial pressure of water vapor in the air-water mixture to the saturated vapor pressure of water at those conditions. The relative humidity of air depends on the temperature and pressure of the system. Relative humidity is often used instead of absolute humidity in situations where the rate of water evaporation is important, as it takes into account the variation in saturated vapor or pressure. Biodegradable starch films generally provide a good barrier against oxygen at low and intermediate relative humidity, and have good mechanical properties, but their barrier against water vapor is poor due to their hydrophilic nature (Kester & Fennema, 1986). When starch films are exposed during storage time in certain environmental conditions, it is possible to obtain both physical and mechanical changes in their nature. Physical changes may be those such as polymer recrystallization (the retrogradation) and those due to the migration of low molecular weight components, such as plasticizers or water contained in film formulation.

This migration of additives can be considered the most important cause of physical instability of starch films. In order to diminish film fragility and increase film flexibility and manageability, plasticizers are added into film formulation (Guilbert, 1986). Water also acts as a plasticizer in hydrophilic films; the plasticizing effect of water is based on the weakening of hydrogen bonds and the dipole–dipole intra and intermolecular interactions due to shielding of these attracting forces by the water molecules. As a consequence, free volume increases which affects their mechanical properties. Stading et al (2001) studied the effect of relative humidity on amylose and amylopectin film properties. They found that when the relative humidity of surrounding films increase this yielded increasing water content and oxygen permeability. However, the storage modulus and glass transition temperature (T_g) showed an inverse effect. The change in surrounding relative humidity affected the water content of the films. When the water content in the films increase this provides an increasing mobility of molecule in the network allows swelling with resulting heterogeneous network structure. Hence, sharply decreased storage modulus and glass transition temperature and increased the oxygen permeability of the resulting films.

Masclaux et al. (2010) reported that relative humidity affected the properties of starch nanocomposite films. They found that at high relative humidity, the water diffusion rate showed higher in the starch nanocomposite films. According to these results, it seemed that it was more water sorption and diffusion in starch matrix due to its initially high swelling capacity and high chain mobility. Beside, the results demonstrated that the oxygen permeability coefficient slightly increased in the range of relative humidity between 30 to 45% and greatly increased at higher relative humidity.

4. Properties enhancement of rice starch-based biodegradable films

4.1. Cross-linking

Cross-linking means that the polymer molecules are interconnected by some sort of bonding. The bonding can be covalent, ionic, or it can result from intermolecular forces such as hydrogen bonding. Cross-linking is used in both synthetic polymer chemistry and in the biological sciences. Although the term is used to refer to the "linking of polymer chains" for both sciences, the extent of cross-linking and the specifics of the cross-linking agents vary. Cross-linking can be accomplished chemically or by irradiation. Cross-linking is a key technique for modifying the properties of starches and can be achieved by adding intra- and inter-molecular bonds at random locations in the starch granules (Figure 1) (Singh et al., 2007). Cross-linking tends to limit the interaction of starch with water and provides a structural integrity of starch-based biodegradable materials during exposure to pressure and moisture (El-Tahlawy et al., 2007). Starch cross-linking is normally performed by treating starches (semi-dry or slurry) with reagents capable of forming either ether or ester linkages between hydroxyl (–OH) groups on starch molecules. Poly-functional chemicals such as phosphorus oxychloride ($POCl_3$), sodium trimetaphosphate (STMP), sodium tripolyphosphate (STPP), epichlorohydrin (EPI), amixture of adipic and acetic anhydrides, and a mixture of succinic anhydride and vinyl acetate have been commonly used to cross-link starches (Singh et al., 2007).

The type of cross-linking agent largely determines the change in functional properties of the treated starches. Starch phosphates may be grouped into two classes: monostarch phosphates and distarch phosphates (cross-linked starches). In general, monostarch phosphates (monoesters) can have a higher degree of substitution than distarch phosphates (diesters) (Singh et al., 2007). Singh et al. (2007) stated that a combination of substitution and cross-linking can provide stability against acid, thermal, and mechanical degradation of starch and delay retrogradation during storage. The mixture of STMP/STPP was a combination of monostarch phosphates (STPP) and distarch phosphates (STMP). STMP is reported to be an efficient cross-linking agent at high temperature with semi-dry starch and at warm temperature with hydrated starch in aqueous slurry (Kerr and Cleveland, 1962). EPI has poor solubility in water and partly decomposes to glycerol, thus water-soluble cross-linking agents such as STMP are preferred. $POCl_3$ is an efficient cross-linking agent in aqueous slurry at pH> 11 in the presence of a neutral salt.

The biodegradable starch film prepared from cross-linked starch provides improved mechanical properties; improved abrasion/cut through; resistance to stress cracking; improved high temperature mechanicals; superior over load characteristics and decrease in flexibility (Rutledge et al., 1974). However, the degree of improvement depends on the type and content of cross- linking agent. Besides, the structure and molecular weight of the starch also affect the intermolecular interactions between molecules. Many researches have been done on the improvement of starch film by using cross linking agent. Kim & Lee (2002) reported that the mechanical properties of starch films prepared with cross-linked corn starch show higher than native corn starch. In addition, Khan et al. (2006) showed that cross-linked sago starch/PVA blend films had mechanical properties higher than native sago starch/PVA blend films. The increase in mechanical properties is due to the increase of cross-linking density. The cross-linking agents react with the -OH groups present in starch and make ether linkages with the available hydroxyl groups. This helps to increase the mechanical properties (Das et al., 2010). This is because cross-linking reinforces the structure of starch granules and limits water absorption of starch thereby restricting the mobility of the starch chain in the amorphous region (Manoi & Rizvi, 2010). According to the results of Das et al. (2010) the moisture absorption was considerably decreased in the a cross-linked starch/poly(vinyl alcohol) (PVA) blend films.

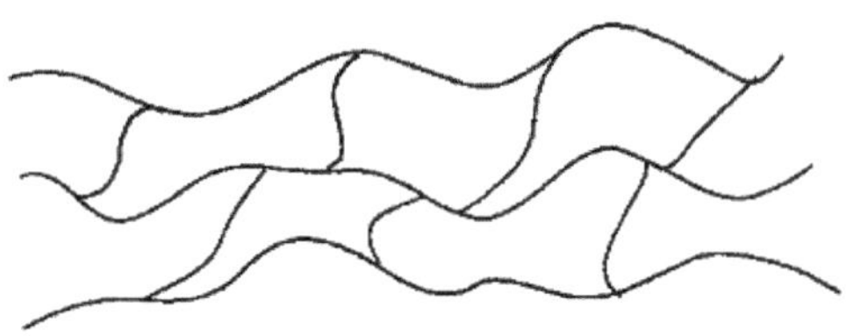

Cross-linking

Figure 1. Schematic of cross-linked polymer. Source: Smiley & Jackson (2002)

4.2. Irradiation treatment

Radioactive treatment, in solid or liquid states such as UV irradiation, gamma irradiation or electron beam have been extensively used to modify starch especially in the grafting process. Radiation processing technology is technically used to improve the properties of starch because of its tendency to undergo the chemical reaction between the polymer molecules under irradiation (Gehring, 2000). Irradiation can reduce sizes of starch molecules. Starch molecules can be broken down by free radicals formed from irradiation which can cause interruption in the molecular bonding structuring of the starch (Woods & Pikaev, 1994).

4.2.1. UV irradiation treatment

Ultraviolet (UV) irradiation as a physical, cost effective, non-thermal, and environmental-friendly technology has received increasing attention during recent years. During this time it has been successfully applied for preservation and decontamination of food products (Bintsis et al., 2000). UV irradiation has been used in medical and pharmaceutical research to crosslink collagen and gelatin films (Bhat & Karim, 2009). UV irradiation requires the presence in the biological medium of certain substances known as photo-sensitizers which induce the changes in the biological substrate after absorbing appropriate radiation (Spielmann et al., 1994). The most efficient photosensitizers, sodium benzoate, are known to be photolysed by UV irradiation (Ghosp & Gangopadhay, 2000).

The properties of biodegradable starch films could be improved by using UV irradiation depending on the intensity, treatment time and content of the photosensitizers. Some researches have been working on the improvement of starch based film by UV irradiation. Delville et al. (2002) studied a new family of cross-linked starches in the solid state by UV irradiation. The original photosensitizers used were water soluble and members of the benzoic acid family. They were able to cross-link starch even at concentrations as low as 0.1%. The mechanical properties showed increases in the cross-linked samples. Follain et al. (2005) studied the impact of the addition of poly vinyl alcohol and photo cross-linking on starch based materials' mechanical properties. The mechanical properties of starch/PVA blends, with or without cross-linking, have been analyzed in order to study how the increase in the linear and branching chains can improve the material's performances. As a result, the mechanical properties of the films obtained were enhanced for both casting and extrusion processes when photo cross-linking was applied.

Khan et al. (2006) prepared and characterized ultra violet (UV) radiation cured bio-degradable films of sago starch/PVA blend. Polymer films of sago starch/polyvinyl alcohol (PVA) were prepared by casting and curing under UV radiation. They found that UV irradiation could promote the degree of cross-linking of sago starch/PVA blend film. The tensile strength values of the films produced increased with UV radiation doses up to a certain limit and then decreased. According to these results, the cross-linking reaction between starch/PVA molecules may be induced while the UV light penetrated into starch/PVA film. The cross-linking reaction occurs through a radical mechanism; the sensitizer (sodium benzoate) is excited or decomposed to produce radicals upon irradiation with UV light (Figure 2). This radical reacted with starch/PVA molecule as a cross-linking reaction to increase the tensile strength of UV treated starch/PVA films . However, too strong a photo-sensitizer would inhibit the UV penetration and decrease the hydrogen abstraction (Delville et al., 2002). When the sensitizer was accessed the tensile strength was decreased. Kumar & Singh (2008) improved the properties of starch biocomposite films by using photo-induced cross-linking. In this study, the starch biocomposite of the films were prepared from the aqueous dispersions of starch with microcrystalline cellulose using glycerol as plasticizer and irradiated under UV light using sodium benzoate as a photo-sensitizer. The results showed that the tensile modulus and strength were found to improve when photo-irradiation was applied. In summary it is clear that treating starch biocomposite films with photo cross-linking under ultraviolet would improve the physical and mechanical properties of the film.

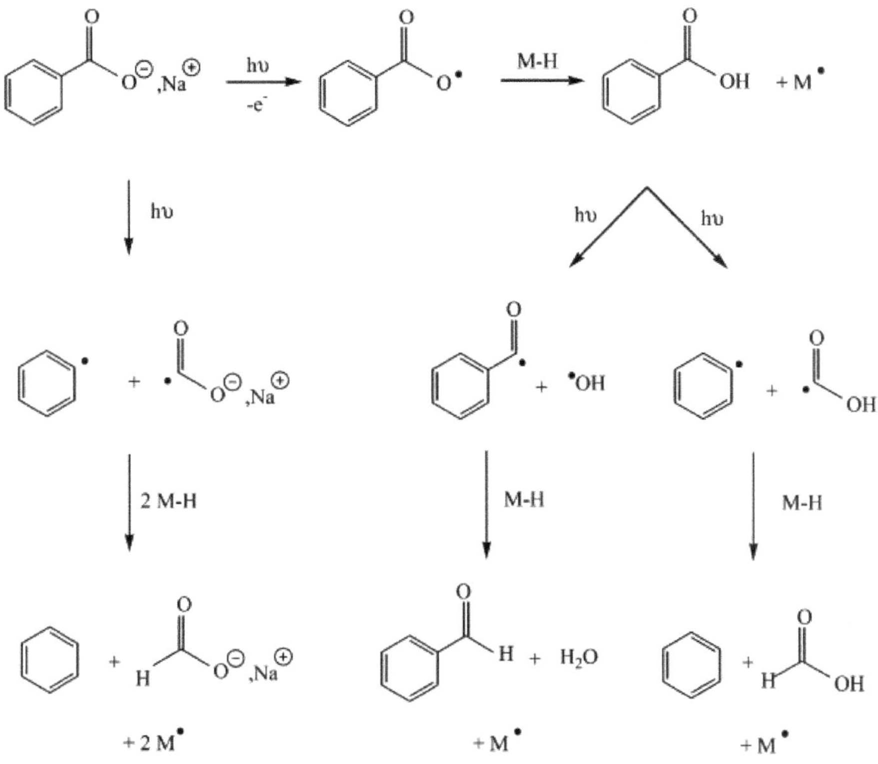

Figure 2. Mechanism involved in sodium benzoate of irradiated rice starch film. Source: Delville et al. (2002)

4.2.2. Gamma irradiation treatment

Gamma irradiation is an ionic, non-heating, processing technology. It has long been used to protect products from insect infestation and microbial contamination during storage and extraction, and to extend the shelf life of foods and medicines (Sabato et al., 2009). Currently, gamma irradiation has become well known as a very convenient tool for the modification of polymer materials through cross-linking, grafting and degradation techniques. This method significantly increases the mechanical properties of films by including cross-links between the starch chains. In addition, gamma irradiation as a cost-lowering and environmentally-friendly alternative method has been widely applied to produce modified starch.

Some researchers have been investigating the effect of gamma irradiation on the properties of starch films. Kim et al. (2008) studied the effect of gamma irradiation on the physicochemical properties of starch based film. They found that using gamma irradiation in aqueous starch based blends resulted in intact and smooth films. The tensile strength of the film increased at the highest value at 3 kGy and its percent elongation was 85% higher than its non-irradiated starch film. Furthermore, water vapor permeability of the starch based film was also improved by gamma irradiation. According to these results, gamma irradiation can be a useful tool as a cross-linking agent of starch to improve the functional properties of a starch based film if the optimum irradiation dose is applied. Akter et al. (2012) found that properties of starch and chitosan blend film could improved by using gamma radiation at optimum doses. The improvement of the mechanical and barrier properties of irradiated blend films resulted from the promoting of the cross-linking of polymer during the exposure to radiation. In addition, they concluded that gamma radiation was found to be an excellent method for grafting and cross-linking of synthetic and biodegradable films for packaging.

4.2.3. Electron beam treatment

Electron beam irradiation is an excitation process in which radicals are produced by the breaking of the H=C bonds of alkene groups. Hence, the monomer molecules, generally multifunctional acrylates, which are in contact with the "activated" material, can copolymerize. The Electron beam irradiation technique is easy to use and control as compared to chemical methods. This technique also displays some advantages over conventional grafting processes including the absence of catalyst residue, complete control over temperature and the absence of a solvent (Peroval et al., 2004). Thus this technique is a green process with potential use in the food field and polymer films. In addition, electron beam irradiation could induce the compatibility of polymers by producing chemical changes in the structure (Olevier et al., 2001). Using this in accord with the starch, it is generally amorphous and mixed with other polymer and/or ingredients to bring about an adequate set of properties for use. The low compatibility between starch polymers in the mixtures is one of the limitations of this approach.

Olivier et al. (2001) reported that the electron beam irradiation of amorphous blends of potato starch and *N*-allylurea (AU) was a fast process that efficiently impedes the spontaneous blooming of mixtures, including large weight fractions of the low molecular weight additive. Pervoral et al. (2004) modified arabinoxylan-based films through grafting of stearyl acrylate by electron beam irradiation. They found that the homogeneous arabinoxylan-based films, which were pre-activated by oxygen plasma and impregnated with a solution of stearyl acrylate before being exposed to an electron beam, had contact angles that increased from 71° (untreated films) to 122 ° (treated films). A decrease of about 24% in the water vapor permeability was obtained and new chemical groupings were observed on the FTIR spectra of these films. These results were explained by the electron beam inducing the grafting of stearyl acrylate monomers on the arabinoxylan-based films surface and also in the film's internal structure resulting a denser film structure.

4.2.4. Ultrasonic treatment

Nowadays, ultrasonic is regarded as an emergent technology in the food, chemical, pharmaceutical and polymer industries. In particular, ultrasonic energy represents a clean way to accelerate and improve the properties of materials. Ultrasonic processing is the use of sound waves beyond the audible frequency range (in general, > 20 kHz) (Chandrapala et al., 2012). When ultrasonic passes through a liquid medium, the interaction between the ultrasonic waves, liquid and dissolved gas leads to an exciting phenomenon known as acoustic cavitation. Acoustic activation generates chemical reactions; and physical forces that include shear forces, shock waves and turbulence (Ashokkumar et al., 2007). In some applications, the physical forces and chemical reactions are needed. For example, the stability of ultrasonically generated microcapsules that encapsulate drugs and high value material components can be substantially increased by chemically cross-linking starch molecules during the formation process (Cavalieri et al., 2008). One of the major issues in using ultrasound in polymer processing is the controlled modification of the interaction between polymers without chemical modification.

Ultrasonic had been extensively performed in the treatment of starch as early as 1933 (Kardos & Luche, 2001). From that point on, many researchers have shown that ultrasonic treatment has a great effect on the behaviors of gelatinized starch dispersions. After treating by ultrasound, the starch dispersion showed a decrease in viscosity and an increase in solubility and clarity due to the increase in number of free mobile macromolecules, rather than the breakage of starch molecules (Iida, 2008; Jenny, 2009). These studies suggest that ultrasonic treatment facilitates the disintegration of starch granules and the formation of homogenous starch solution. Compared to conventional dissolution method for starch, ultrasonic treatment is relatively inexpensive and efficient (Liu et al., 2007). The application of ultrasonic treatment to starch film solution could improve the properties of starch films. In addition, ultrasonic provides dispersion capacity and increases the number of free mobile macromolecules resulting in promotion of chain to chain interaction. Cheng et al. (2010) studied the impact of ultrasonic treatment on the properties of maize starch films. They found that the property of maize starch films was affected remarkably by ultrasound treatment procedure previous to casting. Ultrasonic treatment produces an excellent film with good transparency, improved moisture resistance and provides stronger structure. The results also showed the resulting films with ultrasound treatment lead to a decrease in strain at break. The improvement of film structure by ultrasonic treatment is probably due to the thorough rupture of starch ghosts and rearrangement of free mobile chains of polymer matrix (mainly amylose and amylopectin) during the drying process is favor of forming a more compact and homogeneous film structure. In addition, the destruction of the starch ghosts and releasing of amylose after ultrasound treatment increases amylose content in casting starch solution. It is known that amylose networks form denser films than amylopectin (Rindlav-Westling et al., 1998) which may be the main reason behind the denser films obtained from ultrasonic treatment.

4.3. Combination with other materials

Rice starch can be used to produce biodegradable films to partially or entirely replace plastic polymers because of its low cost and renewability, as well as possessing good mechanical properties (Xu et al., 2005). However, wide application of starch film is limited by its mechanical properties and efficient barrier against low polarity compounds. This constraint has led to the development of improved properties of rice films by modifying its starch properties and/or blending it with other materials. Polymer blending is one of the effective techniques for providing new desirable polymeric materials for a variety of applications. The advantages of blending with other materials versus developing new polymer structure are that it can offer the advantage of reducing research and development expense compared to the development of new monomers and polymers to provide a similar profile (Robeson, 2008). Other advantages of polymer blends are that they can offer properties such as profile combinations not easily obtained with new polymer structures. The properties of polymer blends are varied and depend on the type of polymer materials, ratio between polymers, miscibility, and phase behavior. Various polymer materials can be blended with starch based film in order to improve their property. These are described in the following sections.

4.3.1. Combination with chitosan

Chitosan is a 1, 4 linked-2-deoxy-2-aminoglucose, prepared by N-deacetylation of chitin (Xu et al., 2005). Chitosan provides unique functional, nutritional, and biomedical properties, and its present and potential uses range from dietary fiber to a functional ingredient and processing aid. Some of the well known applications of chitosan include its use for prevention of water pollution, treatment against hypertension, antimicrobial and hypocholesterolemic activity, flavor encapsulation, seed coating, film-forming, and controlled release of food ingredients and drugs (Muzzarelli et al., 1997). Its relatively low cost, widespread availability from a stable renewable source such as shellfish waste of the sea food industry, along with chitosan's ability to form a good film, are primary reasons to seek new applications of this polymer. Numerous investigations have been reported on the studies of films made from chitosan (Park et al., 2002) and chitosan blends with starch.

Zhai et al. (2004) investigated the effect of chitosan on corn starch film properties. The tensile strength and the flexibility of starch film were improved largely after incorporation of 20% chitosan into starch film. In addition, X-ray diffraction and scanning electron microscope analysis of starch and chitosan blend films indicated that there was interaction between starch and chitosan molecules. Xu et al. (2005) prepared and assessed the starch and chitosan blend films. The film's tensile strength, elongation at break and water vapor transmission rate was affected by the chitosan content. According to the results, the mechanical properties of the composite films increased with the addition of chitosan. The increase in tensile strength with the addition of chitosan may occur because inter-molecular hydrogen bonds between starch and chitosan are formed. However, too high a chitosan content provided inferior mechanical properties. This was because intra-molecular hydrogen bonds occurred rather than inter-molecular hydrogen bonds between polymers.

Bangyekan et al. (2005) investigated the properties of chitosan-coated cassava starch films. They found that coating of chitosan solutions led to an improvement of the mechanical and physical properties of cassava starch films. The results of the evaluation of the mechanical properties showed that an increase in chitosan coating concentration resulted in a significant increase in tensile stress at maximum load and tensile modulus, and a decrease in percent elongation at break. According to the results, it can be concluded that the brittle characteristic of chitosan film may be responsible for an increment of tensile strength and tensile modulus of the coated cassava films.

Bourtoom & Chinnan (2008) studied the effect of chitosan on biodegradable films made from rice starch. The biodegradable rice starch films showed an increase in tensile strength, water vapor permeability, a decreasing elongation at the break, and film solubility after incorporation of chitosan. Addition, the introduction of chitosan increased the crystalline peak structure of starch film; however, too high a concentration of chitosan yielded phase separation between starch and chitosan. The increasing of the mechanical properties of the biodegradable blend rice starch films, with the addition of chitosan, are attributable to the formation of intermolecular hydrogen bonding between NH_3^+ of the chitosan backbone and OH^- of the rice starch. The amino groups (NH_2) of chitosan were protonated to NH_3^+ in the acetic acid solution, whereas the ordered crystalline structures of starch molecules were destroyed with the gelatinization process, resulting in the OH^- groups being exposed to readily formed hydrogen bonds with NH_3^+ of the chitosan. This reaction can be confirmed by the results of the amino group band of the chitosan molecule in the FTIR spectrum which shifted from 1541.15 cm^{-1} in the chitosan film to 1621.96 cm^{-1} in the biodegradable blend films. This phenomenon pointed out that interactions were present between the hydroxyl group of rice starch and the amino group of chitosan.An addition of chitosan to rice starch films resulted in increases in the water vapor permeability of the films. This tendency could be explained by the higher hydrophilicity (NH_3^+ groups) of the biodegradable blend films.

4.3.2. *Combination with polyvinyl alcohol*

Polyvinyl alcohol (PVA) provides excellent film forming, emulsifying, and adhesive properties, which has resulted in its broad industrial use, such as for paper adhesives, a textile sizing agent, paper coatings and water soluble films (Ibrahim et al., 2010). PVA is a water soluble polymer; hence PVA can be formed by solution casting and orientation to make high performance PVA films. In addition, PVA can also be resistant to oil, grease and solvents. It is odorless and nontoxic. It has high tensile strength and flexibility, as well as high oxygen and aroma barrier properties. PVA materials show a broad and versatile range of physical properties and other advantageous characteristics at acceptable cost and biodegradation rate and can be employed in a wide range of applications.

The addition of PVA into the starch films to improve their properties (such as mechanical properties, moisture sensitivity) has been reported. Shorgen et al. (1998) reported that the factors that most limit the use of starch in biopolymer material applications are brittleness and loss of strength at low humidity and poor resistance to water. By adding PVA, starch could improve strength and flexibility. However, the desired strength, flexibility, or stiffness can be tailored by formulating with different PVA levels, PVA types, better level of solids and cross-linking agents. Liu et al. (1999) found that incorporation of PVA into starch could enhance the mechanical properties. However, improvement in the mechanical properties is limited, mainly due to the poor interface adhesion between the fibrous PVA structure and the starch matrix. Chen et al. (2008) investigated the effect of PVA on the properties of starch films. It was found that the tensile strength, elongation at break of starch films increased with the addition of PVA. Because both starch and PVA are polar substances having hydroxyl groups (–OH) in their chemical structure, these highly polar hydroxyl groups tend to form inter molecular and intra-molecular hydrogen bonds which improve the integrity of starch–PVA films. On the other hand, the solubility parameters of starch and PVA modeling complexes have been found to be close to each other, which confirmed that starch and PVA are compatible blends. In addition, the vibration frequency analysis of these molecular complexes has shown that the hydroxyl group shifted to lower wave numbers due to formation of hydrogen bonds between starch and PVA.

4.3.3. Combination with polylactic acid

Polylactic acid (PLA) is a rigid degradable thermoplastic polymer which can be semi-crystalline or totally amorphous, depending on the stereochemistry of the polymer backbone. PLA is a unique polymer that has, in many ways, properties like PET, but also performs much like polyolefin. PLA have been used in biomedical fields, for example, as a surgical implant material and drug delivery system, as well as in textile applications. This is because of their excellent properties such as biocompatibility, nontoxicity, high mechanical strength, and thermal plasticity (Lertworasirikul et al., 2008). In addition, PLA also exhibits excellent characteristics and is suitable for food contact surface and related packaging applications.

PLA can be formed into transparent films and fibers. Certain factors suggest the incorporation of PLA into rice starch films would bring benefits and improve their properties. These include relatively reasonable costs and specifically, biocompatibility, nontoxicity, high mechanical strength, and thermal plasticity. Research has been reported, such as Ke & Son (2003) that suggests that PLA performed as a reinforcement in the starch.

These results showed that the tensile strength and elongation of the starch blend films increased when PLA was added. Yew et al. (2005) studied the effect of PLA on the properties of rice starch films and found that the tensile strength of the rice starch increased with the incorporation of PLA. This may be due to the polar interaction between the starch and carboxyl groups of the PLA. In addition, the interaction between PLA and rice starch may be attributed to the possible hydrogen bonding that occurs between the carbonyl group (that is from ester linkage) in PLA and the hydroxyl group in starch. A proposed possible site for interaction between starch and PLA is depicted in Figure 3.

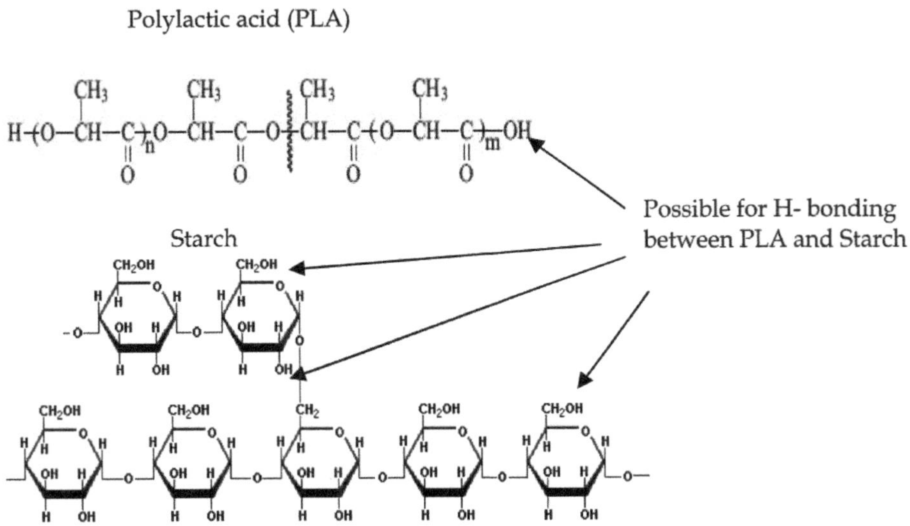

Figure 3. Proposed chemical interactions between starch and PLA. Source: Modified from Yew et al. (2005)

4.3.4. Combination with algae polysaccharides

Algae are photosynthetic microorganisms containing chlorophyll. They can be single cell or multicell, motile or non-motile. Algae and their extracts are currently used in medicine, food, and the nutraceutical, pharmaceutical and packaging industries. Most of the world's annual seaweed harvest is used to produce the algal hydrocolloids and those that show potential for application in biodegradable films are alginate, agar, and carrageenan. Alginates are natural substances extracted from brown seaweed. In polymer chains, monomers are arranged alternately in GG and MM blocks, together with MG blocks (Grant, 1973). The most interesting property of alginates is their ability to react with polyvalent metal cations, specifically calcium ions. The alginate films show hydrophilic matrices, and the cross-linking process with polyvalent cations has been used to improve their water barrier properties, mechanical resistance, cohesiveness and rigidity (Rhim, 2004). Due to the fast cross-linking process between alginate and calcium ions, localized gelling areas are produced, compromising the uniformity and quality of films. While, agar is composed of alternating 1, 3-linked-D-galactose and 1,4- linked 3,6- anhydro-L-galactose units. It is substituted by sulfate esters and methoxyl, and may also carry pyruvic acid residues. Because of its ability to form very hard gels at very low concentrations, agar has been used extensively as a gelling agent in the food industry. Because of its combination of renewability and biodegradability, its enormous gelling power, and the simplicity of the extraction process, agar has been singled out as a promising candidate for future use in plastic materials. Currently, agar has been incorporated in materials such as foams, films and coatings (Phan et al., 2009), and added into starch films to improve the puncture strength and water barrier (Letendre et al., 2002).

Carrageenan is biopolymers extracted from algae, are used extensively in the food and packaging industry. The three main carrageenans differ only in the number of sulphate groups. k-Carrageenan is the most sulphated one and adopts a coil conformation whatever the ionic and temperature conditions. However, kappa and iota carrageenan form gels, but lambda is unable to do this and is used as a pure thickener (Lizarraga et al., 2006). k-Carrageenan itself is known for its good film forming ability (Choi et al., 2005).

The algae polysaccharide could provide good film and shows the potential to improve the properties of starch. Hence, the addition of algae polysaccharide into starch films for improvement of properties has been suggested. Wu et al. (2009) assessed the effect of agar on potato starch films; it was found that the addition of agar could enhance the tensile strength and water barrier of the resulting films. The increase in their properties was attributable to the formation of inter-molecular hydrogen bonds between starch and agar and the compact structure of composite film. In addition, the domains occurred of the three-dimensional network structure formed by entanglement among agar chains. The water barrier properties of the starch film were promoted when agar was added; this was due to the strong inter-molecular interaction formed between starch and agar. It minimized the free volume and inter-molecular distances in the film structures. Consequently, water molecules were diffused with greater difficulty in the compact network and a higher water barrier value was obtained.

In addition, Phan et al. (2009) reported that adding agar into cassava starch film provided an improvement in elongation at break and the tensile strength of cassava starch film. This indicates that incorporation of agar into starch film could provide a very good cohesive matrix, which contributes to enhancing the mechanical properties of starch based films. Cordoba et al. (2008) studied the effect of alginate on the properties of thermoplastic starch. According to the experimental evidence presented in their work, a decrease in the elastic properties and an increase in elongation at break and impact resistance was observed when alginate was added into the thermoplastic starch. These results could be explained by the addition of alginate to thermoplastic starch causing an obvious plasticizing effect. However, the effect of the amount of alginate content on thermoplastic starch when used for specific needs and applications should be determined. Lafargue et al. (2007) determined the influence of k-Carrageenan on film forming and film properties of modified starch. This study demonstrated that the addition of a low amount of the carrageenan into starch resulted in an enhancement of film solution properties and thermo-reversibility without modifying the film properties. According to the results, it might be beneficial to design tailor-made films from starch by using carrageenan.

4.4. Reinforcement

4.4.1. Reinforcement with natural fibers

Narural fibers can be considered as naturally occurring composites, consisting of cellulose fibrils embedded in lignin matrix. The cellulose fibrils are aligned along the length of the fiber, which render maximum tensile and flexural strengths, in addition to providing rigidity. Natural fibers are, in general, suitable to reinforce materials due to their relatively high strength and stiffness and low density. Natural fibers can be processed in different ways to yield reinforcing elements with different mechanical properties. Natural fiber-reinforced starch film has many advantages such as light weight, reasonable strength, stiffness and thermal resistance (Demir et al., 2006; Phataraporn et al., 2010). This behavior has been attributed to the resistance exerted by the fiber itself and also due to 3D hydrogen bonds network formed by intermolecular interactions between starch and fiber, which reduces the flexibility of molecular chains of starch (Lu et al., 2006). Furthermore, there is compatibility between the starch matrix and fiber and performance (such as the mechanical properties). This behavior was confirmed by observing different polysaccharides reinforced with cellulose fiber. Various types of fiber have been applied in starch based polymer films, such as bleached leaf wood fibers, bleached eucalyptus pulp fibers, wood pulp softwood aspen, jute, hemp and flax fibers, tunicin whiskers, and oil palm fibers. It is known that fiber related parameters such as their dispersion, length, type, size, content of natural fiber and their orientation along with their adhesion with the matrix all determine the properties of their composites (Goda et al., 2006).

4.4.2. Reinforcement with modified natural fibers

Reinforcing fibers can be modified by physical and chemical methods. Physical methods, such as stretching, calendaring, thermo-treatment and the production of hybrid yarns do not change the chemical composition of the fibers. Physical treatment changes structure and surface properties of the fiber and thereby influence the mechanical bondings to polymers. With regards to the chemical methods, the strongly polarized cellulose fibers are inherently incompatible with hydrophobic polymers. When two materials are incompatible, it is often possible to bring about compatibility by introducing a third material that has properties intermediate between those of the other two. There are several mechanisms for coupling in materials: (1) weak boundary layers - coupling agents eliminate weak boundary layers; (2) deformable layers – coupling agents produce a tough, flexible layer; (3) restrained layers – coupling agents develop a highly cross-linked inter-phase region, with a modulus intermediate between that of substrate and of the polymer; (4) chemical bonding – coupling agents form covalent bonds with both materials; and (5) acid – base effect – coupling agents alter the acidity of the substrate surface. In relation to the coupling agent used in fiber modification, organosilanes are the main groups of coupling agents. They have been developed to couple virtually any polymer to the minerals used in reinforced composites. Most of the silane coupling agents can be represented by the following formula:

$$R - (CH_2)_n - Si(OR')_3$$

where n = 0-3, OR′ is the coupling agent that causes the reaction with the polymer. This could be co-polymerization, and/or the formation of an interpenetrating network.

The curing reaction of a silane treated substrate enhances the wetting by the resin. Herrera-Franco & Valadez-Gonzalez (2005) investigated the properties of short natural fiber reinforced composites as a function of chemical modification. Short natural fibers were subjected to silane treatment (vinyltris or 2-methoxy-ethoxy silane). The authors observed that the increase in the mechanical properties ranged between 3 and 30%, for the tensile and flexural properties. The shear strength of this composite showed an increase of the order of 25%. From the micrographs, obtained from failure surfaces from the SEM, it was observed that with increasing fiber-matrix interaction the failure mode changed from interfacial failure to matrix failure. The interface failure was mainly a frictional type failure, and matrix tearing and shearing was observed only for the pre-impregnated and silane treated fibers. The silane surface treated fibers also showed a layer of polymer covering the fibers even after failure.

Shih (2007) has also developed epoxy composites from waste water bamboo husk fiber. The fibers were chemically modified by coupling agents and untreated fibers were added to epoxy resin to form new reinforced composites. The results show that the morphology analysis reveals that the fibers modified by coupling agent exhibited better compatibility with the polymer matrices than the untreated fiber. In addition, the mechanical properties were also enhanced due to the addition of fibers treated with coupling agents and untreated fibers. The increments of storage modulus of epoxy were about 16.4 and 36.1% with the addition of 10% in the fibers treated with coupling agents and untreated fibers, respectively.

Phattaraporn et al. (2011) studied the effect of palm pressed fiber (PPF) surface treatment on the properties of rice starch films. They reported that higher tensile strength, water vapor permeability and thermal properties of biodegradable rice starch films were obtained as silane treated PPF was applied. Increasing the concentration of silane and content of silane treated (PPF) resulted in increased tensile strength and water vapor permeability but decreased elongation at break. In addition, the glass temperature (T_g) shifted towards higher temperatures with an increasing concentration of silane. This could be a restriction of the mobility of the starch chain due to the establishment of strong interactions between rice starch films and treated PPF. The maximum improvement of rice starch films in the mechanical and thermal properties was obtained when 40% of silane treated PPF films was applied. These results pointed out that the interfacial interactions improved the filler compatibility, and the mechanical and thermal properties. In addition, the treatment of fiber with alkaline (NaOH) leads to a decrease in the spiral angle, that is it is closer to the fiber axis, and and increase in molecular orientation.

4.4.3. Reinforcement with nanoparticles

4.3.3.1. Nanoclays

Nanoclays are potential filler. It is a naturally abundant mineral that is toxin-free and can be used as one of the components for food, medical use, cosmetics, healthcare products and packaging. Moreover, nanoclays are also environmentally friendly and inexpensive. Nanoclays have also been reported to improve the mechanical strength of biopolymers, making their use feasible. The structure and properties of the resulting material is dependent on the state of the nanoclay in the nanocomposite, that is if it is exfoliate or intercalate. Intercalation is the state in which polymer chains are present between the clay layers, resulting in amultilayered structure with alternating polymer/inorganic layers. Exfoliation is a state in which the silicate layers are completely separated and dispersed in a continuous polymer matrix (Weiss et al., 2006). The most common nanoclay used in composites with starch films is montmorillonite (MMT). Because of the presence of sodium cations between the interlayer spaces, natural MMT is hydrophilic and is miscible with hydrophilic polymers, such as starch. Addition, starch–MMT nanocomposites have gained prominence due to the mechanical reinforcement achieved, even at low concentrations (Ardakani et al., 2010).

Many report that nanoclays show improvement in the mechanical strength of starch films, making their use a possibility. Chung et al. (2010) improved the properties of starch film by adding nanoclay (montmorillonite) into the film solution. The addition of nanoclay in starch film show improved modulus and strength without a decrease in elongation at break. The increase in modulus and strength is 65% and 30%, respectively for the starch film containing 5 wt.% nanoclay compared to the unfilled starch materials. However, further increases in clay result in deterioration in properties that are most likely due to poorer clay dispersion and lower polymer crystallinity.

Muller et al. (2011) studied the effect of nanoclay on mechanical and water vapor barrier properties of starch films. The nanocomposites exhibited a remarkable improvement in mechanical properties and reduced the water vapor permeability when compared with pure starch films. This behavior may be associated with the interaction between nanoclay and the molecules of starch. Gao et al. (2012) studied the effect of nanoclays on the properties of starch film. It was found that the addition of clays, significantly greater tensile strength and lower water vapor permeability were obtained. Besides, starch–clay composite films exhibited a lower glass transition temperature (T_g, −23.8 °C) and better heat endurance than unfilled starch film. They suggested that the coupling between the tremendous surface area of the clay and the starch matrix facilitated the stress transfer to the reinforcement phase, allowing for the tensile and toughening improvements.

4.3.3.2. Nanocrystalline cellulose

Crystalline cellulose is obtained by hydrolysis of the amorphous portion of cellulose until a level-off in the degree of the product of polymerization is obtained. Crystalline cellulose is comprised of highly crystalline regions of cellulose. It is useful for a number of different applications. Pieces of crystalline cellulose easily bond together without the use of an adhesive. Furthermore, crystalline cellulose can be mixed with other substances so as to hold an additive while bonding to itself. It can be made into pharmaceutical-grade tablets, such as those used for vitamins, pain relievers and other medicines. It also may be used as a substitute for starch where starch is used as a smoothener, such as in suntan lotion.

Due to its high aspect ratio and a high modulus, the use of cellulose crystallites for preparation of high performance composite materials has been explored extensively. When the cellulose crystallites were homogeneously dispersed into polymer matrices, they gave a remarkable reinforcing effect, even at concentrations of a few percent (Favier et al., 1996). The hydrolysis of cellulose to obtain crystalline cellulose can be accomplished using mineral acid, enzymes or microorganisms. Although such methods are desirable because glucose, a useful by-product is created, these methods are more expensive and create crystalline cellulose products with a lower crystallinity. Thus, acid hydrolysis is the conventional method of choice for manufacturing crystalline cellulose. Acid that can be used in this process includes hydrochloric acid, sulfuric acid and phosphoric acid. At higher temperatures, sulfuric acid and phosphoric acid can peptize and modified surface crystalline cellulose. Nanocrystalline cellulose is inherently a low cost ingredient and provides a strength of about 10 GPa values which are only about seven times lower than those of single-walled carbon nanotubes (Podsiadlo et al., 2005). Thus, nanocrystalline cellulose shows a high potential application for starch film improvement.

Lu et al. (2006) investigated the morphological and mechanical properties of ramie crystallites (RN) reinforced plasticized starch (PS) biocomposites. The ramie cellulose nanocrystalites, having lengths of 538.5 ± 125.3 nm and diameters of 85.4 ± 25.3 nm on average, were prepared from ramie fibers by acid hydrolysis. The results indicate that the synergistic interactions between fillers and between filler and matrix play a key role in reinforcing the composites. The PS/RN composites, composites conditioned at 50% relative humidity, increases in tensile strength and Young's modulus from 2.8 MPa for PS film to 6.9 MPa and from 56 MPa for PS film to 480 MPa with increasing RN content from 0 to 40 wt% respectively. Furthermore, incorporating RN fillers into PS matrix also leads to a decrease in water sensitivity for the PS based biocomposites. Alemdar & Sain (2007) later studied the morphology, thermal and mechanical properties of wheat straw nanofibers as reinforced starch biocomposites. The nanocomposites from the wheat straw nanofibers and the thermoplastic starch were prepared by the solution casting method. They founded that the tensile strength and modulus of the nanocomposite films showed significantly enhanced properties compared to the pure thermoplastic starch. The glass transition of the nanocomposites was shifted to higher temperatures with respect to the pure thermoplastic starch.

Wittaya et al. (2009) reinforced the rice starch films with microcrystalline cellulose (average length of about 0.480 ± 0.023 μm) from palm pressed fiber. The strong interactions between microcrystalline cellulose from palm pressed fiber and between microcrystalline cellulose from palm pressed fiber and rice starch film matrix played a key role in reinforcing the resulting rice starch film composites. The rice starch film/ microcrystalline cellulose from palm pressed fiber biocomposites increased in tensile strength from 5.16 MPa for pure rice starch film to 44.23 MPa but decreased in elongation at the breaking point of the composites. In addition, the incorporation of microcrystalline cellulose from palm pressed fiber into rice starch films provided an improvement of the water resistance for the rice starch films.

4.3.3.3. Starch nanocrystals

Starch nanocrystal is prepared by submitting native starch granules to an extended time of hydrolysis at a temperature below the gelatinization temperature. This is when the amorphous regions are hydrolyzed allowing the separation of crystalline lamellae, which are more resistant to hydrolysis. The starch crystalline particles show platelet morphology with thicknesses of 6-8 nm (Kristo & Billiaderis, 2007). The use of starch nanoparticles is receiving a significant amount of attention because of the abundant availability of starch, low cost, renewability, biocompatibility, biodegradability and non-toxicity. The latter properties make them excellent candidates for implant materials and drug carriers. Kristo & Biliaderis (2007) studied the addition of starch nanocrystals on the properties of pullulan film. The water uptake of pullulan-starch nanocomposites decreased with increasing filler content whereas water vapor permeability remained constant up to 20% (w/w). This then decreased significantly with further addition of starch nanocrystals. The thermo-mechanical behavior of nanocomposite films was also investigated by means of dynamic mechanical thermal analysis (DMTA) and large deformation mechanical tests. The glass transition temperature (T_g) shifted towards higher temperatures with increasing amounts of nanocrystals. This can be attributed to a restriction of the mobility of pullulan chains due to the establishment of strong interactions not only between starch nanocrystals but also between the filler and the matrix. Moreover, the addition of nanocrystals caused strong enhancement of the Young modulus and the tensile strength, but led to a drastic decrease of the strain at break in the samples

Chen et al. (2007) prepared pea starch nanocrystal (PSN) dispersion containing nanocrystals within a range of 30-80 nm from native pea starch (NPS) granules by acid hydrolysis. The addition of 5 and 10 wt% of PSN content gave improved physical properties compared to the PVA film. The PVA/PSN films showed higher tensile strength, elongation at break and lower moisture uptake than the corresponding PVA/NPS films with the same components. The results revealed that PSN, comparing with NPS, had much smaller sizes and dispersed more homogeneously in the PVA matrix, resulting in stronger interactions with PVA. New applications of native pea starch and its nanocrystals as low-cost fillers were explored in this work, and PSN showed greater potential than NPS to improve the properties of PVA-based composites.

Kaewpool et al. (2010) improved the rice starch films properties by addition of starch nanocrystals. The results showed that the mechanical properties of the rice starch film were enhanced by the addition of starch nanocrystals. This indicated that introduction of starch nanocrystals increased the crystalline peak structure of rice starch film. Furthermore, by increasing the starch nanocrystals content, the water barrier properties of the rice starch films were also improved. The glass transition temperature (T_g) shifted towards higher temperatures when increasing the amount of nanocrystals and the char yield of rice starch films was enhanced as the starch nanocrystals increased. Therefore, the addition of starch nanocrystals to rice starch films can improve the thermal stability of rice starch films. The results demonstrated that the addition of starch nanocrystals by 15% provided better properties, including mechanical, water barrier and thermal properties.

5. Conclusion

Biodegradable film made from rice starch shows potential for use as packaging material. However, compared to the common thermoplastics, biodegradable rice starch films still reveal some disadvantages, such as low mechanical properties and lack of efficient barriers against polarity compounds. This constraint has led to the development of the improved properties of biodegradable film from rice starch by modifying its starch properties and/or incorporating other materials. Chemical treatments and physical treatment, in combination with other biodegradable materials and fiber reinforcement could improve the properties of biodegradable film made from rice starch. However, the potential application, in terms of properties and limitations on the use of facilities, should be taken into account when selecting techniques for improvement.

Author details

Thawien Wittaya
Prince of Songkla University, Department of Material Product Technology, Thailand

6. References

Akter, N.; Lacroix, M.; Salmieri, S.; Dussault, D.; Khan, R. & Sharmin, N. (2012). Effectiveness of silane monomer and gamma radiation on chitosan films and PCL-based composites. *Physics and Chemistry,* In press. (January 2012), ISSN: 0969-806X.

Alemdar, A. & Sain, M. (2008). Biocomposites form wheat straw nanofiber: Morphology, thermal and mechanical properties. *Composite Science and Technology,* Vol.68, No. 2, (February 2008), pp. 557-565, ISSN: 0266-3538.

Alves, D. V.; Mali, S.; Beleia, A. & Grossmann, E. M. (2007). Effect of glycerol and amylose enrichment on cassava starch film properties. *Journal of Food Engineering,* Vol. 78, No. 3 (February 2007), pp. 941-946. ISSN: 0260-8774.

Ardakani, M. M.; Mohseni, M. A.; Beitollahi, H.; Benvidi, A. & Naeimi, H. (2010). Electrochemical determination of vitamin C in the presence of uric acid by a novel TiO2 nanoparticlesmodified carbon paste electrode. *Chinese Chemical Letters,* Vol.21, No.12, (December 2010), pp. 1471–1474. ISSN: 1001-8417.

Ashokkumar, M.; Sunartio, D.; Kentish, S.; Mawson, R.; Simons, L.; Vilkhu, K. &Versteeg, C. (2008). Modification of food ingredients by ultrasound to improve functionality: a preliminary study on a model system. *Innovative Food Science Emerging Technology*, Vol.9, No.2, (April 2008), pp. 155-160. ISSN: 1466-8564.

Bangyekan, C.; Aht-Ong, D. & Srikulkit, K. (2006). Preparation and properties evaluation of chitosan-coated cassava starch films. *Carbohydrate Polymers*, Vol.63, No.1, (January 2006), pp. 61–71. ISSN: 0144-8617.

Banker, G. S. (1966). Film coating theory and practice. *Pharmaceutical Sciences*, Vol.55, No.1, (January 1966), pp. 81-89. ISSN: 1520-6017.

Bhat, R. & Karim, A. A. (2009). Impact of radiation processing on starch. *Comprehensive Reviews in Food Science and Food Safety*, Vol. 8, No. 2 (March 2009), pp. 44-58. ISSN: 1541-4337.

Bintsis, T.; Litopoulou-Tzanetaki, E. & Robinson, R. K. (2000). Existing and potential applications of ultraviolet light in the food industry– A critical review. *Science of Food and Agriculture*, Vol.80, No.6, (May 2000), pp. 637–645. ISSN: 1097-0010.

Bourtoom, T. & Chinnan, M. S. (2008). Preparation and properties of rice starch-chitosan blend biodegradable film. *Lebensmittel-Wissenschaft Und-Technologie*, Vol.41, No.9, (November 2008), pp. 1633-1641. ISSN: 0023-6438.

Bourtoom, T. & Chinnan, M. S. (2009). Improvement of water barrier property of rice starch-chitosan composite film incorporated with lipids. *Food Science and Technology International*, Vol. 15, No. 2, (June 2009), pp. 149-158. ISSN: 1082-0132.

Cavalieri, F.; Ashokkumar, M.; Grieser, F. & Caruso, F. (2008). Ultrasonic synthesis of stable functional lysozyme microbubbles. *Langmuir*, Vol.24, No.18, (September 2008), pp. 10078-10083. ISSN: 1520-5827.

Chandrapala, J.; Zisu, B.; Palmer, M.; Kentish, S. & Ashokkumar, M. (2011). Effects of ultrasound on the thermal and structural characteristics of proteins in reconstituted whey protein concentrate. *Ultrasonics Sonochemistry*, Vol.18, No.5, (September 2011), pp. 951–957. ISSN: 1350-4177.

Chen, Y.; Stipanovic, A. J.; Winter, W. T.; Wilson, D. B. & Kim, Y. J. (2007). Effect of digestion by pure cellulases on crystallinity and average chain length for bacterial and microcrystalline celluloses. *Cellulose*, Vol.14, No.4, (August 2007), pp. 283–293. ISSN: 1572-882X.

Chen, L. S. & Lai, C. H., (2008). Mechanical and water vapour barrier properties of tapioca starch/decolorized hsian-tsao leaf gum films in the presence of plasticizer. *Food Hydrocolloids*, Vol.22, No.8, (December 2008), pp. 1584–1595. ISSN: 0268-005X.

Cheng, W. J.; Chen, J. C.; Liu, D. H.; Ye, X. Q. & Ke, F. S. (2010). Impact of ultrasonic treatment on properties of starch film-forming dispersion and the resulting films. *Carbohydrate Polymers*, Vol.81, No.3, (July 2010), pp. 707–711. ISSN: 0144-8617.

Choi, J. H.; Choi, W. Y.; Cha, D. S.; Chinnan, M. S.; Park, H. J. & Lee, D. S. (2005). Diffusivity of potassium sorbate in kappa-carrageenan based antimicrobial film. *Lebensmittel-Wissenschaft Und-Technologie*, Vol.38, No.4, (June 2005), pp. 417–423. ISSN: 0023-6438.

Chung, Y.; Ansari, S.; Estevez, L.; Hayrapetyan, S.; Giannelis, E. P. & Lai, H. M. (2010). Preparation and properties of biodegradable starch–clay nanocomposites. *Carbohydrate Polymers,* Vol.79, No.2, (January 2010), pp. 391–396. ISSN: 0144-8617.

Clopeau, M. & Prunet-Foch, B. (1990). Electrostatic spraying of liquids. Main functioning modes. *Electrostatics,* Vol.25, No.2, (October 1990), pp. 165–184. ISSN: 0304-3886.

Cordoba, A.; Cuellar, N.; González, M. & Medina, J. (2008). The plasticizing effect of alginate on the thermoplastic starch/glycerin blends. *Carbohydrate Polymers,* Vol.73, No.3, (August 2008), pp. 409–416. ISSN: 0144-8617.

Dai, H.; Chang, P.R.; Geng, F.; Yu, J. & Ma, X. (2010). Preparation and properties of starch-based film using N,N-bis(2-hydroxyethyl) formamide as a new plasticizer. *Carbohydrate Polymers,* Vol.79, No.2, (January 2010), pp. 306–311. ISSN: 0144-8617.

Das, K.; Ray, D.; Bandyopahyay, N. R.; Gupta, A.; Sengupta, S.; Sahoo, S.; Mohanty, A., & Misra M. (2010). .Preparation and Characterization of Cross-Linked Starch/Poly(vinylalcohol) Green Films with Low Moisture Absorption. *Industrial and Engineering Chemistry Research,* Vol.49, No.5, (March 2010), pp. 2176-2185. ISSN :1520-5045.

Delville, J.; Joly, C.; Dole, P. & Bliard, C. (2002). Solid state photo-crosslinked starch based films: a new family of homogeneous modified starches. *Carbohydrate Polymers,* Vol.49, No.1, (July 2002), pp. 71-81. ISSN: 0144-8617.

Demir, H.; Atikler, U.; Balkose, D. & Tihminlioglu, F. (2006). The effect of fiber surface treatmentson the tensile and water sorption properties of polypropylene luffa fiber composites. *Composite Part. A,* Vol.37, No.3, (March 2006), pp. 447-456. ISSN: 1359-835X.

El-Tahlawy, K.; El-Nagar, K. & Elhendawy, A. G. (2007). Cyclodextrin-4 hydroxy benzophenone inclusion complex for UV protective cotton fabric. *Textile Institute,* Vol.98, No.5, (2007), pp. 453–462. ISSN: 1754-2340.

Favier, V.; Chanzy, H. & Cavaille, J. Y. (1995). Polymer nanocomposites reinforced by cellulose whiskers. *Macromolecules,* Vol.28, No.18, (August 1995), pp. 6365–6367. ISSN: 1520-5835.

Follain, N.; Joly, C.; Dole, P.; & Bliard, C. (2005). Properties of starch based blends. Part 2. Influence of poly vinyl alcohol addition and photo-crosslinking on starch based materials mechanical properties. *Carbohydrate Polymers,* Vol.60, No.2, (May 2005), pp.185–192. ISSN: 0144-8617.

Gao, W.; Dong, H.; Hou, H. & Zhang, H. (2012). Effects of clays with various hydrophilicities on properties of starch–clay nanocomposites by film blowing. *Carbohydrate Polymers,* Vol.88, No.1, (March 2012), pp. 321–328. ISSN: 0144-8617.

Gehring, J. (2000). With radiation cross linking of engineering plastics into the next millennium. *Radiation Physics and Chemistry,* Vol. 57, No.3-6 (March 2000), pp. 361-365. ISSN: 0969-806X.

Ghosp, P. & Gangopadhay, R. (2000). Photo functionalisation of cellulose and lignocelluloses fibres using photoactive organic acids. *European Polymer,* Vol.36, No.3, (March 2000), pp. 625-634. ISSN: 0014-3057.

Goda, K.; Sreekala, M.S.; Gomes, A.; Kaji, T. & Ohgi, J. (2006). Improvement of Plant Based Natural Fibers for Toughening Green Composites–effect of Load Application During

Mercerizationof Ramie Fibers. *Composites: Part A,* Vol.37, No.12, (December 2006), pp. 2213–2220. ISSN: 1359-835X.

Grant, G. T.; Morris, E. R.; Rees, D. A.; Smith, P. J. C. & Thom, D. (1973). Biological interactions between polysaccharides and divalent cations: the egg-box model. *FEBS letters,* Vol 32, No. 1 (May 1973), pp. 195–198. ISSN: 0014-5793.

Greener, J. K. & Fennema, O. (1989). Barrier properties and surface characteristics of edible bilayer films. *Food Science,* Vol.54, No.6, (November 1989), pp. 1393-1399. ISSN: 1750-3841.

Guilbert, S. (1986). Technology and application of edible protective films. In: Food packaging and preservation, Theory and Practice, Matholouthi, M., pp. 371-394. Elsevier Applied Science Publishing, ISSN: 0853343772, London.

Haggenmaier, R. D. & Shaw, P. E. (1990). Moisture permeability of edible films made with fatty acid and (hydrocypropyl) methylcellulose. *Agricultural and Food Chemistry,* Vol.38, No.9, (September 1990), pp.1799-1803. ISSN: 0021-8561.

Herrera-Franco, P. J. & Valadez-Gonzalez, A. (2005). A. study of the mechanical properties of short naturalfiber reinforced composites. *Composite. Part B Engineering,* Vol. 36, No.8, (December 2005), pp. 597-608. ISSN: 1359-8368.

Hernandez, R. J. (1994). Effect of water vapor on the transport properties of oxygen through polyamide packaging materials. *Food Engineering,* Vol.22, No.1-4, (1994), pp. 495–507. ISSN: 0260-8774.

Hoseney, R. C. (1986). Principles of Cereal Science and Technology. *Cereal Chemistry,* Vol. 63, No. 3 (May 1986), pp. 285-286. ISSN: 0009-0352

Iida, Y.; Tuziuti, T.; Yasui, K.; Towata, A. & Kozuka, T. (2008). Control of viscosity in starch and polysaccharide solutions with ultrasound after gelatinization. *Innovative Food Science and Emerging Technologies,* Vol.9, No.2, (April 2008), pp. 140–146. ISSN: 1466-8564.

Ibrahim, M. M.; El-Zawawy, W. K. & Nassar, M. A. (2010). Synthesis and characterization of polyvinyl alcohol/nanospherical cellulose particle films. *Carbohydrate Polymers,* Vol.79, No.3, (February 2010), pp. 694–699 ISSN: 0144-8617.

Jenny, Y. Z.; Kai, K.; Raymond, M.; Sandra, K. & Muthupandian, A. (2009). The pasting properties of sonicated waxy rice starch suspensions. *Ultrasonics Sonochemistry,* Vol.16, No.4, (April 2009), pp. 462–468. ISSN: 1350-4177.

Jordan, J.; Jacob, K. I.; Tannenbaum, R.; Sharaf, M. A. & Jasiuk, I. (2005). Experimental trends in polymer nanocomposites: a review. *Materials Science and Engineering A,* Vol. 393, No.1-2, (February 2005), pp. 1-11. ISSN: 0921-5093.

Kaewpool, P. & Witthaya, T. (2010). Properties and Characteristics of Rice Starch Films Reinforced with Rice Starch Nanocrystals, *Proceeding of International Conference on Applied Science Engineering and Technology,* Conference Venue River View Hotel. Singapore, August 2010.

Kardos, N. & Luche, J. L. (2001). Sonochemistry of carbohydrate compounds. *Carbohydrate Research,* Vol.332, No.2, (May 2001), pp. 115–131. ISSN: 0008-6215.

Kasemsuwan, T. & Jane, J. (1996). A quantitative method for the analysis of phosphorus structures and contents in starch by P-31 nmr spectroscopy. *Cereal Chemistry,* Vol. 73, No.6, (1996), pp. 702. ISSN: 0009-0352.

Ke, T. & Sun, X.S. (2003). Starch, poly(lactic acid), and poly(vinyl alcohol) blends. *Polymers and the Environment,* Vol.11, No.1, (January 2003), pp.7–14. ISSN: 1572-8900.

Kerr, R. W. & Cleveland Jr., F. C. (1962). Thickening agent and method of making the same. *United States Patent, 3,021,222.* ISSN: 0161-9470.

Kester, J. J. & Fennema, O. R. (1986). Edible films and coatings: Areview. *Food Technology,* Vol. 40, No.12, (December 1986), pp.47-59. ISSN: 1365-2621.

Khan, M. A.; Bhattacharia, S. K.; Kader, M. A. & Bahari, K. (2006). Preparation and characterization of ultra violet (UV) radiation cured bio-degradable films of sago starch/PVA blend. *Carbohydrate Polymers,* Vol.63, No.3, (March 2006), pp. 500–506. ISSN: 0144-8617.

Kim, J. Y.; Han, S. II. & Hong, S. (2008). Effect of modified carbon nanotube on the properties of aromatic polyester nanocomposites. *Polymers,* Vol.49, No.15, (July 2008), pp. 3335-3345. ISSN: 0032-3861.

Kim, M. & Lee, S. J. (2002). Characteristics of cross linked potato starch and starch-filled linear low-density polyethylene films. *Carbohydrate Polymers,* Vol.50, No.4, (December 2002), pp. 331–337. ISSN:0144-8617.

Kristo, E. & Biliaderis, C. G. (2007). Physical properties of starch nanocrystal reinforced pullulanfilms. *Carbohydrate Polymers,* Vol.68, No.1, (March 2007), pp. 146–158. ISSN: 0144-8617.

Kumar, A. P. & Singh, R. P. (2008). Biocomposites of cellulose reinforced starch: Improvement of properties by photo-nducedcrosslinking. *Bioresource Technology,* Vol.99, No.8, (December 2008), pp. 8803–8809. ISSN: 0960-8524.

Lafargue, D.; Lourdin, D. & Doublier, J. L. (2007). Film-forming properties of a modified starch/k-carrageenan mixture in relation to its rheological behaviour. *Carbohydrate Polymers,* Vol.70, No.1, (August 2007), pp. 101–111. ISSN: 0144-8617.

Lertworasirikul, A.; Noguchi, K.; Ogawa K. & Okuyama, K. (2004). Plausible molecular and crystal structures of chitosan/HI type II salt. *Carbohydrate Research,* Vol.339, No.4, (March 2004), pp. 835-843. ISSN: 0008-6215.

Letendre, M.; D'Aprano, G.; Lacroix, M.; Salmieri, S. & St-Gelais, D. (2002). Physicochemical properties and bacterial resistance of biodegradable milk protein films containing agar and pectin. *Agricultural and Food Chemistry,* Vol.50, No.21, (October 2002), pp. 6017-6022. ISSN: 0021-8561.

Lieberman, E. R. & Gilbert, S. G. (1973). Gas Permeation of Collagen Films as Affected by Cross- linkage, Moisture, and Plasticizer Content. *Journal of Polymer Science, Symposium,* Vol.41, No.1, (1973), pp.33-43. ISSN: 1936-0959.

Liu, H. ; Du, Y.-M. & John, F. K. (2007). Hydration energy of the 1,4-bonds of chitosan and their breakdown by ultrasonic treatment. *Carbohydrate Polymers,* Vol.68, No.3, (April 2007), pp.598–600. ISSN: 0144-8617.

Liu, Z.; Li, X.; Peng, Z.; Tong, B. & Liu, G. (1999). The mechanism of agglomeration and control in the process of ultrafine powder prepared by wet chemical method. *Chemistry,* Vol.7, No. 7, (August, 1999), pp. 54–57. ISSN: 0441-3776

Lizarraga, M. S.; De Piante Vicin, D.; Gonzalez, R.; Rubiolo, A. & Santiago, L. G. (2006). Rheological behaviour of whey protein concentrate and k-carrageenan aqueous mixtures. *Food Hydrocolloids,* Vol.20, No.5, (July 2006), pp. 740–748. ISSN: 0268-005X.

Lourdin, D.; Della Valle, G. & Colonna, P. (1995). Influence of amylose content on starch films and foams. *Carbohydrate Polymer,* Vol.27, No.4, (December 1995), pp. 261-270. ISSN: 0144-8617.

Lu, Y.; Weng, L. & Cao, X. (2006). Morphological, thermal and mechanical properties of ramie crystallites reinforced plasticized starch biocomposites. *Carbohydrate Polymers,* Vol.63, No.1, (January 2006), pp. 1-7. ISSN: 0144-8617.

Manners, D. J. (1989). Recent developments in our understanding of amylopectin structure. *Carbohydrate Polymers,* Vol.11, No.2, (1989), pp. 87-112. ISSN 0144-8617.

Manoi, K. & Rizvi, S. S. H. (2010). Physicochemical characteristics of phosphorylated cross-linked starch produced by reactive supercritical fluid extrusion. *Carbohydrate Polymers,* Vol.81, No.4, (July 2004), pp. 687–694. ISSN 0144-8617.

Masclaux, C.; Gouanve, F. & Espuche, E. (2010). Experimental and modelling studies of transport in starch nanocomposite films as affected by relative humidity. *Membrane Science,* Vol.363, No.1-2, (November 2010), pp. 221–231. ISSN: 0376-7388.

McPherson, A. E. & Jane, J. (1999). Comparison of waxy potato with other root and tuber starches *Carbohydrate Polymers,* Vol.40, No.1, (September 1999), pp.57-70. ISSN 0144-8617.

Ming, L.; Penf, L.; Wei, Z.; Long, Y.; Fengwei, X.; Huanyin, P.; Hongshen, L. & Ling, C. (2011). Extrusion processing and characterization of edible starch films with different amylose contents. *Food Engineering,* Vol.106, No.1, (September 2011), pp. 95–101. ISSN: 0260-8774.

Morrison, W. R. (1995). Starch lipids and how they relate to starch granule structure and functionality. *Cereal Foods World,* Vol.40, No.6 (November 1995), pp. 437-446. ISSN: 0146-6283

Muller, C. M. O. ; Laurindo, J. B. & Yamashita, F. (2011). Effect of nanoclay incorporation method on mechanical and water vapor barrier properties of starch-based films. *Industrial Crops and Products,* Vol.33, No.3, (May 2011), pp. 605–610. ISSN: 0926-6690.

Muzzarelli, R. A. A.; Biagini, G.; Mattioli-Belmonte, M.; Talassi, O.; Gandolfi, M. G.; Solmi, R.; Carraro, S.; Giardino, R.; Fini, M. & Nicoli-Aldini, N. (1997). Osteoinduction by chitosan-complexed BMP. Morpho-structural response in an osteoporotic model. *Bioactive and Compatible Polymers,* Vol.12, No.2, (October 1997), pp. 321–329. ISSN: 1530-8030.

Olivier, A. ; Cazaux, F. ; Gors, C. & Coqueret, X. (2001). Compatibilization of Starch-Allylurea Blends by Electron Beam Irradiation: Spectroscopic Monitoring and Assessment of Grafting Efficiency. *Biomacromolecules,* Vol.2, No.4, (September 2001), pp. 1260-1266. ISSN:1526-4602.

Pareta, R. & Edirisinghe, M. J. (2006). A novel method for the preparation of biodegradable microspheres for protein drug delivery. *Carbohydrate Polymers,* Vol.63, No.3, (March 2006), pp. 425-431. ISSN: 0144-8617.

Park, S. Y.; Marsh, K. S. & Rhim, J. W. (2002). Characteristics of different molecular weight chitosan films affected by the type of organic solvents. *Food Sciences,* Vol.67, No.1, (January 2002), pp. 194–197. ISSN: 1750-3841.

Peroval, C.; Debeaufort, F.; Despre, D. & Voilley, A. (2002). Edible arabinoxylan-based films. I. Effects of lipids types on water vapor permeability, film structure and other physical characteristics. *Agricultural and Food Chemistry,* Vol.50, No.14, (July 2002), pp. 3977-3983. ISSN: 1520-5118.

Phan T. D.; Debeaufort ,F.; Voilley. & A., Luu, D. (2009). Biopolymer interactions affect the functional properties of edible films based on agar, cassava starch and arabinoxylan blends. *Food Engineering,* Vol.90, No.4, (February 2009), pp. 548–558. ISSN: 0260-8774.

Phattaraporn, T.; Waranyou, S.; Fazilah, A. & Thawien, W. (2010). Characteristics and properties of rice starch films reinforced with palm pressed fibers. *International Food Research Journal,* Vol.17, No.3, (2010), pp. 535-547. ISSN: 22317546.

Phattaraporn, T., Waranyou, S. & Thawien, W. 2011. Effect of palm pressed fiber (PPF) surface treatment on the properties of rice starch films. *International Food Research Journal.* Vol.18, No. 1 (January 2011), pp. 535-547. ISSN: 22317546.

Podsiadlo, P.; Paternel, S.; Rouillard, J. M.; Zhang, Z.; Lee, J.; Lee, J. W.; Gulari, E. & Kotov, N. A. (2005). Layer-by-layer assembly of nacre-like nanostructured composites with antimicrobial properties. *Langmuir,* Vol.21, No.25, (December 2005), pp.11915-11921. ISSN :1520-5827.

Rhim, W. J. & Shellhammer, H. T. (2005). Lipid based edible films and coating. *Innovations in food packaging,* Academic Press, ISBN: 978-0-12-311632-1.

Rhim, J. W. (2004). Physical and chemical properties of water resistant sodium alginate films. *Lebensmittel-Wissenschaft Und-Technologie,* Vol.37, No.3, (May 2004), pp. 323–330. ISSN: 0023-6438.

Rindlav- Westling, A.; Stading, M. & Gatenholm, P. (2002). Crystallinity and morphology in films of starch, amylase and amylopectin blends. *Biomacromolecules,* Vol.3, No.1, (January 2002), pp. 84-91.ISSN: 1525-7797.

Rindlav- Westling, A.; Stading, M.; Hermansson A. M. & Gatenholm, P. (1998). Structure, mechanical and barrier properties of amylose and amylopectin films. *Carbohydrate. Polymer,* Vol.36, No.2-3, (July 1998), pp. 217-224. ISSN: 0144-8617.

Robeson, L. M. (2008). The upper bound revisited. *Membrane Science,* Vol.320, No.1-2, (July 2008), pp. 390–400. ISSN: 0376-7388.

Rutledge, J. E.; Islam, M. N. & James, W. H. (1974). Improved canning stability of parboiled rice through cross-linking *Cereal Chemistry,* Vol.51, No.1, (January-February 1974), pp. 46-50. ISSN: 0009-0352.

Sabato, S.; Ouattara, B.; Yu, H.; D'Aprano, G.; Le Tien, C. & Mateescu, M. (2001). Mechanical and barrier properties of cross-linked soy and whey protein based films. *Agricultural and Food Chemistry.* Vol.49, No.3, (March 2001), pp. 1397–1403.ISSN: 0021-8561.

Shih, Y. F. (2007). Mechanical and thermal properties of waste water bamboo husk fiber reinforced epoxy composites. *Material Science and Engineering Part A,* Vol. 445-446, (February 2007), pp. 289-295. ISSN: 0921-5093.

Shogren, R. L. ; Lawton, J. W.; Tiefenbacher, K. F. & Chen, L. (1998). Starch poly(vinyl alcohol) Foamed Articles Prepared by Baking Process. *Applied Polymer Science,* Vol.68, No.13, (June 1998), pp. 2129-2140. ISSN: 1097-4628.

Singh, J.; Kaur, L. & McCarthy, O. J. (2007). Factors influencing the physico-chemical, morphological, thermal and rheological properties of some chemically modified starches for food applications-A review. *Food Hydrocolloids,* Vol.21, No.1, (January 2007), pp. 1–22. ISSN:0268-005X.

Smiley, R. A. & Jackson, H. L. 2002. *Chemistry and the Chemical Industry: A practical guide for non-chemists,* CRC press, ISBN: 1-58716-054-4, Florida, USA.

Spielmann, H.; Lovell, W. W.; Holzle, E.; Johnson, B. E.; Maurer, T.; Miranda, M. A.; Pape, W. J. W.; Sapora, O. & Sladowski, D. (1994). *In Vitro Phototoxicity testing.* The report and recommendations of ECVAM workshop 2. ATLA. Alternative to laboratory animals, Vol.22, No.5, pp.314-348. ISSN: 0261-1929.

Srinivasa, P. C.; Ramesh, M. N. & Tharanathan, R. N. (2007). Effect of plasticizers and fatty acids on mechanical and permeability characteristics of chitosan films. *Food Hydrocolloids,* Vol.21, No.7, (October 2007), pp. 1113-1122. ISSN: 0268-005X.

Stading, M.; Rindlav-Westling, A. & Gatenholm P. (2001). Humidity-induced Structural Transitions in Amylose and Amylopectin Films, *Carbohydrate Polymers,* vol.45, No.3, (July 2001), pp.209-217. ISSN: 0144-8617.

Weiss, J.; Takhistov, P. & McClements, D. J. (2006). Functional materials in food nanotechnology. *Food Science,* Vol.71, No.9, (November/December 2006), pp. R107-R116. ISSN: 1750-3841.

Wittaya, T. (2009). Microcomposites of rice starch film reinforced with microcrystalline cellulose from palm pressed fiber. *International Food Research Journal,* Vol.16, No.4, (2009), pp. 493-500. ISSN: 22317546.

Woods, R. J. & Pikaev, A. K. (1993). *Applied Radiation Chemistry: Radiation Processing.* Wiley, ISBN: 978-0-471-54452-4. New York

Wu, Y.; Geng, F.; Chang, P. R.; Yu, J. & Ma, X. (2009). Effect of agar on the microstructure and performance of potato starch film. *Carbohydrate Polymers,* Vol.76, No.2, (March 2009), pp. 299-304. ISSN: 0144-8617.

Xu, X.Y.; Kim, K.M.; Hanna, M.A. & Nag, D. (2005). Chitosan-starch composite film: preparation and characterization. *Industrial Crops and Products an International Journal,* Vol.21, No.2, (March 2005), pp.185-192. ISSN: 0926-6690.

Yang, L. & Paulson, A. T. (2000). Effects of lipids on mechanical and moisture barrier properties of edible gellan film. *Food Research International,* Vol.33, No.7, (August 2000), pp. 571-578. ISSN: 0963-9969.

Yew, G. H.; Mohd Yusof, A. M.; Mohd Ishak , U.S. & Ishiaku , Z.A. (2005). Water absorption and enzymatic degradation of poly(lactic acid)/rice starch composites. *Polymer Degradation and Stability,* Vol.90, No.3, (December 2005), pp. 488-500. ISSN: 0141-3910.

Zhai, M.; Zhao,L.; Yoshii, F. & Kume, T. (2004). Study on antibacterial starch/chitosan blend film formed under the action of irradiation. *Carbohydrate Polymers,* Vol.57, No.1, (August 2004), pp. 83–88. ISSN: 0144-8617.

Section 2

Foodborne Botulism Poisoning

Chapter 6

Botulinum Toxin Complex: A Delivery Vehicle of Botulinum Neurotoxin Traveling Digestive Tract

Yoshimasa Sagane, Ken Inui, Shin-Ichiro Miyashita, Keita Miyata, Tomonori Suzuki, Koichi Niwa and Toshihiro Watanabe

Additional information is available at the end of the chapter

1. Introduction

Botulinum neurotoxin (BoNT) produced by an anaerobic bacterium *Clostridium botulinum* causes highly fatal disease called botulism. BoNT is a zinc-dependent metalloprotease [1] with a molecular mass of 150 kDa, and classified into seven distinct serotypes A through G. Serotype A, B, E and F BoNTs dominantly cause human botulism, whereas serotype C and D BoNTs are the causative substance of the animal and avian botulism [2,3]. The BoNT invades into the human or animal body via one of three ways; toxin production by the bacterium colonized on the digestive tract in early childhoods (infant botulism), entrance of the toxin from the wound (wound botulism), and the oral ingestion of the toxin-contaminated foods (food-borne botulism). The food-borne botulism is the most frequent among the three cases. In any of these cases, the BoNT ultimately reaches the neuromuscular junction and enters nerve cell via receptor-mediated endocytosis. Once in the nerve cell, the metalloprotease activity of the toxin cleaves the specific site on the target proteins associated with the intracellular vesicle transport, such as synapse-associated protein of molecular weight 25,000 (SNAP-25; serotypes A, C and E), syntaxin (serotype C) and vesicle-associated membrane protein (VAMP; serotypes B, D and F), and inhibits the release of acetylcholine to extracellular space [2,4,5]. These process cause muscular paralysis in humans and animals.

When pure BoNT is exposed to the digestive conditions with acidic fluid and proteases, the BoNT easily degrades into inactive short peptides [6-15] and thus the pure BoNT exerts the only weak or no oral toxicity. This implies that the pure BoNT seems unlikely to cause the food-borne poisoning, and conflicts with the previous description that the food-borne botulism is the most frequent among three botulism diseases. The answer that resolves this discrepancy is the toxin complex (TC). In the culture supernatant and polluted foods, the BoNT is a part of the TC. Serotype A–D strains produce both of hemagglutination-negative and -positive TC, whereas serotype E and F strains produce only hemagglutination-negative TC. The hemagglutination-negative TC is 280-kDa M-TC consists of BoNT and nontoxic nonhemagglutinin (NTNHA). On the other hand, hemagglutination-positive TC is 750-kDa L-TC composed of M-TC and three types of hemagglutinins (HAs; HA-70, HA-33 and HA-17). Serotype A strain produces additional LL-TC, which might be a dimer of L-TC [16,17].

The oral toxicities of the M-TC, L-TC and LL-TC are much greater than pure BoNT. Therefore it seems that the auxiliary nontoxic proteins play a role for the delivery of the toxin through the animal digestive system so that the botulinum TC exerts the oral toxicity. In our chapter, we describe our findings in the series of studies on the structure and function of botulinum TC, especially focused on the nontoxic proteins NTNHA and HAs.

2. Assembly pathway of botulinum toxin complex

On the genome of the *C. botulinum* strains, the gene encoding BoNT forms one or two gene clusters along with other genes (**Figure 1**) [16,18-22]. In serotypes A–D, cluster 1 contains the genes coding the BoNT and NTNHA, whereas the cluster 2 contains three genes coding HAs. The open reading frames of the genes on the cluster 2 run opposite orientation against those on the cluster 1. Serotypes E and F lack cluster 2. In the serotypes C and D, the gene designated as *botR* is located on the downstream of the cluster 2 with opposite directions to reading frames of HA genes. Similar genes are also found in the serotypes A and B, and they are located between the cluster 1 and 2. The gene products of *botR* do not participate as a constituent of the TC, but control the expression of the genes on the cluster 1 and 2 [23,24].

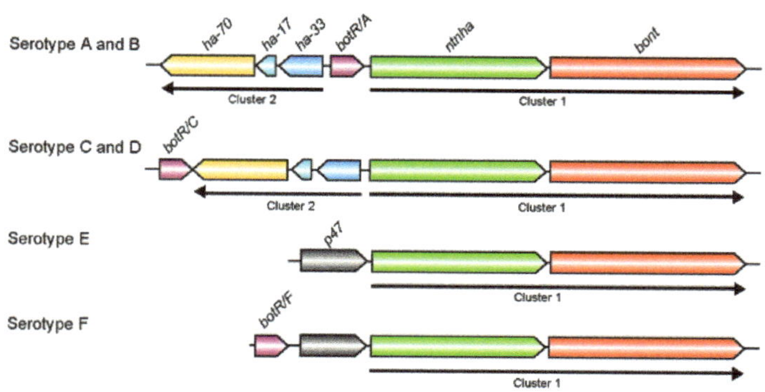

Figure 1. Genetic organization of botulinum neurotoxin related genes of *C. botulinum*. This scheme indicates the BoNT gene clusters in serotype A–F strains. Botulinum TCs are encoded by two gene clusters; cluster 1 encodes BoNT (*bont*) and NTNHA (*ntnha*), and cluster 2 encodes three HAs (*ha-70, ha-17* and *ha-33*). Each gene cluster is transcribed polycistronically, as indicated by arrows. Genetic organization of serotype A resembles that of serotype B. Similarly, gene organizations of serotype C and D BoNT clusters are much alike. On the other hand, serotypes E and F do not possess any HA gene. The gene product of *botR* controls the expression of the genes on the cluster 1 and 2. In serotypes E and F, *p47* gene is arranged sequentially in the upstream region of the *bont* gene cluster and its gene product has been uncharacterized.

Although the *C. botulinum* serotype C and D strains hold five genes involved in the TC constituents, they produces two types of TCs, M-TC (a complex of BoNT and NTNHA) and L-TC (a complex of M-TC and HAs). Ever since the discovery of TC produced by *C. botulinum* strains, discrepancy in the coexistence of two types of the TCs in the same culture had been a longstanding mystery. In the culture supernatant, the 150-kDa BoNT of the L- and M-TC are split into a 50-kDa light chain (Lc) and a 100-kDa heavy chain (Hc) by the excision of several amino acid residues [25], and the HA-70 in the L-TC is also split into 22–23-kDa and 55-kDa fragments by proteolytic processing after translation [26]. The NTNHA of the M-TC is always found nicked at their N-terminus leading to a 15-kDa N-terminal fragment and a 115-kDa C-terminal fragment, whereas the NTNHA of the L-TC remains intact [20,27]. Thus, the components of the TCs are nicked, leading to the appearance of many fragments on the SDS-PAGE. This may complicate the consideration of botulinum TC structure, and hamper the resolution of the discrepancy in coexistence of dual form of TC.

In 2002, we finally found one of the answers to solve this discrepancy [28]. Before that, we needed two breakthroughs to find the answer. First, we serendipitously found a unique serotype D strain 4947 (D-4947), producing the M-TC and L-TC without any nicking in the components of the complex. Second, we have established the method to isolate viable HA components (HA-70, HA-33/HA-17 complex) from the L-TC in the presence of 4 M guanidine hydrochloride [29]. By using the M-TC (BoNT/NTNHA) and isolated HA-70 and HA-33/HA-17 without any nicking obtained from serotype D-4947 strain, we performed reconstitution experiments to construct the L-TC (**Figure 2**) [28]. Mixture of the M-TC and HA-70 yielded a *de novo* M-TC/HA-70 complex, whereas a mixture of the M-TC and HA-33/HA-17 did not produce any complex. Further the M-TC/HA-70 complex converted to the mature L-TC by mixing with the HA-33/HA-17. On the other hand, M-TC, which contained nicked NTNHA prepared by treatment with exogenous trypsin, could no longer be reconstituted to mature L-TC with HA subcomponents, whereas the L-TC treated with trypsin was not degraded into M-TC and HA subcomponents. Consequently, we concluded that the association of BoNT and NTNHA produces M-TC, and thereafter is converted to the L-TC by assembly with HA-70 and HA-33/HA-17.

3. Subunit structure of botulinum toxin complex with three extended arms

C. botulinum serotype C and D strains produce 280-kDa M-TC and 750-kDa L-TC. In addition to these TCs, we found three intermediate TC species in the culture supernatants of serotype C and D strain [30,31]. One of them is M-TC/HA-70 complex (490 kDa). Remaining two is 610- and 680-kDa TC, corresponding to the intermediate products in the pathway leading from the 490-kDa M-TC/HA-70 to the mature 750-kDa L-TC, which has a smaller number of HA-33/HA-17 complexes than mature L-TC. All of the TC species, except for the 750-kDa L-TC, demonstrated no hemagglutination activity. When the intermediate TC species were mixed with an isolated HA-33/HA-17 complex, all TC species were maturated to 750-kDa L-TC with full hemagglutination activity and had the same molecular composition of native 750-kDa L-TC [30]. Therefore, these findings indicated that the mature L-TC contains multiple HA-33/HA-17 sub-complexes.

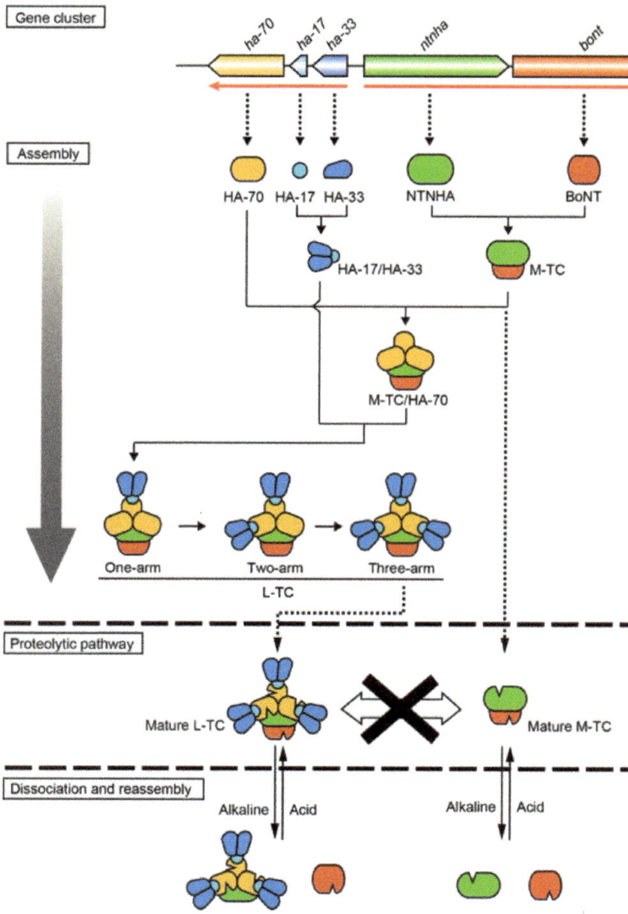

Figure 2. Proposed model for the botulinum toxin complex assembly pathway. Genetic organization of botulinum serotype C and D TCs and their expressed products are represented based on nucleotide sequences and N-terminal amino acid sequences. The assembly pathway from each gene product and proteolytic pathway is indicated by solid and dotted arrows, respectively. The upper panel represents the assembly pathway of the components to TCs, and the middle panel represents putative proteolytic pathway of the nicked TCs usually observed in serotype C and D strains. The lack of mutual conversion between L- and M-TC is represented by the X symbol. The lower panel represents dissociation and reassembly of the TCs depending on pH

To characterize the HA-33/HA-17 complex, we determined the X-ray crystal structure of the isolated HA-33/HA-17 complex from D-4947 L-TC at 1.85 Å resolution [17]. As shown in **Figure 3**, the final model of the complex includes two HA-33 molecules and one HA-17 molecule in the asymmetric unit. This model is consistent with the molecular composition of HA-33/HA-17 complex as determined from molecular mass of the complex by nanoLC/ESI-TOF-MS yielded a mass of 84,118 Da (calculated molecular mass of two HA-33 plus one HA-17 is 84,239 Da).

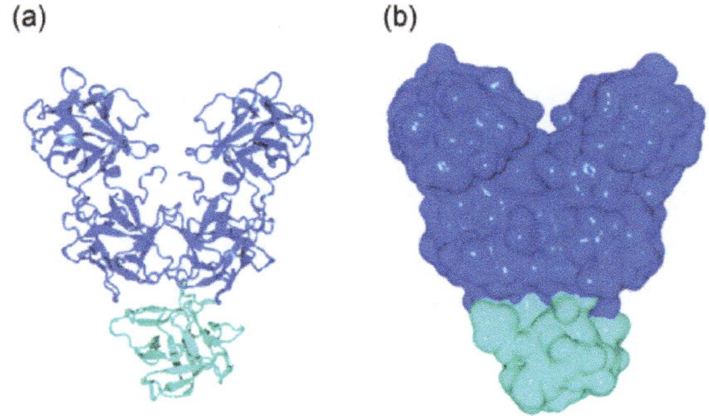

Figure 3. Crystal structure of the HA-33/HA-17 trimer complex. (a) Structure of HA-33/HA-17 trimer complex isolated from serotype D-4947 L-TC as represented by a ribbon diagram. (b) Surface representation of HA-33/HA-17 trimer complex: HA-33 molecules in blue and HA-17 molecule in cyan. Figures were prepared using MolFeat version 3.0 (FiatLux Corp.). The crystal structure of HA-33/HA-17 trimer complex was resolved at 1.85 Å.

To clarify the TC structure, the purified TC species, *i.e.* M-TC, M-TC/HA-70, and 610-, 680- and 750-kDa L-TC, were visualized by negative stained transmission electron microscopy (TEM) [17] (**Figure 4**). As a result, the M-TC (BoNT/NTNHA) appears as an approximately 13-nm spherical or ellipsoidal particle. The M-TC/HA-70 displayed an acorn-like shape with the HA-70 "cone" lying on the M-TC. Interestingly, the 610- and 680-kDa TC and mature 750-kDa L-TC revealed unique "arm" attachments that displayed to be rod-like structure. Number of the arm (approximately 10 nm length) in the TCs was one, two and three for the 610- and 680-kDa TC and mature 750-kDa L-TC, respectively. We concluded that the arm-like structure is the HA-33/HA-17 that attached to the M-TC/HA-70 complex via HA-70 molecule. At this point,

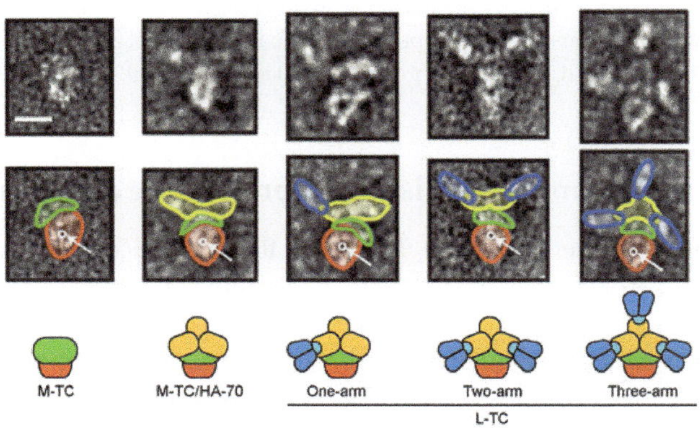

Figure 4. Electron micrographs of the botulinum serotype D-4947 TC species. The upper panel represents the TCs [M-TC, M-TC/HA-70, and one-, two- and three-arm L-TCs, respectively]. Scale bar in the upper left panel indicating 10 nm applies to all panels. The middle panel represents tracing of images of the subunit components: BoNT in red, NTNHA in green, HA-70 in yellow and HA-33/HA-17 complex in blue. A black spot, which is assumed to be a zinc atom or cavity, is indicated by the circle with an arrow in each BoNT image. The lower panel represents corresponding schematic structure model of each TC.

the number of the HA-70 molecule in the L-TC is still unclear. Therefore, we examined a densitometric analysis of Coomasie Brilliant blue staining bands on SDS-PAGE gels to understand the number of the HA-70 molecule in the complex. The result indicated that the number of the molecule is three. After our report regarding the number of the HA-70 molecule in the L-TC, the X-ray crystallographic structure of the HA-70 was published [32]. They displayed a three-leaved propeller-like structure, which is consistent with our experiment.

After the TEM observations of the TCs, we named the 610- and 680-kDa TC and mature 750-kDa L-TC, as one-, two- and three-arm L-TC [33]. Based on the TEM images and the crystallographic structure of the HA-33/HA-17, we constructed a 14-heteromer model of the mature three-arm L-TC [17]. That is, the three-arm L-TC is composed of a single BoNT molecule, a single NTNHA molecule, and an HA complex. The HA complex is comprised of three HA-70 molecules and three arm structure of HA-33/HA-17 that consists of two HA-33 molecules and a single HA-17 molecule (**Figure 5**). This model suggests that the six HA-33 molecules exposed to outer of the mature three-arm L-TC, where they play a role in anchoring the complex at the epithelial cell surface.

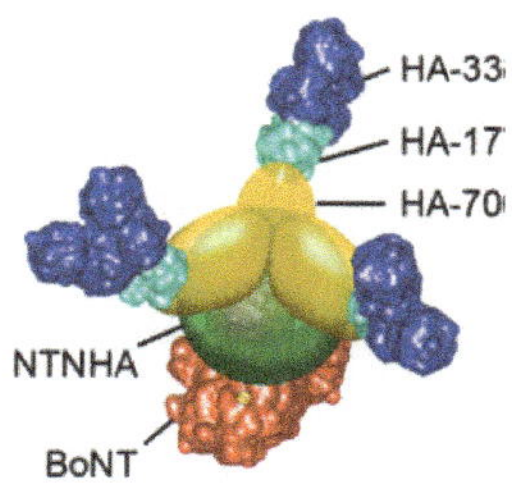

Figure 5. Subunit structure of the botulinum large toxin complex (L-TC). L-TC possesses 14-mer subunit structure and is composed of five types of protein; BoNT, NTNHA and three types of HAs. BoNT molecule is shown in red, NTNHA in green, HA-70 in yellow, HA-33 in blue and HA-17 in cyan, respectively.

4. Botulinum toxin complex resistant to proteolytic digestion

BoNT dissociates from the botulinum TC under the alkaline condition, while under the acidic condition the BoNT re-associates with the nontoxic component complex generating the L-TC. Botulinum TC exerts the grater oral toxicity than BoNT detached from the complex. Thus the nontoxic components play a vital role to exert the oral toxicity of the botulinum TC. Of the nontoxic components of the botulinum TC, the NTNHA displays unique property. That is, the NTNHA components in both isolated form and the M-TC (BoNT/NTNHA) complexed form are spontaneously converted to the nicked form leading 15-kDa N-terminal and 115-kDa C-terminal fragments with excision of several amino acid residues at specific sites during long-term incubation [34]. To clarify the role of the NTNHA on oral toxicity of the TC, we constructed the overexpression system that produces the recombinant NTNHA (rNTNHA) in the *Escherichia coli* strain [15]. The NTNHA gene was amplified by the PCR using the gene specific primer, ligated to pET200-D/TOPO, and then transformed into a large-scale *E. coli* BL21 cell expression system to produce the rNTNHA with His-tagged N-terminus. The rNTNHA was also spontaneously converted to nicked form during long-term incubation generating N-terminal 18-kDa N-terminal (containing His-tag) and 115-kDa C-terminal fragments. Using the rNTNHA, crystallization and preliminary X-ray analysis was successfully performed [35]. Further the rNTNHA was successfully reconstituted with isolated BoNT generating M-TC by mixing these proteins at a molar ratio 1:1 followed by incubation at pH 6.0. The reconstituted M-TC reversibly dissociated into rNTNHA and BoNT at pH 8.8. During reconstitution experiment, the intact rNTNHA in the reconstituted M-TC was spontaneously converted into nicked form indicated by the 18- and 115-kDa bands on the SDS-PAGE. These features including spontaneous cleavage occurred in the molecule and reconstitution with the BoNT are consistent with those of the native NTNHA.

During the oral delivery of the botulinum TC, the orally ingested TC is exposed to acidity and proteolysis in the gastrointestinal tract. To examine the stability of the rNTNHA, BoNT and reconstituted M-TC, these proteins were exposed to harsh conditions mimicking environment of the stomach (30 U pepsin in pH 2.7) and small intestine (1250 U trypsin in pH 6.0) (**Figure 6**). The rNTNHA was digested into several fragments after incubation with pepsin for 60 min, whereas trypsin treatment for 360 min yields just nicking in rNTNHA. The BoNT was highly sensitive to both trypsin and pepsin digestion, producing no bands on the SDS-PAGE. In contrast, when the BoNT formed complex with rNTNHA generating the reconstituted M-TC, both proteins exhibited amazing resistance to proteolysis. When the reconstituted M-TC was treated with the pepsin, the BoNT displayed no sign of the fragmentation, while the rNTNHA was converted to nicked form yielding 18- and 115-kDa fragments on the SDS-PAGE. Trypsin treatment of the M-TC produced specific nicked form BoNT and rNTNHA. Because only the reconstituted M-TC displayed high tolerance against both trypsin and pepsin attacks, we presumed that the NTNHA component can alter its conformation to assemble with BoNT, forming an oral toxin that protects BoNT from harsh conditions in the digestive tract. The presumption was certified very recently by the X-ray crystal structure of serotype A M-TC [36].

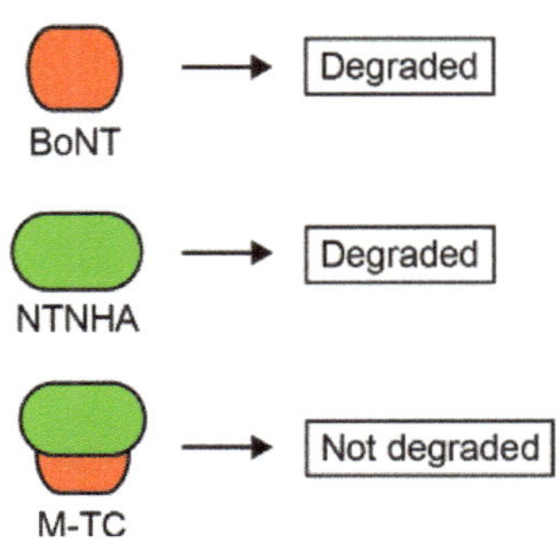

Figure 6. Stability of BoNT, NTNHA and M-TC (BoNT/NTNHA complex) to protease attack. Both BoNT and NTNHA molecules in free form are easily degraded by protease attack, while both proteins in M-TC shows the resistance to pepsin and trypsin. These protease tolerance represents the NTNHA could protect BoNT from protease attack in the digestive tract.

5. HA-33 facilitates transport of toxin complex across intestinal epithelial cell monolayer

For the oral delivery, not only the acidity and proteases in the digestive tract, the physical barrier presented by the intestinal wall is also large obstacle. Of the nontoxic components of the botulinum TC, HA-33 component exposed outermost of the complex, and possesses an ability to recognize sugar chains on intestinal microvilli. Additionally a series of the investigations indicated that the L-TC containing HA-33 components transports across the intestinal epithelial cell monolayer more effectively than pure BoNT [14,37,38]. Therefore the HA-33 component appears to play a role in the transport of the TC across the intestinal epithelium. However, some serotype A and serotype E and F strains do not possess the genes that encode the HA components [18,39,40], implying that the absorption of the botulinum TC into intestinal epithelial cells does not depend on the presence of HA components.

In the study of [33], BoNT, M-TC (BoNT/NTNHA), M-TC/HA-70 and three types of L-TC (one-, two- and tree-arm L-TC) with different numbers of the HA-33/HA-17 arm attached were purified from the culture of the D-4947, to examine the cell binding and monolayer transport of serotype D toxins in the rat intestinal epithelial cell line IEC-6. The TCs including pure BoNT were incubated with IEC-6 cells at 4 °C for 1 h. Toxins bound to the cells and cytoskeletal actins were visualized by immunohistochemistry using anti-BoNT and phalloidin (**Figure 7a**). As a result, fluorescence from the BoNT-, M-TC- and M-TC/HA-70-treated cells demonstrated similar intensities, whereas higher intensities were observed after cells were treated with one-, two- or three-arm L-TCs. To quantitatively assess binding of the TCs, toxins bound to the cells were extracted into the sample buffer containing SDS, applied to SDS-PAGE and subjected to Western blot analysis using anti-BoNT. Staining intensities were compared to determine the relative amount of the toxins bound to the cells (**Figure 7b**). BoNT, M-TC and M-TC/HA-70 displayed similar cell-binding potencies. In contrast, the one-, two- and three-arm L-TCs exhibited two, four and five times greater binding than pure BoNT.

Toxin transport through the IEC-6 cell monolayer was examined using the Transwell two-chamber system. Toxins suspended in the medium were applied to apical side of the cell monolayer. After 1 to 48 h incubation at 37 °C, toxins transported through the cell layer were collected from the basal side of the layer, applied to SDS-PAGE and then detected by Western blot analysis using anti-BoNT, anti-NTNHA, anti-HA-70 and anti-HA-33/HA-17 (**Figure 8a**). The result indicated all TCs including pure BoNT transported through the IEC-6 cell monolayer. L-TC demonstrated greater transport potency than the smaller TCs, whereas pure BoNT, M-TC and M-TC/HA-70 exhibited similar transport efficiencies at all time points (**Figure 8b**). After a 24-h incubation, the one-, two- and three-arm L-TCs displayed 2, 2.5 and 3 times greater cell monolayer transport, respectively.

To determine the responsible components on the cell binding and cell monolayer transport, pure BoNT, M-TC, M-TC/HA-70 and three-arm L-TC were preincubated with one of the antibodies against their constituents before binding or transport assay (**Figure 9**). The binding and transport of the pure BoNT were significantly inhibited by preincubation with anti-BoNT. Binding and transport of M-TC was significantly inhibited by anti-BoNT, and slightly by anti-NTNHA. As for the M-TC/HA-70, anti-BoNT and anti-HA-70 significantly reduced both the binding and transport of the toxin, while anti-NTNHA had very little effect. Of the antibodies for three-arm L-TC, the HA-33 significantly interfered with both the cell binding and the cell monolayer transport.

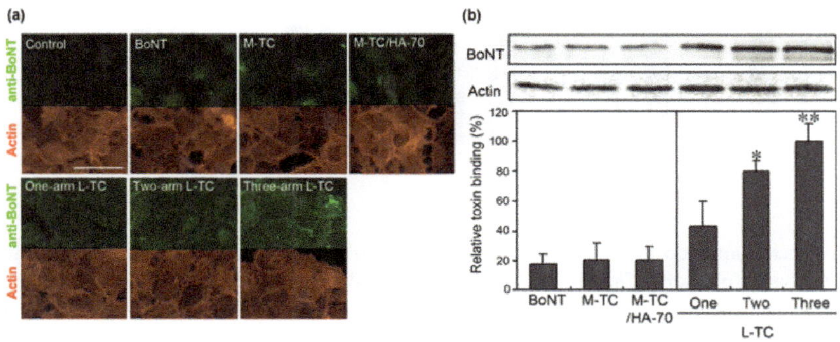

Figure 7. Binding of the BoNT and botulinum TCs to IEC-6 cells. (a) Cells were incubated with BoNT, M-TC, M-TC/HA-70, and one-, two- and three-arm L-TCs at 100 nM for 1 h at 4° C. Toxins bound to the cells and actins were visualized by immunostaining using BoNT antibody followed by Alexa Fluor 488-conjugated secondary antibody (green) and Alexa Fluor 546-conjugated Phalloidin (red). Cells without toxin treatment was employed as a control. Scale bar indicating 50 μm at upper left panel applies to all images. (b) Western blot analysis of the binding of the BoNT and TCs to IEC-6 cells. Cells were incubated with 20 nM TCs. To detect the TCs, BoNT in the cell lysate was detected by using anti-BoNT. Representative data (upper panel) and calculated relative amounts (lower panel) are shown. The relative amounts of the BoNT were calculated based on the intensities of the signals on the Western blot. Double and single asterisk denote a significant increase in binding ($P<0.01$ and $P<0.05$) compared to the BoNT, respectively.

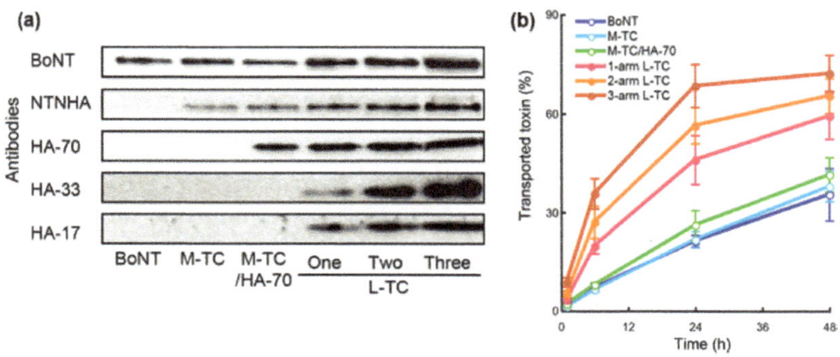

Figure 8. Transport of the BoNT and botulinum TCs through the IEC-6 cell monolayer. (a) Western blot analysis of the transport of TCs through the cell monolayer. Each 20 nM toxin suspension containing 0.2 mg/ml FITC-dextran (M.W. 500 kDa) was applied to apical side of the cell monolayer. Transported TCs through the cell monolayer were collected from the medium in the basal side of the layer, and then applied to Western blot analysis. Each component was detected by using corresponding antibody. (b) Time course of toxin transport through the cell layer. Toxin concentration in the basal side medium was estimated from the band intensities of the BoNT on the Western blot. Amounts of the transported toxins are indicated as percentages of the original toxin amounts applied to apical side medium.

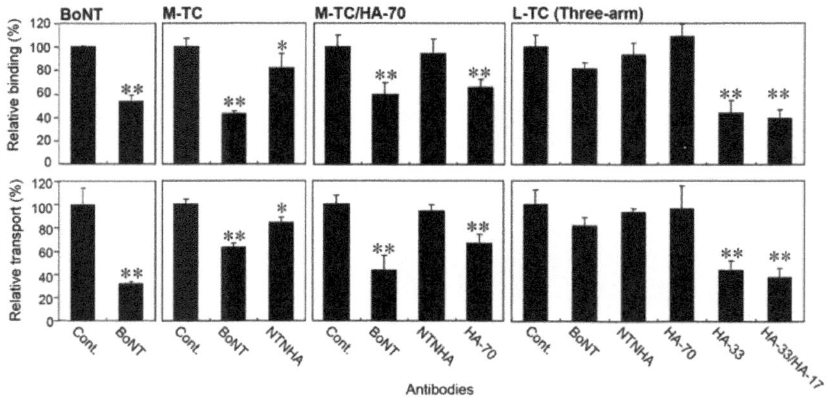

Figure 9. Effect of antibodies against each TC constituents on the cell binding and cell monolayer transport of the botulinum TCs. The toxins (40 nM) were preincubated with antibodies (diluted to 1:10 in incubation buffer) for 1 h at 37 °C before cell binding (upper panel) and cell monolayer transport (lower panel) assay. Relative binding and transport were calculated in the same manner for the Figure 8 and compared with the control without antibody preincubation. Double and single asterisk denote a significant increase in binding and transport ($P<0.01$ and $P<0.05$) compared to the control.

Consequently, all TCs including pure BoNT can transport across the intestinal epithelial cell monolayer via BoNT, NTNHA, HA-70 and HA-33 depending on the TC structure. Nonetheless binding and transport markedly increased with the number of HA-33/HA-17 arms in the TC. We therefore concluded that the HA-33/HA-17 arm is not necessarily required for, but facilitates, transport of the botulinum TC.

6. Conclusion

BoNT forms TCs by assembling with nontoxic proteins that include NTNHA and three types of HAs. We provide definitive evidence that NTNHA plays a crucial role in protecting BoNT, which is an oral toxin, from digestion by proteases common in the stomach and intestine. Furthermore, we concluded that the HA-33 is not necessarily required for, but facilitates, transport of botulinum TC. In the food-borne botulism poisoning, the nontoxic components of the TC play a role as a "delivery vehicle" of the unwanted toxic protein. However, we are expecting that the nontoxic proteins would be available for the "delivery vehicle" of the drug and vaccine by substituting the BoNT into functional peptides and proteins.

Author details

Yoshimasa Sagane, Ken Inui, Shin-Ichiro Miyashita, Keita Miyata,
Koichi Niwa and Toshihiro Watanabe

Department of Food and Cosmetic Science, Faculty of Bioindustry, Tokyo University of Agriculture, Abashiri, Japan

Tomonori Suzuki
Department of Bacteriology, Okayama University Graduate School of Medicine, Dentistry and Pharmaceutical Sciences, Okayama, Japan

7. References

[1] Schiavo G, Rossetto O, Santucci A, DasGupta BR, Montecucco C (1992) Botulinum neurotoxins are zinc proteins. J. biol. chem. 267: 23479–23483.

[2] Montecucco C, Schiavo G (1994) Mechanism of action of tetanus and botulinum neurotoxins. Mol. microbiol. 13: 1–8.

[3] Li L, Singh BR (1999) Structure-function relationship of clostridial neurotoxins. J. toxicol. toxin rev. 18: 95–112.

[4] Schiavo G, Rossetto O, Catsicas S, Polverino de Laureto P, DasGupta BR, Benfenati F, Montecucco C (1993) Identification of the nerve terminal targets of botulinum neurotoxin serotypes A, D, and E. J. biol. chem. 268: 23784–23787.

[5] Dolly JO, Black J, Williams RS, Melling J (1984) Acceptors for botulinum neurotoxin reside on motor nerve terminals and mediate its internalization. Nature 307: 457–460.

[6] Halliwell G (1954) The action of proteolytic enzymes on *Clostridium botulinum* type A toxin. Biochem. j. 58: 4–8.

[7] Ohishi I, Sugii S, Sakaguchi G (1977) Oral toxicities of *Clostridium botulinum* toxins in response to molecular size. Infect. immun. 16: 107–109.

[8] Sugii S, Ohishi I, Sakaguchi G (1977) Correlation between oral toxicity and in vitro stability of *Clostridium botulinum* type A and B toxins of different molecular sizes. Infect. immun. 16: 910–914.

[9] Sugii S, Ohishi I, Sakaguchi G (1977) Intestinal absorption of botulinum toxins of different molecular sizes in rats. Infect. immun. 17: 491–496.

[10] Bonventre PF (1979) Absorption of botulinal toxin from the gastrointestinal tract. Rev. infect. dis. 1: 663–667.

[11] Ohishi I, Sakaguchi G (1980) Oral toxicities of *Clostridium botulinum* type C and D toxins of different molecular sizes. Infect. immun. 28: 303–309.

[12] Ohishi I (1984) Oral toxicities of *Clostridium botulinum* type A and B toxins from different strains. Infect. immun. 43: 487–490.

[13] Chen F, Kuziemko GM, Stevens RC (1998) Biophysical characterization of the stability of the 150-kilodalton botulinum toxin, the nontoxic component, and the 900-kilodalton botulinum toxin complex species. Infect. immunol. 66: 2420–2425.

[14] Niwa K, Koyama K, Inoue S-I, Suzuki T, Hasegawa K, Watanabe T, Ikeda T, Ohyama T (2007) Role of nontoxic components of serotype D botulinum toxin complex in permeation through a Caco-2 cell monolayer, a model for intestinal epithelium. FEMS immun. med. microbiol. 49: 346–352.

[15] Miyata K, Yoneyama T, Suzuki T, Kouguchi H, Inui K, Niwa K, Watanabe T, Ohyama T. (2009) Expression and stability of the nontoxic component of the botulinum toxin complex. Biochem. biophys. res. commun. 384: 126–130.

[16] Inoue K, Fujinaga Y, Watanabe T, Ohyama T, Takeshi K, Moriishi K, Nakajima H, Inoue K, Oguma K (1996) Molecular composition of *Clostridium botulinum* type A progenitor toxins. Infect. immun. 64: 1589–1594.

[17] Hasegawa K, Watanabe T, Suzuki T, Yamano A, Oikawa T, Sato Y, Kouguchi H, Yoneyama T, Niwa K, Ikeda T, Ohyama T (2007) A novel subunit structure of *Clostridium botulinum* serotype D toxin complex with three extended arms. J. biol. chem. 282: 24777–24783.

[18] East AK, Bhandari M, Stacey JM, Campbell KD, Collins MD (1996) Organization and phylogenetic interrelationships of genes encoding components of the botulinum toxin complex in proteolytic *Clostridium botulinum* types A, B, and F: evidence of chimeric sequences in the gene encoding the nontoxic nonhemagglutinin component. Int. j. syst. bacteriol. 46: 1105–1112.

[19] Hauser D, Eklund MW, Boquet P, Popoff MR (1994) Organization of the botulinum neurotoxin C1 gene and its associated non-toxic protein genes in *Clostridium botulinum* C 468. Mol. gen. genet. 243: 631–640.

[20] Ohyama T, Watanabe T, Fujinaga Y, Inoue K, Sunagawa H, Fujii N, Inoue K, Oguma K (1995) Characterization of nontoxic-nonhemagglutinin component of the two types of progenitor toxin (M and L) produced by *Clostridium botulinum* type D CB-16. Microbiol. immun. 39: 457–465.

[21] Fujinaga Y, Inoue K, Shimazaki S, Tomochika K, Tsuzuki K, Fujii N, Watanabe T, Ohyama T, Takeshi K, Inoue K, Oguma K (1994) Molecular construction of *Clostridium botulinum* type C progenitor toxin and its gene organization. Biochem. biophys. res. commun. 205: 1291–1298.

[22] Nakajima H, Inoue K, Ikeda T, Fujinaga Y, Sunagawa H, Takeshi K, Ohyama T, Watanabe T, Inoue K, Oguma K (1998) Molecular composition of the 16S toxin produced by a *Clostridium botulinum* type D strain, 1873. Microbiol. immunol. 42: 599–605.

[23] Marvaud JC, Gibert M, Inoue K, Fujinaga Y, Oguma K, Popoff MR (1998) *botR/A* is a positive regulator of botulinum neurotoxin and associated non-toxin protein genes in *Clostridium botulinum* A. Mol. microbiol. 29: 1009–1018.

[24] Raffestin S, Dupuy B, Marvaud J, Popoff MR (2005) BotR/A and TetR are alternative RNA polymerase sigma factors controlling the expression of the neurotoxin and associated protein genes in *Clostridium botulinum* type A and *Clostridium tetani*. Mol. microbiol. 55: 235–249.

[25] Sagane Y, Watanabe T, Kouguchi H, Sunagawa H, Inoue K, Fujinaga Y, Oguma K, Ohyama T (1999) Dichain structure of botulinum neurotoxin: identification of cleavage sites in types C, D, and F neurotoxin molecules. J. protein chem. 18: 885–892.

[26] Watanabe T, Sagane Y, Kouguchi H, Sunagawa H, Inoue K, Fujinaga Y, Oguma K, Ohyama T (1999) Molecular composition of progenitor toxin produced by *Clostridium botulinum* type C strain 6813. J. protein chem. 18: 753–760.

[27] Sagane Y, Watanabe T, Kouguchi H, Sunagawa H, Inoue K, Fujinaga Y, Oguma K, Ohyama T (2000) Characterization of nicking of the nontoxic-nonhemagglutinin

components of *Clostridium botulinum* types C and D progenitor toxin. J. protein chem. 19: 575–581.

[28] Kouguchi H, Watanabe T, Sagane Y, Sunagawa H, Ohyama T (2002) *In vitro* reconstitution of the *Clostridium botulinum* type D progenitor toxin. J. biol. chem. 277: 2650–2656.

[29] Kouguchi H, Watanabe T, Sagane Y, Ohyama T (2001) Characterization and reconstitution of functional hemagglutinin of the *Clostridium botulinum* type C progenitor toxin. Eur. j. biochem. 268: 4019–4026.

[30] Mutoh S, Kouguchi H, Sagane Y, Suzuki T, Hasegawa K, Watanabe T, Ohyama T (2003) Complete subunit structure of the *Clostridium botulinum* type D toxin complex via intermediate assembly with nontoxic components. Biochemistry 42: 10991–10997.

[31] Mutoh S, Suzuki T, Hasegawa K, Nakazawa Y, Kouguchi H, Sagane Y, Niwa K, Watanabe T, Ohyama T (2005) Four molecules of the 33 kDa haemagglutinin component of the *Clostridium botulinum* serotype C and D toxin complexes are required to aggregate erythrocytes. Microbiology 151: 3847–3858.

[32] Nakamura T, Kotani M, Tonozuka T, Ide A, Oguma K, Nishikawa A (2009) Crystal structure of the HA3 subcomponent of *Clostridium botulinum* type C progenitor toxin. J. mol. biol. 385: 1193–1206.

[33] Ito H, Sagane Y, Miyata K, Inui K, Matsuo T, Horiuchi R, Ikeda T, Suzuki T, Hasegawa K, Kouguchi H, Oguma K, Niwa K, Ohyama T, Watanabe T (2011) HA-33 facilitates transport of the serotype D botulinum toxin across a rat intestinal epithelial cell monolayer. FEMS immunol. med. microbiol. 61: 323–331.

[34] Sagane Y, Watanabe T, Kouguchi H, Sunagawa H, Obata S, Oguma K, Ohyama T. (2002). Spontaneous nicking in the nontoxic-nonhemagglutinin component of the *Clostridium botulinum* toxin complex. Biochem. biophys. res. commun. 292: 434–440.

[35] Miyata K, Inui K, Miyashita S-I, Sagane Y, Hasegawa K, Matsumoto T, Yamano A, Niwa K, Ohyama T, Watanabe T. (2012) Crystallization and preliminary X-ray analysis of the *Clostridium botulinum* type D nontoxic nonhaemagglutinin. Acta cryst. F68: 227–230.

[36] Gu S, Rumpel S, Zhou J, Strotmeier J, Bigalke H, Perry K, Shoemaker CB, Rummel A, Jin R (2012) Botulinum neurotoxin is shielded by NTNHA in an interlocked complex. Science 335: 977–981.

[37] Fujinaga Y, Inoue K, Watanabe S, Yokota K, Hirai Y, Nagamachi E, Oguma K (1997) The haemagglutinin of *Clostridium botulinum* type C progenitor toxin plays an essential role in binding of toxin to the epithelial cells of guinea pig small intestine, leading to the efficient absorption of the toxin. Microbiology 143: 3841–3847.

[38] Inui K, Ito H, Miyata K, Matsuo T, Horiuchi R, Ikeda T, Watanabe T, Ohyama T, Niwa K (2010) Involvement of sialic acid in transport of serotype C1 botulinum toxins through rat intestinal epithelial cells. J. vet. med. sci. 72: 1251–1255.

[39] Sakaguchi G, Sakaguchi S, Kamata Y, Tabita K, Asao T, Kozaki S (1990) Distinct characters of *Clostridium botulinum* type A strains and their toxin associated with infant botulism in Japan. Int. j. food microbiol. 11: 231–241.

[40] Johnson EA, Bradshaw M (2001) *Clostridium botulinum* and its neurotoxins: a metabolic and cellular perspective. Toxicon 39: 1703–1722.

Chapter 7

Mycotoxin Decontamination Aspects in Food, Feed and Renewables Using Fermentation Processes

Grazina Juodeikiene, Loreta Basinskiene,
Elena Bartkiene and Paulius Matusevicius

Additional information is available at the end of the chapter

1. Introduction

Mycotoxins are secondary metabolites produced by a wide variety of filamentous fungi, including species from the genera *Aspergillus, Fusarium* and *Penicillium*. They cause nutritional losses and represent a significant hazard to the food and feed chain. Humans have long been exposed to mycotoxins by several different routes: directly, via foods of plant origin; by air (both indoors and outdoors); or indirectly, through foods of animal origin. The economically most important mycotoxins in foods and feeds are aflatoxins, ochratoxin A and patulin produced mainly by *Aspergillus* and *Penicillium* spp. and *Fusarium* toxins: type A and B trichothecenes, zearalenone and fumonisins (Table 1) [1, 2]. With the exception of T-2 and HT-2 and ergot or the sum of its individual alkaloids in unprocessed cereals and cereal products all mentioned mycotoxins have been regulated in the EU quite detailed in regulations (EC) No. 1881/2006 and 105/2010.

Occurrence of mycotoxin contamination in foods is more prevalent in the tropical and subtropical countries resulting in acute and chronic mycotoxicoses in humans and animals. Wild [3] reported that many West African countries, over 98% of the tested people were positive to aflatoxin-DNA adducts indicating aflatoxin exposure in the population. The importance of this situation is highlighted with an outbreak of aflatoxicosis as 2004 in Kenya [4] and a report on impaired child growth in Benin caused by post weaning exposure to aflatoxins [5]. Many of the developed countries have regulations for mycotoxins in grains and its products, because at least 60% of the food produced and consumed in the world originates from cereal crops such as rice, wheat, corn, barley,rye, sorghum and oats. However, the risk of mycotoxin exposures continues in the developing countries due to lack of food security, poverty and malnutrition [5, 6].

Mycotoxin		Commodity	Genus	Species	Disease/mode of action
Ergot alkaloids	Ergotamine, ergometrine Ergosine, ergocristine Ergocryptine, ergocornine	Cereals Pearl millet	*Claviceps*	*purpurea* *fusiformis*	Necrose of limbs – St. Anthony's fire Vasoconstrictive properties Gangrenous and colvulsive ergotism
Aflatoxin (AFT) (AFM)	B1, B2 G1, G2 M1, M2	Cereals, nuts Spices, figs Milk	*Aspergillus*	*flavus* *parasiticus* *nominus* *pseudotamarii*	Carcinogenic Liver cancer Immune suppressive
Ochratoxin (OTA)	A B	Cereals Wine Coffee Spices	*Aspergillus* *Penicillium*	*ochraceus* *carbonarius* *verrucosum*	Neurotoxic Nephrotoxic Kidney damage and cancer Immune suppressive
Type B trichothecenes	DON (Vomitoxin) ADON, NIV ANIV	Cereals	*Fusarium*	*graminearum* *culmorum* *acuminatum* *crookwellence* *arenaceum*	ATA (alimentary toxic leukopenia) Immunotoxic Acute toxicity
Type A trichothecenes	T-2 HT-2	Small grains Cereals	*Fusarium*	*poae* *sporotrichioides*	Acute toxicity linked to ATA Immunotoxic Immune system and hematological disorders
Zearalenone (ZEA)		Cereals	*Fusarium*	*graminearum* *culmorum* *cerealis* *avenaceum* *equiseti*	Estrogenic effects Reproductive disorders Effects endocrine system
Fumonisin (FUM)	FB1 FB2	Corn Cereals Multiple crops	*Fusarium* *(formerly)* *F.moniliforme*	*verticilloides* *fujikuroi* *proliferatum* *subglutinans*	Esophageal cancer Sphingolipid methabolism disruption Immune suppression

Table 1. Most important parent mycotoxins for cereal products, their occurrence, primary fungus producer and mode of action

Several approaches have been developed for decontamination of mycotoxins in foods [7, 8]. Principally, there are three possibilities to avoid the harmful effect of contamination of food and feed caused by mycotoxins:

1. prevention of contamination;
2. decontamination of mycotoxin-containing food and feed; and
3. inhibition or absorption of mycotoxin content of consumed food into the digestive tract.

The most obvious measure as to prevent mycotoxin production is in general to reduce the moisture content of the commodity to the equivalent of less than 0.65 water activity (e.g. for cereals <14.5% moisture by weight) direct after harvest. To minimize mycotoxin contamination can also be achieved as to limit bird and insect damage. Create also anaerobic conditions by storage and promote crop rotation to minimize the carry-over of moulds from one year to the next. Many of these recommendations can be found in documents of the Codex Alimentarius (CODEX STAN 193). Other useful documents are the codes of practice for the prevention and reduction of mycotoxin contamination in cereals, including annexes on OTA, ZEA, FUM and trichothecenes. Together with a suitable Hazard Analysis Critical Control Points (HACCP) and reliability procedures and following properly reporting systems such as RASFF (Rapid Alert System for Food and Feed) in Europe and RFR (Reportable Food Registry) in the US, health and safety measures in food and feed are optimum served by preventing and controlling the formation of mycotoxin.

The approach of prevention is doubtless also to breed cereals and other food and feed plants for resistance to mould infection and consequently exclude mycotoxin production. Particularly in breeding wheat and corn, significant improvement of resistance has been achieved. The slow-moving research however in the breeding field to culture resistant crops even through marker-assisted breeding (MAS),seems to be an endless task since up until now moulds have been a step ahead of breeders. The identification of microbial species (and genes coding enzymes degrading mycotoxins) allows transfer of these genes into plants and production of such enzymes by transgenic plants. In this way, the safety problems connected with the use of live microorganisms may be avoided. Another practical approach to prevention of mycotoxin contamination is the inhibition of the growth of molds and their production of mycotoxins. First, optimal harvesting, storage and processing methods, and conditions may be successful in prevention of mold growth. Although the primary goal is the prevention of mycotoxin contamination, mycotoxin formation appears to be unavoidable under certain adverse conditions. Treatment of grains by some chemicals to prevent mycotoxin formation is also possible but not desirable in view of a.o. the chemophobia of Europeans and the far-reaching legislation in the EU to restrict herbicides and pesticides with 'REACH' (REACH = Registration, Evaluation, Authorisation and restriction of Chemicals), effective as of 1 June 2007.

Most of these compounds work by inhibiting fungal growth. For example, approximately one hundred compounds have been found to inhibit aflatoxin production. Two extensively studied inhibitors of aflatoxin synthesis are dichlorvos (an organophosphate insecticide) and caffeine. As reported in reference [8] some surfactants have been found to suppress the growth of *Aspergillus flavus* and aflatoxin synthesis. When contamination cannot be prevented, physical and chemical decontamination methods have been employed (such as for ergot physically) with varying success in the past, principally for feed. Whichever decontamination strategy is used, it must meet some basic criteria:

1. the mycotoxin must be inactivated or destroyed by transformation to non-toxic compounds;
2. fungal spores and mycelia should be destroyed, so that new toxins are not formed;
3. the food or feed material should retain its nutritive value and remain palatable;
4. the physical properties of raw material should not change significantly; and it must be economically feasible / the cost of decontamination should be less than the value of contaminated commodity.

Mycotoxin decontamination by physical and chemical methods has been reviewed extensively in several papers [7, 9, 10]. Partial removal of mycotoxin may be achieved by dry cleaning of the grain and in the milling process, as well. Milling led to a fractionation, with increased level of mycotoxin in bran and decreased level in flour. The majority of mycotoxins are heat-stable so heat treatment, usually applied in food technology, does not have significant effect on the mycotoxin level. Efforts were made in several countries to find an economically acceptable way of destruction of mycotoxins into non-toxic products using different chemicals such as alkali and oxidative agents. Although such treatment reduces nearly completely the mycotoxin concentration, these chemicals also cause losses of some nutrients and such treatment is too drastic for e.g. grain destined for food uses. Many physical adsorbents have been studied and are available as commercial preparations as animal feed additives. However, many of these adsorbents can bind to only a small group of toxins while showing very little or no binding to others [9]. Although the different methods used at present have been to some extent successful, most methods have major disadvantages, starting with limited efficacy to losses of important nutrients and generally with high costs. Because the EU does not allow chemical treatment methods anyway, manoeuvrability is limited.

More recently, biological decontamination and biodegradation of mycotoxins with microorganisms or enzymes have been used [11, 12]. Many species of bacteria and fungi have been shown to enzymatically degrade mycotoxins [13-19]. In this case no harmful chemicals were used, so no significant losses in nutritive value and palatability of decontaminated food and feed occurred. Today, ruminants appear to be a promising potential source of microbes or enzymes for use in the biotransformation of fungi and/or there mycotoxins they create. Biological decontamination of fungi/mycotoxins by microorganisms is reviewed in some papers [8, 13, 16, 20, 21], however, there are no many reviews on decontamination of fungi/mycotoxins by microorganisms involved in food fermentation and its implications. Fermentation is one of the easiest and cheapest means of food preservation in addition to imparting nutritional and organoleptic benefits to fermented foods. Fermentation is effected by the natural microbiota of raw materials, microorganisms attached to the fermentation equipments or from externally added starter cultures. Yeasts, especially *S. cerevisiae* and *Candida krusei,* and lactic acid bacteria (LAB) occur as part of natural microbial population in spontaneous food fermentation and as starter cultures in the food and beverage industry [22]. In addition, yeasts have been fed to animals for more than a century and commercial yeast products are being specifically produced on a commercial scale for animal feeding [23]. Hence, yeasts and LAB have immense potential as tools in tackling the problem of fungi/mycotoxins in cereal-based foods and in animal feed.

However, question remains on the toxicity of products of enzymatic degradation and undesired effects of fermentation with non-native microorganisms on the quality of food [21]. In this chapter the biocontrol of aflatoxins, trichothecenes (type A and B), zearalenone, fumonisins, ochratoxins and patulin by bacteria, fungi and yeasts will be discussed more in depth to fill the existing gaps and to develop further proper management practices using biocontrol agents to ensure food safety and to protect consumer's health.

2. Possible approaches of microorganisms and enzymes to biodegradation of mycotoxins

One of the most frequently used strategies for biodegradation of mycotoxins includes isolation of microorganisms able to degrade the given mycotoxin and treatment of food or feed in an appropriate fermentation process. From a food safety point-of-view, fermentation with microorganisms commonly used in food production (fermentation with lactic acid bacteria, alcoholic fermentation, traditional fermentation of vegetable protein used in South Asia, etc.) should be preferred.

Knowledge of enzymes that take part in degradation of mycotoxins opens some new approaches:

1. the production of genetically modified species of microorganisms commonly used in food production and their use for production of enzymes mentioned above; or
2. the transfer of genes coding for these enzymes to transgenic plants and use the plants for production of mycotoxin degrading enzymes.

In staple food such as bread and bakery products in the flour sector, yeast and lactobacilli now play an important role. It thus stands to reason that the same microbes and enzymes are the first to have been considered for use as detoxifying or decontaminating agents. This type of biodegradation could therefore prove a useful strategy for partially overcoming the problem of some mycotoxins. Indeed, this already takes place in bread and in sourdough processes [24]; and OTA in food can also undergo biodegradation [25] and certain antagonistic yeast strains can substantially degrade OTA. This might offer new possibilities for reducing this mycotoxin in bread and bakery products and their raw materials [15]. The use of enzymes or engineered micro-organisms (provided that these are allowed by legislation) as processing aids in the bread and bakery sector would also prove beneficial. Genetic engineering technologies will improve the efficiency with which enzymes can be produced from these organisms, and will allow the production of engineered organisms which have the target genes. They will additionally increase the availability and bioavailability, and will improve the quality of the end product.

2.1. Inhibition of mycotoxins biosynthesis by lactic acid bacteria

Several papers dealing with the inhibition of mycotoxin biosynthesis by LAB have focused on aflatoxins [26]. During cell lysis, it is possible that LAB releases molecules that potentially inhibit mould growth and therefore lead to a lower accumulation of their mycotoxins [27]. These "anti-mycotoxinogenic" metabolites could also be produced during LAB growth. Gourama [28], using a dialysis assay, demonstrated the occurrence of a metabolite that inhibits aflatoxin accumulation in *Lactobacillus* cell-free extracts. It was suggested that this inhibition of aflatoxin biosynthesis was not the result of a hydrogen peroxide production or a pH decrease [29]. These findings were consistent with those of Gourama, who suggested that inhibition of aflatoxin biosynthesis by *Lactobacillus* cell free supernatants was probably due to specific bacterial metabolites. Coallier-Ascah and Idziak [30] reported a significant reduction of aflatoxin biosynthesis by *Lactobacillus* cell free supernatants and suggested that this inhibition was related to a heat stable, low-molecular-weight inhibitory compound. Although *Lactobacillus* spp. were found to delay aflatoxin biosynthesis, other lactic strains such as *L. lactis* were found to stimulate aflatoxin accumulation [31].

2.2. Decontamination of mycotoxins using microorganisms by binding or degradation

Biological detoxification of mycotoxins works mainly *via* two major processes, sorption and enzymatic degradation, both of which can be achieved by biological systems. Recently a critical review on biological detoxification by Dalié *et al.* [32] summarized different and interesting aspects of the biological detoxification of mycotoxins. Microorganism detoxification can be performed in many different ways [33]:

1. the entire organism can be used as a starter culture, as in the fermentation of beer, wine and cider, or in lactic acid fermentation of vegetables, milk and meat;
2. the purified enzyme can be used in soluble or immobilized (biofilter) forms;
3. the gene encoding the enzymatic activity can be transferred and overexpressed in a heterologous system; interesting candidates for this application include yeasts, probiotics and plants.

2.2.1. Binding by yeast and LAB

Live microorganisms can absorb either by attaching the mycotoxin to their cell wall components or by active internalization and accumulation. Dead microorganisms too can absorb mycotoxins, and this phenomenon can be exploited in the creation of biofilters for fluid decontamination or probiotics (which have proven binding capacity) to bind and remove the mycotoxin from the intestine.

Yeast and LAB cells are known to bind different molecules such as killer toxins and metal ions on complex binding structures on the cell wall surface [34-36]. Yeast cell wall is known to bind sterols from the medium and the binding molecule was identified as the cell wall mannan [37]. It is confirmed that removal of mycotoxins is by adhesion to cell wall components rather than by covalent binding or by metabolism, as the dead cells do not lose their binding ability [38-41]. Reported literature indicates that mannan components of the cell wall play a major role in aflatoxin binding by *S. cerevisiae*. However, more kinetic studies with live and physico chemically modified intact *S. cerevisiae* cells are needed to assess the role of different components of cell wall in mycotoxin binding. Raju and Devegowda [40] have shown that mannans can also bind other mycotoxins such as ochratoxin A and T-2 toxin. Yiannikouris *et al.* [42, 43] mentioned the zearalenone binding to β-D-glucans and Freimund *et al.* [44] shows that modified yeast β-1,3-glucan showed excellent binding with T-2 toxins in addition to zearalenone mycotoxins. This indicates the possibilities of more than one target for mycotoxin binding on the cell wall. The fact that binding of ochratoxins is also enhanced when yeast cells are replaced with physically extracted cell walls [9] or heat treated cells and a very rapid nature of toxin removal from liquid medium [45] indicates the adsorption physical nature of ochratoxin binding also. A comprehensive study of key interactions between zearalenone and β-D-glucans, based on NMR and X-ray diffraction studies, was reported in reference [46]. Interestingly, β-1,3-D-glucan chains favor a very stable intra-helical association with zearalenone, nicely stabilized by β-1,6-D-glucans side chains (Figure 1) [47]. Both hydrogen bonds and van der Waals interactions were precisely identified in the complex and could thus be proposed as driving interactions to monitor the association between the two molecules. It appears that the carbohydrate components are common sites for binding, with different toxins having different binding sites.

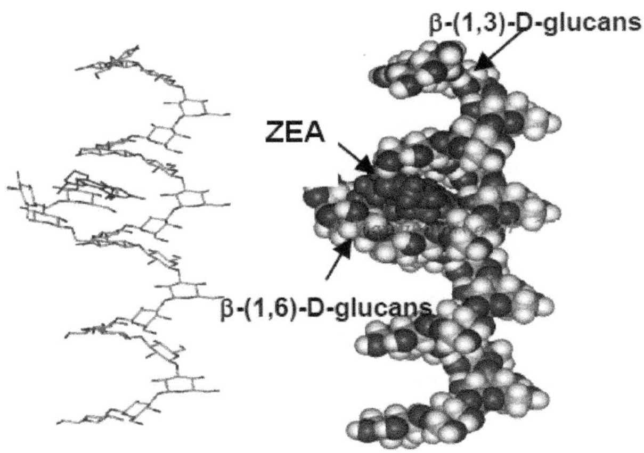

Figure 1. Computer-generated views of the energy-minimized structure of the docking of the most favorable conformation of zearalenone (ZEA) into the singlehelix of β-1,3-D-glucan chain branched with β-1,6-D-glucan side chain [47]

Differences between strains of LAB with respect to aflatoxin binding indicates that binding ability is highly strain specific [48]. In some of the earlier studies, LAB are considered to be inefficient binders of aflatoxin B_1 [26, 30]. This may be due to the strains used in those studies were binding low amounts of aflatoxins occurs. Binding of aflatoxin B_1 by viable or heat and acid treated bacteria depend on the initial concentration of the toxin [49]. Haskard *et al.* [50] studied the mechanism of binding of aflatoxins to *L. rhamnosus* using enzyme treatments and showed that binding is predominantly to carbohydrate and to some extent protein components in the cell wall. Treatment with urea considerably decreased the binding indicating the major role of hydrophobic interactions in binding. Electrostatic interactions are shown to play only a minor role as shown by treatments with NaCl and $CaCl_2$. Binding of aflatoxin B_1 was not affected by pH, but had considerable effect on binding of aflatoxin B_{2a}, indicating that different metabolites of the same mycotoxins with minor differences may show considerable differences with respect to binding mechanisms. Recently, it has been shown that peptidoglycan or the structures closely associated with peptidoglycan might be the most likely carbohydrate involved in aflatoxin B_1 binding process [51]. Treatment with different specific proteases did not have any specific effect on the binding and a possible explanation for the effect seen earlier with pronase may be due to the release of other components associated with the proteins, thus affecting the binding process. Another mechanistic study [52] showed that binding of aflatoxins to the cell surface is considerably strong. Viable cells of *L. rhamnosus* strains LGG and LC105 retained 38 and 50% (w/w) respectively of the bound toxins after repeated washings with water. However, non-viable (heat and acid treated) cells retained 66–71% (w/w) of the toxin, indicating a higher stability of the complex. This higher binding of heat and acid treated cells was attributed to the better access of groups in the treated cells. Autoclaving and sonication did not release any detectable toxin from pre-washed cells indicating a high stability of the complex. Binding of aflatoxin B_1 was found to be unaffected at a pH scale ranging from 2.5 to 8.5, suggesting the absence of a cation exchange mechanism and when the cell was treated with organic solvents, bound toxin was quickly extracted suggesting a major role of hydrophobic interaction in the binding [50]. Accessibility of bound aflatoxin to specific monoclonal antibody indicates the surface nature of binding [52].

2.2.2. Degradation or biotransformation

Another approach to the biological decontamination of mycotoxins involves their degradation or conversion into less toxic molecules by enzymes and selected microorganisms. Initial research in the field of mycotoxin biotransformation started 40 years ago. It has been demonstrated that some microorganisms produce enzymes that could alter the structure of mycotoxins and/or proteins that can conjugate these compounds, making them less active as pathogenic agents [53]. However, only few microorganisms have shown the capacity of degrading mycotoxins.

Karlovsky [20] has shown that the 12,13-epoxide ring of trichothecenes is responsible for their toxicity and that a reductive de-epoxidation caused by specific enzymes (deepoxidases) entails a significant loss of toxicity. For the elimination of the toxic effects of trichothecenes (deoxynivalenol or DON and T-2 toxin are the most well known members) *Eubacterium* BBSH 797 was isolated. The enzymes produced by this organism (for example epoxidases) play an important role by enabling the specific disruption of the toxic epoxy ring possessed by this group of mycotoxins (Figure 2). The destruction of deoxynivalenol may also be correlated to the action of oxidative enzymes that would catalyze the opening of the epoxide ring [54, 55].

Garvey *et al.* [56] have reported that enzymes known as trichothecene 3-*O*-acetyltransferases have the ability to modify DON by converting it to an acetylated derivative. These enzymes are also produced by fungi in the genus *Fusarium,* and are encoded by the genes TRI101 or TRI201 [57, 58]. The enzymatic modification involves the attachment of an acetyl group to the C-3 hydroxyl moiety of the trichothecene molecule [56], forming the derivative 3-acetyldeoxynivalenol (3ADON) [59]. Expression of TRI101 has been shown to reduce the phytotoxic effects of trichothecenes in tobacco and rice [60, 61], and to decrease the inhibitory effects of trichothecenes on the growth of *Saccharomyces pombe* [59] and *Chlamydomonas reinnardtii* [62].

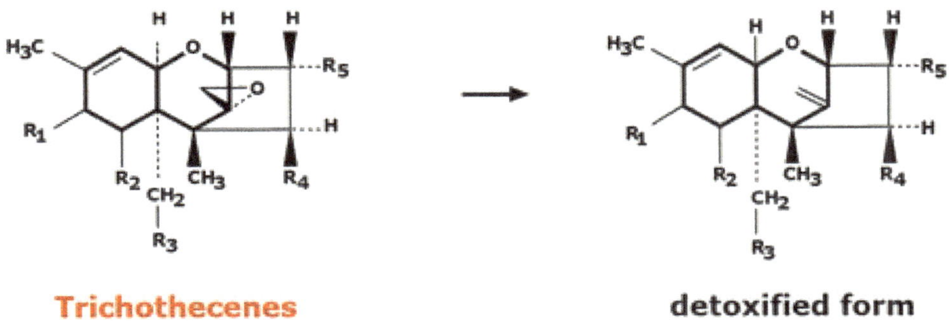

Figure 2. Biotransformation of trichothecenes into the detoxified forms (de-epoxy structures)

Very few scattered reports are available on the use of yeast and fungal strains for degradation zearalenone. *S. cerevisiae* strains were tested for their ability to degrade zearalenone in Sabouraud broth. Two strains were capable to degrade zearalenone totally, one strain decreased the mycotoxin concentration up to 25% and one strain up to 75% of the original amount. Later, the non-pathogenic yeast *T. mycotoxinivorans* MTV was isolated, described and patented for its ability to degrade zearalenone and ochratoxin A. For the elimination of zearalenone's negative effects it is vital that the lactone ring within the molecule is destroyed. This reaction is once again mediated by enzymes (e.g. esterases) (Figure 3). In doing so, zearalenone's resemblance with the sexual female hormone estradiol is lost and therefore impairment of the reproduction system is avoided.

In the case of ochratoxin A (OTA), two pathways may be are involved in OTA microbiological degradation [25]. First, OTA can be biodegraded through the hydrolysis of the amide bond that links the L-β-phenylalanine molecule to the ochratoxin alpha (OTα) moiety (Figure 4). Since OTα and L-β-phenylalanine are virtually non-toxic, this mechanism can be considered to be a detoxification pathway. Second, a more hypothetical process involves OTA being degraded via the hydrolysis of the lactone ring. In this case, the final degradation product is an opened lactone form of OTA, which is of a similar toxicity to OTA when administered to rats. However, it is less toxic to mice and *Bacillus brevis.* Although this is hypothetical, it is likely to occur since microbiological lactonohydrolases, which undertake a similar transformation, are common.

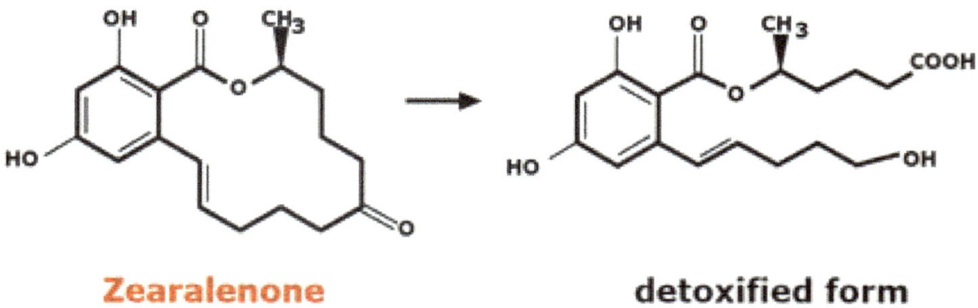

Figure 3. Biotransformation of zearalenone into a detoxified form (ZOM-1)

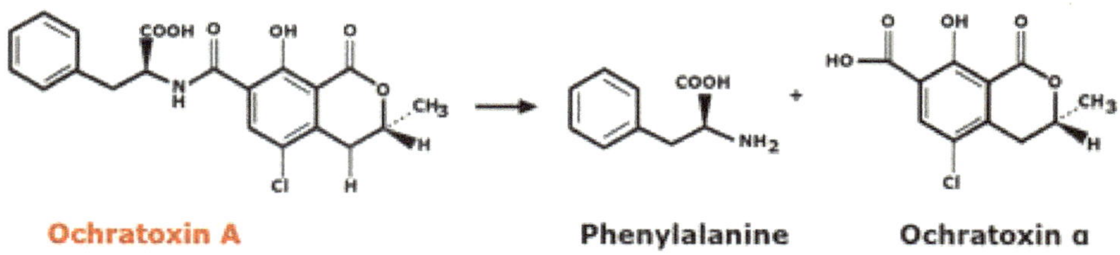

Figure 4. Biotransformation of ochratoxin A into a detoxified form (ochratoxin alpha)

As a conclusion, mycotoxin risk management strategies must comprise several components. The elimination of adsorbable mycotoxins is possible to be done through adsorption; however, for the elimination of the toxicity of non-adsorbable mycotoxins, such as zearalenone, ochratoxins and trichothecenes biotransformation is crucial. Biotransformation, which is enabled by enzyme-producing microorganisms, allows the conversion of the toxic structure of mycotoxins into non-toxic, harmless metabolites.

3. Possible approaches of microorganisms and enzymes for biodegradation of mycotoxins in food

Mycotoxins frequently contaminate the food raw materials such as cereals, fruits, nuts, spices, milk and meat at various levels. Hence, for the food industry, it has always been an uphill task to keep the mycotoxin levels under check in the products, because mixing high contaminated commodities with low contaminated commodities to reach a level below the regulatory maximum limit is standard not allowed in the EU. Normally, low mycotoxin levels in the food are ensured by using mycotoxin-free, or raw materials with low levels of mycotoxins. In spite of these efforts, sporadically mycotoxin contamination is reported in the food products such as wine, beer, milk and milk products [63-65]. Strains of *S. cerevisiae* and LAB with high mycotoxin binding abilities can be used as part of the starter cultures in the fermentation of food and beverages, or heat treated cell walls or purified components of *S. Cerevisiae* and LAB can be used as additives in small quantities without compromising the characteristics of the final product. More importantly, *S. cerevisiae* and LAB are the major microorganisms involved in food fermentations in tropical countries with high levels of mycotoxin contamination in their foods. Strains of *S. cerevisiae* and LAB isolated from native fermented foods can be used as starter cultures with additional capacities to decontaminate mycotoxins in the food.

3.1. Aflatoxins (AF)

As the first mycotoxins being discovered were the aflatoxins (AF), also the first being targeted to be screened for microbial degradation were the aflatoxins. Several examples of the detoxification of the most common and important mycotoxins are reviewed. Almost 40 years ago, several species of microorganisms – including yeasts, moulds, bacteria, actinomycetes and algae – were screened for detoxification activity; based on this studies only one isolate was found, *Flavobacterium auranotiacum,* which significantly removed aflatoxin from a liquid medium.

Later AF decontamination during fermentation was reported in several cases. About 50% reduction in aflatoxins B_1 and G_1 (AFB_1 and AFG_1) has been reported during an early stage of miso fermentation. It was attributed to the degradation of the toxin by microorganisms. Significant losses of AFB_1 and OTA were observed during beer brewing [66]. Detoxification of AFB_1 occurred during the fermentation of milk by LAB and in dough fermentation during breadmaking. Govaris *et al.* [67] showed that aflatoxin M_1 (AFM_1)concentrations fell between 13 and 22 % when cows' milk was fermented to produce yoghurt and by 16 and 34 % after storage for yoghurts of pH 4.6 and 4.0 respectively.

Digestive tract microorganisms are able to reduce mycotoxin levels not only by binding and removal but also by detoxification. Most data dealing with the effects of LAB on the accumulation of mycotoxins are related to aflatoxin-producing moulds. Wiseman and Marth [68] revealed the existence of an amenable relationship between *L. lactis* and *A. parasiticus*. When these authors added the spores of *A. parasiticus* to a 13-day-old culture of *L. lactis,* they observed the entire repression of AF production. When the fungal spore suspension and the LAB were inoculated simultaneously, an increase in AF production was observed. In contrast, Coallier-Ascah and Idziak [30] showed an inhibition of AF accumulation when both microorganisms were simultaneously cultivated in Lab-Lemco tryptone broth (LTB). Addition of glucose to the cultivation medium during the conidiation phase of the mould did not restore the production of AF. Several LAB have been found to be able to bind AFB_1 *in vitro* [69, 70], with an efficiency depending on the bacterial strain [71]. El-Nezami with co-workers [72] have evaluated the ability of five *Lactobacillus* strains (*L. rhamnosus* GG, *L. rhamnosus* LC705, *L. acidophilus, L. gasseri,* and *L. casei*) to bind aflatoxins *in vitro* and have shown that probiotic strains such as *L. rhamnosus* GG and *L. rhamnosus* LC705 were very effective for removing AFB_1, removing as much as 80% (w/w) of toxin immediately. Similar observations were also made with *Flavobacterium aurantiacum* (now known as *Nocardia corynebacterioides*), although there was a small amount of toxin which was degraded in live cells [73]. Later, many other strains of LAB were shown to bind AF in a strain specific manner [74]. *L. rhamnosus* strains GG and LC705 most effectively bound AFB_1 than AFB_2 and AFG_1 [75]. In addition, the two strains showed similar AFB_1 binding, even though they showed differences with respect to other metabolites. According to Coallier-Ascah and Idziak [30], the inhibition of AF accumulation was not related to a pH decrease but rather to the occurrence of a low-molecular-weight metabolite produced by the LAB at the beginning of its exponential phase of growth. Inhibition of aflatoxin production by other LAB belonging to the genus *Lactobacillus* was also reported [29]. It was assumed that this inhibition resulted from the production of a metabolite different from hydrogen peroxide or organic acid [28]. Haskard wich co-wokers [52] demonstrated that *L. rhamnosus* GG and *L. rhamnosus* LC705 were able to eliminate AFB_1 from the culture medium by a physical process.

Several studies have suggested that the antimutagenic and anti-carcinogenic properties of probiotic bacteria can be attributed to their ability to non-covalently bind hazardous chemical compounds such as AF in the colon [70, 76]. Both viable and non viable forms of the probiotic bacterium *L. rhamnosus* GG effectively removed aflatoxin B_1 from an aqueous solution [76]. Since metabolic activation is not necessary, binding can be attributed to weak, non-covalent, physical interactions, such as association to hydrophobic pockets on the bacterial surface [50].

3.2. Ochratoxin A (OTA)

The major OTA producers in food and feed products are considered to be *A. alliaceus, A. carbonarius, A. ochraceus, A. steynii, A. westerdijkiae, P. nordicum* and *P. verrucosum* [77]. These are mainly associated with agricultural crops preharvest, or in post harvest storage situations. Biological methods use microorganisms, which can decompose, transform or adsorb OTA to detoxify contaminated products or to avoid the toxic effects when mycotoxins are ingested. Several bacterial, yeast and filamentous fungal species are able to biodegrade OTA [25]. After success in clarifying the mechanism and degradation products of OTA, three directions in recent research may be observed (1) possibilities of bacterial and yeast degradation, (2) study of molds able to degrade this mycotoxin and (3) identification and isolation of enzymes taking part in the degradation process.

Lactobacillus strains were demonstrated to eliminate 0.05 mg OTA l^{-1} added to culture medium – in particular, *L. bulgaricus, L. helveticus, L. acidophillus,* eliminated up to 94%, 72% and 46%, respectively, of OTA [78]; *L. plantarum, L. brevis* and *L. sanfrancisco* were reported to eliminate 54%, 50% and 37%, respectively, of 0.3 mg OTA l^{-1} after 24 h of incubation [79]. It is now generally accepted that OTA adsorption to the cells walls is the predominant mechanism involved in this OTA detoxification phenomenon by LAB. For example, adsorption effects were claimed by Turbic *et al.* [48], who found that heat and acid treated cells from two *L. rhamnosus* strains were more effective at removing OTA from phosphate buffer solutions than viable cells. The strains removed 36% to 76% in the buffer solution (pH

7.4) after 2 h at 37°C. Similarly, Piotrowska and Zakowska [80] verified that *L. acidophilus* and *L. rhamnosus* caused OTA reductions of 70% and 87% of 1 mg OTA l^{-1} after five days at 37°C, and that significant levels of the OTA were present in the centrifuged bacteria cells. Other LAB (*L. brevis, L. plantarum* and *L. sanfranciscencis*) also produced smaller decreases on OTA (approximately 50%). Del Prete *et al.* [81] tested 15 strains of oenological LAB in order to determine the *in vitro* capacity to remove OTA, and reported *Oenococcus oeni* as the most effective, with OTA reductions of 28%. The involvement of cell-binding mechanisms was confirmed as (1) up to 57% of the OTA absorbed by the cells was recovered through methanol extraction from the bacteria pellets; (2) crude cell-free extracts were not able to degrade OTA; and (3) degradation products were not detected. Nevertheless, some authors consider that metabolism may also be involved. Fuchs *et al.* [82] confirmed that viable cells of *L. acidophilus* removed OTA more efficiently then unviable. *L. acidophilus* strain was able to decrease ≥95% the OTA in buffer solutions (pH 5.0) containing 0.5 and 1 mg OTA l^{-1} when incubated at 37°C for 4 h. In addition, a detoxification effect was also demonstrated since pre-incubation of OTA with this strain reduced OTA toxicity to human derived liver cells (HepG2) [83]. Other *L. acidophilus* strains demonstrated only a moderate reduction in OTA contents suggesting that the effect was strain specific. In summary, some LAB adsorbs OTA by a strain specific cell-wall binding mechanism, although some undetected catabolism can also be involved. The detection of this OTA catabolism may only be possible with radiolabeled OTA. The potential of LAB as mycotoxin decontaminating agents has been studied in different fermentation processes and reviewed. The OTA content, its fate during wine-making and possibilities of its degradation have been intensively studied. Overviews concerning the presence and fate of this mycotoxin in grapes, wine and beer were published by Mateo *et al.* [84] and Varga and Kozakiewicz [85]. Although the decrease of OTA content in liquid phase during vinification process is observed by the majority of researchers, reports are controversial regarding the mechanism of OTA removal. Is it a result of malolactic fermentation due to the action of lactic acid bacteria [86], or is it adsorption to yeast cell walls [87]. *Streptococcus salivarius* subsp. *thermophilus, Bifidobacterium bifidum,* and yogurt bacteria have completely reduced OTA levels in milk samples containing 0.05 and 0.1 mg l^{-1}; *L. delbrueckii* subsp. *bulgaricus* completely reduced OTA level in milk samples with 0.5 mg l^{-1} [88].

Several reports of OTA biodegradation by *S. cerevisiae* and other yeast have been published, and most of the effects detected and reported are from wall adsorption mechanisms. *S. cerevisiae* was claimed to biodegrade 41% of 0.3 mg OTA l^{-1} after 24 h at 30 °C, but details were not provided about the mechanism involved [73]. Similarly, Böhm *et al.* claimed that some strains degraded up to 38% of 0.05 mg OTA l^{-1} without describing any resulting degradation metabolites [78]. On the other hand, the adsorption of OTA by oenological *Saccharomyces* strains was demonstrated by Bejaoui and co-authors, since they verified that heat and acid treated cells could bind significantly more OTA than viable ones [45]. Viable yeast bound up to 35% and 45% of the OTA, depending on the medium and strain, while heat and acid treated cells bound a maximum of 75%. Additionally, yeast are reported to reduce OTA in alcoholic fermentation processes such as brewing or vinification. During wort fermentation, yeasts adsorbed a maximum of 21% of the added OTA [25]. Also, almost 30% of the added OTA was removed after extended contact with yeast biomass [89]. Cecchini and co-authors verified during vinification trials that up to 70% of OTA could be removed from wine and that a significant percentage of the removed OTA was found in yeast lees [90]. Adsorption assays that used several yeasts products or fractions were also carried out in order to understand and explain the mechanisms involved. Moruno and co-authors tested the capacity of active dried yeasts and yeast lees to remove OTA from wines and reported a reduction of approximately 70% when yeast lees were used [91]. The *in vitro* biosorption of OTA by vinasse containing yeast cell walls, purified yeast β-glucan and dried yeast cell wall fractions was studied [92]. Dried yeast cell wall fractions were reported to be the most efficient at adsorbing OTA. Several reports explained this phenomena by relating it to yeast β-D-glucans [93], glucomannans [94] and mannanoligosaccharide [95]. On the other hand, some studies emphasized the involvement of biodegradation mechanisms. *Trichosporon*, *Rhodotorula* and *Cryptococcus* demonstrated an ability to biodegrade OTA through the cleavage of the amide bond and releasing OTα [96]. In this study, the most effective strain degraded up to 100% of 0.2 mg OTA l^{-1} after five hours of incubation at 35 °C. This yeast was classified subsequently as the novel species *Trichosporon mycotoxinivorans* due to its excellent ability to detoxify OTA and ZEA [97]. However, a recent study recognized *T. mycotoxinivorans* as a novel human pathogen associated with cystic fibrosis and the death of a patient with histologically documented *Trichosporon* pneumonia: this obviously raises safety issues on its practical use [98]. A *Phaffia rhodozyma* strain was also able to degrade 90% of 7.5 mg OTA l^{-1} after 15 days at 20 °C [99]. In this study, the conversion of OTA into OTα and the adsorption of OTA into viable and heat-treated cells was observed. More recently, *Aureobasidium pullulans* was reported to degrade OTA through the hydrolysis of the amide bond since OTα was detected [100]. The use as a biocontrol agent was also assessed as a reduction of OTA in grapes and wine was reported. However, the fungus appears to be involved in human disease and this issue needs to be resolved before more general use can be recommended [101].

Reports about the capacity of some filamentous fungi to biodegrade OTA can also be found. Some strains of *Aspergillus fumigatus*, *A. japonicus*, and *A. niger* have been reported to be able to degrade OTA to less toxic products such as ochratoxin R [102]. *A. fumigatus*, *A. japonicus* and *A. niger* degraded 2 mg OTA l^{-1} after 10 days of incubation at 30 °C. OTα was detected and further degradation into an unknown compound was observed. *A. niger* and other filamentous fungi have also shown to biodegrade OTA completely or partially, after growth in 1 mg OTA l^{-1} for six days at 25 °C [103]. OTα was detected, particularly in the assays performed with *A. niger* and other black aspergilli. An unidentified biodegradation metabolite was observed in the assays carried out with *A. ochraceus* which did not produce OTA, and some *A. wentii* strains. Additionally, *R. homothallicus*, *R. oryzae*, *R. stolonifer* and other *Rhizopus* species degraded more than 95% of 7.5 mg OTA l^{-1} after 16 days of incubation at 25 °C [18]. OTα was also detected in this study. Later, the excellent capacity of some black aspergilli to degrade OTA was confirmed: some *A. carbonarius*, *A. japonicus*, and *A. niger* strains degraded more than 80% of 2 mg OTA l^{-1} [104]. More recently, the capacity of *Botrytis cinerea* to degrade OTA was confirmed with reductions of 24.2% to 26.7% [105]. This provided an explanation for the low OTA contamination of noble rot and late-harvest wines. The white rot fungus *Pleurotus ostreatus* could degrade OTA (77%) and OTB (97%) when growth on contaminated barley by solid state fermentation, with OTα being detected from OTA biodegradation [106]. *Rhizopus japonicus* and *Phanerochaete chrysosporium* were also shown to biodegrade OTA to the lesser extents of 38% and 36%, respectively.

Several enzymes may be involved in the microbiological degradation of OTA. However, little information is available and very few have been purified and characterized. The first reported protease able to hydrolyze OTA was carboxypeptidase A (CPA) from bovine pancreas [107]. Subsequently, a screening study which included several commercial hydrolases, verified that a crude lipase product from *A. niger* was able to hydrolyze OTA via the amide bond [108]. The enzyme was purified by anion exchange chromatography and was demonstrated to cleave OTA and p-nitrophenyl palmitate, a specific lipase substrate. Several proteolytic preparations were also studied, which were involved in the hydrolysis of OTA to OTα. These included protease A from *A. niger*, pancreatin from porcine pancreas and to a lesser extent, prolyve PAC from *A. niger* [109]. Additionally, the production and purification of an *A. niger* cell-free crude enzyme preparation that demonstrated a significant capacity to cleave the amide bond of OTA was reported. The OTA-degrading enzyme involved was purified by anion exchange chromatography and characterized [110]. This enzyme showed higher OTA-degrading activity then CPA at pH 7.5 and 37 °C, and was inhibited by EDTA, which is a specific inhibitor of metalloproteases. It was found that carboxypeptidase Y (CPY) from *S. cerevisiae* is also able to hydrolyze OTA with optimal activity at pH 5.6 and 37 °C. However, the specific activity of CPY is very low as indicated by the OTA hydrolyzation reaction being very slow. Nevertheless, after five days of incubation, CPY converted 52% of the OTA present in the reaction assay into OTα. This activity is sufficient to reduce significantly levels of OTA during wine or beer fermentation, since these processes take several days to complete. Hence, a biodegradation pathway is possible for *S. cerevisiae* in addition to the OTA adsorption phenomenon. It is necessary to consider that CPY is a vacuolar exopeptidase where OTA enters the yeast cells before it is catabolized. However, the *S. cerevisiae* wall-binding properties can make difficult OTA uptake. Although the results of these studies look very promising for reducing OTA contamination, studies on model systems do not guarantee the degradation of OTA *in situ*, using food. Further studies are needed to characterize the products of degradation and to investigate the activity of these microorganisms and enzymes in foods.

3.3. Patulin (PAT)

PAT contamination of apple and other fruit-based foods and beverages is an important food safety issue due to the high consumption of these commodities. PAT contamination is considered of greatest concern in apples and apple products; however, this mycotoxin has also been found in other fruits, such as pears, peaches, strawberries, blueberries, cherries, apricots and grapes as well as in cheese [8]. A number of studies have shown that PAT is generally unstable during fermentation so that products such as cider are usually free of PAT. It is likely that when PAT is reported in cider this is the result of the addition of apple juice to produce 'sweet cider'.

The initial studies concerning degradation of PAT by actively fermenting yeasts were reported by Stinson and co-authors [111]. However, authors were not able to chemically characterize the products of degradation. More recently, Moss and Long [112] reported that under fermentative conditions, the commercial yeast *S. cerevisiae* transformed PAT into ascladiol. They also showed that PAT was unstable in a study of the fate of [14C]-labelled PAT during the alcoholic fermentation of apple juice with *S. cerevisiae.* High-performance liquid chromatography (HPLC) analysis of the fermentations showed the appearance of two major metabolites, probably E- and Z-ascladiol. In a recent study [113] the ability of *Gluconobacter oxydans* to degrade PAT was investigated and the degradation products of this mycotoxin determined. More than 96% of PAT was degraded after 12 h treatment, due to change of chemical structure (opening of the pyran ring). The degradation product was confirmed to be ascladiol. The genus *Gluconobacter,* whose taxonomy is made up of five different species which have no health risk, are commonly used in food manufacturing [114]. Apple juice inoculated with this bacterium and incubated for 3 days still tasted like juice and was drinkable. However, keeping in mind the toxicity of ascladiol and eventual unsatisfactory organoleptic properties of alcoholic apple (fruit) juice (apple wine), the use of this bacterium at the industrial level needs additional investigation. In screenings for PAT detoxifying bacteria, a bacterium from fermented sausage has been isolated; it was identified as *L. plantarum,* and it significantly reduced PAT levels via an intracellular enzime. [8]

3.4. Trichothecenes (type A and B) – *Fusarium* toxins such as DON and T-2/HT-2

Several microorganisms have been found that can degrade deoxynivalenol (DON) and T-2 toxin. On the basis of morphological and phylogenetic studies, the degrader strain was classified as a bacterium belonging to the *Agrobacterium Rhizobium* group. *L. rhamnosus* strains LGG and LC 705 and *Propionibacterium freudenreichii* (PJ) were also shown to effectively bind some of the *Fusarium* toxins such as DON, 3-acetyldeoxynivalenol (3ADON), nivalenol (NIV), fusarenon-X (FUS-X), diacetoxyscirpenol (DAS), T-2 toxin, and HT-2 toxin [115]. The strains showed considerable differences in binding.

Studies on the effect of detoxification procedures for *Fusarium* mycotoxins by yeast fermentation have been mentioned by several authors. Using yeast in bread making from wheat contaminated with DON, levels of this mycotoxin were shown to be reduced. Samar *et al.* [116] found that fermentation reduced naturally occurring DON in Argentinean bread processing technology using a pilot scale plant. French bread and Vienna bread were prepared from wheat flour naturally contaminated with DON at 150 $\mu g\ kg^{-1}$ in which dough was fermented at 30–50 ºC. The maximum reduction obtained in dough at 50 ºC was 46 % for the Vienna bread and 41 % for French bread. This agreed with a study of Neira *et al.* [117], that showed a significant reduction of DON during the bread-making process. According to Garda *et al.* [118], alcoholic fermentation with *S. cerevisiae* can be considered as a promising method of detoxification of different levels of DON and T-2 toxin. It was found that the fermentation process of malt contaminated with DON and T-2 caused a decontamination of 53% for these mycotoxins, taking into account both the wort and the filtered sample. Yiannikouris and Jouany [119] noted that glucomannans extracted from the external cell walls of the yeast *S. cerevisae* are able to bind effectively aflatoxins and, to a lesser degree, ochratoxins and fusarium toxins. In contrast, the results mentioned by Hanschmann and Krieg [120], Bennet and Richard [121] and Schaafsma *et al.* [122] showed that DON was not destroyed by alcoholic fermentation and high levels in both the solid residue and the fermented liquid could be detected. Scott [123] mentioned that DON was stable after 7–9 days of wort fermentation by *S. cerevisae*. Böhm-Schraml *et al.* [124] demonstrated increasing DON levels within the first 20 h of fermentation of wort, which then subsequently decreased up to 100 h. These results suggest that other procedures should be carried out on *Fusarium*-contaminated grains used as raw materials for fermentation and, also, that studies of the effect of fermentation on the decontamination process should be better evaluated.

Considerable amounts of other *Fusarium* mycotoxin zearalenone (ZEA) and its derivative α-zearalenol were bound effectively (up to 55% w/w) to the probiotic bacteria *L. rhamnosus* GG and *L. rhamnosus* LC705. Both heat-treated and acid-treated bacteria were capable of removing the toxins, indicating that binding, not metabolism is the mechanism by which the toxins are removed from the media [72]. A few other microbial activities that transform ZEA have been published but are protected by patents. Interactions between LAB and ZEA and its derivative, α-zearalenol were also investigated. It was suggested that the yeast epihydroxylase might be involved. A significant proportion (38–48%) of both toxins was trapped in the bacterial pellet and no degradation product of ZEA or α-zearalenol was detected [72], leading to the conclusion that binding and not metabolism was the mechanism by which the toxins were removed from the media. Similar results were obtained with other mycotoxins including OTA [82, 125] and fumonisins B_1 (FB_1) and B_2 (FB_2) [126]. Therefore, two specific processes such as binding and inhibition of biosynthesis may be involved in the interaction between LAB and the accumulation of some mycotoxins.

3.5. Fumonisins

Concerning the mechanisms of action involved in the removal of fumonisins by LAB, Niderkorn [126] suggested that peptidoglycans were the most plausible fumonisin binding sites. The quenching ability of LAB was increased when bacteria were killed using different physical and chemical treatments, while lysozyme and mutanolysin enzymes that target peptidoglycans partially inhibited it. It was also reported that tricarballylic acid chains found in fumonisin molecules played an important role in the binding process since hydrolysed fumonisin had less affinity for LAB, and free amine group inactivation had no effect on the binding process. The same article attempted to explain the low affinity of FB_1 using a molecular modelling approach. In fact, an additional hydroxyl group in FB_1 could form a hydrogen bond with one of the tricarballylic acid chains, resulting in a spatial configuration where the tricarballylic acid chain is less available to interact with bacterial peptodoglycans. Removal of fumonisins by LAB was ascribed to adhesion to cell wall components rather than covalent binding or metabolism, since the dead cells fully retained their binding ability. Peptidoglycans probably play a key rule in this binding process. Therefore, elucidating the differences between bacterial cell wall components of LAB strains might make it possible to select LAB species with the potential to act as biopreservative agents capable of reducing exposure from fumonisins that occur in foods.

4. Detoxification of mycotoxins in animal feed

Different mycotoxins are more commonly found in or associated with certain feedstuffs. Some develop in the growing crop due to its being susceptible to certain toxigenic fungi, while infection and toxin production by others is facilitated by the preservation and storage system used if insufficient care is taken to prevent this.

Most feeds are produced from crops at the farm and consumed by the animals some time later. However, some straight feeds and particularly mixed feeds are also produced and sold by feed mills. Feed raw materials are often divided into (1) cereals and by-products; (2) oilseed by-products; (3) leguminous seeds; (4) roots and tubers; (5) animal by-products; (6) green crops/pasture; (7) silages; (8) hays; (9) straws. They can be used straight, single or in combination in the feeding regime. Component types 1–5 are often combined and used as mixed feed and concentrates. The other raw material groups (6–9) are combined as roughage and mainly used for ruminants and horses.

4.1. Ensiling

Ensiling of mycotoxin-contaminated crops for detoxification has been proposed as an interesting and possible method for elimination or reduction of mycotoxins. Normal ensiling has, however, only rarely been studied for its mycotoxin degrading potential. A study by Rotter *et al.* [127] showed that ensiling of ochratoxin-contaminated barley could reduce the toxin by approximately 68 %. Yeasts in grass silage have been found to degrade patulin in silage inoculated with *Paecilomyces* sp. to induce patulin production [128]. Both bacteria and yeasts from maize silage have also been shown to be able to degrade fumonisins [129]. Stimulation of mycotoxin degradation by naturally occurring micro-organisms in silage or the addition of yeasts or bacteria with known mycotoxin degradation ability to silage may in the future become practical means to detoxify mycotoxins in certain crops.

4.2. Yeasts as feed additives

Yeasts have been fed to animals for more than a century and commercial yeast products are being specifically produced in commercial scale for animal feeding [39]. Hence, yeasts have immense potential as tools in tackling the problem of mycotoxins in animal feed.

In the poultry industry, *S. cerevisiae* has been used as general performance promoter in poultry feeds and has recently been shown to have beneficial effects against AFB_1 exposure [39, 130]. There are many reports on use of physically separated yeast cell walls obtained from brewery as feed additive in poultry diet resulting in amelioration of toxic effects of aflatoxins [41]. When dried yeast and yeast cell walls were added to rat-ration along with AFB_1, a significant reduction in the toxicity was observed [38]. In an *in vitro* study with the cell wall material, there was a dose dependent binding of as much as 77% (w/w) and modified mannan-oligosaccharides derived from the *S. cerevisiae* cell resulted in as much as 95% (w/w) binding [131]. In addition, compound also showed considerably high binding to ZEN and FB_1, but only little binding to DON. Esterified glucomannan was later shown to provide protection in broilers exposed to OTA and T-2 toxin in addition to aflatoxin [94] and reduced the toxic effects of *Fusarium* toxins in horses [132]. A later study also confirms the protective effects of yeast glucomannan on aflatoxicosis in broilers [133].

In a recent study, seven different trichothecene 3-*O*-acetyltransferases transformed into the yeast strain RW2802 were analyzed for their ability to modify DON into 3ADON during a series of feeding assays; conversion levels ranged from 50.5% to 100%, depending on the source of the acetyltransferase [134].

4.3. Impact of enzymes on alcohol fermentation of mycotoxin contaminated grains and safety of distiller's dried grains with solubles (DDGS)

A valuable co-product of fuel ethanol production, known as distillers dried grains with solubles (DDGS), is increasingly being used as a feed source for domestic animals. DDGS contains high levels of protein, fiber, minerals and vitamins [135, 136]. An increase in the supply and demand for DDGS [137] is expected to coincide with the increased production of fuel ethanol in commercial plants [138], which rely on the sale of DDGS to turn a profit [139]. One of the challenges facing the fuel ethanol industry is the management of mycotoxins such as DON in DDGS.

The degradation of mycotoxins during alcohol fermentation for the production of ethanol has been investigated only in a few studies, but many papers have appeared on the fate of mycotoxins during the production of beer and wine. Bennett *et al.* used zearalenone-contaminated maize for ethanol production, but the toxin was not degraded and remained in the fermentation residues [140]. Mycotoxins can often be found in beer and wine, but they are partly degraded and partly transferred to beer, with part remaining in the draff during the fermentation process [141-144]. Studies on alcohol fermentation by *S. cerevisiae* using grains (barley, corn or wheat) contaminated with DON [122, 123, 145, 146] showed that DON was stable to the alcoholic fermentation process and not metabolised by *Saccharomyces* yeasts.The mycotoxin concentration remaining in the draff (DDGS) may be relatively high and so traditional alcohol fermentation is not a very effective means of detoxifiying feed material.

Mycotoxin-contaminated cereals may in the future be used more often in industrial ethanol production. Enhanced degradation of mycotoxins by eventual addition of microorganisms with mycotoxin-degradation ability is needed if the DDGS are to be utilized for animal feed.

The impact of β-xylanase on the fermentation of *Fusarium* contaminated wheat was studied in our lab [12]. It was found that *Fusarium*-contaminated wheat with a high concentration of

DON (3.95 mg kg^{-1}) negatively affected the fermentation by *S. cerevisiae* process and reduced the ethanol concentration by 13.5% in the fermented wort (Figure 5). The explanation for the decrease in the ratio of fermentation by *S. cerevisiae* might be the inhibitory effect of high concentrations of *Fusarium* mycotoxins on yeast growth [147]. Whitehead and Flannigan [148] suggested that trichothecenes inhibit mitochondrial function, causing slower oxygen utilisation by the yeast resulting in a slower growth rate. In addition, damage by *Fusarium* could result in loss of sugars and other components which could affect the ethanol productivity [149].

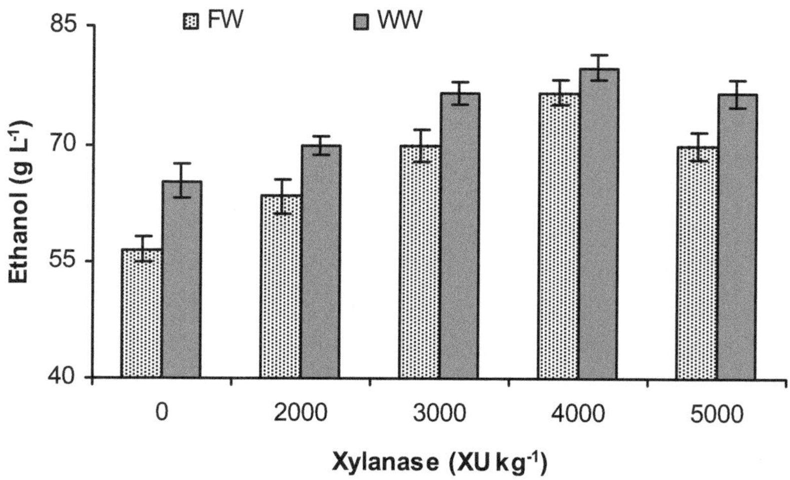

Figure 5. Effect of *T. reesei* xylanase on ethanol concentration in the fermented wort produced from *Fusarium* contaminated wheat (FW) *and wholesome wheat (WW)*; 0 – sample prepared by using only amylolytic enzymes

The complex of amylolytic enzymes with xylanase from *T. reesei,* used for wort saccharification, increased the efficiency of the fermentation process of *Fusarium*-contaminated wheat: the ethanol concentration increased by 35.3%. The addition of xylanase also improved the quality of bioethanol by decreasing the concentrations methanol, methyl acetate, isoamyl and isobutyl alcohols (Figure 6). The increase of the ethanol concentration could be influenced by synergetic action of the different side glycoside hydrolases,which renders the solubilisation and depolymerisation of non-starch polysaccharides to their monomeric constituent sugars and the higher concentration of hexoses in the medium. A higher activity of yeast enzymes and biomass formation are obtained, herewith fostering the carbohydrate metabolism to ethanol and carbon dioxide [150]. The formation of methanol and methyl acetate during the fermentation indicated that *Fusarium*-contaminated wheat was a pure growing medium for yeasts. Studies by Reinehr and Furlong demonstrated that unpredictable productions of methanol and higher alcohols could be obtained during the fermentation process of malt, depending on the presence of fungal contamination in cereals [151].

The results indicated that DON concentrations in the DDGS were significantly higher than in the starting material (*Fusarium* contaminated wheat). DON concentrations increased by a factor of 2.1–2.4 on a dry matter basis in DDGS in compare with starting wheat (Table 2). Increase in concentration of mycotoxins in DDGS is due to the reduction in total solid mass during fermentation, together with the loss of moisture during drying of the DDGS, and therefore a mass balance was used. The balance of mycotoxin contents showed the reduction of DON in the process. Traditional mash preparation and alcoholic fermentation by *S. cerevisiae* decreased the initial contamination by 26%. By using the complex of amylolytic enzymes and *T. reesei* xylanase for mash saccharification, decontamination of DDGS was 31–41%. The highest level of decontamination was obtained by a xylanase activity of 4000 XU kg^{-1}, and it was possible to decrease DON level in the DDGS by 15% in compare with the sample with only amylolytic enzymes used.

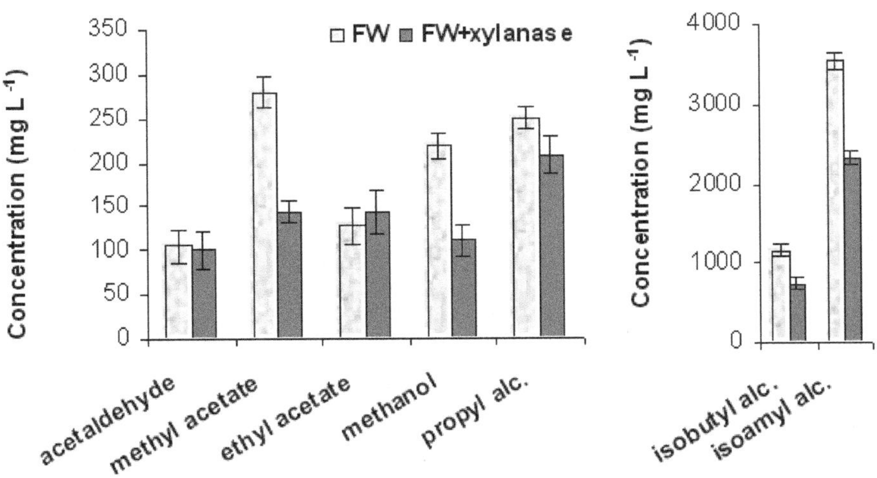

Figure 6. Effect of *T. reesei* xylanase on the concentrations of fusel oils and methanol in the distillates from *Fusarium* contaminated wheat (FW)

Sample	Moisture content (g kg^{-1})	Dry weight (kg)	DON concentration, dry matter		DON recovery (%)
			Absolute (mg)	mg kg^{-1}	
Input wheat	118 ± 2	0.882 ± 0.002	3.48 ± 0.01	3.95	
DDGS, without xylanase	68 ± 5	0.266 ± 0.005	2.57 ± 0.01	9.66	74
DDGS, 2000 XU kg^{-1}	82 ± 9	0.253 ± 0.003	2.41 ± 0.01	9.53	69
DDGS, 3000 XU kg^{-1}	79 ± 8	0.246 ± 0.006	2.34 ± 0.01	9.51	67
DDGS, 4000 XU kg^{-1}	74 ± 7	0.245 ± 0.003	2.05 ± 0.01	8.37	59
DDGS, 5000 XU kg^{-1}	80 ± 9	0.245 ± 0.005	2.10 ± 0.01	8.57	60

Table 2. DON assessment and mass balance on wheat and its bioethanol co-product DDGS

The decrease in DON levels may have occurred due to absorbtion by yeast cells or extracellular metabolism resulting in protection from mycotoxin-induced toxicities [124]. According to Yiannikouris and Jouany [119], glucomannans from the external part of the cell wall of S. *cerevisae* are able to bind DON by 12.6%. The potential effects of ethanol fermentation by yeasts on the decomposition of *Fusarium* mycotoxins (trichothecenes) were studied by Flesch and Voight-Scheuerman [152] during alcoholic fermentation of grape juice. Derivatives that did not contain epoxide groups were found, suggesting that a yeast epihydroxylase may have been present. It was also suggested that the yeast probably produced ligases and also a keto–enol tautomerase. It was established that a significant amount of mycotoxins (about 20%) were taken up by yeast. With reference to our investigations, by using xylanase suplementation for saccharification of contaminated wheat the synthesis of yeast biomass could be increased, herewith fostering the absorbtion or biotransformation of mycotoxins and reduction of DON levels (up to 41%) in DDGS. Therefore, saccharification using the complex of amylolytic enzymes with xylanase and alcoholic fermentation by *S. cerevisiae* can be considered as a promising method of DON detoxification.

Another approach to reduce DON in DDGS might be to add an exogenous trichothecene 3-*O*-acetyltransferase preparation to the mash at the start of fermentation. In a recent study [153], Two *Fusarium* trichothecene 3-*O*-acetyltransferases (FgTRI101 and FfTRI201) were cloned and expressed in yeast (*Saccharomyces cerevisiae*) during a series of small-scale ethanol fermentations using barley (*Hordeum vulgare*). During the fermentation process, FgTRI101 converted 9.2% to 55.3% of the DON to 3ADON, resulting in DDGS with reductions in DON and increases in 3ADON. Employing yeast to express mycotoxin-detoxification genes represents a potential strategy to reduce mycotoxin levels in fuel ethanol co-products. However, a number of issues must be addressed before this process is commercialized. First, the composition of DDGS in future work using transformed yeast would need to be evaluated. Second, the use of a transgenic yeast strain for fuel ethanol production will need to be accepted by policy makers and ethanol production facilities in order to be implemented on a commercial scale.

5. Conclusion and future concerns

Prevention of mycotoxin formation is the best defence for protecting the consumer's health. However, prevention is not always possible, especially for those mycotoxins formed under field conditions. Introduction of further legislation for a wider range of mycotoxins in more food commodities means that there is a much greater need to determine how mycotoxins survive processing so that this can be taken into account when setting statutory or guideline limits. It is thus expected that there will be a trend towards further study of the fate of those mycotoxins that pose the greatest potential risk for humans. In some instances it may then be possible to introduce modifications to commercial processes that result in a significant reduction of mycotoxin content in the retail product.

Contaminated crops condemned as food can otherwise be diverted for use as animal feed. Therefore, prevention and reduction of mycotoxin contamination during feed production will become more important. Mycotoxin-contaminated crops must be used in more cases for purposes other than direct food and feed and their utilization will be further investigated. Decontamination procedures will be further studied. Physical methods will be studied, but biological methods probably more so. Mycotoxin-contaminated cereals may in the future be used more often in industrial ethanol production. An enhanced degradation of mycotoxins by eventual addition of microorganisms or enzymes with mycotoxin-degradation ability is needed if the fermentation remains are to be utilized for animal feed.

In this chapter, we have tried to provide information on decontamination of various mycotoxins during fermentation procceses using bacteria, yeasts, fungi and enzymes. Based on the available reports, we can conclude that microorganisms are the main living organisms applicable for mycotoxin decontamination in foods. Results of various researchers showed that yeasts and bacterial strains had differences in decontamination of mycotoxins. For example, *Kluyveromyces marxianus* was sensitive to all trichothecene toxins, but this yeast was not inhibited by other mycotoxins, and *B. brevis* was sensitive to eight mycotoxins, including ZEA and OTA, but was not affected by high concentrations of trichothecene toxins. However, the results achieved in microbiological decontamination of mycotoxins until today may be treated as a first step in development of practical commercial technologies. Further intensive screening of mycotoxin-degrading microorganisms may lead to detection of efficient and applicable ones. Based on the available reports of mycotoxin-degrading microorganisms in the digestive tract of animals, the activity of these microorganisms may be increased and they may be used *in vivo* for degradation of mycotoxins in food. With the application of molecular biology techniques, the potential mycotoxin degrading microbial strains can be engineered to significantly improve the quality and safety of foods from mycotoxins contamination to protect consumer's health.

Author details

Grazina Juodeikiene and Loreta Basinskiene
Department of Food Technology, Kaunas University of Technology, Kaunas, Lithuania

Elena Bartkiene and Paulius Matusevicius
Department of Food Safety and Quality, Lithuanian University of Health Sciences, Kaunas, Lithuania

6. References

[1] Reddy KRN, Salleh B, Saad B, Abbas HK, Abel CA, Sheir WT (2010) An overview of mycotoxin contamination in foods and its implications for human health. Toxin Rev. 29: 3-26.

[2] Chassy BM (2010) Food safety risk and consumer health. New Biotechnol. 27(5): 534-544.

[3] Wild CP (1996) Summary of data on aflatoxin exposure in West Africa. In: Cardwell KF, editor. Proceedings of the workshop on mycotoxins in food in Africa, November 6-10, 1995, Cotonou, Benin. Benin: International Institute of Tropical Agriculture. p. 26.

[4] Anonymous (2004) Outbreak of aflatoxin poisoning. Eastern and central provinces, Kenya, Januar-July 2004. Morb. Mortal Wkly. Rep. 53: 790-793.

[5] Gong Y, Hounsa A, Egal S, Turner PC, Sutcliffe AE, Hall AJ, Cardwell K, Wild CP (2004) Post weaning exposure to aflatoxin results in impaired child growth: A longitudinal study in Benin, West Africa. Environ. Health Perspect. 112: 1334-1338.

[6] Williams JH, Phillips TD, Jolly PE, Stiles JK, Jolly CM, Agarwal D (2004) Human aflatoxicosis in developing countries: A review of toxicology, exposure, potential health consequences, and interventions. Am. J. Clin. Nutr. 80: 1106-1122.

[7] Diaz DE, Hagler WM, Blackwelder JT, Eve JA, Hopkins BA, Andersen KL (2004) Aflatoxin binders II: Reduction of aflatoxin M1 in milk by sequestering agents of cows consuming aflatoxin in feed. Mycopathol.157: 233-241.

[8] Halász A, Lásztity R, Abonyi T, Bata A (2009) Decontamination of mycotoxin-containing food and feed by biodegradation. Food Rev. Int. 25(4): 284–298.

[9] Huwig A, Freimund S, Käppeli O, Dutler H (2001) Mycotoxin detoxication of animal feed by different adsorbents. Toxicol. Lett. 122: 179-188.

[10] Piva G, Galvano F, Pietri A, Piva A (1995) Detoxification methods of aflatoxins. A review. Nutr. Res. 15: 767-776.

[11] He J, Zhou T, Young JC, Boland GJ, Scott PM (2010) Chemical and biological transformations for detoxification of trichothecene mycotoxins in human and animal food chains: a review. Trends Food Sci. Technol. 21(2): 67-76.

[12] Juodeikiene G, Basinskiene L, Vidmantiene D, Makaravicius T, Bartkiene E (2012) Benefits of β-xylanase for wheat biomass conversion to bioethanol. J. Sci. Food Agr. 92(1): 84-91.

[13] Bata A, Lasztity R (1999) Detoxification of mycotoxin contaminated food and feed by microorganisms. Trends Food Sci. Technol. 10: 223-228.

[14] Reddy KRN, Reddy CS, Muralidharan K (2009) Potential of botanicals and biocontrol agents on growth and aflatoxin production by Aspergillus flavus infecting rice grains. Food Control. 20: 173-178.

[15] Patharajan S, Reddy KRN, Karthikeyan V, Spadaro D, Gullino ML, Garibaldi A (2011) Potential of yeast antagonists on *in vitro* biodegradation of ochratoxin A. Food Control. 22: 290-296.

[16] Styriak I, Concova E (2002) Microbial binding and biodegradation of mycotoxins. Vet. Human Toxicol. 44: 358-361.

[17] La Penna M, Nesci A, Etcheverry M (2004) *In vitro* studies on the potential for biological control on *Aspergillus* section Flavi by *Kluyveromyces* spp. Lett. Appl. Microbiol. 38: 257-264.

[18] Varga J, Péteri Z, Tábori K, Téren J, Vagvolgyi C (2005) Degradation of ochratoxin-A and other mycotoxins by *Rhizopus* isolates. Int. J. Food Microbiol. 99: 321-328.
[19] Pereira P, Nesci A, Castillo C, Etcheverry M (2010) Impact of bacterial biological control agents on fumonisin B1 content and *Fusarium verticillioides* infection of field-grown maize. Biol. Control. 53: 258-266.
[20] Karlovsky P (1999) Biological detoxification of fungal toxins and its use in plant breeding, feed and food production. Nat. Toxins. 7: 1-23.
[21] Shetty PH, Jespersen L (2006) Saccharomyces cerevisiae and lactic acid bacteria as potential mycotoxin decontamination agents. Trends Food Sci. Technol. 17: 48-55.
[22] Jespersen L (2003) Occurrence and taxonomic characteristics of strains of *Saccharomyces cerevisiae* predominant in African fermented foods and beverages. FEMS Yeast Res. 3: 191-200.
[23] Celyk K, Denly M, Savas T (2003) Reduction of toxic effects of aflatoxin by using baker yeast (*Saccharomyces cerevisiae*) in growing broiler chicken diets. Rev. Bras. Zootecn. 32: 615-619.
[24] Bartkiene E, Juodeikiene G, Vidmantiene D (2008) Evaluation of deoxynivalenol in wheat by acoustic method and impact of starter on its concentration during wheat bread baking process. Food Chem. Technol. 42(1): 5-12.
[25] Abrunhosa L, Paterson RRM, Venâncio A (2010) Biodegradation of ochratoxin A for food and feed decontamination. Toxins. 2: 1078-1099.
[26] Thyagaraja N, Hosono A (1994) Binding properties of lactic acid bacteria from 'Idly' towards food-borne mutagens. Food Chem. Toxicol. 32: 805-809.
[27] Gourama H, Bullerman LB (1995) Inhibition of growth and aflatoxin production of *Aspergillus flavus* by *Lactobacillus* species. J. Food Protect. 58: 1249-1256.
[28] Gourama H (1991) Growth and aflatoxin production of *Aspergillus flavus* in the presence *Lactobacillus species*. Ph.D. thesis, University of Nebraska-Lincoln.
[29] Karunaratne A, Wezenberg E, Bullerman LB (1990) Inhibition of mold growth and aflatoxin production by *Lactobacillus* spp. J. Food Protect. 53: 230-236.
[30] Coallier-Ascah J, Idziak E (1985) Interaction between *Streptococcus lactis* and *Aspergillus flavus* on production of aflatoxin. Appl. Environ. Microb. 49: 163-167.
[31] Luchese RH, Harrigan WF (1990) Growth of and aflatoxin production by *Aspergillus parasiticus* when in the presence of either *Lactococcus lactis* or lactic acid and at different initial pH values. J. Appl. Bacteriol. 69: 512-519.
[32] Dalié DKD, Deschamps AM, Richard-Forget F (2009) A review: Lactic acid bacteria – Potential for control of mould growth and mycotoxins. Food Control. 21(4): 370-380.
[33] Moreno-Aribas MV, Polo MC (2010) Wine Chemistry and Biochemistry. In: Magan N, Olsen M, editors.Mycotoxins in food Detection and control: Biological decontamination of mycotoxins. New York: Springer, Woodhead Publishing Ltd and CRC Press: LLC. pp. 2006-2211.

[34] Brady D, Stoll A D, Strake L, Dunkan J R (1994) Chemical and enzymatic extraction of heavy metal binding polymera from isolated cell walls of *Saccharomyces cerevisiae.* Biotechnol. Bioeng. 44: 297-302.

[35] Bolognani F, Rumney CJ, Rowland IR (1997) Influence of carcinogen binding by lactic acid-producing bacteria on tissue distribution and in vitro mutagenecity of dietary carcinogens. Food Chem. Toxicol. 35: 535-545.

[36] Santos A, Marquina D, Leal JA, Peinado JM (2000) (1-6)-β-D-glucan as cell wall receptor for *Pichia membranifaciens* killer toxin. Appl. Environ. Microb. 66: 1809-1813.

[37] Thompson ED, Knights BA, Parks LW (1973) Identification and properties of a sterol-binding polysaccharide isolated from *Saccharomyces cerevisiae.* Biochim. Biophys. Acta. 304: 132-141.

[38] Baptista AS, Horii J, Calori-Domingues MA, da Gloria EM, Salgado JM, Vizioli MR (2004) The capacity of mannooligosaccharides thermolysed yeast and active yeast to attenuate aflatoxicosis.World J. Microb. Biot. 20: 475-481.

[39] Celyk K, Denly M, Savas T (2003) Reduction of toxic effects of aflatoxin by using baker yeast (*Saccharomyces cerevisiae*) in growing broiler chicken diets. Rev. Bras. Zootecn. 32: 615-619.

[40] Raju MVLN, Devegowda G (2000) Influence of esterifiedglucomannan on performance and organ morphology, serum biochemistry and haematology in broilers exposed to individual and combined mycotoxicosis (aflatoxin, ochratoxin and T-2 toxin). Brit. Poultry Sci. 41: 640-650.

[41] Santin E, Paulilo AC, Maiorka A, Nakaghi LSO, Macan M, de Silva AVF (2003) Evaluation of the efficiency of *Saccharomyces cerevisiae* cell wall to ameliorate the toxic effects of aflatoxin in broilers. Int. J. Poultry Sci. 2: 241-344.

[42] Yiannikouris A, François J, Poughon L, Dussap CG, Bertin G, Jeminet G, Jouany JP (2004) Adsorption of zearalenone by beta-D-glucans in the *Saccharomyces cerevisiae* cell wall. J. Food Prot. 67: 1195-1200.

[43] Yiannikouris A, François J, Poughon L, Dussap CG, Bertin G, Jeminet G, Jouany JP (2004) Alkali extraction of beta-D-glucans from *Saccharomyces cerevisiae* cell wall and study of their adsorptive properties toward zearalenone. J. Agric. Food Chem. 52: 3666-3673.

[44] Freimund S, Sauter M, Rys P (2003) Efficient adsorption of mycotoxins zearalenone and T-2 toxin on a modified yeast glucan. J. Environ. Sci. Health B. 38(3): 243–255.

[45] Bejaouii H, Mathieu F, Taillandier P, Lebrihi A (2004) Ochratoxin A removal in synthetic and natural grape juices by selected oenological *Saccharomyces* strains. J. Appl. Microbiol. 97: 1038-1044.

[46] Yiannikouris A, Andre G, Buleon A, Jeminet G, Canet I, François J, Bertin G, Jouany JP (2004) Comprehensive conformational study of key interactions involved in zearalenone complexation with beta-D-glucans. Biomacromolecules. 5: 2176-2185.

[47] Jouany JP,. Yiannikouris A, Bertin G (2005) The chemical bonds between mycotoxins and cell wall components of *Saccharomyces cerevisiae* have been identified. Arch. Zootech. 8: 26-50.

[48] Turbic A, Ahokas JT, Haskard CA (2002) Selective in vitro binding of dietary mutagens, individually or in combination, by lactic acid bacteria. Food Addit. Contam. 19: 144-152.

[49] Haskard CA, El-Nezami HS, Peltonen KD, Salminen S, Ahokas JT (1998). Sequestration of aflatoxin B1 by probiotic strains: Binding capacity and localization. Rev. Med. Vet.-Toulouse. 149: 571.

[50] Haskard CA, Binnion C, Ahokas J (2000) Factors affecting the sequestration of aflatoxin by *Lactobacillus rhamnosus* strain GG. Chem-Biol. Interact. 128: 39-49.

[51] Lahtinen SJ, Haskard CA, Ouwehandz AC, Salminenz SJ, Ahokasy JT (2004) Binding of aflatoxin B1 to cell wall components of *Lactobacillus rhamnosus* strain GG. Food Addit. Contam. 21: 158-164.

[52] Haskard, C. A., El-Nezami, H. S., Kankaanpaa, P. E., Salminen, S.,& Ahokas, J. T. (2001). Surface binding of alfatoxin B1 by lactic acid bacteria. Appl. Environ. Microb. 67: 3086-3091.

[53] Westby A, Reilly A, Bainbridge Z (1997) Review of the effect of fermentationon naturally occurring toxins. Food Control. 8(5/6): 329-339.

[54] Sweeney MJ, Dobson ADW (1998) Mycotoxin prodution by *Aspergillus, Fusarium* and *Penicillium* species. Int. J. Food Microbiol. 43: 141-158.

[55] Moss MO, Thrane U (2004) *Fusarium* taxonomy with relation to trichothecene formation. Toxicol. Lett. 153: 23-28.

[56] Garvey GS, McCormick SP, Rayment I (2008) Structural and functional characterization of the TRI101 trichothecene 3-O-acetyltransferase from *Fusarium sporotrichioides* and *Fusarium graminearum*: kinetic insights to combating Fusarium head blight. J. Biol. Chem. 283: 1660-1669.

[57] Kimura M, Tokai T, Matsumoto G, Fujimura M, Hamamoto H, Yoneyama K, Shibata T, Yamaguchi I (2003) Trichothecene nonproducer Gibberella species have both functional and nonfunctional 3-*O*-acetyltransferase genes. Genetics. 163: 677-684.

[58] Khatibi PA, Newmister SA, Rayment I, McCormick SP, Alexander NJ, Schmale DG (2011) Bioprospecting for trichothecene 3-*O*-acetyltransferases in the fungal genus fusarium yields functional enzymes with different abilities to modify the mycotoxin deoxynivalenol. Appl. Environ. Microbiol. 77: 1162-1170.

[59] Kimura M, Kaneko I, Komiyama M, Takatsuki A, Koshino H, Yoneyama K, Yamaguchi I (1998) Trichothecene 3-*O*-acetyltransferase protects both the producing organism and transformed yeast from related mycotoxins. Cloning and characterization of Tri101. J. Biol. Chem. 273: 1654-1661.

[60] Muhitch MJ, McCormick SP, Alexander NJ, Hohn TM (2000) Transgenic expression of the TRI101 or PDR5 gene increases resistance of tobacco to the phytotoxic effects of the trichothecene 4,15-diacetoxyscirpenol. Plant. Sci. 157: 201-207.

[61] Ohsato S, Ochiai-Fukuda T, Nishiuchi T, Takahashi-Ando N, Koizumi S, Hamamoto H, Kudo T, Yamaguchi I, Kimura M (2007) Transgenic rice plants expressing trichothecene 3-*O*-acetyltransferase show resistance to the *Fusarium* phytotoxin deoxynivalenol. Plant. Cell. Rep. 26: 531-538.

[62] Alexander NJ, McCormick SP, Ziegenhorn SL (1999) Phytotoxicity of selected trichothecenes using *Chlamydomonas reinhardtii* as a model mystem. Natural Toxins. 265-269.

[63] Blesa J, Soriano JM, Molto JC, Manes J (2004) Concentration of ochratoxin A in wines from supermarkets and stores of Valencian Community (Spain). J. Chromatogr. A. 1054: 397-401.

[64] Elgerbi AM, Aidoo KE, Candlish AA, Tester RF (2004) Occurrence of aflatoxin M1 in randomly selected North African milk and cheese samples. Food Addit. Contam. 21: 592-597.

[65] Odhav B, Naicker V (2002) Mycotoxins in South African traditionally brewed beers. Food Addit. Contam. 19: 55-61.

[66] Chu FS, Chang CC, Ashoor SH, Prentice N (1975) Stability of aflatoxin B, and ochratoxin A in brewing. Appl. Microbiol. 29: 313-316.

[67] Govaris A, Roussi V, Koidis PA, Botsoglou NA (2002) Distribution and stability of aflatoxin M1 during production and storage of yoghurt. Food Addit. Contam. 19: 1043-1050.

[68] Wiseman DW, Marth EH (1981) Growth and aflatoxin production by *Aspergillus parasiticus* when in the presence of *Streptococcus lactis*. Mycopathologia. 73: 49-56.

[69] Kankaanpää P, Tuomola E, El-Nezami H, Ahokas J, Salminen SJ (2000) Binding of aflatoxin B1 alters the adhesion properties of *Lactobacillus rhamnosus* strain GG in Caco-2 model. J. Food Protect. 63: 412-414.

[70] Gratz S, Mykkänen H, Ouwehand AC, Juvonen R, Salminen S, El-Nezami HS (2004) Intestinal mucus alters the ability of probiotic bacteria to bind aflatoxin B1 *in vitro*. Appl. Environ. Microb. 70: 6306-6308.

[71] Shah N, Wu X (1999) Aflatoxin B1 binding abilities of probiotic bacteria. Biosci. Microflora. 18: 43-48.

[72] El-Nezami HS, Polychronaki N, Salminen S, Mykkänen H (2002) Binding rather metabolism may explain the interaction of two food-grade *Lactobacillus* strains with zearalenone and its derivative α-zearalenol. Appl. Environ. Microb. 68: 3545-3549.

[73] Line JE, Brackett RE (1995) Role of toxin concentration and second carbon source in microbial transformation of aflatoxin B1 by Flavobacterium aurantiacum. J. Food Protect. 58: 1042-1044.

[74] Peltonen K, El-Nezami H, Haskard C, Ahokas J, Salminen S (2001) Aflatoxin B1 binding by dairy strains of lactic acid bacteria and bifidobacteria. J. Dairy Sci. 84: 2152-2156.

[75] El-Nezami H, Haskard C, Salminen E, Mykkanen H, Ahokas J, Salminen S (2002) Lactic acid bacteria and bifidobacteria can reduce dietary exposure to aflatoxins. Brit. J. Nutr. 88: 119-121.

[76] El-Nezami H, Kankaanpää P, Salminen S, Ahokas J (1998) Physicochemical alterations enhance the ability of dairy strains of lactic acid bacteria to remove aflatoxin from contaminated media. J. Food Protect. 61: 466-468.

[77] Frisvad JC, Thrane U, Samson RA, Pitt JI (2006) Important mycotoxins and the fungi which produce them. Adv. Exp. Med. Biol. 571: 3-31.

[78] Böhm J, Grajewski J, Asperger H, Rabus B, Razzazi E (2000) Study on biodegradation of some trichothecenes (NIV, DON, DAS, T-2) and ochratoxin A by use of probiotic microorganisms. Mycot. Res. 16: 70-74.

[79] Piotrowska M, Zakowska Z (2000) The biodegradation of ochratoxin A in food products by lactic acid bacteria and baker's yeast. In: Bielecki S, Tramper J, Polak J, editors. Progress in Biotechnology (Food Biotechnology). Vol. 17. Amsterdam: Elsevier. pp. 307-310.

[80] Piotrowska M, Zakowska Z (2005) The elimination of ochratoxin A by lactic acid bacteria strains. Pol. J. Microbiol. 54: 279-286.

[81] Del Prete V, Rodriguez H, Carrascosa AV, Rivas BDL, Garcia-Moruno E, Munoz R (2007) In vitro removal of ochratoxin A by wine lactic acid bacteria. J. Food Protect. 70: 2155-2160.

[82] Fuchs S, Sontag G, Stidl R, Ehrlich V, Kundi M, Knasmuller S (2008) Detoxification of patulin and ochratoxin A, two abundant mycotoxins, by lactic acid bacteria. Food Chem. Toxicol. 46: 1398-1407.

[83] Shetty PH, Jespersen L (2006) *Saccharomyces cerevisiae* and lactic acid bacteria as potential mycotoxin decontaminating agents. Trends Food Sci. Techn. 17: 48-55.

[84] Mateo R, Medina A, Mateo EM, Mateo F, Jimenez M (2007) An overview of ochratoxin A in beer and wine. Int. J. Food Microbiol. 119 (1–2): 79-83.

[85] Varga J, Kozakiewicz Z (2006) Ochratoxin-A in grapes and grape-derived products. Trends Food Sci. Techn. 1: 72-81.

[86] Kozakiewicz Z, Battilani P, Cabanes I, Venancio A, Mule G, Tjamos E (2003) Making wine safer. In: van Egmond H, van Osenbruggen T, Lopez Garcia R, Visconti A, editors. Meeting the Mycotoxin Menace. Wageningen: Wageningen Academic Publisher. pp.131-140.

[87] Binder EM, Heidler D, Schatzmayr G, Thimm N, Fuchs E, Schuh M, Krska R, Binder J (2000) Microbial detoxification of mycotoxins in animal feed. In: De Koe WJ, Samson RA, Van Egmond HP, Gilbert J, Sabino M, editors. Mycotoxins and Phytotoxins in Perspective at the Turn of the Millenium. Proceedings of the 10-th International IUPAC Symposium on Mycotoxins and Phytotoxins, Garuja, Brazil, May 20-25. IUPAC. pp. 271-277.

[88] Sÿkrinjar M, Rašić JL, Stojičić V (1996) Lowering of ochratoxin A level in milk by yoghurt bacteria and bifidobacteria. Folia Microbiol. 1: 26-28.

[89] Bizaj E, Mavri J, Cus F, Raspor A (2009) Removal of ochratoxin A in *Saccharomyces cerevisiae* liquid cultures. S. Afr. J. Enol. Vitic. 30: 151-155.

[90] Cecchini F, Morassut M, Garcia Moruno E, Di Stefano R (2006) Influence of yeast strain on ochratoxin A content during fermentation of white and red must. Food Microbiol. 23: 411-417.

[91] Moruno EG, Sanlorenzo C, Boccaccino B, Di Stefano R (2005) Treatment with yeast to reduce the concentration of ochratoxin A in red wine. Am. J. Enol. Vitic. 56: 73-76.

[92] Ringot D, Lerzy B, Bonhoure JP, Auclair E, Oriol E, Larondelle Y (2005) Effect of temperature on *in vitro* ochratoxin A biosorption onto yeast cell wall derivatives. Process Biochem. 40: 3008-3016.

[93] Yiannikouris A, Andre G, Poughon L, Francois J, Dussap CG, Jeminet G, Bertin G, Jouany JP (2006) Chemical and conformational study of the interactions involved in mycotoxin complexation with β-D-glucans. Biomacromolecules. 7: 1147-1155.

[94] Raju MVLN, Devegowda G (2002) Esterified-glucomannan in broiler chicken diets-contaminated with aflatoxin, ochratoxin and T-2 toxin: Evaluation of its binding ability (*in vitro*) and efficacy as immunomodulator. Asian Australas. J. Anim. Sci. 15: 1051-1056.

[95] Oguz H, Parlat SS (2004) Effects of dietary mannanoligosaccharide on performance of Japanese quail affected by aflatoxicosis. S. Afr. J. Anim. Sci. 34: 144-148.

[96] Schatzmayr G, Heidler D, Fuchs E, Mohnl M, Täubel M, Loibner AP, Braun R, Binder EM (2003) Investigation of different yeast strains for the detoxification of ochratoxin A. Mycot. Res. 19: 124-128.

[97] Molnar O, Schatzmayr G, Fuchs E, Prillinger H (2004) *Trichosporon mycotoxinivorans* sp nov., a new yeast species useful in biological detoxification of various mycotoxins. Syst. Appl. Microbiol. 27: 661-671.

[98] Hickey PW, Sutton DA, Fothergill AW, Rinaldi MG, Wickes BL, Schmidt HJ, Walsh TJ (2009) *Trichosporon mycotoxinivorans*: A novel respiratory pathogen in patients with cystic fibrosis. J. Clin. Microbiol. 47: 3091-3097.

[99] Peteri Z, Teren J, Vagvolgyi C, Varga J (2007) Ochratoxin degradation and adsorption caused by astaxanthin-producing yeasts. Food Microbiol. 24: 205-210.

[100] de Felice DV, Solfrizzo M, De Curtis F, Lima G, Visconti A, Castoria R (2008) Strains of *Aureobasidium pullulans* can lower ochratoxin A contamination in wine grapes. Phytopathology. 98: 1261-1270.

[101] Hawkes M, Rennie R, Sand C, Vaudry W (2005) *Aureobasidium pullulans* infection: Fungemia in an infant and a review of human cases. Diagn. Microbiol. Infect. Dis. 51: 209-213.

[102] Varga J, Rigó K, Téren J (2000) Degradation of ochratoxin A by *Aspergillus* species. Int. J. Food Microbiol. 59: 1-7.

[103] Abrunhosa L, Serra R, Venâncio A (2002) Biodegradation of ochratoxin A by fungi isolated from grapes. J. Agric. Food Chem. 50: 7493-7496.

[104] Bejaoui H, Mathieu F, Taillandier P, Lebrihi A (2006) Biodegradation of ochratoxin A by *Aspergillus* section Nigri species isolated from French grapes: A potential means of

ochratoxin A decontamination in grape juices and musts. FEMS Microbiol. Lett. 255: 203-208.

[105] Valero A, Sanchis V, Ramos AJ, Marin S (2008) Brief *in vitro* study on *Botrytis cinerea* and *Aspergillus carbonarius* regarding growth and ochratoxin A. Lett. Appl. Microbiol. 47: 327-332.

[106] Engelhardt G (2002) Degradation of ochratoxin A and B by the white rot fungus *Pleurotus ostreatus*. Mycot. Res. 18: 37-43.

[107] Pitout MJ (1969) The hydrolysis of ochratoxin A by some proteolytic enzymes. Biochem. Pharmacol. 18: 485-491.

[108] Stander MA, Bornscheuer UT, Henke E, Steyn PS (2000) Screening of commercial hydrolases for the degradation of ochratoxin A. J. Agric. Food Chem. 48: 5736-5739.

[109] Abrunhosa L, Santos L, Venâncio A (2006) Degradation of ochratoxin A by proteases and by a crude enzyme of *Aspergillus niger*. Food Biotechnol. 20: 231-242.

[110] Abrunhosa L, Venâncio A (2007) Isolation and purification of an enzyme hydrolyzing ochratoxin A from *Aspergillus niger*. Biotechnol. Lett. 29: 1909-1914.

[111] Stinson EE, Osman SF, Bills DD (2006) Water soluble products from patulin during alcoholic fermentation of apple juice. J. Food Sci. 44(3): 788-789.

[112] Moss MO, Long MT (2002) Fate of patulin in the presence of the yeast *Saccharomyces cerevisiae*. Food Addit. Contam. 19: 387-399.

[113] Richelli A, Baruzzi F, Solfrizzo M, Morea M, Fanizzi FP (2007) Biotransformation of patulin by *Gluconobacter oxydans*. Appl. Environ. Microb. 73: 785-792.

[114] Tanasupawat S, Thawai C, Yukphan P, Moonmangmee D, Itoh T, Adachi O, Yamada Y (2004) *Gluconobacter thailandicus* sp. nov., an acetic acid bacterium in the alpfa-proteobacteria. J. Gen. Appl. Microb. 50, 159–167.

[115] El-Nezami HS, Chrevatidis A, Auriola S, Salminen S, Mykkanen H (2002) Removal of common Fusarium toxins in vitro by strains of Lactobacillus and Propionibacterium. Food Addit. Contam. 19: 680-686.

[116] Samar MM, Neira MS, Resnik SL, Pacin A (2001) Effect of fermentation on naturally occurring deoxynivalenol (DON) in Argentinean bread processing technology. Food Addit. Contam. 18: 1004-1010.

[117] Neira MS, Pacin AM, Martinez EJ, Molto G, Resnik SL (1997) The effects of bakery processing on natural deoxynivalenol contamination. Int. J. Food Microbiol. 37: 21-25.

[118] Garda J, Macedo RM, Faria R, Bernd L, Dors GC, Badiale-Furlong RE (2005) Alcoholic fermentation effects on malt spiked with trichotecenes. Food Control. 16: 423-428.

[119] Yiannikouris A, Jouany J (2002) Mycotoxins in feed and their fate in animals: a review. Anim. Res. 51: 81-99.

[120] Hanschmann G, Krieg D (2006) The fate of *Fusarium* toxins in the course of the synthesis of bioethanol from contaminated grain. Mycotoxin Res. 22: 174-177.

[121] Bennett GA, Richard JL (1996) Influence of processing on *Fusarium* mycotoxins in contaminated grains. Food Technol. 50: 235-238.

[122] Schaafsma AW, Limay-Rios V, Paul DE, Miller JD (2009) Mycotoxins in fuel ethanol co-products derived from maize: a mass balance for deoxynivalenol. J Sci. Food Agric. 89: 1574-1580.

[123] Scott PM (1992) Fermentation of wort containing deoxynivalenol and zearalenone. Mycotoxin Res. 8: 58-66.

[124] Böhm-Schraml M, Stettner G, Geiger E (1997) Studies into the influence of yeast on *Fusarium* toxins in wort. Cereal Res. Comm. 25: 729-730.

[125] Del Prete V, Rodriguez H, Carrascosa AV, Rivas BDL, Garcia-Moruno E, Munoz R (2007) In vitro removal of ochratoxin A by wine lactic acid bacteria. J. Food Protect. 70: 2155-2160.

[126] Niderkorn V, Boudra H, Morgavi DP (2006) Binding of *Fusarium* mycotoxins by fermentative bacteria *in vitro*. Appl. Environ. Microb. 101: 849-856.

[127] Rotter RG, Marquardt RR, Frohlich AA, Abramson D (1990) Ensiling as a means of reducing ochratoxin A concentrations in contaminated barely. J. Sci. Food. Agric. 50: 155-66.

[128] Dutton MF, Westlake K, Anderson MS (1984) The interaction between additives, yeasts and patulin production in grass silage. Mycopathologia. 87: 29-33.

[129] Camilo SB, Ono CJ, Ueno Y, Hirooka EY (2000) Anti-Fusarium moniliforme activity and fumonisin biodegradation by corn and silage microflora. Arquivos de Biologia e Tecnologia. 43: 159-64.

[130] Devegowda G, Raju MVLN, Afzali N, Swamy HVLN (1998) Mycotoxin picture worldwide: Novel solutions for their counteraction. In: Lyons TP, Jacques KA, editors. Biotechnology in feed industry. Nottingham: Nottingham University Press. pp. 241-255.

[131] Devegowda G, Arvind BIR, Morton MG (1996) Saccharomyces cerevisiae and mannanoligosaccharides to counteract aflatoxicosis in broilers. Proceedings of Australian poultry science symposium. Sydney. pp. 103-106.

[132] Raymond SL, Smith TK, Swamy HVLN (2003) Effects of feeding a blend of grains naturally contaminated with *Fusarium* mycotoxins on feed intake, serum chemistry, and hematology of horses, and the efficacy of polymeric glucomannan mycotoxin adsorbent. J. Anim. Sci. 81: 2123-12130.

[133] Karaman M, Basmacioglu H, Ortatatli M, Oguz H (2005) Evaluation of the detoxifying effect of yeast glucomannan on aflatoxicosis in broilers as assessed by gross examination and histopathology. Brit. Poultry Sci. 46: 394-400.

[134] Khatibi PA, Newmister SA, Rayment I, McCormick SP, Alexander NJ, Schmale DG (2011) Bioprospecting for trichothecene 3-o-acetyltransferases in the fungal genus fusarium yields functional enzymes with different abilities to modify the mycotoxin deoxynivalenol. Appl. Environ. Microbiol. 77: 1162-1170.

[135] Ingledew WM (1999) Yeast-Could you base a business on this bug? In: Lyons TP, Jacques KA, editors. Biotechnology in the Feed Industry. Nottingham: Nottingham University Press. pp.27-47.

[136] Stone CW (1998) Yeast Products in the Feed Industry: A Practical Guide for Feed Professionals. Cedar Rapids, IA: Diamond V Mills Inc.

[137] Wu F, Munkvold GP (2008) Mycotoxins in ethanol co-products: modeling economic impacts on the livestock industry and management strategies. J. Agric. Food. Chem. 56:3900-3911.

[138] Mielenz JR (2001) Ethanol production from biomass: technology and commercialization status. Curr. Opin. Microbiol. 4:324-329.

[139] Madson PW, Monceaux DA (1995) Fuel ethanol production. The Alcohol Textbook. pp. 227-238.

[140] Bennett GA, Lagoda AA, Shotwell OL, Hesseltine CW (1981) Utilization of zearalenone-contaminated corn for ethanol production. J. Am. Oil. Chem. Soc. 58: 974-976.

[141] Niessen L, Donhauser S (1993) Fate of deoxynivalenol in the process of brewing and its prevalence in commercial beer. In: Scudamore KA, editor. Proceedings of UK Workshop Occurrence and Significance of Mycotoxins, Slough, Central Science Laboratory, MAFF. pp. 203-207.

[142] Scott PM, Lawrence GA (1995) Analysis of beer for fumonisins, J. Food Prot. 58: 1379-1382.

[143] Scott PM, Lawrence GA (1997) Determination of aflatoxins in beer. J. AOAC Int. 80: 1229-1234.

[144] Baxter ED (1996) The fate of ochratoxin A during malting and brewing. Food Addit. Contam. 13(suppl): 23-24.

[145] Hanschmann G, Krieg D (2006) The fate of *Fusarium* toxins in the course of the synthesis of bioethanol from contaminated grain. Mycotoxin Res. 22: 174-177.

[146] Böswald C, Engelhardt G, Vogel H, Wallnöfer PR (1995) Metabolism of the *Fusarium* mycotoxins zearalenone and deoxynivalenol by yeast strains of technological relevance. Nat. Toxins. 3: 138-144.

[147] Boeira LS, Bryce JH, Stewart GG, Flannigan B (1999) Inhibitory effect of *Fusarium* mycotoxins on growth of brewing yeasts. 2. Deoxynivalenol and nivalenol. J. Inst. Brew. 105: 376-381.

[148] Whitehead MP, Flannigan B (1989) The *Fusarium* mycotoxin deoxynivalenol and yeast growth and fermentation. J. Inst. Brew. 95: 411-413.

[149] Wang J, Pawelzik E, Weinert J, Wolf GA (2005) Impact of Fusarium culmorum on the polysaccharides of wheat flour. J. Agric. Food. Chem. 53: 5818-5823.

[150] Maloney DH, Foy JJ (2003) Yeast Fermentations. In: Kulp K, Lorenz K, editors. Handbook of Dough Fermentations. New York: Marcel Dekker Inc. pp.43-102.

[151] Reinehr CO, Furlong EB (2003) Profile of the alcohols produced in fermentations with malt contaminated with trichothecenes. Braz. Arch. Biol. Technol. 46: 587-593.

[152] Flesch P, Voight-Scheuerman I (1994) Über den Abbau der Pilztoxine Trichotecin und Iso-Trichotecin bei der alkoholischen Garung in Traubensaft. Wein und Weinwissenschaft. 49: 180-184

[153] Khatibi PA, Montanti J, Nghiem NP, Hicks KB, Berger G, Brooks WS, Griffey CA, Schmale DG (2011) Conversion of deoxynivalenol to 3-acetyldeoxynivalenol in barley-derived fuel ethanol co-products with yeast expressing trichothecene 3-*O*-acetyltransferases. Biotechnol. Biofuels. 4:26 doi:10.1186/1754-6834-4-26

Chapter 8

Climate Change: Implication for Food-Borne Diseases (Salmonella and Food Poisoning Among Humans in R. Macedonia)

Vladimir Kendrovski and Dragan Gjorgjev

Additional information is available at the end of the chapter

1. Introduction

Using the classic epidemiologic triad (host, agent, and environment), it is clear that climate, which impacts all three sectors of the triad, can have a dramatic effect on infectious disease. Evidence of the impact of climate change on the transmission of food and waterborne diseases comes from a number of sources, e.g. the seasonality of foodborne and diarrhoeal disease, changes in disease patterns that occur as a consequence of temperature, and associations between increased incidence of food and waterborne illness and severe weather events as "in reference [1,2]. There are also theoretical and unintended consequences of global climate change on food safety. Climatic factors influence the growth and survival of pathogens, as well as transmission pathways as "in [3]". Higher ambient temperatures increase replication cycles of food-borne pathogens, and prolonged seasons may augment the opportunity for food handling mistakes - in 32% of investigated food-borne outbreaks in Europe "temperature misuse" is considered a contributing factor [4].

Climate change associated diseases are estimated already to comprise 4.6% of all environmental risks. It has been estimated that climate change in the year 2000 contributed to about 2.4% of all diarrhoea outbreaks in the world, 6% of malaria outbreaks in certain developing countries and 7% of the episodes of dengue fever in some industrial countries. In total, the estimates show that mortality due to climate change has been 0.3%, whereas the related burden of disease has been 0.4% [5]. Climate change is linked to human health in a complex manner. There are direct impacts, such as diseases and conditions that may result in morbidity or mortality related to extreme temperatures, and other, more indirect health effects such as diseases related to consumption of contaminated drinking water, foodborne or vector-borne diseases and zoonoses, or health conditions related to lack of food and water. There are projections regarding the expansion of diseases from the southern to the northern latitudes, especially re-emerging diseases that had already been eradicated, such as malaria, yellow fever, etc. Changes have also been detected in the distribution of rodent borne diseases, such as the hantavirus disease and leptospirosis. Geographical, weather and environmental changes are likely to affect the vectors of disease and to have a corresponding impact on the distribution of diseases such as leishmaniasis, Lyme disease, tick-borne encephalitis, malaria (in endemic regions), dengue, etc. An increased burden of disease related to drinking water and food may be expected due to inadequate distribution at a global level and the projections for decreased availability of drinking water and food production (cholera and food poisoning). Exposure to extremely high temperatures may lead to cardiovascular or respiratory diseases, whereas extreme disturbances in climate conditions (floods, warm winds) may lead to injuries, choking, respiratory disorders, diarrhoea, etc. Increased temperature and floods are the cause of` an increase in water contamination and resulting food- and waterborne diseases. Climate change is also likely to have an impact on the distribution of aeroallergens, especially pollen, and thereby cause changes in the distribution of allergic diseases. On the positive side, health conditions related to extreme low temperatures will decrease [6,7]

Weather effects, especially related to temperature, act in an indirect manner as regards transmission of infectious diseases. Temperature may affect both the causes of infectious diseases and their carriers (vectors) or water supplies. The link between weather impacts and infectious diseases has led to the development of scenario models to predict the expansion of infectious diseases due to climate change. Changed lifestyles, food production, modern urban planning, climate change and variations in the quality of the environment increase the danger of expansion of zoonoses.

Before the prospect of anthropogenic climate change emerged, epidemiologists were not greatly interested in climate-health relations [8]. Modern epidemiology has focused mainly on studying risk factors for noncommunicable diseases in individuals, not populations. Meanwhile, there have been occasional studies examining deaths due to heatwaves, some epidemiological studies of air pollution incorporating temperature as a covariate, and a continuation of the longer standing research interest in meteorological effects on microbes, vectors, and infectious disease transmission. Overall, the health risks of climate-related thermal stress, floods, and infectious diseases have been the most amenable to conventional epidemiological studies. Can current effects be estimated, if not yet directly observed? The current burden of disease attributable to climate change has been estimated by WHO as part of the Global Burden of Disease (2000) project as "in reference [5]", as a comprehensive standardised risk assessment exercise that underwent critical review. The estimation of the attributable burden was a statistical exercise that entailed three steps: (i) estimation of the baseline average annual disease burden in 1961–90; (ii) specification (from published work) of the increase in disease risk per unit increase in temperature or other climate variable; and (iii) estimation, by geographic region, of the current and future global distributions of population health effects of the change in climate. The extent of climate change (relative to the 1961–90 average climate) by the year 2000 is estimated to have caused in that year around 160 000 deaths worldwide and the loss of 5 500 000 disability-adjusted life-years (from malaria, malnutrition, diarrhoeal disease, heatwaves, and floods). This exercise was conservative in several respects, including being limited to quantifiable health outcomes. Nevertheless, is it reasonable to attribute a proportion of global deaths from malaria, malnutrition, or other such outcomes in 2000, to the global warming that has taken place since around 1975? The fact that equivalent estimations are routinely made for other such relationships involving a disease with known multivariate causation—eg, the proportion of all stroke deaths in 2000 attributable to hypertension—suggests that, in principle, wherever a well documented exposure-effect relation exists, the incremental change in health outcome can legitimately be estimated for an incremental exposure (eg, temperature) [8]. Climate change is perhaps the most significant environmental problem which mankind will face in the coming century. Efforts to reduce the extent of climate change are of course important, but it is likely that we will have to deal with at least some impacts on health. Climate change is linked to human health in a complex manner. There are direct impacts, such as diseases and conditions that may result in morbidity or mortality related to extreme temperatures, and other, more indirect health effects, such as diseases related to consumption of contaminated drinking water, foodborne or vector-borne diseases and zoonoses, or health conditions related to lack of food and water. Cimate change will have consequences for the health of Macedonian citizens as well. In 2006 a report on the Vulnerability and Adaptation of Climate Change in Health Sector has been published [9]. This report was the first of its kind in the country, insofar as it sought to provide quantitative estimates of the possible impacts of climate change on health.

2. Climate change and food security and utilization

The Food and Agriculture Organization (FAO) defines food security as a "situation that exists when all people, at all times, have physical, social, and economic access to sufficient, safe, and nutritious food that meets their dietary needs and food preferences for an active and healthy life" [10]. This definition comprises four key dimensions of food supplies: availability, stability, access, and utilization. The first dimension relates to the availability of sufficient food, i.e., to the overall ability of the agricultural system to meet food demand. Its subdimensions include the agro-climatic fundamentals of crop and pasture production [11] and the entire range of socio-economic and cultural factors that determine where and how farmers perform in response to markets.

Climate change affects agriculture and food production in complex ways. It affects food production directly through changes in agro-ecological conditions and indirectly by affecting growth and distribution of incomes, and thus demand for agricultural produce. Impacts have been quantified in numerous studies and under various sets of assumptions [12].

Depending on the SRES emission scenario and climate models considered, global mean surface temperature is projected to rise in a range from 1.8°C (with a range from 1.1°C to

2.9°C for SRES B1) to 4.0°C (with a range from 2.4°C to 6.4°C for A1) by 2100. In temperate latitudes, higher temperatures are expected to bring predominantly benefits to agriculture: the areas potentially suitable for cropping will expand, the length of the growing period will increase, and crop yields may rise. A moderate incremental warming in some humid and temperate grasslands may increase pasture productivity and reduce the need for housing and for compound feed. These gains have to be set against an increased frequency of extreme events, for instance, heat waves and droughts in the Mediterranean region or increased heavy precipitation events and flooding in temperate regions, including the possibility of increased coastal storms [13]; they also have to be set against the fact that semiarid and arid pastures are likely to see reduced livestock productivity and increased livestock mortality. In drier areas, climate models predict increased evapotranspiration and lower soil moisture levels. As a result as "in [10]", some cultivated areas may become unsuitable for cropping and some tropical grassland may become increasingly arid. Temperature rise will also expand the range of many agricultural pests and increase the ability of pest populations to survive the winter and attack spring crops.

Global and regional weather conditions are also expected to become more variable than at present, with increases in the frequency and severity of extreme events. By bringing greater fluctuations in crop yields and local food supplies and higher risks of landslides and erosion damage, they can adversely affect the stability of food supplies and thus food security.

Climate change will also affect the ability of individuals to use food effectively by altering the conditions for food safety and changing the disease pressure from vector, water, and food-borne diseases.

The main concern about climate change and food security is that changing climatic conditions can initiate a vicious circle where infectious disease causes or compounds hunger, which, in turn, makes the affected populations more susceptible to infectious disease. The result can be a substantial decline in labor productivity and an increase in poverty and even mortality. Essentially all manifestations of climate change, be they drought, higher temperatures, or heavy rainfalls have an impact on the disease pressure, and there is growing evidence that these changes affect food safety and food security

Extreme rainfall events can increase the risk of outbreaks of water-borne diseases particularly where traditional water management systems are insufficient to handle the new extremes [11]. Likewise, the impacts of flooding will be felt most strongly in environmentally degraded areas, and where basic public infrastructure, including sanitation and hygiene, is lacking. This will raise the number of people exposed to water-borne diseases and thus lower their capacity to effectively use food.

Vulnerability refers to the degree to which a system or societies are susceptible to, and unable to cope with, adverse effects of climate change, including climate variability and extremes. Vulnerability is a function of the character, magnitude and rate of climate change and variation to which a system is exposed, its sensitivity and its adaptive capacity [12]. Since impacts and adaptive capacity of systems may vary substantially over the next decades and within countries, vulnerabilities can be highly dynamic in space and time. Consequently, there is a strong need to build resilient agricultural systems that have a high capacity to adapt to stress and changes and can absorb disturbances. Impacts of climate change on food security are global and local. Climate change will affect agricultural food systems in all countries, including exporters and importers as well as those at subsistence level. Many impacts, such as increased land degradation and soil erosion, changes in water availability, biodiversity loss, more frequent and more intense pest and disease outbreaks as well as disasters need to be addressed across sectors. To describe the effect of climate change on a more global scale, the World Health Organization has released data regarding the estimated effects on human health as of the year 2004. What is readily apparent from these data is that developing regions of the world have been disproportionately affected by climate change relative to developed regions. This imbalance stands in stark contrast to the imbalance in greenhouse gas emissions, which are almost entirely attributable to developed countries, and countries with rapidly developing economies. The WHO report also includes estimates of the future global burden of disease that will result from climate change. It is predicted that by 2030 there will be 10% more diarrhoeal disease than there would have been with no climate change and that it will primarily affect the health of young children; indeed, the impact on children might well be amplified by the effects of such diseases on malnutrition, development and cognition. If global temperatures increase by 2–3°C, as expected, it is estimated that the population at risk for malaria will increase by 3–5%, which means that millions of additional people would probably become infected with malaria each year [14].

3. Current impacts of climate and weather

Nowadays, a wide range of events shape the behaviour and social interactions of the human host. The spread of childhood communicable diseases mirrors school calendars and community activities. Holidays promote travel and new social interactions, increasing the spatial distribution of disease transmission, even more efficiently vectored through packed planes and other modes of mass transportation. Seasonal shifts in immunity and host susceptibility, exacerbated by increased exposure through crowds during the colder months, will also increase patterns of infectious disease spread [15]. The first detectable changes in human health may well be alterations in the geographical range (latitude and altitude) and seasonality of certain infectious diseases – including vector-borne infections such as malaria and dengue fever, and foodborne infections (e.g. salmonellosis) which peak in the warmer months. Warmer average temperatures combined with increased climatic variability, would alter the pattern of exposure to thermal extremes and the resultant health impacts, in both summer and winter. By contrast, the public health consequences of the disturbance of natural and managed food-producing ecosystems, rising sea levels and population displacement for reasons of physical hazard, land loss, economic disruption and civil strife, may not become evident for up to several decades. New challenges associated with the emergence of large epidemics related to food consumption are arising as a result of globalization, increased trade in food products, increased consumption of fast food, international travel, environmental contamination by human faecal matter in areas with poor sanitation, the increased frequency of natural disasters related to climate change, the introduction of new technologies in food production processes, etc. There are different ways in which weather conditions can affect the incidence of foodborne diseases. Firstly, the prevalence of specific pathogenic organisms in animals may increase with higher temperatures. Secondly, the food cooling chain is harder to maintain in higher temperatures and prolonged warm weather increases the risk of mistakes in food handling. Thirdly, higher air temperatures may speed up the replication cycles of foodborne pathogenic organisms, which leads to a higher degree of contamination. Higher temperatures, in interaction with inadequate hygiene conditions, improper food handling, and lack of hand-washing, may lead to an increased number of epidemics resulting from consumption of unsafe food. In the Republic of Macedonia, foodborne and climate-sensitive pathogenic organisms causing the greatest concern in the context of climate change include the following:

Alimentary toxic infections (ATI) – These diseases were reported throughout the period 1991–2008, with fairly uniform prevalence each year. During the period there were a total of 26 092 cases of ATI, an average of 1450 cases a year. Total morbidity for the entire period was 1304.6 per 100 000, a yearly average of 72.4 per 100 000, with a clear tendency to maintenance. During the period, ATI continually ranked between fourth and sixth among the ten most frequently reported infectious diseases in the Republic of Macedonia, depending on whether ATI epidemics had been more common in any specific year. Syndromes related to ATI tend to be seasonal (with an increase during the summer months), with a few very large outbreaks reported in 2008, connected to specific closed communities and having one common source.

Shigellosis – In the Republic of Macedonia during 1991–2008, a total number of 2652 cases of shigellosis were reported, or 147 cases a year, with a total morbidity of 132.6 per 100 000 inhabitants for the entire period. The trend has significantly decreased over the last eight years, with the average being 35 reported cases each year. This is most likely due to improved access to safe food and drinking water as well as other provisions, proper and hygienic disposal of liquid and solid waste substances, and increased levels of health education and information among the general population regarding hygiene, safe food preparation, etc.

Campylobacter – The risk of infections caused by *Campylobacter* is directly proportional to the increase in temperature. Recent studies show increased incidence of campylobacteriosis at 2–5% per each degree Celsius rise of temperature, based on weekly temperature data. Notwithstanding that it is mandatory to report cases of campylobacteriosis in the Republic of Macedonia, there is currently no reliable information on its distribution, although estimates indicate that its incidence exceeds 18 000 cases annually.

Other foodborne pathogenic organisms causing concern in the context of climate change – These include *Brucella, Hepatitis A, E. coli* O157 H7 (EHEC) and bacteria causing bacterial food poisoning (e.g. *Clostridium perfringens*). As far as these pathogenic organisms are concerned, the effect of climate change remains within the area of speculation. However, due to their possible sensitivity to climate conditions and their importance for public health in the Republic of Macedonia, they have been included in the programmes for monitoring and prevention of climate-change-related infectious diseases. Such diseases are subject to mandatory reporting under the current legislation. Hepatitis A is constantly present in the Republic of Macedonia and there were 290 reported cases in 2009, 243 reported cases in 2008 and 257 reported cases in 2007.

Waterborne communicable diseases – About 10% of the population in the Republic of Macedonia still lacks access to clean and safe water, be it for drinking or for meeting their basic needs. In addition, there are year-on-year growing trends for certain groups of communicable diseases, especially those associated with contaminated food and water (salmonellas, alimentary toxic infections, shigelloses). Climate change will most probably have an impact on the incidence of waterborne infections, not only as a result of changing average meteorological parameters (e.g. rainfall), but also as a result of the increased frequency of extreme weather events, such as heavy rainfall, flash floods and droughts. Such extreme weather events will have an impact on the available quantity of water, on the quality of the water or on the availability of clean and safe water.

Waterborne pathogens include viral (*Hepatitis A*), bacterial (*Cryptosporidiae, E. coli*) and protozoan (*Giardia lamblia*) agents, which cause gastroenteritis. Waterborne diseases may even occur following adequate treatment of water. An example of this is the epidemic of cryptosporidiosis associated with the urban drinking water supply of Milwaukee, Wisconsin, USA in 1989, which resulted in 400 000 cases. Heavy rainfall may contaminate watercourses by bringing human and animal faecal products and other waste substances into surface waters. There is evidence of contamination of the water during heavy rainfall by *Cryptosporidium, Giardia* and *E. coli*. Contamination most commonly occurs in the event of high saturation of the soil due to a more efficient transport of microorganisms. Floods and low water levels may both lead to contamination of water and higher disease incidence and mortality due to diarrhoea. Warming and the higher variability of rainfall increase the risk of greater burden of these diseases. Pathogenic organisms identified as relevant for the former Yugoslav Republic of Macedonia in this context include:

Cryptosporidium – This has only recently been added to the list of infections that are mandatory to report; therefore, no details on incidence are available yet. No cases have been registered in the Republic of Macedonia so far.

Giardia lamblia – This has recently been added to the list of infections that are mandatory to report. At the moment no incidence data is available, other than information on laboratory isolates.

Leptospirosis – There is firm evidence showing that leptospirosis is affected by climate conditions. In the Republic of Macedonia, eight cases were reported in the period 1991–2008. Due to the lack of diagnostic facilities, it is assumed that a large number of cases have not been reported. Regions at high risk might include the rice fields in the region of Kocani, in addition to urban areas, riverbanks and lakes.

Vector-borne communicable diseases – Vector-borne infections are passed onto humans from arthropods or mammals, including rodents. Arthropod vectors, such as mosquitoes and ticks, are cold-blooded and thus especially sensitive to climatic factors. Climate change might have an impact on the distribution and the activity of arthropods. In addition, rodents are reservoirs of a large number of human diseases and the population of rodents is subject to the impact of weather conditions. Warm winters and warm springs may increase the population of rodents, a phenomenon that has been reported over the last few years.

Climate is an important factor for the distribution of vectors, in addition to other factors such as the destruction of their habitats, pest control and the density of hosts. Some vector-borne diseases sensitive to climate change have already been reported in the Republic of Macedonia (e.g. Lyme disease) and 4 human cases of West Nile fever in 2011 [16].

4. Temperature as a function of salmonella food poisoning cases in the Republic of Macedonia

The second largest number of human foodborne diseases is caused by the *Salmonella* spp. bacteria. In 2007, the European Union incidence was 31.1 cases per 100 000 population (151 995 confirmed cases), with eggs being the biggest contributors to these outbreaks, followed by fresh poultry and pork. Higher ambient temperatures have been associated with 5–10% higher salmonellosis notifications for each degree increase in weekly temperature, for ambient temperatures above 5°C. Roughly one-third of the transmission of salmonellosis (population attributable fraction) in England and Wales, Poland, the Netherlands, the Czech Republic, Switzerland and Spain can be attributed to temperature influences. Temperature has the most noticeable effect on salmonellosis and food poisoning notifications one week before disease onset, indicating inappropriate food handling and storage at the time of consumption. Indeed, an analysis of foodborne illnesses from England and Wales showed that the impact of the temperature of the current and preceding week has decreased over the past decades, indicating that the potential risk from elevated temperatures related to climate change can be counteracted through concerted public health action. Thus, regardless of climatic factors, health-behaviour interventions and food-safety regulations should be able to attenuate possible negative consequences on public health. Indeed, bacterial enteric infections have recently started to decrease throughout Europe, in part due to control measures [17,18].

Between 1971 and 2000, the annual mean temperature in the Republic of Macedonia increased by -0.1°C to 0.2°C in comparison to the period 1961–1991. Recorded values for the period 1996–2005 are 1.3°C higher for Demir Kapija and Prilep, 1.4°C higher for Stip and Bitola and 1.5°C higher for Skopje. Significantly higher mean annual temperatures were also recorded in 1999, 2002, 2003 and 2007, with the most dramatic variations of temperature recorded during the summer period. According to climate change scenarios, the average increase of temperature may reach 3.8°C in 2100 and the average decrease in precipitation may be 13% compared with 1970–1990 averages [16].

There are different ways in which weather conditions can affect the incidence of foodborne diseases. Firstly, the prevalence of specific pathogenic organisms in animals may increase with higher temperatures. Secondly, the food cooling chain is harder to maintain in higher temperatures and prolonged warm weather increases the risk of mistakes in food handling. Thirdly, higher air temperatures may speed up the replication cycles of foodborne pathogenic organisms, which leads to a higher degree of contamination. Higher temperatures, in interaction with inadequate hygiene conditions, improper food handling, and lack of hand-washing, may lead to an increased number of epidemics resulting from consumption of unsafe food. In the Republic of Macedonia, foodborne and climate-sensitive pathogenic organisms causing the greatest concern in the context of climate change include the following:

Salmonellosis – Recent studies on foodborne diseases show that disease episodes caused by *Salmonella* bacteria increase by 5-10% per each degree Celsius rise in temperature. During 1991–2008, 6969 cases of salmonellosis were reported in the Republic of increasing trend Macedonia, with total morbidity of 340.3 per 100 000, or an average of 387 cases a year, with an in recent years [19].

5. Materials and methods

Data on reported cases of Salmonella infection for the period 1998-2008 were obtained from the national surveillance centre, i.e., Institute for Public Health of R. Macedonia for the city of Skopje (capital) and countrywide. Data of average maximum weekly temperature for the same period were obtained from the National Hydrometeorological Office. The following age groups were also modelled: young children (0–6 years), children (7–14 years); adults (15–59 years); and the elderly (60+years). We investigated the epidemiological characteristics of salmonellosis both at the national level and in the city of Skopje using a retrospective research as a method of analysis. We created a Seasonal Index for the same period for monthly distribution of the reported cases for Skopje and for the entire country. The monthly number of reported Salmonella cases for Skopje was related to the average monthly temperature on the same month using Regression statistical analyses. For the analytical approach we made use of Poisson regression model. These techniques helped us in assessing any short-term effects of temperature on the disease. The Statgraphics Centurion software package was used.

For the prediction of the burden of the ambient Skopje temperature in relation with salmonella cases among humans and for determining the current burden for the period 1998-2008 as attributable factor, the following model was used:

5.1. PB=FB-CB

CB_{months} (^{0}C) = MT_{months} (2001…2008) - MT_{months} (1991…2000)

PB-Predicted burden as estimation of difference between future burden and attributable current burden; **FB**-Future burden of the monthly mean temperature due to Climate change estimated by scenario , **CB**- Current burden as attributable fraction of the monthly mean temperature due to Climate change, **MT**- Monthly mean temperature

Then the current burden of the weather temperature was compared by forecasting the future burden due to climate change in the period ranging 2025-2050 and 2075-2100 for the central region of Macedonia where the city of Skopje belong. The scenario from the Second National communication on climate change was used [20].

6. Results

Skopje, the capital, is the most populous Macedonian city. According to an official estimate from 2009, 20.5% of the total population registered in the country (2,052,722) lives in Skopje [21].

During the period 1998-2008, nationally, 3,890 salmonella human cases were registered; 1,951 (50.1%) males and 1,939 (49.9%) females. S. Entiritidis with 90% and S. Typhimurium with 8% are predominant serovars causing human infections in the R. Macedonia. 1,085 salmonella cases were reported in Skopje for the same period with an average of 8.2 patients per month (28% of the total national average; about 29 patients per month). Specific morbidity distribution of salmonellosis (rate per 100,000) in Skopje and countrywide are shown as follows (Figure 1).

While a decreasing tendency was registered at the national level, the analysis for Skopje showed increasing tendency of salmonella incidence. The salmonellosis morbidity rate for Skopje in 1998 was 8.3/100,000; in 2000, the rate was 24.6/100,000 and in 2008 was 41.2/100,000.

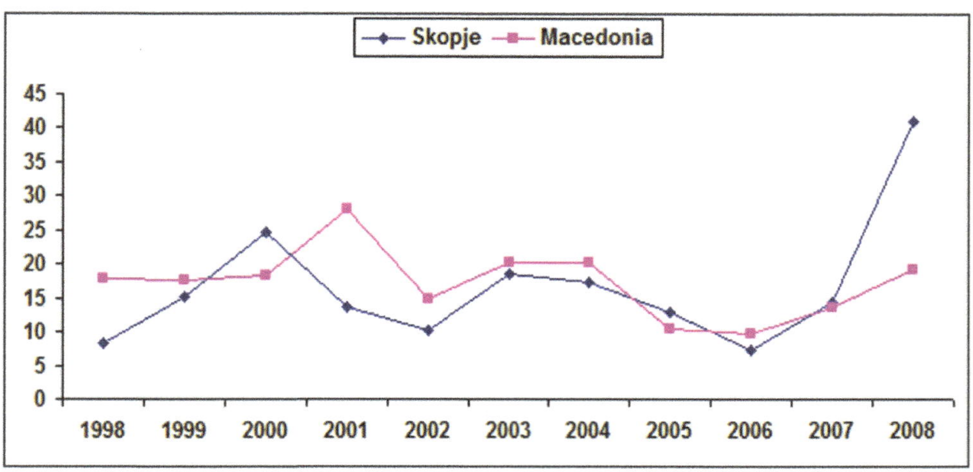

Figure 1. Reported Salmonella cases in humans in Macedonia and Skopje 1998-2008 (Mb/100.000).

Health Indicators	Total	Age groups								
		0 - 6	7 - 9	10 - 14	15 - 19	20 - 29	30 - 39	40 - 49	50 - 59	60 >
Skopje (Number of Salmonella cases)	1085	327	88	96	85	131	81	95	85	97
Specific Mb./ 100 000	142,4	646,1	400,9	239	198,7	141,8	90,3	117,6	108	103,8
Macedonia (Number of Salmonella cases)	3890	1023	334	459	297	416	394	385	272	310
Specific Mb./ 100 000	190,7	622,1	431,6	307	183,2	128,5	131,6	131,3	107,4	97,8

Table 1. Specific salmonella morbidity among humans (per 100,000) by age groups for Skopje and Macedonia during the period 1998-2008

The analysis of specific salmonellosis morbidity by age groups in Skopje showed higher morbidity among 0 to 6 year old children with 646.1/100,000 and lowest among adults from 30 to 39 years old with 90.3/100,000 (Table 1). At the national level, the higher morbidity was registered also among 0 to 6 years old children with 622.1/100,000 but the lowest was recorded among the elderly 60 years and above with 97.8/100,000.

The highest values of the Seasonal Index for Salmonella cases were registered in the summer months, i.e. June with 160.1%, July with 188.6%, August with 171.3% and September with 182.5%. The lowest reported Salmonella cases were registered in February with 20.7% (Figure 2).

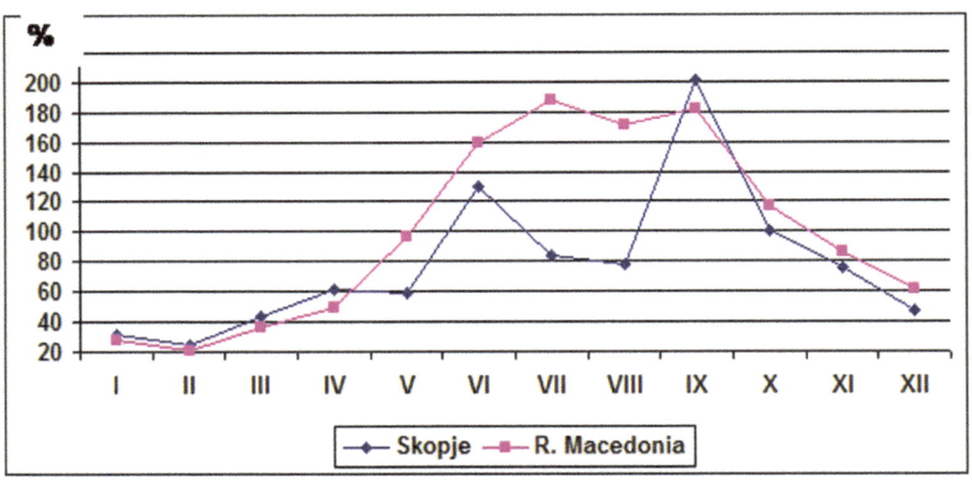

Figure 2. Seasonal Indexes for reported Salmonella cases in Macedonia and Skopje for the period 1998-2008 -distribution by months

Furthermore, in Skopje, the Seasonal Index for Salmonella cases showed two peaks in the summer months (September with 201.6%, and June with 130.5%). The lowest reported Salmonella cases were registered in February with 24.3%. The largest percentage of outbreaks of salmonellosis in the review period, were registered in the months with the highest seasonal index. A total of 42 outbreaks of salmonellosis or an average of 5 outbreaks per year was registered for the same period with a total 6,015 exposed persons. In these outbreaks, according to data obtained from epidemiological surveys, 1,871 persons (31.1% from total number of exposed) were registered as salmonella cases and 608 patients were hospitalized (32.4% from total registered patients). No deaths were recorded in those outbreaks.

The estimated correlation coefficient (0.54), indicates a moderately strong relationship between the monthly number of reported Salmonella cases for Skopje, and the average monthly maximum temperature at $p<0.05$. The 1 month lag time shows Pearson Correlation coefficient = 0.51 and 2 month lag shows Pearson Correlation coefficient = 0.49. Our investigation indicates that higher and sustained temperatures for longer periods of time are likely to lead to increasing cases of salmonellosis. The 1 month lag time of rising salmonella cases suggests that temperatures might be influential earlier in the production phase. The largest increase of air temperature in the next decades for the Republic of Macedonia is expected in the summer season, associated with a strong decrease in precipitation, due to climate change [22]. It is anticipated that there will be a corresponding rise in the incidence of salmonellosis.

The plot of Poisson distribution with 95% Confidence limits, between the monthly number of reported Salmonella cases for Skopje, and average monthly maximum temperature for the period 1998-2008 has been estimated (Figure 3).

The estimated rate ratio for Skopje is 1.052, which means that under conditions of increasing maximum monthly mean temperature of 1°C, salmonellosis incidence will increase 5.2% per month.

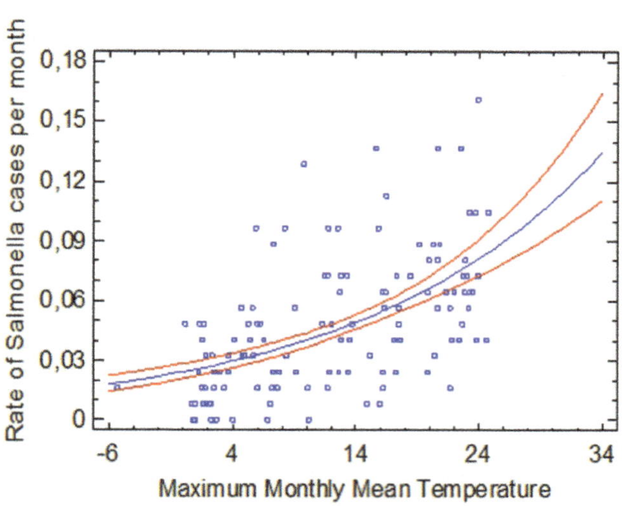

Figure 3. Plot of fitted Poisson Regression Model for Skopje

We detected the current burden of the weather monthly mean temperature for the city of Skopje in particular as follows: January (0,4^{0}C), February (-0,3 ^{0}C), March (2,1 ^{0}C), April (0,5 ^{0}C), May (0,4 ^{0}C), June (-0,5 ^{0}C), July (0,3 ^{0}C), August (0,7 ^{0}C), September (-0.7 ^{0}C) and October (1,1 ^{0}C), November (-0,7^{0}C), December (1,1^{0}C). According to the scenario the largest increase of air temperature in the Republic of Macedonia is expected in the summer season (1,4-5,4°C). We projected the changes of future burden of mean monthly air temperature (°C) by seasons using predicted burden as attributive fraction by estimation of difference between future and currant burden for Skopje in 21st Century (Tab.3)

	2025		2050		2075		2100	
	FB	PB	FB	PB	FB	PB	FB	PB
Winter	0,8	0,4	1,7	1,3	2,3	1,9	3	2,6
Spring	0,8	1	1,5	0,5	2,2	1,2	3,2	2,2
Summer	1,4	0,17	2,5	2,33	4,1	3,93	5,4	5,23
Autumn	0,9	0,37	1,7	1,33	2,8	2,43	3,7	3,33

Table 2. Projected changes of future burden of mean monthly air temperature (°C) by seasons as predicted burden by estimation of currant burden for Skopje in 21st Century

A projection of the Seasonal Index for the year 2030 in relation to increase of average monthly temperatures under various scenarios for climate change, show that in addition to two peaks in the summer months which were not that significant, there is also a possible peak in the colder months as a result of increase of average monthly temperatures in the future. The largest increase of air temperature in the next decades for the Republic of Macedonia is expected in the summer season, associated with a strong decrease in precipitation. Climate change in the Republic of Macedonia will have an impact in terms of higher air temperatures and reduced rainfall during the summer period. The scenarios show that the total available amounts of water (the river basin of the Vardar river) for the year 2100 will most probably be 18% less than today (estimates vary between 13% and 23%). In addition, more frequent flash flood and floods may be expected. Various parts of the country will suffer different impacts. The regions of the Republic of Macedonia with a Mediterranean climate are likely to experience reduced availability of water, increased number of dry periods and increased health-related impacts resulting from heat waves. Those regions with a continental climate are likely to suffer an increased number of floods and impacts resulting from extreme weather conditions. The local projections of climate change indicate that different climatic regions of Macedonia will respond slightly different on large-scale climate changes. The difference between a strong increase in temperature in summer season and weaker in winter season is not that evident as in sub-Mediterranean climate region where the city of Skopje belong. Although empirical downscaling projections of climate change on a local level contain uncertainties relating to the results, they present a step forward towards the need for implementing adaptation measures now. The Canadian Study showed that, for Alberta, the log relative risk of Salmonella weekly case counts increased by 1.2% for every degree increase in weekly mean temperature [23]. In our Study under conditions of increasing maximum monthly mean temperature for 1°C, the salmonellosis incidence increase for 5.2% per month. Similar higher ambient temperatures have been associated with 5-10% higher salmonellosis notifications for each degree increase in weekly temperature. In other 10 European countries, for ambient temperatures above 5°C, the estimated change in incidence above a common 6°C threshold ranged from 0.3% in Denmark to 12.5% in England and Wales [18]. The strongest effects were found for temperatures 1 week before the onset of illness rather than the longer lag of 1 month found in the Australian study. A significant positive association between mean temperature of the previous month and the number of salmonellosis notifications in the current month, with the estimated increases for a 1°C in temperature ranging from 4% to 10% in five Australian cities were reported [24]. In our Study, food poisoning by salmonellosis, was positively associated with ambient maximum temperature in the previous month, i.e., for each increase in temperature for 1°C resulted in 5.2% increase in salmonellosis notifications in the current month. In the UK, the monthly incidence of food poisoning was most strongly associated with the temperatures occurring in the previous two to five weeks [25]. The time lag of 1 month of rising salmonella cases suggests that temperatures might be influential earlier in the production phase [26]. Roughly one-third of the transmission of salmonellosis (population attributable fraction) in England and Wales, Poland, the Netherlands, the Czech Republic, Switzerland and Spain can be attributed to temperature influences [27]. In our investigation the higher and sustained temperatures for longer periods of time are likely to lead to increasing cases of salmonellosis. Indeed, an analysis of foodborne illnesses from England and Wales showed that the impact of the temperature of the current and preceding week has decreased over the past decades, indicating that the potential risk from elevated temperatures related to climate change can be counteracted through concerted public-health action [28].

7. Conclusion

The climate change process has already started and efforts should be concentrated towards the assessment of the current and future vulnerability of the population in the Republic of Macedonia, with the aim of identifying the necessary interventions and adaptation. The influence of global climate change, including its effects on people's health, can be reduced or avoided by undertaking measures for adaptation. The primary goal of the climate change adaptation measures in the health sector is to reduce the burden of disease, injury, invalidity, suffering and morbidity. The adaptation measures are not solely intended for the health sector, they are also relevant to other fields and sectors such as energy, sanitation and water supply, education, agriculture, economy, tourism, transport, development and housing, etc.

The incidence of Salmonella cases among humans in the Macedonian population varies seasonally, and may be expected to be change in response to global climate changes. During the review period, the highest values of the Seasonal Index for Salmonella cases were registered in the summer months, i.e. June, July, August and September. An understanding of how specific environmental factors influence human disease may improve disease forecasting; enhance the design of integrated warning systems; advance the development of efficient adaptation action plans; and, underline the need for implementing adaptation measures now.To demonstrate this technique, we conducted a comparative study of seasonality in *Salmonella* cases as reported by the state surveillance system in relation to seasonality in ambient temperature, and found that the incidence in *Salmonella* infection peaked two weeks after a peak in temperature. The limitations of such study are small numbers and under-reporting or late notifications. The results suggest that ambient temperature can be a potential predictor of *Salmonella* infections at a seasonal scale. It is clear that one overall challenge is the generation and maintenance of constructive dialogue and collaboration between public health, veterinary and food safety experts, bringing together multidisciplinary skills and multi-pathogen expertise. Such collaboration is essential to monitor changing trends in the well-recognised diseases and detect emerging pathogens. It will also be necessary understand the multiple interactions these pathogens have with their environments during transmission along the food chain in order to develop effective prevention and control strategies. Reducing the effects of communicable diseases related to climate change requires continuous epidemiological surveillance, as well as preparedness to take immediate epidemiological measures to respond to the threats. Furthermore, consideration should be given to investigating the routes of transmission and improving the safety of drinking water and food, controlling the insects and vectors that transmit disease, as well as providing a rapid response by the public health sector in the event of outbreaks.

8. Action in place

Climate change research needs to be properly coordinated and the benefits optimized to meet the needs of policy-makers in the country. Attention needs to be focused on data that will assist with mitigation of, and adaptation to, climate change and address specific areas of vulnerability. Further, national data are required to show the advantages and acceptability of a variety of technologies related to climate change.

A variety of methodologies of assessments of the potential health effects of climate variability and change in the Republic of Macedonia have been used. Both qualitative and quantitative approaches were used, as appropriate, depending on the data availability, level and type of knowledge.

The National Committee for Climate Change and Health within the Ministry of Health identified the most important climate-sensitive diseases and conditions that will be included in the health vulnerability assessment process.

Before the health impact assessment process started, a management structure was established to supervise each stage of the assessment. Identifying experts from the National Public Health Institute (the leading institution) was useful to ensure that the assessment was supervised through to completion. The Ministry of Health has primary responsibility for assessing and promoting the health of the population. The Second National Communication document was compiled by the Ministry of Environment and Physical Planning, with contributions from a wide range of experts, whose efforts are all gratefully acknowledged.

Funding for the Vulnerability Assessment Studies was provided by the German Federal Ministry for the Environment, Nature Conservation and Nuclear Safety, through the WHO Regional Office for Europe project "Protecting health from climate change" and the WHO Country Office in Skopje.

A basic assessment was conducted using readily available information and data, such as previous assessment, literature reviews by the IPCC and others and available region-specific health data. Limited analysis was conducted of regional health data, such as plotting the data against weather variables over time. A more comprehensive assessment included a literature search focused on the goals of the assessment, some quantitative assessment using available data (such as the incidence or prevalence of weather-sensitive diseases), more involvement by experts, some quantification of effects and a formal peer review of results. Some comprehensive assessment included a detailed literature review, collecting new data (i.e. mosquito survey in 2010) and/or generating new models to estimate impact, extensive analysis of quantification and sensitivity.

Interactions between weather and climate and health are location-specific – using epidemiological evidence based on local data, if they are available, is therefore important. The first step was to specify the scope of the assessment in the country in relation to climate change and health and community security issues of concern today and of potential risk in the future, both nationally (floods, droughts) and, where applicable, in the geographical region to be covered by the assessment (temperature-related mortality in Skopje, salmonella poisoning distribution in five Macedonian cities), and over time (projection for pollen distribution in future).

Health effects related to communicable diseases in the context of climate change are generally preventable, provided that the health care system is prepared and the population informed. The health care system should strengthen its functions as a leading sector that needs to have the capacity to protect the population and to work together with other government sectors, to establish a proactive, multisectoral and multidisciplinary approach. The activities encompassed by the health care sector should include strengthening the capacities of health care practitioners and strengthening the laboratory diagnostic system for identification and diagnosis; obtaining knowledge; adaptation; and health promotion. Generally, it is expected (with some presumption of uncertainty) that the effect of rapid climate change on human health will be negative. Adverse effects are expected to include:

The variations in rainfall will **most probably** compromise the supply of fresh drinking water, thereby increasing the risk of waterborne diseases.

The higher temperatures and the variability of rainfall will **most probably** reduce food production in the least developed regions, thereby increasing the risk of malnutrition.

Climate change will **most probably** prolong the season of transmission of certain significant vector-borne diseases and will tend to change their geographical distribution, potentially allowing them to spread into regions characterized by lack of immunity among the population and/or lack of well-organized health care infrastructure.

The link between weather impacts and infectious diseases has led to the development of scenario models to predict the expansion of infectious diseases due to climate change. Changed lifestyles, food production, modern urban planning, climate change and variations in the quality of the environment increase the danger of expansion of zoonoses. The health care system has an important role in establishing adaptation, health promotion, prevention and response measures against the health risks related to climate change and communicable diseases, such as:

- strengthening existing public health capacities for early detection and adequate response to communicable disease outbreaks;
- anticipating the consequences of emerging communicable diseases possibly related to climate change; and
- raising awareness among the general population about the possible links between climate change and communicable diseases.

Measures for adaptation to climate-change-related health risks are aimed at reducing the effects of climate change on human health and they can be categorized as follows:

- **primary adaptation measures:** measures aimed at preventing the initiation of disease occuring as a consequence of certain environmental conditions among the exposed population;
- **secondary adaptation measures:** preventive measures aimed at providing a response to the early evidence of impacts on health (e.g. strengthening disease control and providing an adequate response to the disease); and
- **tertiary adaptation measures:** health care measures aimed at reducing the mortality or morbidity caused by disease (e.g. improved diagnostics and treatment of certain infectious diseases).

Adaptation to potential consequences of climate change on communicable diseases at local and regional levels encompasses public health measures in the following fields:

- establishing early warning systems;
- systematic control and surveillance of foodborne, waterborne and vector-borne diseases;
- upgrading existing facilities for laboratory diagnosis and expertise;
- promoting and improving the health education of the general population, promoting hygiene measures among the population and enforcing environmental protection measures.

Adaptation measures form part of the National Climate Change Health Adaptation Strategy. As regards the surveillance of communicable diseases, the Republic of Macedonia has in place a syndrome-based early warning system (EWARN system that includes reporting upon eight syndromic diseases, such as diarrhoea, outbreaks, acute haemorrhagic fevers, etc.) and a system for mandatory reporting of diseases under the Infectious Diseases Protection Law. In addition, it is necessary to provide adequate laboratory confirmation capacity as well as capacity for continuous education of medical staff regarding the health risks associated with climate change. In this regard, the key activity for the health sector must be health promotion and improvement of health education for the general population, as well as the promotion of good hygiene practices, HACCP system in whole food chain. Health education campaigns should promote good hygiene, and include guidance on the safe preparation of food, education about avoiding certain foods in specific climate conditions, and sanitary-hygienic knowledge for individuals with their own water supply and food production facilities. Education and information of the public should especially be targeted to those parts of the country that are at higher risk due to shortage or lack of water. Key activities for health sector institutions should include health education and information for the public; preparation of health advocacy materials, such as posters and leaflets providing information about infectious diseases, and distribution thereof; and media campaigns for health promotion [16] On the basis of analysis of the Vulnerability Assessment Report in the Republic of Macedonia, the Climate Change Adaptation Strategy [29] was endorsed in 2011 by the government with the following priority domains of action:

- raising awareness of climate change and the effect on health in the former Yugoslav Republic of Macedonia;
- identifying, registering and monitoring risks connected with climate change and their influence on people's health; and
- improving the health system in its promotion, prevention and timely response to climate change risks for people's health.

The following specific goals are envisaged as part of the implementation of the Climate Change Adaptation Strategy in relation with food:

- Provide a coordinated approach and functional cooperation between the sectors and the relevant institutions in terms of effective and efficient use of the available resources.
- Raise public awareness about climate change and its effect on health
- Establish an integrated, efficient and effective approach for prevention, early warning, management and overcoming of the effects of climate change due to heat-waves including food safety advices .
- Establish an integrated, efficient and effective approach for prevention, early warning, management and overcoming of the effects of floods and fires.
- Protect people from climate-change-related communicable diseases.

Author details

Vladimir Kendrovski* and Dragan Gjorgjev
Medical Faculty, University "St. Cyrile and Methodius",
Department for Hygiene and Environmental Health, Skopje, R. Macedonia

Acknowledgement

This study was conducted under the activities for Macedonia within the project "Protecting health from climate change" implemented by the World Health Organization, Regional

* Corresponding Author

Office for Europe with support by the German Federal Ministry for the Environment and the Nature Conservation and Nuclear Safety.

9. References

[1] Hall GV, D'Souza RM, Kirk MD (2002) Foodborne disease in the new millennium: out of the frying pan and into the fire? Med. J. Aust. 177:614-618.
[2] Rose JB, Epstein PR, Lipp EK, Sherman BH, Bernard SM, Patz JA (2001) Climate variability and change in the United States: potential impacts on water- and foodborne diseases caused by microbiologic agent. Environ. Health Perspectives 109:211-221.
[3] Food and Agriculture Organization (2008) Climate Change: Implications for Food Safety. Food and Agriculture Organization, Rome
[4] Tirado C, Schmidt K (2001). WHO surveillance programme for control of foodborne infections and intoxications: preliminary results and trends across greater Europe. J Infect. 43(1):80-4.
[5] World Health Organization (2002). World Health Report. Geneva
[6] Patz, JA et al. (2000), Effects of environmental change on emerging parasitic diseases. Int J Parasitol, .30 p 12-13
[7] Wilson ML (2001) Ecology and infectious disease, in Ecosystem Change and Public Health: A Global Perspective, Aron JL, Patz JA Editors. Johns Hopkins University Press: Baltimore. p. 283-324.
[8] McMichael JA, Woodruff ER, Hales S (2006) Climate change and human health: present and future risks. Lancet 367: 859–69
[9] Kendrovski V (2006). Vulnerability and Adaptation of Climate Change in Health Sector, under the Second National Communication to UNFCCC. UNDP, MoEPP. Skopje
[10] Food and Agriculture Organization (2002) The State of Food Insecurity in the World 2001.Food and Agriculture Organization, Rome
[11] Schmidhuber J, Tubiello NF (2007) Global food security under climate change. PNAS; Vol 104: 50: 19703-19708
[12] Intergovernmental Panel on Climate Change. Climate Change (2007) Impacts, Adaptation and Vulnerability, Contribution of Working Group II to the Fourth Assessment Report of the Intergovernmental Panel on Climate Change, Cambridge Univ Press, Cambridge, UK
[13] Rosenzweig C, Tubiello FN, Goldberg RA, Mills E, Bloomfield J (2002) Global Environ Change 12: 197–202.
[14] World Health Organization (2004). World Health Report. Geneva
[15] Altizer S, Dobson A, Hosseini P, Hudson P, Pascual M, Rohani P (2006). Seasonality and the dynamics of infectious diseases. Ecol Lett 9:467–484.
[16] Kendrovski V, Milenkovik Z, Karadzovski Z, Spasenovska M, Kostovska J, Kisman M (2011) Guidelines for prevention of communicable diseases related to climate change in the former Yugoslav Republic of Macedonia. World Health Organization, Copenhagen

[17] European Food Safety Authority (2009). The Community Summary Report on Trends and Sources of Zoonoses and Zoonotic Agents in the European Union in 2007. The EFSA Journal 223:1–215.
[18] Kovats RS, Edwards SJ, Hajat S, Armstrong BG, Ebi KL, Menne B (2004). The effect of temperature on food poisoning: a time-series analysis of salmonellosis in ten European countries. Epidemiol Infect 132(3):443–453.
[19] Kendrovski V, Karadzovski Z, Spasenovska M (2011). Ambient maximum temperature as a function of Salmonella food poisoning cases in the Republic of Macedonia North Am J Med Sci 3: 264-267
[20] Ministry of Environment and Physical Planning of the Republic of Macedonia (2008). Second national communication under the United Nations Framework convention on climate change. Skopje
[21] State Statistical Office (2010). Macedonia in figures, 2010. State Statistical Office, Skopje
[22] Alcinova SM, Ristevski P, Pavlovska V, Todorovska S (2010) Projection of Temperature and Precipitation Changes in the XXI Century on the Territory of Republic of Macedonia, Fourth International Scientific Conference BALWOIS 2010, Ohrid, Republic of Macedonia.
[23] Fleury M, Charron DF, Holt JD, Allen OB, Maarouf AR (2006) A time series analysis of the relationship of ambient temperature and common bacterial enteric infections in two Canadian provinces. Int J Biometeorol. 50(6): 385-391.
[24] D'Souza RM, Becker NG, Hall G, Moodie KBA (2004) Does Ambient Temperature Affect Foodborne Disease? Epidemiology 15(1): 86-92.
[25] Bentham G, Langford IH (2001) Environmental temperatures and the incidence of food poisoning in England and Wales. Int J Biometeorol 45: 22–26.
[26] Kovats RS, Campbell-Lendrum D, McMichael AJ, Woodward A, Cox J (2001) Early effects of climate change: do they include changes in vector-borne disease? Philos Trans R Soc Lond B Biol Sci 356: 1057–1068.
[27] Lake IR, Gillespie IA, Bentham G, Nichols GL, Lane C, Adak GK, Threlfall EJ (2009) A Re-evaluation of the Impact of Temperature and Climate Change upon Foodborne Illness. Epidemiol Infect 137(11): 1538-1547.
[28] Semenza JC, Menne B (2009) Climate Change and Infectious Diseases in Europe. Lancet Infec Dis 9: 365-75.
[29] Ministry of Health of the Republic of Macedonia (2011) Adaptation Strategy in Health Sector from Climate Change of the Republic of Macedonia. Skopje

Chapter 9

Possible Risks in Caucasians by Consumption of Isoflavones Extracts Based

Maria Graça Campos and Maria Luísa Costa

Additional information is available at the end of the chapter

1. Introduction

With the increase in human life expectancy many health concerns start to be important, which raises the need for more and different therapeutic solutions.

One of the emerging issues that attracted the attention of research teams looking for new and better medicines is prevention of hormonal dependent tumors in both genders and the relief of climacteric symptoms in post-menopausal women, as the prevention of cardiovascular diseases (CVD).

For both of the above situations testosterone, estrona and estradiol metabolism are fundamental for a possible cardiovascular effect or a desoxinucleic acid (DNA) damage in breast cancer. Phytoestrogens, as isoflavones, in a first approach, seems to be potentially useful solutions. These compounds have the ability to mimic estrogens and induce an estrogen-like effect dependent on their affinity for alpha and beta estrogen receptors (ERs) [1], and this is related to their chemical structures.

Phytoestrogens were discovered by Bennetts *et al,* who were trying to identify the cause of a specific sheep breeding problem in Western Australia. They implicated equol, a metabolite of daidzin that is an isoflavone existing in subterranean clover pastures [2]. Four decades later, in 1984, this compound was identified in human Urine by Setchell *et al* [3] and related to consumption of soy and further implicated in a possible prevention of hormonal dependent tumors, cardiovascular diseases and osteoporosis.

At that time, the world was focused on the *French Paradox* (data collected with French population) that claimed the ingestion of antioxidants, mainly from red wine as the alcohol itself, could be a panacea for cancer prevention and cardiovascular diseases. According to data from the world's largest study of heart disease, conducted by the World Health Organization (WHO), during the past decade in 21 countries with 10 million men and women, French heart disease statistics appeared to have been under-estimated and the "*French Paradox*" overestimated. The French rate of heart disease was actually similar to that of the Italians, Spanish, and the Germans (mainly southern Germany), but still lower than many other countries.

Despite that, other researchers looking in a different way for the same reasons made epidemiological studies on Asian populations in the '90s that revealed a possible relationship between the ingestion of soy and the prevention of hormonal dependent tumors and also a lower incidence of climacteric symptoms in postmenopausal women. These results, known as the "*Japanese Paradox*" (data collected with Japonese population), induced Western people to consume soy and soy derivatives rich in isoflavones as genistein, daidzein, glycitein and the respective glucosidic forms (genistin, daidzin and glycitin). A large number of studies have attempted to demonstrate that soy consumption decreases the risk of developing several chronic diseases, in particular, cancer, osteoporosis [4, 5], cardiovascular diseases [6], and also the relief of climacteric symptoms [7]. Up until now however, the majority of these benefits have not been proven [for review see 9].

These two epidemiologic studies are mainly focus on prevention of cardiovascular diseases and cancer, using phenolic compounds as the targets for this bioactivity. The first one, *French paradox,* fail and leave behind an important side effect, cirrhosis, in people that increase the intake of wine all over this last decades. Related to the "*Japanese Paradox*", the scientific community starts to have data enough to think that this concept needs to be evaluated with caution to prevent a future failure.

Until now, in Europe isoflavones are considered as food compounds, nevertheless they are antinutrient compounds, and is the European Food Safety Authority (EFSA) that is responsible for the risk assessment evaluation of them. They recognize the potential importance for human health of the issue of isoflavones from food digestion, but should the alleged beneficial or detrimental health effects be scientifically proven. The Isoflavones ESCO working group evaluates the relevant scientific information available. This includes *inter alia*: to assess the potential of isolated isoflavones to trigger adverse human health effects; the possible human health benefits of the use of isolated isoflavones for the general population and particularly for women with complaints during and after the menopausal period; whether there is any scientific basis for differences concerning the hazard assessment of isolated isoflavones from soy and/or red clover in comparison with soy or red clover extracts. The final document will be public and available at EFSA web site soon (http://www.efsa.europa.eu/en/esco/escoisoflavones.htm).

For this reason in the present chapter we'll not focus an overview of the isoflavones, as that is the current work of, but we'll discuss other points of view that can contribute for understanding of the topic especially for cancer and relief of climacteric symptoms. In this last point, recently (2010) Rebbeck *et al* [12] presented a study were they evaluated whether genes involved in the metabolism of steroid hormones are associated with hormone levels or menopausal symptoms. They used a population-based prospective sample of 436 African American (AA) and European American (EA) women who were premenopausal at enrollment and were followed longitudinally through menopause. Evaluation of the relationship between steroid hormone metabolism genotypes at catechol O-methyltransferase (COMT), cytochrome P450 (CYP) as the isoformes CYP1A2, CYP1B1, CYP3A4 and CYP19, Sulfotransferase 1A1 (SULT1A1), and SULT1E1 with hormone levels and menopausal features were carried out. The results show in EA women, SULT1E1 variant carriers had lower levels of dehydroepiandrosterone sulfate, and SULT1A1 variant carriers had lower levels of estradiol, dehydroepiandrosterone sulfate, and testosterone compared with women who did not carry these variant alleles. In AA women, CYP1B1*3 genotypes were associated with hot flashes (odds ratio [OR], 0.62; 95% CI, 0.40-0.95). Interactions of CYP1A2 genotypes were associated with hot flashes across menopausal stage ($P = 0.006$). Interactions of CYP1B1*3 ($P = 0.02$) and CYP1B1*4 ($P = 0.03$) with menopausal stage were associated with depressive symptoms. In EA women, SULT1A1*3 was associated with depressive symptoms (OR, 0.53; 95% CI, 0.41-0.68) and hot flashes (OR, 2.08; 95% CI, 1.64-2.63). There were significant interactions between SULT1A1*3 and hot flashes ($P < 0.001$) and between SULT1A1*2 and depressive symptoms ($P = 0.007$) on menopausal stage, and there were race-specific effects of SULT1A1*2, SULT1A1*3, CYP1B1*3, and CYP3A4*1B on menopause. These results suggest that genotypes are associated with the occurrence of menopause-related symptoms or the timing of the menopausal transition [10].

Isoflavones as we'll explain later in this chapter will interfere with most of these enzymes and Caucasians and Asians present polymorphic changes, in some of them, that can change the bioactive response to various situations, from compounds metabolism to cancer induction or prevention.

2. Chemistry, main food sources, bioactivity and possible toxicity of isoflavones related to ethnic differences

Isoflavones are biologically active heterocyclic phenolic compounds (subgroup of isoflavonoids) (Figure 1) that are absorbed by the intestine, circulate systemically, and are eliminated by the kidneys and liver [11]. In plants isoflavones occur predominantly as β-glucosides (genistin, daidzin, glycitin), or as acetyl-β-glucosides and malonyl-β-glucosides [12, 13]. Genistein is the more abundant isoflavone in the majority of soy products and also the most active of these compounds, and being able to interact with the estrogenic receptors [14].

Traditionally the main food sources of isoflavones are soy and other beans and pulses, and also fermented soy foods, where the glucosides have been transformed into aglycones which are absorbed more efficiently than glucosides.

In the past these have been more commonly consumed by Asian populations, but are growing in popularity in Europe. Similarly, in recent decades a new generation of soy products have entered the market (e.g. yogurts, cheeses, soy milk drinks, infant formula's) and commonly consumed food products incorporating soy flour (e.g. bakery products) and protein isolates (e.g. meat products and soy meatless products such as soy burgers). More recently, the development of nutritional supplements rich in isoflavones has targeted niche markets in response to scientific research, that is still controversial, but that suggests a beneficial effect from these food components.

Isoflavones were first discovered in the 1930's, as a bioactive agent, following the disruption of estrogen action and increased infertility in sheep that had been grazing on red clover, thereby earning the often used name 'phytoestrogens'. Subsequently isoflavones have been shown to bind to, or indirectly interact with several key nuclear receptors, including hormonal (estrogen receptors alpha and beta [ERs], progesterone and androgen receptors), xenobiotic sensing receptors (Pregnane X receptors [PXR] and Peroxisome proliferator activated receptors [PPARs], and steroidogenic and hypothalamus-pituitary-thyroid (HPT) axis pathways. Isoflavones are structurally similar to the endogenous estrogen 17-beta estradiol, but much less potent on binding to the ERs, although genistein, the more active isoflavone, have a greater binding affinity for ER beta (43.9% related to 100% for estradiol [15]).

These compounds, as other phytoestrogens, are molecules with the ability to mimic estrogen pharmacological action through the linkage with ERs, and because of that are called *estrogen-like* molecules [15].

The benefits and the inherent risks associated to the ingestion of *estrogen-like* molecules are related to their binding affinity to beta-ERs and to alpha-ERs.

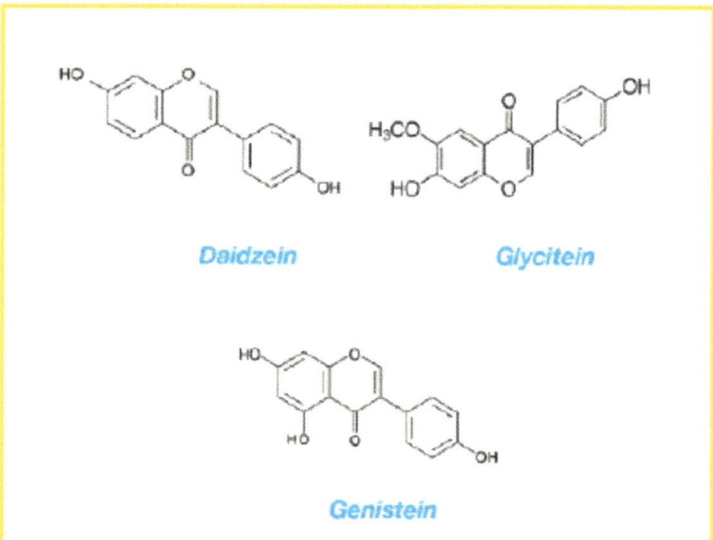

Figure 1. Chemical structures of iso-flavone equivalents

Beta-ERs are mainly located in bones, brain, thymus, bladder, and the cardiovascular system. Its activation by estrogenic compounds or estrogen-like compounds can probably improve or prevent conditions such as osteoporosis and cardiovascular diseases [16, 17]. In the other hand, alpha-ERs appear to predominate in breast, uterus and ovary; some authors report that linkage between estrogen or estrogen-like molecules and alpha-ERs is potentially dangerous. In breast cancer, for example, it could promote proliferation of damaged DNA [18,19]; and it can also lead to endometrial hyperplasia, as concluded in a randomized, double-blind, placebo-controlled study evaluating the safety of a long-term treatment (up to 5 years) with soy isoflavones. Although no cases of malignancy were detected, the hyperplasia called into question the long-term safety of phytoestrogens with regard to the endometrium [8].

Another important issue indicates a possible relationship between low doses of isoflavones and an increase in tumor proliferation, as it was pointed out by some authors.The same authors also concluded that, genistein and daidzein could not only stimulate pre-existing tumors, but also antagonize the effects of the tamoxifen (antitumour agent). Thus, women with breast cancer or with a history of a previous breast cancer event should be made aware of the risks of potential tumor proliferation while taking soy products [20]. Previously, Hsieh *et al*, also pointed out the possibility that lower amounts of genistein could stimulate alpha – receptors and influence the proliferation of breast cancer cells [22].

It is furthermore of particular relevance the fact that genistein can be estrogenic and antiestrogenic in human body. Clarify these mechanisms, as it was done for tamoxifen, can be crucial to the understanding of a possible benefit/risk ratio. For example, although tamoxifen has been useful both in treating breast cancer patients and in decreasing the risk of getting breast cancer in women at high risk, it also has some serious side effects. These side effects arise from the fact that while tamoxifen acts as an antiestrogen that blocks the effects of estrogen on breast cells, it mimics the actions of estrogen in other tissues such as the uterus. Its estrogen-like effects on the uterus stimulate proliferation of the uterine endometrium and increase the risk of uterine cancer [23]. The same was refered before in the data from, Unfer *et al*, (2004) [8].

The relevance of ethnic metabolism will be crucial to clarify some misunderstandings. As an example of the ethnic importance of polymorphic enzymes in metabolic pathways a meta-analysis conducted combining the data with 34 published studies that included 11 962 cancer cases and 14 673 controls in diverse cancers was carried out. The SULT1A1*2 revealed contrasting risk association for UADT cancers (OR=1.62, 95% CI: 1.12, 2.34) and genitourinary cancers (OR=0.73, 95% CI: 0.58, 0.92). Furthermore, although SULT1A1*2 conferred significant increased risk of breast cancer to Asian women (OR=1.91, 95% CI: 1.08, 3.40), it did not confer increased risk to Caucasian women (OR=0.92, 95% CI: 0.71, 1.18). Thus risk for different cancers in distinct ethnic groups could be modulated by interaction between genetic variants and different endogenous and exogenous carcinogens [24].

Even all these results were carried out with Europeans (which don´t consume soy very often) and others with Asians (which included this food many years ago in their normal diet plan), it isn't frequent to discuss their metabolic pathways.

As it is known, populations vary genetically and differ in the occurrence and frequency of particular genetic polymorphism depends on the time when the mutation occurred in relation to human migrations. If it occurred after the populations split, it will be unique to one population and its descendants. If it happened in the ancestor population, it is possible present in all descendent groups. A well-known example is the frequent ALDH2 deficiency in Asian individuals which is rarely found in Caucasians. As metabolizing enzymes are involved in the detoxification on endogenous and xenobiotic toxins, in the activation of procarcinogens, in the formation of reactive intermediates and in their neutralization, many studies have been undertaken concerning the correlation between genetic polymorphisms and metabolizing enzymes and cancer risk [25].

Ethnic differences in metabolism are a consequence of various factors as, for example, adaptation to different environments, differences in nutrition, behavior and cultural differences. This genetic polymorphism of metabolizing enzymes cause differences in effects and toxicity of the ingested compounds between individuals and all population. The so called "idiosyncratic drug interactions" are now explained on the basis of genetic polymorphisms. Beside the genetic variability differences can appear after enzyme induction or inhibition by the affected compound itself or by others or xenobiotics, e.g., from foods [25].

In Phase I metabolization, genistein, daidzein and equol (important metabolite of daidzein-7-*O*-glucoside) inhibit the Cytochrome P450 (CYP) 1A2, 1B2, 2E1, 3A4 and in phase II they induce UDP glucuronosyl transferase (UGT), Glutathione-S-Transferase (GST) and Quinone Redutase (QR); SULT1A1, SULT1A3 and SULT1E1 are inhibited for all of them [26].

CYP 1A2, for example, is important for estrogen metabolization (E1 and 2) in 2- or 4-OH E1 and 2 followed of the 4-OH E1 and 2 sulphate metabolite by SULT1A1*2 and *3, or to 2- and 4-OCH3 E1 and 2 by COMT. The inhibition of CYP 1A2 and SULT1A1, SULT1A3, probably can induce E1 and E2 to be metabolized by CYP 3A4 (*1B) to 16-alpha OH E1 and 2.

Rebbeck *et al*, also observed race-specific associations with CYP1B1, CYP1A2, and SULT1A1 on menopausal symptoms; race-specific effects of SULT1A1*2, SULT1A1*3, CYP1B1*3, and CYP3A4*1B on time to late premenopause, early menopausal transition, and menopause; and interactions of race with SULT1A1*2 and SULT1A1*3 on time to menopause [10].

The involvement of all these enzymes in the different ethnic groups constitutes a gap in the understanding of what can be used with benefit from other continents. The controversy persists regarding the role of a low ratio of 2-hydroxyestrone /16alpha-hydroxyestrone (2-OHE_1)/(16 -OHE_1) as a potential estrogen metabolism marker of increased risk for breast cancer [27]. Was suggested that soy consumption increases this ratio only in women who are equol producers given a possible protection against breast cancer [28].

The European American women that carry the CYP3A4*1B allele important in these last steps of metabolization of estrogens were indicated for a early menopause independent of the SULT1A1*2 or *3 type. The SULT1A1*2 (Arg213His) have a frequency of *ca* 30% for Caucasians and African Americans but only 8-17% for Asians [25].

GST present important genetic differences and is induce by isoflavones. The consequences of polymorphic GST isoenzymes are probably more relevant for carcinogenesis and for the detoxication of toxic xenobiotics and of chemotherapeutics than for drug metabolism in general. For example, 44-64% of the Asians present a GSTT1*0 polymorphism against 10-36% in Caucasians with the consequent difference in the metabolizing behavior [25].

This is also important for possible interactions between those molecules and some medicines, special in polimedicated patients. Daidzein, one of the principal isoflavones in soybean, can inhibit CYP1A2 activity and alter the pharmacokinetics of theophylline in healthy volunteers. Theophylline is a bronchodilator with a narrow therapeutic index (5–20 mg/L), and it is primarily eliminated by hepatic metabolism mediated by CYP1A2. [26].

Steroid hormone metabolism genes are not generally responsible for interindividual variation in steroid hormone levels or with changes in these levels across the menopausal transition. However a better understanding of all these mechanisms will be important to prevent future damage if we ingest compounds that will change the natural equilibrium of ethnic groups.

If these associations are confirmed, they may provide information about the prediction of menopausal symptoms and allow clinicians to individualize and target hormone therapy in women experiencing menopausal symptoms. Because hormone exposures, genotypes involved in hormone metabolism, and the phenotypic manifestations of these factors on symptoms are all associated epidemiologically with risk of cancer and other diseases, a better understanding of the role of genotypes and intermediate phenotypes such as hormone levels may ultimately assist our understanding of steroid hormone– related disease etiology and prevention [10].

So although, until now, the risk assessment for soy products is not clear enough to consider their consumption safe, humans are increasingly being exposed to isoflavones in soy and soy derivatives (other products containing isoflavones from different sources aren't so rich). Thus better information about the safety of these soy phytoestrogens is urgently needed.

3. Examples of products available to be consumed by European population

In this section is provided an overview (not exhaustive but presenting some discuss points) of the potential benefit/risk impact of nutritional products and supplements, that include isoflavones in their composition and that are recommended mainly for climacteric relief symptoms in post-menopausal women and/or prostatic cancer prevention in man and CVD prevention.

The European Prospective Investigation into Cancer and Nutrition (EPIC) study evaluated in 2002 the consumption of soy products in 10 European countries [29]. Results from this study revealed that soy consumption at that time was low in Western Europe and that non-dairy substitutes were the most frequently consumed items.

In this section we'll demonstrate the wide variation of isoflavone levels that these products may provide to the consumer. Because there are not recommendations of the optimal isoflavone consumption, the information displayed in these products should be clearly specified, especially in the case of health supplements. This is essential for accurate risk assessment studies, where knowing the exposure level of the population to these products and more specifically the exposition to each isoflavone is fundamental.

Nowadays, over the counter (OTC) tablet preparations [30, 31], nutraceuticals and various supplements with isoflavones extracted from soy [32] and other plants are sold in various countries. These products are often used by postmenopausal women to do hormonal therapy replacement as another option to allopathic medicines with oestrogen derivatives, and also with the claim of cancer prevention. However, the impact of that exposure in Caucasian women health is underestimated.

As a result of the promising scientific findings related to isoflavones, the soy-processing industry has grown worldwide from manufacturing alternative dairy products and milk-free infant formulas to provide a wider range of products embracing all types of tastes and consumer health concerns. These new products, sometimes named as second generation soy foods, are based in the inclusion of soy ingredients (soy pieces, soy powder, soy flour, soy protein, etc) among the ingredient list of food items commonly consumed by western populations.

Development of new soy products is especially important in Europe where the population does not accept the characteristic soybean flavor. Some years ago, only a few traditional soy foods were common items in European markets, and among those, soy nuts, soy sauce, tofu and soymilk were easily found, but the recent appearance of the so-called second-generation of soy foods; products made by adding soy ingredients (soy protein) to a wide variety of manufactured foods resulting in products which generally simulate familiar dishes [32].

A number of soy protein-based products are included, mainly meat substitutes (texturised soy protein) in a number of forms, as well as dairy-products substitutes. Same protein is added to enrich a number of foods commonly present in European diets: bread, cookies and soups or dietary supplements (powders) and energy bars used to complement sport diets. The latest version to use soy protein or isoflavone isolates appeared in the form of capsules or tablets directed, and labelled, to promote the claimed health benefits of soy isoflavones between the consumers [32].

It is well known that isoflavone levels in soy-based products depend primarily upon the soybean variety used. Isoflavone content within a single variety can differ three times from year to year [33]. In some studies, this variability has been attributed to climatic and environmental factors [34]. Moreover, soybean processing and storage conditions usually lead to a significant isoflavone loss in soy-based products [35]. In addition, isoflavone glucoside conjugates are easily alterable during extraction and cooking[36], and it has been even shown variations between products included in the same batch [32].

In some cases the presence of isoflavones is intended and therefore its concentration is known, but in the majority of the soy foods the levels of isoflavones need to be quantified in order to evaluate the possible health effect of these products.

Results from our previous work with some of the above cited products, selected soy-foods produced and distributed within Europe (Table 1), provide total isoflavone values for a possible evaluation of the potential exposure to these molecules [32].

Group	Designation	Examples	N
I	Traditional soy foods	Tofu, Soy beans, Soy nuts, Soy sauce	9
II	Non-dairy soy products	Soy-enriched plain yogurt, Soy milk yogurt, Soy milk dessert, Plain soy milk, Cream substitute	18
III	Meat analogues	Soy meat strips, Soy escallops, Meat pieces, Soy minced meat, Flavoured minced meat	5
IV	Second-generation soy foods	Pasta with broccoli and cheese, Energy bar with fruit and muesli, Powdered tomato soup, Soy-enriched orange cookies, Soy-enriched rice cookies, Crisp bread, Snack with almonds	24
V	Health supplements		59

Table 1. Groups of soy based foods and health supplements

In this work a total of 115 soy-based products were purchased at local retailers and natural health food stores in Finland, Spain, UK and Portugal during the years 2002-2005. Commercial availability and European manufacture were the only selection criteria. The selected items belong to different groups of soy foods and a classification was needed in order to allow the comparison between levels of isoflavones provided by the different products. Items were divided into four different groups: 1) Traditional soy foods, 2) Non-dairy soy products, 3) Meat analogues, 4) Second-generation soy foods and 5) Health supplements. Food samples were (when necessary) freeze-dried, homogenized and stored at -70 °C until analysis [32].

Soy isoflavones in samples were quantified as previously reported [30, 38, 39] using a HPLC systems equipped with a Coularray detector or diode array detection. Peaks corresponding to soy isoflavones were confirmed by LC/MS-MS as previously reported [39]. Synthetic standards were used for quantification through calibration curves, and control samples, were introduced between run to assure repeatability was acceptable (CV <15%) over the analysis time. All reagents were from major suppliers and HLPC-grade. Results are mean values of triplicate analyses, and were only accepted when coefficient of variation (CV) between the replicates was <15%. Only values above the limits of quantification of the method for each isoflavone (~1 ppm, depending on the analyte) are reported [39]. All values are expressed as aglycone equivalents.

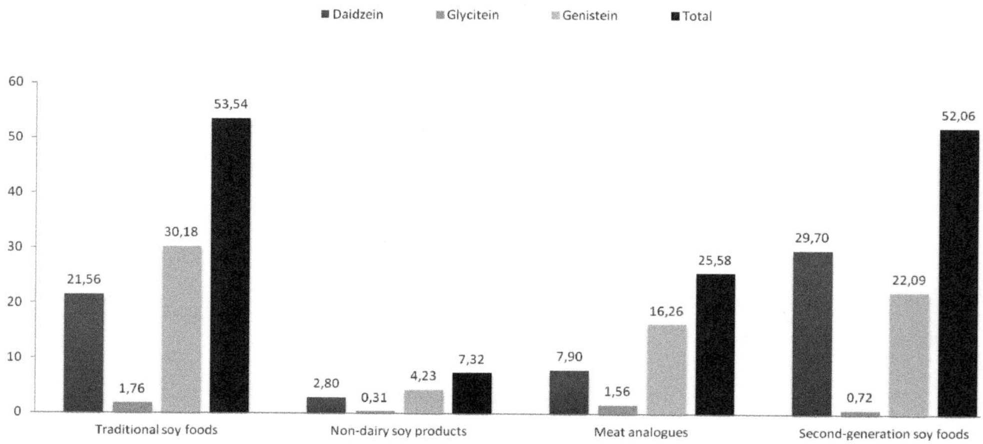

Figure 2. Isoflavone means in Traditional soy foods, no-dairy soy products and meat analogues analised as examples of European consume.

In Figure 2, from our previous data [32], genistein was found to be the most prevalent isoflavone in Traditional soy foods, no-dairy soy products and meat analogues. In second generation soy foods, daidzein was the most prevalent isoflavones. However differences between the various products were considerable.

Diverse studies have been conducted in different countries in order to analyze the isoflavone content in soy products. Chan *et al* determined the concentrations and distribution of isoflavones in 47 soy-based foods [40]. They also studied the conjugation pattern of isoflavones and found that varied within and between food groups as influenced by the types of soybeans and the processing or cooking techniques used. The authors found very high values in certain foods and concluded that might not be safe, supporting the relevance of the risk assessment for the recommendation of safe intake levels. Similar results were obtained by Boniglia et al [41].

Three years ago, Boniglia *et al*, achieved similar results. These authors presented a study of the isoflavone content of 14 soy-based Italian dietary supplements – indicated for the improvement of perimenopausal and menopausal symptoms. The objective was to quantify soy isoflavones fraction after hydrolysis into free aglycones. They noticed that, in the examined products, the amounts of isoflavones were frequently expressed ambiguously. None of the products stated whether the isoflavone content of the product was expressed as aglycones or as conjugates. Even worse, each product revealed a different aglycone concentration profile. They also concluded that this difference was probably related to the different sources of raw materials and methods used in the processing and preparation of extracts. They also observed that in more than half the supplements tested, the actual values contained were below those stated and below those expected to relief perimenopausal and menopausal symptoms [41].

Similarly, our study of a 59 selected soy health supplements shown a large variation in the composition, and a wide difference between the content of isoflavones analyzed and that reported in the product [32].

Our study demonstrates, as the others cited before, the wide variation of isoflavone levels that these products may provide to the consumer. Because there are not recommendations of the optimal isoflavone consumption, the information displayed in these products should be clearly specified, especially in the case of health supplements. This is essential for accurate risk assessment studies, where knowing the exposure level of the population to these products and more specifically the exposition to each isoflavone is fundamental.

Despite the predominant isoflavone forms in soybeans [38] and non-fermented soy products (like soy protein or soy milk derivates) are glucosides [42]; fermented soy foods (like soy sauce) contain mainly aglycones [43]. The form in which isoflavones are present in soy-based foods and health supplements is important to assess the bioavailability of these products [36], Considering that genistein is, in a theoretical point of view [15,44], about 1000 times more active that daidzein the heath benefit and/or the toxicity could be different in two apparently identical products.

For this confusing data our previous proposition was, and still is, applying the calculation of "Theoretical Efficacy (of isoflavones) Related to Estradiol (TERE) [44]. Like that it is possible evaluate the theoretical impact of exposure to estrogen-like activity of isoflavones in various countries from all the data of different studies already published [44]. This theoretical calculation estimates the "Theoretical Efficacy (TE)" of a mixture with different bioactive compounds in a way to obtain a "Theoretical Efficacy Related to Estradiol (TERE)". The theoretical calculation that was proposed for some of the authors of this chapter integrates different knowledge about this subject and sets methodological boundaries that can be used to analyse data already published. The outcome should set some consensus for new clinical trials using isoflavones (isolated or included in mixtures) that will be evaluated to assess their therapeutically activity.

To do the theoretical calculation [44] the amounts of isoflavones were multiply for the "ERs binding affinities" based on the values obtained in literature [15]. After applying the proposed model, "Theoretical efficiency related to estradiol (TERE), for each mixture of isoflavones, the" Theoretical Efficiency (TE) was estimated which can be used to compare the potential bioactivity [44].

Daidzein and genistein do not have the same binding affinity to alpha- and/or beta-ERs [15]. Also known and frequently mentioned by several authors is that the linkage of oestrogens or oestrogenic compounds to alpha-ERs could be dangerous with breast cancers because it could aid the proliferation of damaged DNA in tumours [45, 46]. The amounts theoretical linked from the different isoflavones to the alpha- oestrogen receptors as an add benefit (when added to the beta-receptors affinity) but also as a potential risk to be under evaluation.

However, it is also accepted that beta-ERs are mainly located in bones, brain, thymus, bladder, cardiovascular system and its activation by estrogenic compounds, or compounds with the ability to mimic estrogenic molecules, such as phytoestrogens, can improve and prevent conditions like osteoporosis and cardiovascular diseases [17, 47] and this will be used as the benefit parameter.

In global terms the total efficacy of the TERE will be determined adding these two values. However the amount linked to the alpha-receptors will be consider as a possible risk limitation, which that needs to be evaluated when the increased dose will induce an improved TERE but the risk assessment for the toxicity can be a handicap. Ultimately, different extracts could be compared, even when the relative amounts present in the extracts are very different.

The example of TERE calculation for samples show in Figure 2 is presented in Table 2, using "Second generation soy foods" data.

	Second generation soy foods	Receptor Type	
	Intake (mg/100g)	Alpha-ERs	Beta-ERs
daidzein	29,2	(29.2 x 0.031/100) 0.009	(29.2 x 0.020/100) 0.006
genistein	22,09	(22.09 x 0.86/100) 0.19	(25 x 43.9/100) 9.7
Total	51.79	0.199	9.706
TE	**9.9**	(= 0.199 + 9.706)	
TERE	(100 total linkage to ERs/9.9 TE) = **1/10.1** of the theoretical activity of the estradiol		
RISK		**≃ 2%**	
BENEFIT			**≃ 98%**

Daidzein binding affinity for alpha–ER is 0.031 and 0.020 for beta-ER [15];
Genistein binding affinity for alpha–ER is 0.86 and 43.9 for beta-ER [15].

Table 2. Example of TERE calculation with mean values for *Second generation soy foods* using the "Estrogen Receptor affinity binding values" with daidzein and genistein.

From the total values in Figure 3 and 4 is possible evaluate the amount of daidzein and genistein linked to both receptors and have an idea about the Theoretical Efficacy (TE) of these components if related to estradiol (TERE), Figure 5.

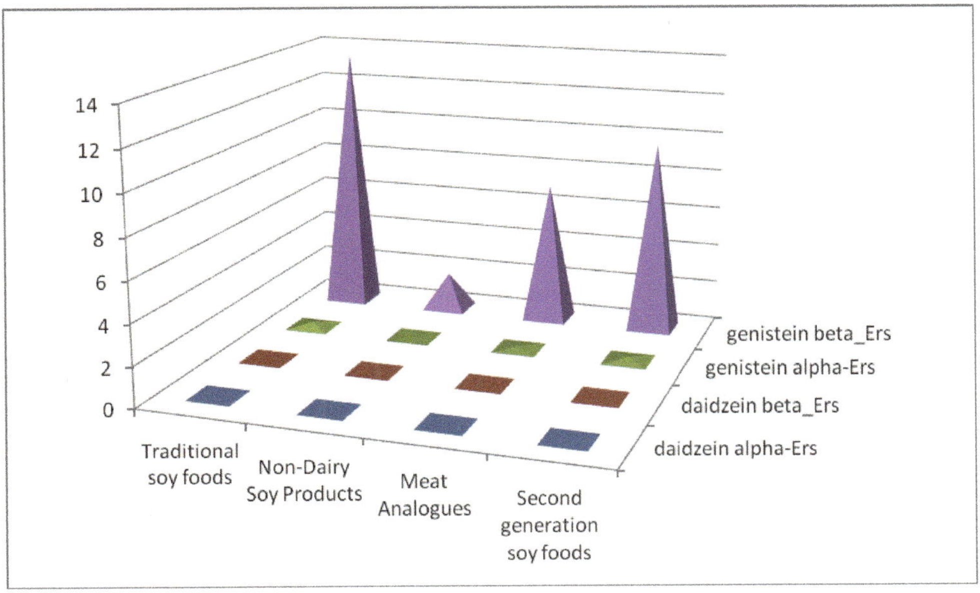

Figure 3. "Estrogen Receptor affinity binding values" for daidzein and genistein content in samples of Group I to IV.

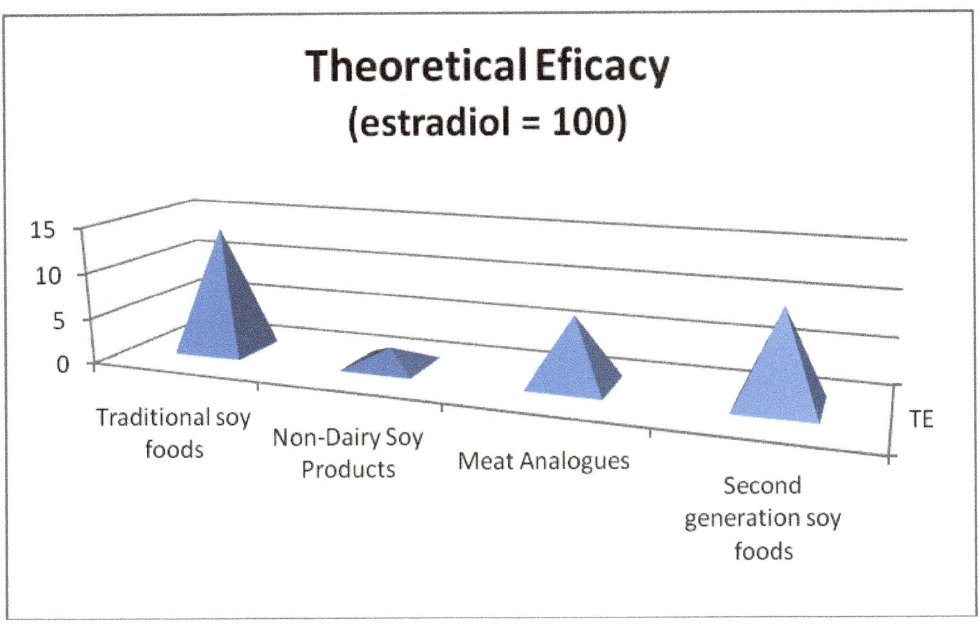

Figure 4. TE values for samples of Group I to IV.

With the exception of Non-dairy soy products that present very low bioactivity related to estradiol the other three groups have values between 7 and almost 14% of the estradiol bioactivity.

Nevertheless the apparent majority of benefice related to risk, as was pointed out above, the low amounts of isoflavones linked to alpha receptors can increase the possibility of proliferation in hormonal dependent tumours and the safe amounts were not evaluated yet.

If the same analysis is made with health supplements the values are absolutely different. In Figure 6 it is shown the TE for the highest values of daidzein and genistein found in our samples (daidzein 371.48 and genistein 172,47 mg/g of pill) [32]. The pill with the content of 172,47mg of genistein also had 208.41mg of daidzein. The TE for this product is almost 80% of the bioactivity of estradiol with a clear uncertainty of influence in the bioability of these compounds in human body.

The exposure risk of all those products is unknown and needs to be evaluated in a near future even no enough data for now make sense.

The data presented herein intend to demonstrate the wide range of isoflavone levels that different products may provide to the consumer, even when recommended for the same therapeutic effect and still if the product keeps the same isoflavones profile all the time. Its relevant to mention that the majority of the soy suppliers breed a wide range of soy cultivars, which results in a different mixture of compounds, and consequently in a variable final product composition and hazard [19, 38].

The Europeans, as can be seen in results presented recently by Konar *et al*, in 2012 consumed legumes with low levels of dietary isoflavones. In this study, 6 legumes (chickpea, red kidney bean, haricot bean, yellow lentil, red lentil and green lentil) were analysed to determine their contents for 10 different isoflavones (both free and conjugated). Methanolic extracts obtained by ultrasound-assisted extraction were analysed by triple quadrupole LC-MS/MS. Chickpeas were the best source of isoflavones (3078 372 mu g/kg total content), with a significant amount of biochanin A and its conjugated form, sissotrin. Kidney beans had the second highest concentration of isoflavones (1076 mu g/kg) and were particularly rich in genistin (946.4 +/- 228.5 mu g/kg). The total isoflavone concentrations of yellow split lentils, green lentils, red lentils and haricot beans were each below 200.0 mu g/kg. However it was determined that the legumes commonly consumed in Western diets (those analyzed in this study) are not so concentrated as soy and soy products as sources of Isoflavones [48].

As it was explained and discussed in the previous section the metabolic pathways and the genetic polymorphic enzymes involved in detoxication of the ingested compounds in our body can conduce to a different bioactivities and consequent risk impact in the various ethnic groups. Only a full risk assessment will prevent the danger or will help to understand the benefit of an increase in the isoflavone intake by Caucasians that for the moment is absolutely out of control.

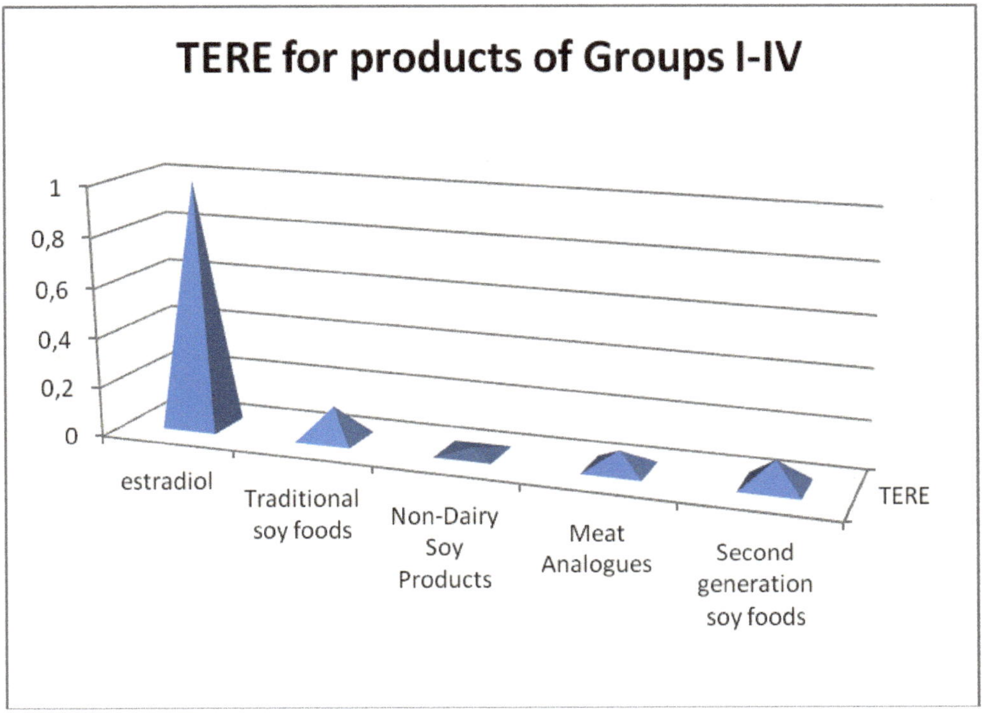

Figure 5. TERE values for samples of Group I to IV.

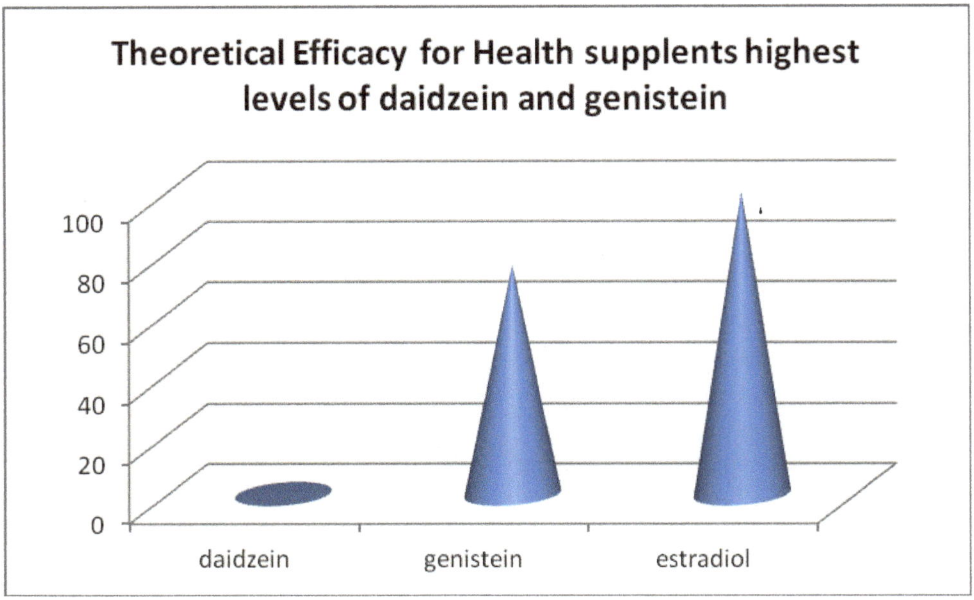

Figure 6. TE values for samples of Group V (only for highest daidzein and genistein content).

4. Conclusions

The above findings emphasize the need for a thorough correct risk assessment to be carried out to evaluate the differences between the various isoflavones, their relative levels in formulations (dietetic supplements, foods soy based, etc) and their safety profile, in order to establish limits for safe therapeutic effects. Various interactions with conventional medicines have already been published and it is important to alert already medicated patients who are also consuming soy or soy based products. The relationship between this intake and cancer and or cardiovascular diseases prevention is unclear moreover the danger of a cancer improvement.

Without such a risk assessment *ad libitum* consumption of these compounds could be hazardous.

Author details

Maria Graça Campos* and Maria Luísa Costa

Observatory of Herb-Drug Interactions & Faculty of Pharmacy, University of Coimbra, Polo III, Azinhaga de Santa Comba, Coimbra, Portugal

Drug Discovery Group, Center of Pharmaceutical Studies, Laboratory of Pharmacognosy, Faculty of Pharmacy, University of Coimbra, Polo III, Azinhaga de Santa Comba, Coimbra, Portugal

5. References

[1] The use of dietary isoflavones and isolated isoflavones from soy or red clover in food and food supplements – 2009. *document available at* http://www.efsa.europa.eu/en/escoisoflavones/docs/escoisoflavonesmandate.pdf (acessed 31 May 2012)

[2] Bennetts, H. N., Underwood, E. J., Shier, F. L. A specific breeding problem of sheep on subterranean clover pastures in Western Australia. *Aust. Vet. J.* 1946, 22: 2-12.

[3] Setchell, K.D.R.; Borriello, S.P.; Hulme, P.; Kirk, D.N.; Axelson, M. Nonsteroidal oestrogens of dietary origin: possible roles in hormone-dependent disease. *Am J Clin Nutr* 1984, 40, 569–578.

[4] Kenny, A. M.; Mangano, K. M.;Abourizk, R.H.; Bruno, R. S.; Anamani, D. E.; Kleppinger,A.; Walsh, S. J.; Prestwood, K. M.; Kerstetter, J. E. Soy proteins and isoflavones affect bone mineral density in older women: a randomized controlled trial. *Am J Clin Nutr* 2009, 90, 234–424.

[5] Atmaca, A.; Kleerekoper, M.; Bayraktar, M.; Kucuk, O. Soy isoflavones in management of postmenopausal osteoporosis. *Menopause* 2008, 15, 1–10.

[6] Lissin, L.W.; Oka, R.; Lakshmi, S.; Cooke, J.P. Isoflavones improve vascular reactivity in post-menopausal women with hypercholesterolemia. *Vascul. Med.* 2004, 9, 26–30.

* Corresponding Author

[7] Crisafulli, A.; Marini, H.; Bitto, A.; Altavilla, D.; Squadrito, G.; Romeo, A.; Adamo, E.B.;Marini, R.; D'Anna, R.; Corrado, F.; Bartolone, S.; Frisina, N.; Squadrito, F. Effects of genistein on hot flushes in early postmenopausal women: A randomized, double-blind EPT- and placebo-controlled study. *Menopause* 2004, 11, 400–404

[8] Unfer, V.; Casini, M.L.; Costabile, L.; Mignosa, M.; Gerli, S.; Di Renzo, G.C. Endometrial effects of long-term treatment with phytoestrogens: a randomized, double-blind, placebo-controlled study. *Fertil Steril* 2004, 82, 145-8

[9] Wuttke, W.; Jarry, H., Seidlova-Wuttke, D. Isoflavones - safe food additives or dangerous drugs? *Ageing Res Rev.* 2007; 6, 150-188.

[10] Rebbeck T. R, Su H I., Sammel M. D, Lin Hui, T. T.V, Gracia C. R, Freeman E. W., Effect of hormone metabolism genotypes on steroid hormone levels and menopausal symptoms in a prospective population-based cohort Menopause. 2010,17(5): 1026–1034.

[11] Fanti, P.; Sawaya, B. P.; Custer, L. J.; Franke, A. A. Serum levels and metabolic clearance of the isoflavones genistein and daidzein in hemodialysis patients. *J. Am. Soc. Nephrol.* 1999, *10*, 864-871.

[12] Coward, L.; Smith, M.; Kirk, M.; Barnes, S. Chemical modification of isoflavones in soyfoods during cooking and processing. *Am. J. Clin. Nutr.* 1998, *68*, 1486S-1491S.

[13] Song, T.; Barua, K.; Buseman, G.; Murphy, P. A. Soy isoflavone analysis: quality control and a new internal standard. *Am. J. Clin. Nutr.* 1998, *68*, 1474S-1479S.

[14] Orgaard, A.; Jensen, L. The effects of soy isoflavones on obesity. *Exp. Biol. Med. (Maywood)* 2008, *233*, 1066-1080.

[15] Fokialakis N, Lambrinidis G, Mitsiou DJ, Aligiannis N, Mitakou S, Skaltsounis AL, Pratsinis H, Mikros E, Alexis MN. A new class of phytoestrogens; evaluation of the estrogenic activity of deoxybenzoins. *Chem Biol.* 2004, 11(3):397-406.

[16] Kreijkamp-Kaspers, S.; Kok, L.; Grobbee, D.E.; De Haan, E.H.; Aleman, A.; Lampe, J.W.; Van der Schouw, Y.T. Effect of soy protein containing isoflavones on cognitive function, bone mineral density, and plasma lipids in postmenopausal women: A randomized controlled trial. *J. Am. Med. Assoc.* 2004, *292*, 65–74.

[17] De Lemos, M. Effects of Soy Phytooestrogens Genistein and Daidzein on Breast Cancer Growth. *Ann Pharmacother* 2001, 35, 1118-1121

[18] Wang, C.; Kurzer, M.S. Phytooestrogen concentration determines effects on DNA synthesis in human breast cancer cells. *Nutr. Cancer* 1997, *28*, 236–247.

[19] De Lemos, M.; Van Patten, C.L.; Gelmon, K.A.; Olivotto, I.A. Safety issues of soy phytooestrogens in breast cancer patients [3] (multiple letters) *A. J. Clin. Oncol.* 2002, *20*, 3040–3042.

[20] Moon, Y. J.; Wang, X.; Morris, M. E. Dietary flavonoids: Effects on xenobiotic and carcinogen metabolism. *Toxicology in Vitro* 2006, 20, 187-210

[21] Seo, H.S.; DeNardo, D.G.; Jacquot, Y.; Laïos, I.; Vidal, D. S.; Zambrana, C. R.; Leclercq, G.; Brown, P. H. Stimulatory effect of genistein and apigenin on the growth of breast

cancer cells correlates with their ability to activate ER alpha. Breast Cancer Res Treat 2006, 99, 121–134.

[22] Hsieh, C.; Santell, R. C.; Haslam, S. Z.; Helferich, W.G. Oestrogenic Effects of Genistein on the Growth of Oestrogen Receptor-positive Human Breast Cancer (MCF-7) Cells in Vitro and in Vivo *Cancer Research* 1998, 58, 3833-3838.

[23] http://www.cancer.gov/cancertopics/understandingcancer/estrogenreceptors/page20 acceded 30 April 2012.

[24] Kotnis A, Kannan S, Sarin R, Mulherkar R. Case-control study and meta-analysis of SULT1A1 Arg213His polymorphism for gene, ethnicity and environment interaction for cancer risk. Br J Cancer. 2008, 21;99(8):1340-7.

[25] Kramer S. and Testa B. The Biochemistry of Drug Metabolism – An Introduction Part 6. Inter-Individual Factors Affecting Drug Metabolism, in *Chemistry & Biodiversity* –, 2008. 5, 2465-2578

[26] Moon Y J, Wang X., Morris M. E., Dietary flavonoids: Effects on xenobiotic and carcinogen metabolism. Mini review. *Toxicology in Vitro* 2006, 20. 187–210

[27] Falk R.T., *et al.* Urinary Estrogen Metabolites and Their Ratio among Asian American *Women Cancer Epidemiology Biomarkers & Prevention*. 2005, 14, 221-226

[28] Nettleton J. A.,Greany K. A, Thomas W, Wangen K. E., Adlercreutz H. and Kurzer M.S. *The Effect of Soy Consumption on the Urinary 2:16-Hydroxyestrone Ratio in Postmenopausal Women Depends on Equol Production Status but Is Not Influenced by Probiotic Consumption* J. Nutr. 2005, 135: 603-608

[29] Keinan-Boker, L.; Peeters, P. H.; Mulligan, A. A.; Navarro, C.; Slimani, N.; Mattisson, I.; Lundin, E.; McTaggart, A.; Allen, N. E.; Overvad, K.; Tjonneland, A.; Clavel-Chapelon, F.; Linseisen, J.; Haftenberger, M.; Lagiou, P.; Kalapothaki, V.; Evangelista, A.; Frasca, G.; Bueno-de-Mesquita, H. B.; van der Schouw, Y. T.; Engeset, D.; Skeie, G.; Tormo, M. J.; Ardanaz, E.; Charrondiere, U. R.; Riboli, E. Soy product consumption in 10 European countries: the European Prospective Investigation into Cancer and Nutrition (EPIC) study. *Public Health Nutr.* 2002, *5*, 1217-1226.

[30] Campos M. G.; Paranhos, A. H.; Matos, M. P.; Câmara, M. T.; Cunha, M. M.; Pinto P., Silvestre, A. J.; Amado, F.; Neto, P. C. Comparative Analysis Of Over the Counter Tablet Preparations Of Isoflavones Extracted From Soy Available In Portugal, *Natural Products Communications*, 2006, 1, 973-980.

[31] Matos, M.; Castilho, C.; Campos, M.G.; Ramos, F.; Silveira, I. Quais os benefícios de uma alimentação rica em Fitoestrogénios. *Med. Int.* 2005, 2, 171–182.

[32] Moreno-Franco, B., Campos M G, Matos, M ; Cunha, M ; Adlercreutz, H Penalvo, J 2011 Isoflavone profile of selected soy-products commercialized in Europe. *Annals Of Nutrition And Metabolism*. 58 (3) 134-135

[33] Wang, H.; Murphy, P. A. Isoflavone Composition of American and Japanese Soybeans in Iowa: Effects of Variety, Crop Year, and Location. *J. Agric. Food Chem.* 1994, *42*, 1674-1677.

[34] Eldridge, A. C.; Kwolek, W. F. Soybean isoflavones: effect of environment and variety on composition. *J. Agric. Food Chem.* 1983, *31*, 394-396.

[35] Rau De Almeida Callou, K.; Sadigov, S.; Lajolo, F. M.; Genovese, M. I. Isoflavones and antioxidant capacity of commercial soy-based beverages: effect of storage. *J. Agric. Food Chem.* 2010, *58*, 4284-4291.

[36] Setchell, K. D.; Brown, N. M.; Desai, P.; Zimmer-Nechemias, L.; Wolfe, B. E.; Brashear, W. T.; Kirschner, A. S.; Cassidy, A.; Heubi, J. E. Bioavailability of pure isoflavones in healthy humans and analysis of commercial soy isoflavone supplements. *J. Nutr.* 2001, *131*, 1362S-1375S.

[37] Setchell, K. D.; Cole, S. J. Variations in isoflavone levels in soy foods and soy protein isolates and issues related to isoflavone databases and food labeling. *J. Agric. Food Chem.* 2003, *51*, 4146-4155.

[38] Campos M. G.; Matos M. P.; Câmara, M. T.; Cunha, M. M. The variability of isoflavones in soy seeds and the possibility of obtaining extracts for over the counter tablet preparations that can be standardized. *Ind. Crop. Prod.*, 2007, 26, 85-92.

[39] Penalvo, J. L.; Nurmi, T.; Haajanen, K.; Al-Maharik, N.; Botting, N.; Adlercreutz, H. Determination of lignans in human plasma by liquid chromatography with coulometric electrode array detection. *Anal. Biochem.* 2004, *332*, 384-393.

[40] Chan, S. G.; Murphy, P. A.; Ho, S. C.; Kreiger, N.; Darlington, G.; So, E. K.; Chong, P. Y. Isoflavonoid content of Hong Kong soy foods. *J. Agric. Food Chem.* 2009, *57*, 5386-5390.

[41] Boniglia, C.; Carratù, B.; Gargiulo, R.; Giammarioli, S.; Mosca, M.; Sanzini, E. Content of phytooestrogens in soy-based dietary supplements. *Food Chemistry* 2009, 115:1389–1392.

[42] Naim, M.; Gestetner, B.; Zilkah, S.; Birk, Y.; Bondi, A. Soybean isoflavones. Characterization, determination, and antifungal activity. *J. Agric. Food Chem.* 1974, 22, 806-810.

[43] Coward, L.; Barnes, N. C.; Setchell, K. D. R.; Barnes, S. Genistein, daidzein, and their b-glycoside conjugates: antitumor isoflavones in soybean foods from American and Asian diets. *J. Agric. Food Chem.* 1993, *41*, 1961-1967.

[44] Campos M. G., Matos M. P. Bioactivity of Isoflavones: Assessment Through a Theoretical Model as a Way to Obtain a "Theoretical Efficacy Related to Estradiol (TERE)", *Int. J. Mol. Sci.* 2010, 11, 480-491.

[45] Wang C. and Kurzer M. S., Phytoestrogen Concentration Determines Effects on DNA Synthesis in Human Breast Cancer Cells *Nutrition and Cancer*, 1997, 28(3): 236-247.

[46] Lemos M. Phytoestrogen Concentration Determines Effects on DNA Synthesis in Human Breast Cancer Cells. *American Society of Clinical Oncology*. 2002 www.jco.org on

[47] Kreijkamp-Kaspers S, Kok L, Grobbee DE, de Haan EH, Aleman A, Lampe JW & van der Schouw YT Effect of soy protein containing isoflavones on cognitive function, bone

mineral density, and plasma lipids in postmenopausal women: a randomized controlled trial. *Journal of the American Medical Association* 2004, 292, 65–74.

[48] Konar, N, Poyrazoglu, ES, Demir, K. Artik, N. Determination of conjugated and free isoflavones in some legumes by LC-MS/MS. *Journal Of Food Composition And Analysis* 2012, 25 (2), 173-178.

Section 3

Food Processing Technology

Chapter 10

Pulsed Electric Fields for Food Processing Technology

Maged E.A. Mohamed and Ayman H. Amer Eissa

Additional information is available at the end of the chapter

1. Introduction

Pulsed electric fields PEF is a non-thermal method of food preservation that uses short pulses of electricity for microbial inactivation and causes minimal detrimental effect on food quality attributes. PEF technology aims to offer consumers high-quality foods. For food quality attributes, PEF technology is considered superior to traditional thermal processing methods because it avoids or greatly reduces detrimental changes in the sensory and physical properties of foods. PEF technology aims to offer consumers high-quality foods. For food quality attributes, PEF technology is considered superior to traditional thermal processing methods because it avoids or greatly reduces detrimental changes in the sensory and physical properties of foods (Quass, 1997).

PEF technology has been presented as advantageous in comparison to, for instance, heat treatments, because it kills microorganisms while better maintaining the original color, flavor, texture, and nutritional value of the unprocessed food . PEF technology involves the application of pulses of high voltage to liquid or semi-solid foods placed between two electrodes. Most PEF studies have focused on PEF treatments effects on the microbial inactivation in milk, milk products, egg products, juice and other liquid foods (Qin et al., 1995). However, whereas a considerable amount of research papers have been published on the microbial aspects of food preservation by PEF, a lesser amount of information is available about the effect of this technology on food constituents and overall quality and acceptability. Recently, the interest in application of pulsed electric fields (PEF) for food processing has revived. The PEF treatment was shown to be very effective for inactivation of microorganisms, increasing the pressing efficiency and enhancing the juice extraction from food plants, and for intensification of the food dehydration and drying (Gulyi *et al.*, 1994; Barbosa-Cánovas *et al.*, 1998; Barsotti and Cheftel, 1998, 1999; Estiaghi and Knorr, 1999; Vorobiev *et al.*, 2000, 2004; Bajgai and Hashinaga, 2001; Bazhal *et al.*, 2001; Taiwo *et al.*, 2002).

(Wouters *et al.*, 2001) mentioned that, Pulsed electric field technology (PEF) is viewed as one of the most promising nonthermal methods for inactivating microorganisms in foods. Electric fields in the range of 5-50 kV/cm generated by the application of short high voltage pulses (μs) between two electrodes cause microbial inactivation at temperatures below those used in thermal processing. The precise mechanisms by which microorganisms are inactivated by pulsed electric fields are not well understood; however, it is generally accepted that PEF leads to the penneabilization of microbial membranes.

Non-thermal processes have gained importance in recent years due to the increasing demand for foods with a high nutritional value and fresh-like characteristics, representing an alternative to conventional thermal treatments. Pulsed electric fields (PEF) is an emerging technology that has been extensively studied for non-thermal food processing. PEF processing has been studied by a number of researchers across a wide range of liquid foods. Apple and orange juices are among the foods most often treated in PEF studies. The sensory attributes of juices are reported to be well preserved, and the shelf life is extended. Yogurt drinks, apple sauce, and salad dressing have also been shown to retain a fresh-like quality with extended shelf life after processing. Other PEF-processed foods include milk, tomato juice (Min et al., 2003), carrot juice, pea soup (Vega-Mercado et al., 1996), liquid whole egg (Martı´n-Belloso et al., 1997), and liquid egg products.

Food preservation technologies are based on the prevention of microbial growth or on the microbial inactivation. In many cases, foods are preserved by inhibiting microbial activity through those factors that most effectively influence the growth and survival of microorganisms such as temperature, water activity, addition of preservatives, pH, and modified atmosphere. In this case, the microorganisms will not be destroyed and will still be metabolically active and viable if transferred to favorable conditions. As estimates of the infection dose of some pathogenic microorganisms are very low, growth of these microorganisms in foods is not necessary to cause infection (Blackburn and McClure, 2002).

To qualify as an alternative method, a new technology should have significant impact on quality while at the same time maintain the cost of technology within feasibility limits. In recent years, several technologies have been investigated that have the capability of inactivating microorganisms at lower temperatures than typically used in conventional heat treatments (Lado and Yousef, 2002).

Application of pulsed electric fields of high intensity and duration from microseconds to milliseconds may cause temporary or permanent permeabilization of cell membranes. The effects of PEF on biomembranes have been thoroughly studied since the use of PEF has attracted great interest in several scientific areas such as cell biology, biotechnology, medicine, or food technology (Zimmermann, 1986; Palaniappan and Sastry, 1990; Ho and Mittal, 1996; Prassanna and Panda, 1997).

Recently published research results will be reviewed and compared with those obtained for other thermal and non-thermal processing technologies, with a special stress on the effect of PEF-processing variables on the bioactive composition of foods throughout their whole shelf-life. Furthermore, different examples will be presented to illustrate not only the potential but also the limitations of PEF technology when aiming at preserving the health-promoting features of plant-based foods. With the use of electric fields, PEF technology enables inactivation of vegetative cells of bacteria and yeasts in various foods. As bacterial spores are resistant to pulsed electric fields, applications of this technology mainly focus on food-borne pathogens and spoilage microorganisms, especially for acidic food products. In addition to the volumetric effect of PEF technology in controlling the microbiological safety of foods in a fast and homogenous manner, successful application provides extended shelf life without the use of heat to preserve the sensory and nutritional value of foods. PEF technology has the potential to economically and efficiently improve energy usage, besides the advantage of providing microbiologically safe and minimally processed foods. Successful application of PEF technology suggests an alternative substitute for conventional thermal processing of liquid food products such as fruit juices, milk, and liquid egg (Mertens and Knorr, 1992; Bendicho et al., 2002; Hodgins et al., 2002).

The objective of this chapter is to provide some basic information about the pulsed electric field technology for preservation of food.

2. Nonthermal technologies for food processing

Nonthermal technologies represent a novel area of food processing and are currently being explored on a global scale; research has grown rapidly in the last few years in particular. The main purpose of thermal processing is the inactivation of pathogenic microorganisms and spores (depending on the treatment) to provide consumers with a microbiologically safe product. However, despite the benefits of thermal treatment, a number of changes take place in the product that alter its final quality, for example, flavor, color, texture, and general appearance. Now, consumers are looking for fresh-like characteristics in their food, along with high sensorial quality and nutrient content. Consumers are more aware of food content and the technologies used to process their food, showing a higher preference for natural products (Evans and Cox, 2006) free of chemicals and/or additives. Thus, the need for processing alternatives that can achieve microbial inactivation, preserve food's fresh like characteristics, and provide environment friendly products, all at a reasonable cost, has become the present challenge of numerous food scientists/technologists around the world.

Nonthermal processing technologies were designed to eliminate the use of elevated temperatures during processing and so avoid the adverse effects of heat on the flavour, appearance and nutritive value of foods (Barbosa-Canovas et al., 1999).

Novel nonthermal processes, such as high hydrostatic pressure (HHP), pulsed electric fields (PEFs), ionizing radiation and ultrasonication, are able to inactivate microorganisms at ambient or sublethal temperatures. Many of these processes require very high treatment intensities, however, to achieve adequate microbial destruction in low-acid foods. Combining nonthermal processes with conventional preservation methods enhances their antimicrobial effect so that lower process intensities can be used. Combining two or more nonthermal processes can also enhance microbial inactivation and allow the use of lower individual treatment intensities. For conventional preservation treatments, optimal microbial control is achieved through the hurdle concept, with synergistic effects resulting from different components of the microbial cell being targeted simultaneously. The mechanisms of inactivation by nonthermal processes are still unclear; thus, the bases of synergistic combinations remain speculative (Ross et al.,2003).

Nonthermal technologies encompass all preservation treatments that are effective at ambient or sub lethal temperatures including antimicrobial additives, pH adjustment and modified atmospheres. The term 'nonthermal processing' is more apt for novel nonthermal technologies, such as high hydrostatic pressure, pulsed electric fields (PEFs), high-intensity ultrasound, ultraviolet light, pulsed light, ionizing radiation and oscillating magnetic fields, which are intended for application as microbe-inactivating processes during food manufacture. Such novel technologies have the ability to inactivate microorganisms to varying degrees (Butz and Tauscher, 2002).

One nonthermal technology, high hydrostatic pressure (HHP), has shown a negligible effect on the nutrient content of food, for example, in processing of fruits and vegetables, where pressure has minimal effect on the anthocyanin content after processing. Anthocyanins are considered phytonutrients, and they not only are responsible for color but also have an important antioxidant effect on human health. However, anthocyanin content in juices after pulsed electric fields (PEF) treatment has shown contradictory results. Some researchers report a minimum effect on the pigment content after processing, while others show that there is degradation in anthocyanin content after pulsing (Tiwari et al., 2009).

The most extensively researched and promising nonthermal processes appear to be high hydrostatic pressure (HHP), pulsed electric fields (PEF) and high intensity ultrasound combined with pressure. Gamma irradiation has high potential although its development and commercialization has been hampered in the past by unfavourable public perceptions (Resurreccion et al., 1995).

Despite the current gaps in understanding, combining nonthermal processes with other nonthermal technologies has been investigated to improve control over food borne microorganisms, with promising results. A better understanding of the antimicrobial mechanisms of emerging nonthermal technologies as well as their effectiveness when combined with traditional food preservation hurdles is needed so that new food preservation strategies can be developed on a sound scientific basis (Barbosa-Canovas et al., 1998).

High-pressure processing applied at room temperature yields a product with most of food's quality attributes intact; for example, pressurization does not affect covalent bonds, avoiding any development of strange flavors in the food (Knorr et al., 2002).

Ultrasound has also been used in milk pasteurization, with important results; milk shows a higher degree of homogenization, whiter color, and better stability after processing. In this method, pasteurization and homogenization are completed in a one-step process (Bermúdez-Aguirre et al., 2009).

3. The principles of pulsed electric field

The basic principle of the PEF technology is the application of short pulses of high electric fields with duration of microseconds micro- to milliseconds and intensity in the order of 10-80 kV/cm. The processing time is calculated by multiplying the number of pulses times with effective pulse duration. The process is based on pulsed electrical currents delivered to a product placed between a set of electrodes; the distance between electrodes is termed as the treatment gap of the PEF chamber. The applied high voltage results in an electric field that causes microbial inactivation. The electric field may be applied in the form of exponentially decaying, square wave, bipolar, or oscillatory pulses and at ambient, sub-ambient, or slightly above-ambient temperature. After the treatment, the food is packaged aseptically and stored under refrigeration. applied to a food product held between two electrodes inside a chamber, usually at room temperature. Food is capable of transferring electricity because of the presence of several ions, giving the product in question a certain degree of electrical conductivity. So, when an electrical field is applied, electrical current flows into the liquid food and is transferred to each point in the liquid because of the charged molecules present (Zhang et al., 1995).

Several nonthermal processing technologies were proposed on the basis of the same basic principle of keeping food below temperatures normally used in thermal processing. This would retain the nutritional quality of food including vitamins, minerals, and essential flavors while consuming less energy than thermal processing. High hydrostatic pressure, oscillating magnetic fields, intense light pulses, irradiation, the use of chemicals and biochemicals, high intensity pulse electric fields, and the hurdle concept were all recognized as emerging nonthermal technologies in recent years (Barbosa-Cánovas et al., 1999).

As a result of this permanent membrane damage, microorganisms are inactivated. Some applications of PEF technology are in biotechnology and genetic engineering for electroporation in cell hybridization (Chang et al., 1992).

Several nonthermal processing technologies were proposed on the basis of the same basic principle of keeping food below temperatures normally used in thermal processing. This would retain the nutritional quality of food including vitamins, minerals, and essential flavors while consuming less energy than thermal processing. High hydrostatic pressure, oscillating magnetic fields, intense light pulses, irradiation, the use of chemicals and biochemicals, high intensity pulse electric fields, and the hurdle concept were all recognized as emerging nonthermal technologies in recent years (Barbosa-Cánovas et al., 1999).

The basis for this prediction is because of PEF ability to inactivate microorganisms in the food, reduce enzymatic activity, and extend shelf-life with negligible changes in the quality of the final product as compared to the original one. According to the intensity of the field strength, electroporation can be either reversible (cell membrane discharge) or irreversible (cell membrane breakdown or lysis), but this effect can be controlled depending on the application (Ho and Mittal, 1996).

(Zimmermann and Benz, 1980) mentioned that, PEF technology is based on a pulsing power delivered to the product placed between a set of electrodes confining the treatment gap of the PEF chamber. The equipment consists of a high voltage pulse generator and a treatment chamber with a suitable fluid handling system and necessary monitoring and controlling devices (Fig. 1.). Food product is placed in the treatment chamber, either in a static or continuous design, where two electrodes are connected together with a nonconductive material to avoid electrical flow from one to the other. Generated high voltage electrical pulses are applied to the electrodes, which then conduct the high intensity electrical pulse to the product placed between the two electrodes. The food product experiences a force per unit charge, the so-called electric field, which is responsible for the irreversible cell membrane breakdown in microorganisms.

This leads to dielectric breakdown of the microbial cell membranes and to interaction with the charged molecules of food (Fernandez-Díaz et al., 2000; Zimmermann, 1986). Hence, PEF technology has been suggested for the pasteurization of foods such as juices, milk, yogurt, soups, and liquid eggs (Vega-Mercado et al.,1997; Bendicho., 2003; Puértolas et al., 2004).

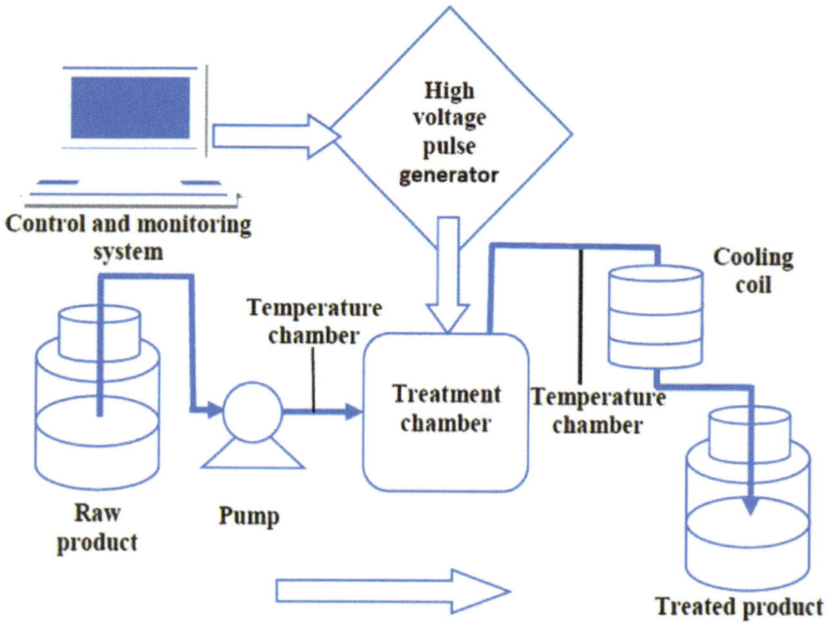

Figure 1. Flow chart of a PEF food processing system with basic component.

4. System components

A pulsed Electric Field processing system consists of a high-voltage power source, an energy storage capacitor bank, a charging current limiting resistor, a switch to discharge energy from the capacitor across the food and a treatment chamber. An oscilloscope is used to observe the pulse waveform. The power source, a high voltage DC generator, converts voltage from an utility line (110 V) into high voltage AC, then rectifies to a high voltage DC.

Energy from the power source is stored in the capacitor and is discharged through the treatment chamber to generate an electric field in the food material. The maximum voltage across the capacitor is equal to the voltage across the generator . The bank of capacitors is charged by a direct current power source obtained from amplified and rectified regular alternative current main source. An electrical switch is used to discharge energy (instantaneously in millionth of a second) stored in the capacitor storage bank across the food held in the treatment chamber. Apart from those major components, some adjunct parts are also necessary. In case of continuous systems, a pump is used to convey the food through the treatment chamber. A chamber cooling system may be used to diminish the ohmic heating effect and control food temperature during treatment. High-voltage and high-current probes are used to measure the voltage and current delivered to the chamber. (Ho et al., 1995; Barbosa-Cánovas et al. 1999; Floury et al. 2005; Amiali et al. 2006). Fig. 2 shows a basic PEF treatment unit (Ortega-Rivas et al. 1998)

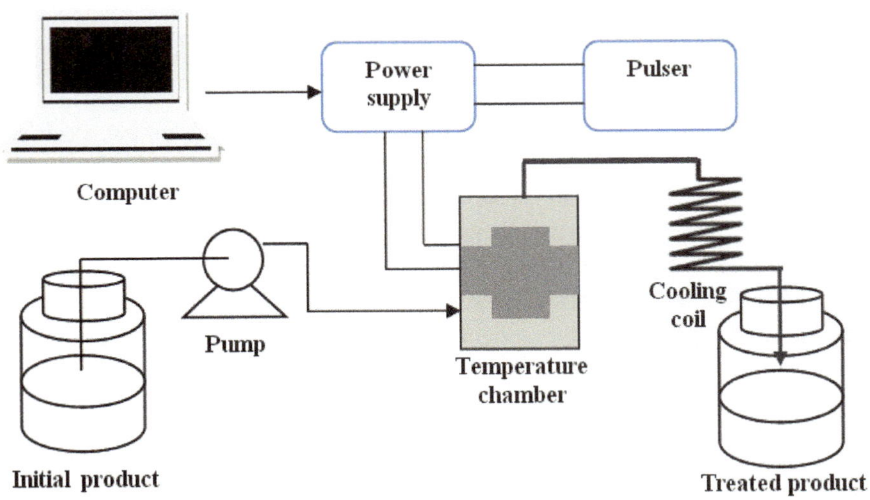

Figure 2. Schematic diagram of a pulsed electric fields operation.

A PEF system for food processing in general consists of three basic components (Fig.3): a high voltage pulse generator, a treatment chamber and a control system for monitoring the process parameters (Loeffler, 2006).

Many successful steps have been taken in the design of system components and inactivation mechanism for different species, however, there are still many points that have not been fully explained. Inactivation kinetics and the effect of PEF on spores are some of the most discussed issues in recent studies. Methods applied to thermal processing technologies by plotting logs of the numbers of survivors against log or treatment time, or number of pulses, have been used to explain inactivation kinetics neglecting the deviations from linearity for these plots (Zhang *et al.*, 1995).

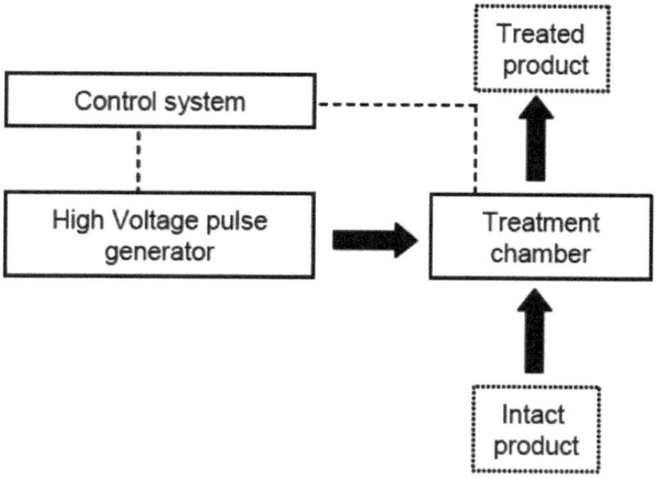

Figure 3. Scheme of a pulsed electric field system for food processing.

The high intensity pulsed electric field processing system is a simple electrical system consisting of a high voltage source, capacitor bank, switch, and treatment chamber. Generation of pulsed electric fields requires a fast discharge of electrical energy within a short period of time. This is accomplished by the pulse-forming network (PFN), an electrical circuit consisting of one or more power supplies with the ability to charge voltages (up to 60 kV), switches (ignitron, thyratron , tetrode, spark gap, semiconductors), capacitors (0.1-10 μF), resistors (2Ω-1O MΩ), and treatment chambers (Gongora-Nieto et al., 2002).

The PEF processing system is composed of a high voltage repetitive pulser, a treatment chamber(s), a cooling system(s), voltage- and current measuring devices, a control unit, and a data acquisition system. A pulsed power supply is used to obtain high voltage from low utility level voltage, and the former is used to charge a capacitor bank and switch to discharge energy from the capacitor across the food in the treatment chamber. Treatment chambers are designed to hold the food during PEF processing and house the discharging electrodes. After processing the product is cooled, if necessary, packed aseptically, and then stored at refrigerated or ambient temperatures depending on the type of food (Qin et al., 1995a; Zhang et al., 1997).

4.1. Power supply

High voltage pulses are supplied to the system via a high voltage pulse generator at required intensity, shape, and duration. The high voltage power supply for the system can either be an ordinary source of direct current (D) or a capacitor charging power supply with high frequency AC inputs that provide a command charge with higher repetitive rates than the DC power supply (Zhang et al., 1996).

High voltage pulses are supplied to the PEF system via a high voltage generator at required electric field intensity, pulse waveform and pulse width. In general, the high voltage power supply is used to charge the capacitor bank and store the energy to the capacitor bank. Liquid food may be processed in a static treatment chamber or in a continuous treatment chamber through a pump. For preliminary laboratory-scale studies, the static treatment chamber is used, but a continuous treatment chamber (s) is desirable for the pilot plant or industrial-scale operations. In order to avoid undesirable thermal effects, cold water of the cooling system is recirculated through the electrodes to dissipate the heat generated by the electric current passing through the food (Barbosa-Cánovas et al., 1999).

Total power of the system is limited by the number of times a capacitor can be charged and discharged in a given time. The electrical resistance of the charging resistor and the number and size of the capacitors determine the power required to charge the capacitor, wherein a smaller capacitor will require less time and power to be charged than a larger one. The capacitance C_o (F) of the energy storage capacitor is given by Eq. (1):

$$C_o = \frac{\tau}{R} = \frac{\tau \sigma A}{d} \quad (1)$$

where τ (s) is the pulse duration, R (Ω) is the resistance, σ (S/m) is the conductivity of the food, d (m) is the treatment gap between electrodes, and A (mm^2) is the area of the electrode surface. The energy stored in a capacitor is defined by the mathematical expression:

$$Q = 0.5 C_o V^2 \quad (2)$$

where Q is the stored energy, C_o is the capacitance, and V is the charge voltage.

More complex PFN systems can provide square pulses, bipolar pulses, and instantaneously reversal pulses, as illustrated in Fig. 4.

4.2. High-power capacitors

The main components of high-power sources are storage capacitors and on- and off-switches. Because of their relatively high ohmic power consumption, inductors in comparison to capacitors play a minor role. The energy stored in capacitors is used to generate electric or magnetic fields. Electric fields are used to accelerate charged particles, leading to thermal, chemical, mechanical, electromagnetic wave, or breakdown effects. Electromagnetic fields transfer energy as electromagnetic waves. xray, microwaves, and laser beam generation are typical examples. Magnetic fields facilitate the generation of extremely high pressures ranging from 0.1 GPa to many GPa. These effects are applied to modify molecules to remodell, compress, weld, segment, fragment, or destroy materials; and t o modify the surface of organic and inorganic parts and particles (Weise and Loeffler, 2001).

4.3. Switches

The discharging switch also plays a critical role in the efficiency of the PEF system. The type of switch used will determine how fast it can perform and how much current and voltage it

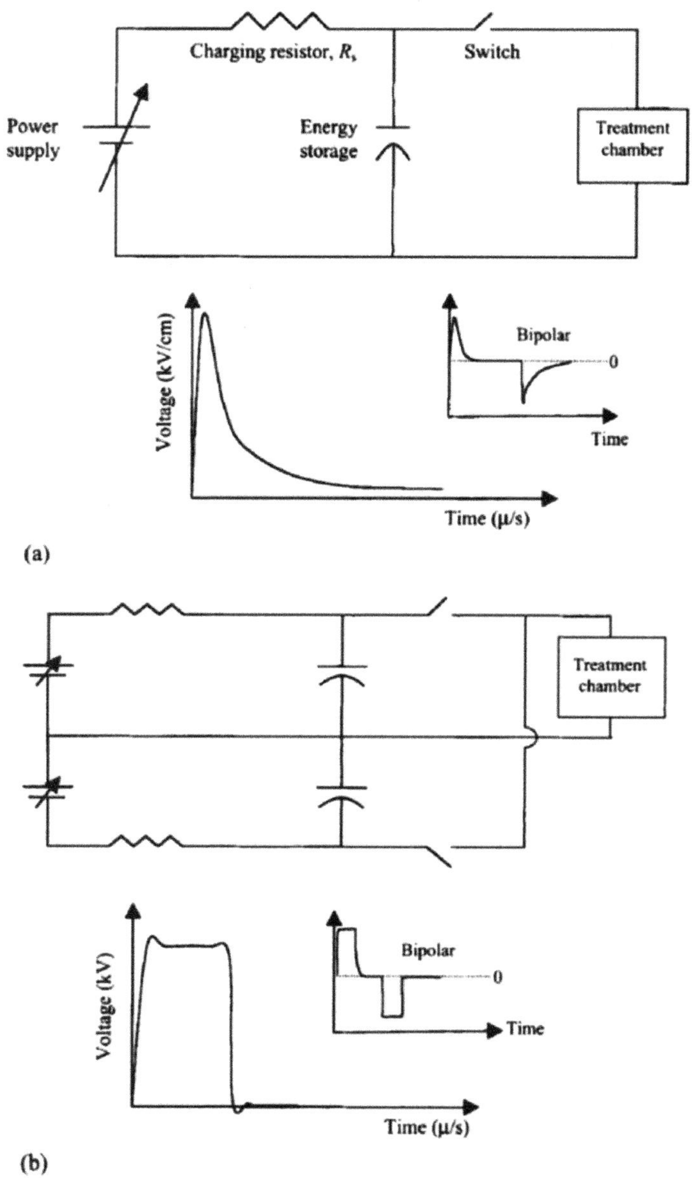

Figure 4. Commonly used pulse wave shapes and the generic electrical circuits: () Monopolar exponential decaying circuits and possible waveform; () Monopolar square circuit and possible waveform.

can withstand. In increasing order of service life, suitable switches for PEF systems include: ignitrons, spark gaps, trigatrons, thyratrons, and semiconductors. Solid-state semiconductor switches are considered by the experts as the future of high power switching (Bartos, 2000).

After the energy storage device, the switch is the most important element of a high-power pulse generator. High-power switching systems are the connecting elements between the storage device and the load. The rise time, shape, and amplitude of the generator output pulse depends strongly on the properties of the switches in the pulse forming elements. Generators with capacitive storage devices need closing switches, while generators with inductive storage devices require opening switches (Bluhm 2006).

There are two main groups of switches currently available: ON switches and ON/OFF switches. ON switches provide full discharging of the capacitor but can only be turned off when discharging is completed. ON switches can handle high voltages with relatively lower cost compared to ON/OFF switches, however, the short life and low repetition rate are some disadvantages to be considered for selection. The Ignitron, Gas Spark Gap, Trigatron, and Thyratron are some of the examples from this group. ON/OFF type switches have been developed in recent years that provide control over the pulse generation process with partial or complete discharge of the capacitors. Improvements on switches, mainly on semiconductor solid-states witches, have resulted in longer life spans and better performance. The gate turn off (GTO) thyristor, the insulated gate bipolar transistor (IGBT), and the symmetrical gate commutated thyristor (SGCT) are some examples from this group (EPRI and Army, 1997; Barbosa-Cánovas *et al.*, 1999; Barsotti *et al.*, 1999; Gongora-Nieto *et al.*, 2002; Sepulveda and Barbosa-Cánovas, 2005).

4.4. High voltage pulse generator

The high voltage pulse generator provides electrical pulses of the desired voltage, shape and duration by using a more or less complex pulse forming network (PFN). More in detail, a PFN is an electrical circuit consisting of several components: one or more DC power supplies, a charging resistor, a capacitor bank formed by two or more units connected in parallel, one or more switches, and pulse-shaping inductors and resistors. The DC power supply charges the capacitors bank to the desired voltage. Using this device, the ac power from the utility line (50-60Hz) is converted in high voltage alternating current (A) power and then rectified to high voltage dc power (Zhang et al., 1995).

A low-energy PEF system, which consists of a high voltage pulse generator (Fig. 5) is used to treat the spoiled grape juice samples. The details are given by Ho and Mittal (2000). The system consists of a 30 kV d.c. high-voltage pulse generator, a circular treatment chamber, and devices for pumping and recording. The 110V a.c. was raised in voltage through a high-voltage transformer, and then rectified. The d.c. high-voltage supply then charges up the 0.12 uF capacitor through a series of 6 MΩ resistors (the time constant =0.72 s). The pulse generator emits a train of 5V pulses, and the trigger circuit serves to convert that to 500V pulses using a silicon control rectifier (SCR). The generation of high voltage pulses relies on the discharge of the 0.12 uF capacitor through the thyratron. The batch unit can generate short duration pulses (2 ms width, 0.5Hz frequency) with a peak-to-peak electric field strength up to 100 kV/cm. The uniqueness of this pulser is that the pulses of low energy (<25 J/pulse) and of instant charge reversal shape are generated.

4.5. Treatment chamber

One of the most important and complicated components in the processing system is the treatment chamber. The basic idea of the treatment chamber is to keep the treated product inside during pulsing, although the uniformity of the process is highly dependent on the characteristic design of the treatment chamber. When the strength of applied electric fields exceeds the electric field strength of the food product treated in the chamber, breakdown of food occurs as a spark. Treatment chambers are mainly grouped together to operate in either a batch or continuous manner; batch systems are generally found in early designs for handling of static volumes of solid or semi-solid foods. Several treatment chambers have been designed. they can be categorized within two types: parallel plate and coaxial. (Fig. 6). parallel plate chambers have been typically used in batch modes while coaxial designs have been used in continuous modes where the medium is pumped through at a known flow rate and pulses are applied at a known pulse frequency. coaxial chambers used in continuous operation have been found to result in higher inactivation rates compared to batch systems sins there is a more uniform diatribution of the electric field in continuously flowing media (Qin et al., 1998). (Fig.7) presents different chamber designs

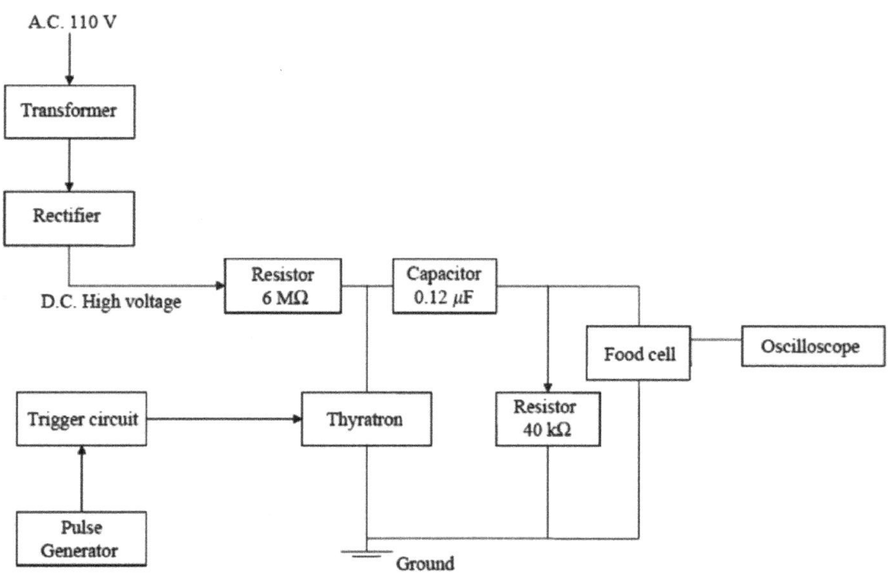

Figure 5. Generalized scheme of pulsed electric field equipment.

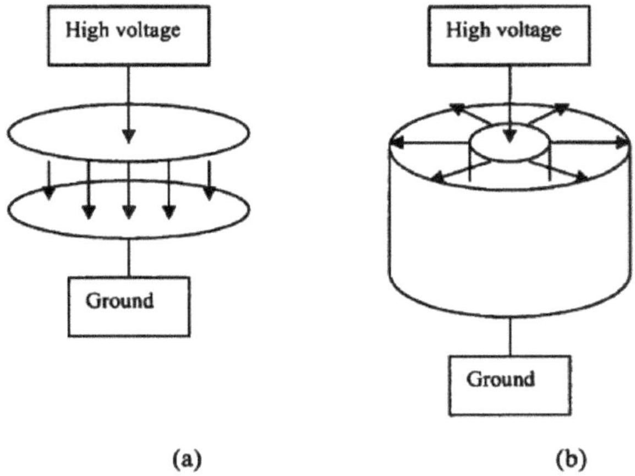

Figure 6. Common electrode configurations in pulsed electric field treatment chambers () parallel-plate () coaxial

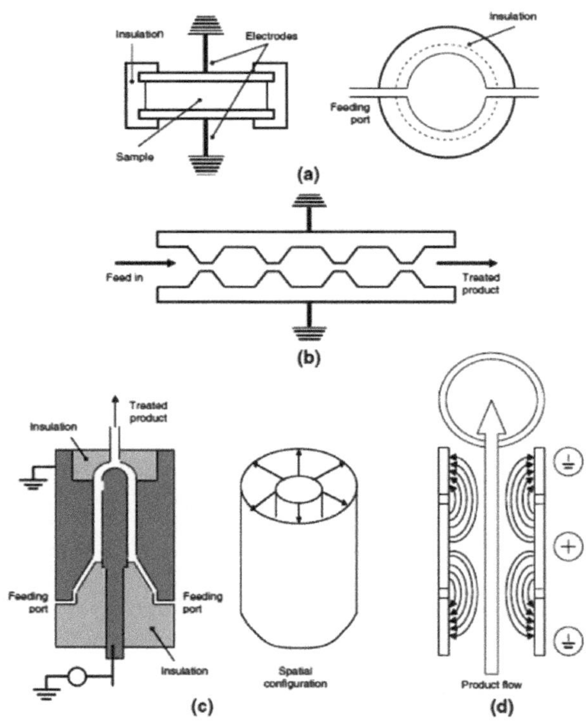

Figure 7. Schematic diagram of a pulsed electric fields operation design of treatment chambers for pulsed electric fields equipment: a static chamber, b side view of a basic continuous design, c coaxial chamber, and d co-linear chamber.

PEF investigators studying inactivation and preservation effects have been highly inventive in treatment chamber design (Fig. 8). Several different designs have been developed through the years for this key component, wherein high voltage delivered by the power supply is applied to the product located between a pair of electrodes. The basic idea of the treatment chamber is to keep the treated product inside during pulsing, although the uniformity of the process is highly dependent on the characteristic design of the treatment chamber. When the strength of applied electric fields exceeds the electric field strength of the food product treated in the chamber, break down of food occurs as a spark. Known as the dielectric breakdown of food, this is one of the most important concepts to be considered in treatment chamber design. Dielectric breakdown of the food is generally characterized as causing damage on the electrode surfaces in the form of pits, a result of arching and increased pressure, leading to treatment chamber explosions and evolution of gas bubbles. Intrinsic electrical resistance, electric field homogeneity, and reduction and generation of enhanced field areas are some other important design criteria for a successful design in terms of energy consumption and low product heating (Barbosa-Cánovas and Sepulveda, 2005).

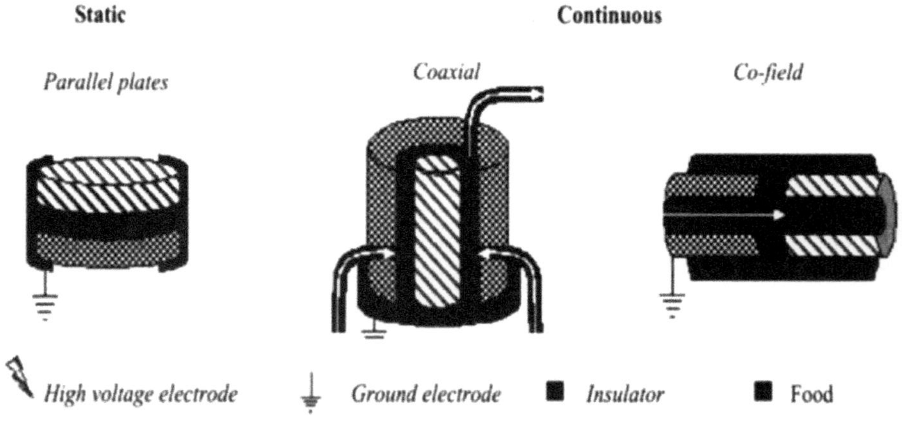

Figure 8. Schematic configurations of the three most used PEF treatment chambers.

Dunn and Pearlman (1987) designed a chamber consisting o f two parallel plate electrodes and a dielectric space insulator (Fig. 9). The electrodes are separated from the food by ion conductive membranes made of sulfonated polystyrene and acrylic acid copolymers, but fluorinated hydrocarbon polymers with pendant groups would also be suitable. An electrolyte is used to facilitate electrical conduction between electrodes and ion permeable membranes. Suitable electrolyte solutions include sodium carbonate, sodium hydroxide, potassium carbonate, and potassium hydroxide. These are circulated continuously to remove the products of electrolysis and replaced in the event of excess concentration or depletion of ionic components.

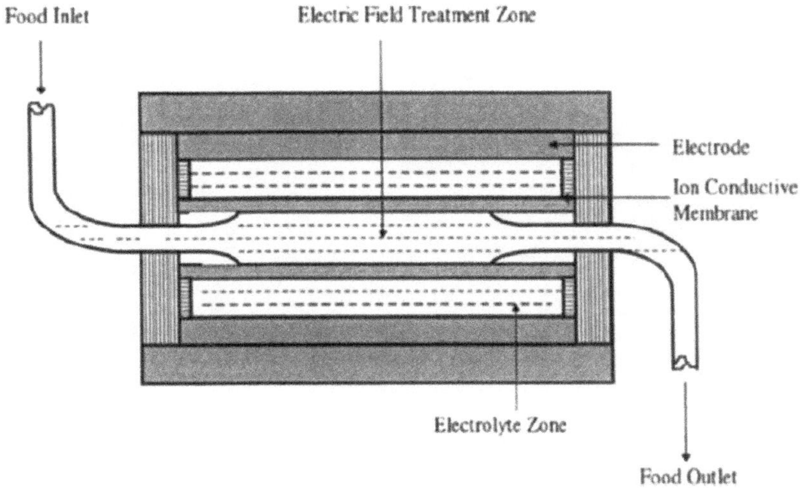

Figure 9. A continuous chamber with ion-conductive memberanes separating the electrodes and food.

From the electrical point of view, the PEF treatment chamber represents the electrical load consisting of two or more electrodes filled with the liquid substance to be treated. The chamber has to be constructed in such a way that the electrical field acting on the liquid is more or less homogeneous across the entire active region. Planar electrode configurations consist of two parallel electrodes fixed by insulators. The insulators and the electrodes form a channel for the streaming liquid. Coaxial electrode configurations consist of two coaxial electrodes. The liquid streams between these electrodes that are fixed by insulators not shown in the Fig.. Axial electrode configurations consist of several electrode rings on alternating potentials separated by insulating rings. Electrode materials also play an essential role. If monopolar voltage waveforms are applied, electrode corrosion can become critical and the substance to be treated can be contaminated. In commercially available electroporation devices with small probes, aluminum, stainless steel, carbon, gold-plated electrodes, and even silver electrodes are used (Barbosa-Cánovas *et al.*, 2000; Puc *et al.*, 2004).

The static parallel plate electrode chamber was modified by adding baffled flow channels inside to make it operate as a continuous chamber (Fig. 10). Two stainless-steel disk-shaped electrodes separated by a polysulfone spacer form the chamber. The designed operating conditions are: chamber volume, 20 or 8 ml; electrode gap, 0.95 or 0.51 cm; and food flow rate, 1200 or 6 ml/min (Qin et al., 1996).

The static parallel plate electrode chamber was modified by adding baffled flow channels inside to make it operate as a continuous chamber (Fig. 2.19). Two stainless-steel disk-shaped electrodes separated by a polysulfone spacer form the chamber. The designed operating conditions are: chamber volume, 20 or 8 ml; electrode gap, 0.95 or 0.51 cm; and food flow rate, 1200 or 6 ml/min (Qin et al., 1996).

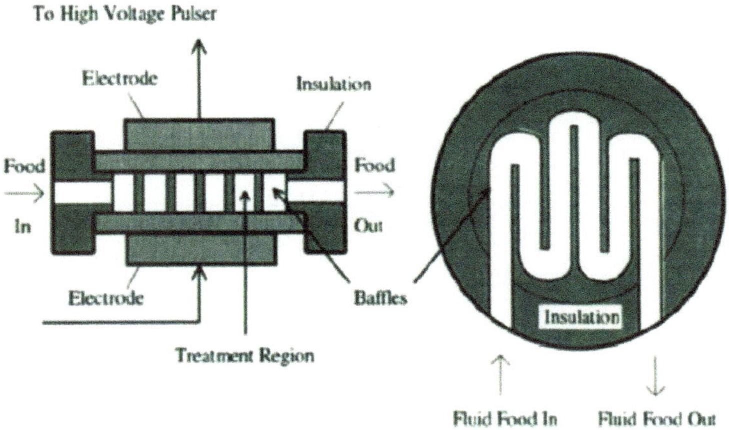

Figure 10. The continuous chamber with baffles

Jemai and Vorobiev, (2007) mentioned that In most runs, one or two chambers were manually filled with grated cossettes (see Fig. 11 for shapes and cross-sections of two types of cossettes). In a few runs, up to six chambers were used at the same time. Each chamber consists of a plate covered by a filter cloth and a flexible electrode (metallic grid) on one side and a rigid electrode on the other. The pressure (compressed air) is applied to the membrane of the plate, which in turn exerts and distributes the pressure over the cossettes placed between the plate and the rigid electrode (Fig. 11). Juice is drained through channels leading to the outlet, where juice accumulation is monitored by a weighing balance connected to a data acquisition system (Bouzrara, 2001).

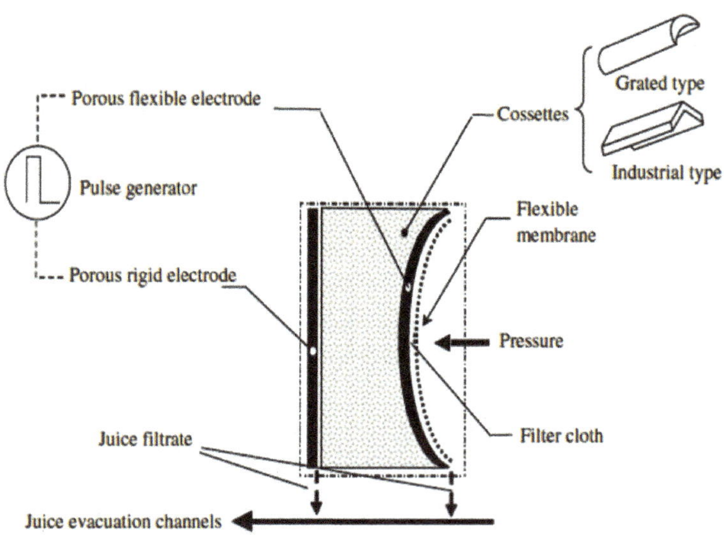

Figure 11. Schematic of a one chamber configuration of plate and frame filter press

5. Applications of pulsed electric fields technology

Application of pulsed electric fields technology has been successfully demonstrated for the pasteurization of foods such as juices, milk, yogurt, soups, and liquid eggs. Application of PEF processing is restricted to food products with no air bubbles and with low electrical conductivity. The maximum particle size in the liquid must be smaller than the gap of the treatment region in the chamber in order to ensure proper treatment. PEF is a continuous processing method, which is not suitable for solid food products that are not pump able. PEF is also applied to enhance extraction of sugars and other cellular content from plant cells, such as sugar beets. PEF also found application in reducing the solid volume (sludge) of wastewater.

PEF processing has been successful in a variety of fruit juices with low viscosity and electrical conductivity such as orange, apple, and cranberry juice. Recent studies reported more than a 3-10g reduction in orange juice (Qin *et al.,* 1998) and apple juice (Evrendilek *et al.,* 2000).

Additionally, the color change in fruit juices (subject to prolonged storage) was reportedly less in juices treated by PEF, as in a recent study of PEF-treated orange juice stored at 4°C for 112 days; there was less browning than thermally pasteurized juice, which was attributed to conversion of ascorbic acid to furfural (Yeom *et al.,* 2000).

Considering the effectiveness of PEF treatment on liquid products, such as milk, fruit juices, liquid egg, and any other pumpable food products, extensive research has been done to implement the process at an industrial level. Flavor freshness, economic feasibility, improvements in functional and textural attributes and extended shelf life are some of the main points of interest besides achievement of microbiological safety of food products (Dunn, 2001). Among all liquid products, PEF technology has been most widely applied to apple juice, orange juice, milk, liquid egg, and brine solutions (Qin *et al., 1995).*

Each of the nonthermal technologies has specific applications in terms of the types of foods that can be processed. Among these, pulsed electric fields (PEF) is one of the most promising nonthermal processing methods for inactivation of microorganisms, with the potential of being an alternative for pasteurization of liquid foods. Comparable to pasteurization, yet without the thermal component, PEF has the potential to pasteurize several foods via exposure to high voltage short pulses maintained at temperatures below 30-40°C. The basic definition of PEF technology relies on the use of high intensity pulsed electric fields (10-80 kV/cm) for cell membrane disruption where induced electric fields perforate microbial membranes by electroporation, a biotechnology process used to promote bacterial DNA interchange. Induction of membrane potentials exceeding a threshold value often result in cell damage and death (Zimmermann,1986).

PEF technology has recently been used in alternative applications including drying enhancement, enzyme activity modification, preservation of solid and semisolid food products, and waste water treatment, besides pretreatment applications for improvement of metabolite extraction. The ability of PEF to increase permeabilization means it can be successfully used to enhance mass and heat transfer to assist drying of plant tissues. Studies conducted on different plant tissues such as potato tissue (Angersbach and Knorr, 1997), coconut (Ade-Omowaye *et al.*, 2000), carrots (Rastogi *et al.*, 1999), mango (Tedjo *et al.*, 2002), and apple slices (Ade-Omowaye *et al.*, 2002).

Considering the effectiveness of PEF treatment on liquid products, such as milk, fruit juices, liquid egg, and any other pump able food products, extensive research has been done to implement the process at an industrial level. Flavor freshness, economic feasibility, improvements in functional and textural attributes and extended shelf life are some of the main points of interest besides achievement of microbiological safety of food products (Dunn, 2001).

Among all liquid products, PEF technology has been most widely applied to apple juice, orange juice, milk, liquid egg, and brine solutions (Qin *et al., 1995).*

Application of PEF is especially promising for the citrus industry, which is concerned with the spoilage microorganisms and resultant production of off-flavor compounds such as lactic acid bacteria (Hendrix and Red, 1995).

Jemai and Vorobiev (2002) stated the enhancing effect of a PEF treatment on the diffusion coefficients of soluble substances in apple slices. The results available in literature clearly indicate that PEF can be successfully applied to disintegrate biological tissue and to improve the release of intracellular compounds, though an industrial application has not been achieved up to now. At Berlin University of Technology a system with a peak voltage of 20 kV, an average power of 7 kW and a production capacity of 2 ton/h has been developed for the treatment of fruit mashes. The power supply and the treatment chamber are shown in Fig. 12. It is noteworthy that avoiding an enzymatic maceration the pectin fractions will remain in a native, highly esterified structure. This provides a potential to extract high quality pectin from the pomace after juice winning and therefore a step toward a more economic and sustainable processing.

PEF treatments are applied in the form of short pulses to avoid excessive heating or undesirable electrolytic reactions. In general, a continuous PEF treatment system is composed of treatment chambers, a pulse generator, a fluid-handling system, and monitoring systems (Fig. 13.) (Elez-Martínez et al., 2006; Min et al., 2007).

An OSU-4D bench-scale continuous unit manufactured in Ohio State University (US) was used to treat the food sample (Fig. 14). Six co-field chambers with a diameter of 2. $3*10^{-3}$m and a gap distance of $2.93 * 10^{-3}$ m between electrodes were connected in series. Two cooling coils were connected before and after each pair of chambers and submerged in a circulating bath (Polystat, Cole Parmer, IL, USA, ±0.05 ^{0}) to maintain the selected temperature at 35 or 55 ^{0}C. The temperature was recorded by thermocouples (T type, ±0.1 ^{0}) at the entrance of the first treatment chamber (initial temperature) and at the edcxit of the last treatment chamber (final temperature) (Sampedro, 2007).

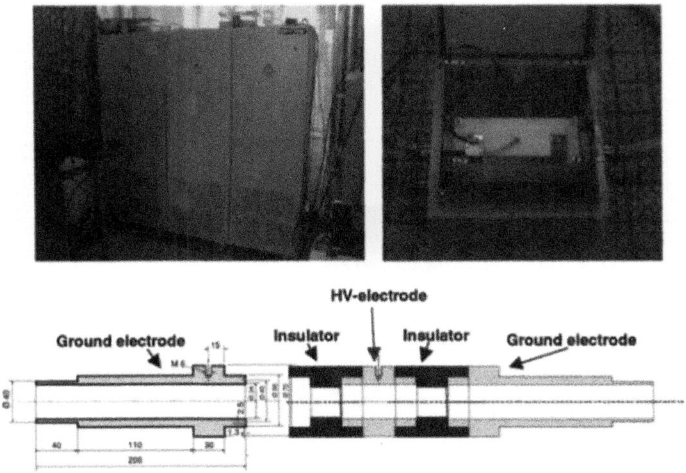

Figure 12. Power supply and treatment chamber of a pilot scale system for fruit mashes. The treatment chamber is a co-linear configuration of 3 cylindrical electrodes separated by insulators.

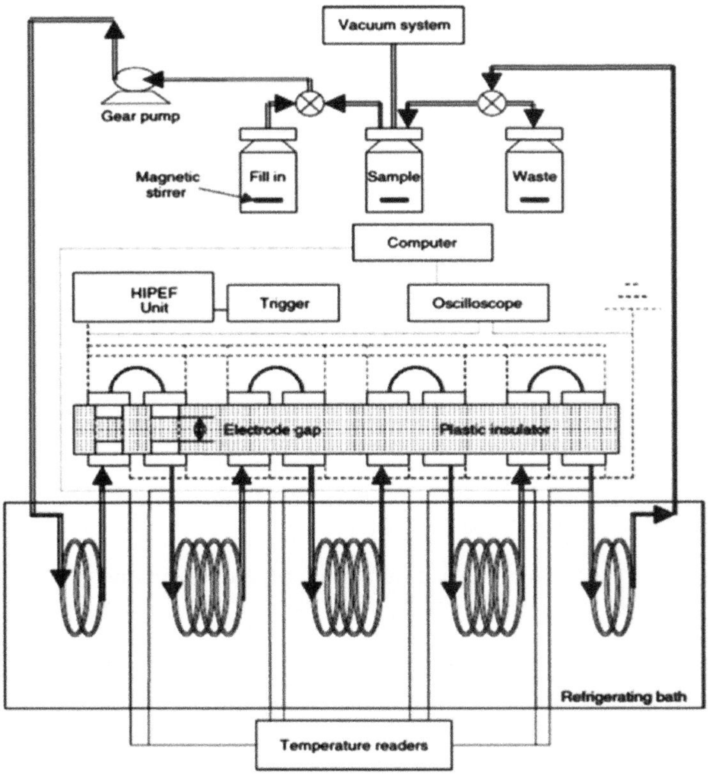

Figure 13. Diagram of the high-intensity pulsed electric field bench-scale processing unit.

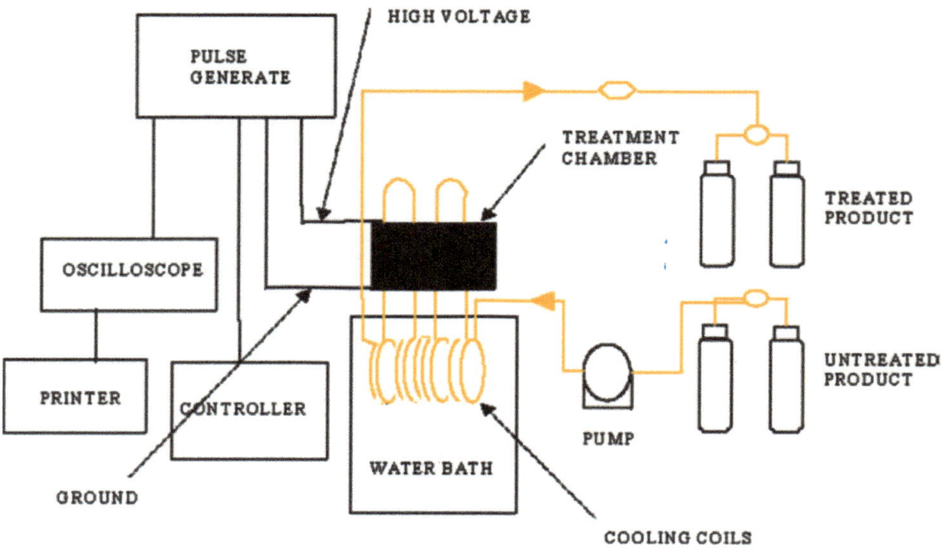

Figure 14. Scheme of the pulsed electric field equipment.

Vorobiev (2008), developed a laboratory device in the University of Technology of Compiegne (UT) (Fig. 15) permits both pretreatment and intermediate treatment by PEF. The treatment cell has a polypropylene frame with a cylindrical cavity compartment (20 mm thick, 56 mm in diameter), which should be initially filled with gratings and then closed from both sides by steel covers. A mobile electrode is attached to the elastic rubber diaphragm. A stationary wire gauze electrode is installed between the filter cloth and the layer of gratings. Both electrodes are connected to the PEF generator, which can provide the monopolar or bipolar pulses of near-rectangular shape.

Studies conducted on the effects of PEF on dairy products such as skim milk, whole milk, and yogurt compromise a major section of PEF applications (Alvarez and Ji, 2003). Milk is very susceptible to both spoilage and pathogenic microorganisms requiring the application of thermal pasteurization under current regulations, which ensures safety but generally results in a cooked flavor (Wirjantoro and Lewis, 1997).

6. Factors affecting the outcome of pulsed electric fields treatments

In order to use PEF technology as a pasteurization process it is necessary to estimate its efficacy against pathogenic and spoilage food-borne microorganisms. To obtain this objective there is a need to accumulate knowledge on the critical factors affecting microbial inactivation, to describe the PEF inactivation kinetics and to understand the mechanisms involved in microbial PEF inactivation. The lethality factors contributing to the effectiveness of pulsed electric field technology can be grouped as technological, biological, and media factors. Each group of determinant factors is related to type of equipment, processing parameters, target microorganism, and type and condition of media used.

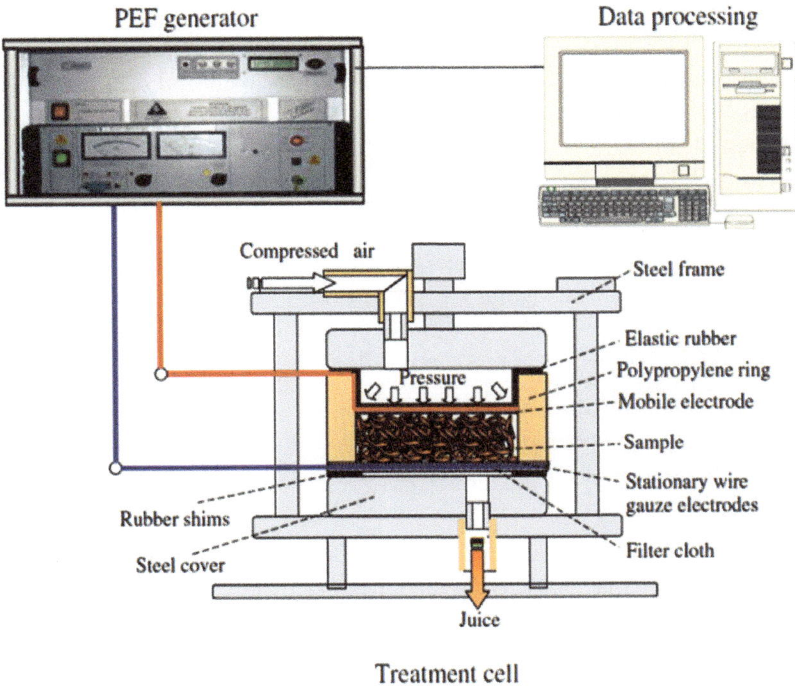

Figure 15. Experimental setup for permits both pretreatment and intermediate treatment by PEF.

6.1. Technological factors

A number of other factors during PEF processing can affect specific microbial inactivation as well. Some of these critical factors include the field strength, treatment time, treatment temperature, pulse shape, type of microorganism, growth stage of microorganism, and characteristics of the treatment substrate.

Microbial inactivation increases with an increase in the electric field intensity, above a critical trans membrane potential (Qin et al., 1998)

It is important that the electric field intensity should be evenly distributed in the treatment chamber to achieve an efficient treatment. Electric field intensities of smaller than 4-8 kV cm"' usually do not affect microbial inactivation (Peleg, 1995).

In general, the electric field intensity required to inactivate microorganisms in foods in the range of 12-45 kV cm"1. The fact that microbial inactivation increases with increases in the applied electric field intensity and can be attributed to the high energy supplied to the cell suspension in a liquid product (Wouters et al., 1999).

An important aspect that differentiates between PEF processing and other microbial inactivation technologies is that the PEF treatment is delivered by pulsing. The pulses commonly used in PEF treatments are usually either exponential or square wave pulses (Jeyamkondan *et al.*, 1999).

There is some controversy with respect to the influence of the pulse width on the PEF microbial lethality. Some authors have indicated that after the same treatment time, inactivation tested in several microorganisms was independent of the pulse width (Hülsheger *et al.*, 1981; Raso *et al.*, 2000; Alvarez *et al.*, 2003b,).

Treatment time could be defined as the effective time during which range microorganisms are subjected to the field strength. It depends on the number of pulses and the width of the pulses applied. This parameter and the electric field strength are the main factors determining the lethal effect of PEF treatments (Sale and Hamilton, 1967; Jayaram *et al.*, 1991; Barsotti and Cheftel, 1999; Wouters *et al.*, 2001).

Studies on microbial inactivation by PEF have been conducted at frequencies ranged from 1 to 500 Hz. If the same number of pulses is applied, microbial inactivation is generally independent of the number of pulses applied per second (Hülsheger *et al.*, 1981; Jeantet *et al.*, 1999; Raso *et al.*, 2000; Alvarez *et al.*, 2003b,).

PEF treatment time is calculated by multiplying the pulse number by the pulse duration. An increase in any of these variables increases microbial inactivation (Sale and Hamilton 1967).

A good understanding of the electrical principles behind PEF technology is essential for a comprehensive analysis of the PEF system. The electrical field concept, introduced by Faraday, explains the electrical field force acting between two charges. When unit positive charge q located at a certain point within the electric field is generated in the treatment gap (Er), it experiences force F identified by position vector r (Blatt, 1989). The electrical field per unit charge is then defined as shown in Eq. (3):

$$E_r = \frac{F_{qr}}{q} \qquad (3)$$

The electrical potential difference *(V)* between voltage across two points, separated by a nonconductive material , results in generation of an electric field between these points , with an electrical intensity *(E)* directly proportional to the magnitude of potential difference *(V)* and inversely proportional to the distance *(d)* between points, as given in Eq. (4):

$$E = \frac{V}{d} \qquad (4)$$

The type of electrical field waveform applied is one of the important descriptive characteristics of a pulsed electric field treatment system. The exponentially decaying or square waves are among the most common waveforms used. To generate an exponentially decaying voltage wave, a DC power supply charges the bank of capacitors that are connected in series with a charging resistor. When a trigger signal is applied, the charge stored in the capacitor flows through the food in the treatment chamber. Exponential waveforms are easier to generate from the generator point of view. Generation of square waveform generally requires a pulse forming network (PFN) consisting of an array of capacitors and inductors. It is more challenging to design a square waveform system compared to an exponential waveform system. However, square waveforms may be more lethal and energy efficient than exponentially decaying pulses since square pulses have longer peak voltage duration compared to exponential pulses (Amiali et al. 2006).

The electric field should be evenly distributed in the treatment chamber in order to achieve an efficient treatment. An electric field intensity of 16 kV/cm or greater is usually sufficient to reduce the viability of Gram negative bacteria by 4 to 5 log cycles and Gram positive bacteria by 3 to 4 log cycles (Pothakamury et al. 1995)

6.2. Biological factors

Biological factors that include the individual characteristics of target microorganisms and their physiological and growth states are determinant factors affecting PEF application. The susceptibility of a microorganism to PEF inactivation is highly related to the intrinsic parameters of the microorganism such as size, shape, species or growth state. Generally, Gram-positive vegetative cells are more resistant to PEF than Gram-negative bacteria, while yeasts show a higher sensitivity than bacteria. Induction of electric fields into cell membranes is greater when larger cells are exposed to PEF treatment (Sale and Hamilton, 1967; Htilsheger et al., 1983; Zhang et al., 1994). Most of the research focuses on the inactivation of vegetative cells of bacteria, while only a few reports are available on the inactivation of spores describ ing a limited effect of PEF. *Bacillus cereus* spores were mostly resistant (approximately I log reduction) to a mild PEF treatment at electric field strength of 20 kV/cm and 10.4 pulses in a study conducted on apple juice (Cserhalmi *et al., 2002).*

Another study conducted by Pagan *et al.* (1998) found that *Bacillus cereus* spores were not affected with PEF treatment of 60 kV/cm for 75 pulses at room temperature. On the other hand, Marquez *et al.* (1997) reported 3.42-log and 5-log reductions of *Bacillus subtiLis* and *Bacillus cereus* spores, respectively, with PEF treatment of 50 kV/cm for 50 pulses at 25°C in salt solution. Additionally, mold Condi spores were reported to be sensitive to PEF in fruit juices whereas *Neosartorya fischeri*asco spores were resistant to PEF treatments (Raso *et al.*, 1998).

Compared to the number of studies reported for enzyme inactivation by PEF, little information is available on the mechanism of inactivation, which may be due to the lack of analysis of enzyme structural data (Yeom et al., 1999).

6.3. Media factors

The effects of PEF on the food system are related to the PEF system and the properties of the liquid food. The most important factors in the PEF system are the electric field intensity, number of pulses, pulse waveform, pulse width, treatment time and treatment temperature. But enzymes and proteins are generally more resistant to electric field intensity and pulses than microorganisms. This requires further investigation, especially on the effects of pH, temperature, resistivity and composition of the enzyme or protein-containing medium or food system (Barsotti & Cheftel, 1999).

The physical and chemical characteristics of food products are known to strongly influence the effectiveness of microbial inactivation during PEF application (Wouters et al., 2001), thus the challenge experienced using real food systems was due to the important role of the media 's chemical and physical characteristics. These factors most likely influence the recovery of injured microbial cells and their subsequent growth following PEF exposure, since the presence of food components, such as fats and proteins , has reportedly had a preventive effect on microorganisms against PEF treatment (Ho et aI., 1995; Grahl and Markl, 1996; Martin et al., 1997).

Similar to the intrinsic parameters of microorganisms, treated media has its own intrinsic factor s such as conductivity, resistivity, dielectric properties, ionic strength, pH, and composition . Each of these parameter sinfluences the PEF treatment either alone or in combination. PEF technology has recently been used in alternative applications including drying enhancement, enzyme activity modification, preservation of solid and semisolid food products, and waste water treatment, besides pretreatment applications for improvement of metabolite extraction. The ability of PEF to increase permeabilization means it can be successfully used to enhance mass and heat transfer to assist drying of plant tissues. Studies conducted on different plant tissues such as potato tissue (Angersbach and Knorr, 1997), coconut (Ade-Omowaye *et al.*, 2000), carrots (Rastogi *et al.*, 1999), mango (Tedjo *et al.*, 2002), and apple slices (Ade-Omowaye *et al.*, 2002) reported increased yield of water removal by 20-30% when exposed to low intensity electric fields.

Temperature is one factor proposed that has been correlated with microbial inactivation, and although PEF application is strictly a nonthermal processing technology, the synergistic effect of temperature on foods (due to changes in the properties of cell membranes) becomes greater when foods are subjected to high intensity pulse electric fields (Jayaram et al., 1993). In general, the lethality of PEF treatments increases with an increase in processing temperature; therefore, a proper cooling device is necessary to maintain temperatures below levels. that affect nutritional, sensory or functional properties of food products (Wouters et al., 1999).

The influence of pH and water activity (aw) on microbial growth was documented by (Jay, 1992).

Sepulveda (2003) proposed that a PEF treatment time between 0.1 to 0.5 ms produced the best results for microbial inactivation. The pulse width is defined as the time where the peak field is maintained for square wave pulses or the time until decay to 37% for exponential decay pulses. Typically, increasing the number of pulses causes an increase in treatment time, as the pulse width is fixed by the impulse generation setup.

Dunn and Pearlman (1987) found that a combination of PEF and heat was more efficient than conventional heat treatment alone. A higher level of inactivation was obtained using a combination of 55°C temperature and PEF to treat milk.

Zhang et al (1995) reported that increasing treatment temperature from 7 to 20°C significantly increased PEF inactivation of *E. coli* in simulated milk ultra-filtrate (SMUF).

However, additional increase in temperature from 20 to 33°C did not result in any further increase in PEF inactivation.

7. Conclusion

The objective of food preservation technologies used by the food industry is to control microorganisms once they are contaminating foods. Food preservation technologies are based on the prevention of microbial growth or on the microbial inactivation.

Pulsed electric field (PEF) is a potential non-thermal food preservation technique to replace conventional thermal processing. When exposed to high electrical field pulses, cell membranes develop pores either by enlargement of existing pores or by creation of new ones. These pores may be permanent or temporary, depending on the condition of treatment. The pulsed electric processing system is composed basically of a high power pulse generator, a treatment cell, voltage and current measuring devices. A traditional treatment cell consists of two electrodes held in parallel by insulating material that form an enclosure containing the food to be treated. The application of high intensity pulsed electric fields consists of the generation of short time pulses of electric fields between two parallel plate electrodes enclosing a dielectric material. Pulsed electric fields technology is the application of very short pulses (micro – to milliseconds), at electric field intensities ranging from 10-80 kV/cm, applied to a food product held between two electrodes inside a chamber, usually at room temperature.

Research of pulsed electric fields technology is ongoing around the world. Most of the research conducted up until now has been in the laboratory and on a pilot plant scale level, and has shown promising results.

The basis for this prediction is because of PEF's ability to inactivate microorganisms in the food, reduce enzymatic activity, and extend shelf-life with negligible changes in the quality of the final product as compared to the original one. According to the intensity of the field strength, electroporation can be either reversible (cell membrane discharge)

The present chapter reviews the current state of the art in microbial inactivation by PEE after discussing critical factors determining microbial inactivation by PEF and mathematical kinetic models used for describing PEF death, the most successful combinations of PEF with other preservation techniques for enhancing the safety of minimally processed foods are presented. The chapter concludes with some aspects that need further investigation for the development of PEF processes to supply safe food products of high organoleptic and nutritional quality.

The chapters focus on the electrical bases, various equipment configurations, and principal components of pulsed electric field systems, including various types of electric circuits and processing chambers, By explaining the following points

1. What is main electrical fundamental parameter of pulsed electric field treatments?
2. How pulsed electric field works?

3. What is pulsed electric field processing?
4. What is a pulsed electric field treatment chamber?
5. Application of pulsed electric field technology in food processing.
6. Pulsed Power Systems for Application of Pulsed Electric Fields.

Author details

Maged E.A. Mohamed
Department of Agriculture Engineering, Faculty of Agricultural, Minoufiya University, Egypt
Date Palm Research Center, King Faisal University, Saudi Arabia

Ayman H. Amer Eissa
Department of Agriculture Engineering, Faculty of Agricultural, Minoufiya University, Egypt
Department of Agriculture Systems Engineering,
College of Agricultural and Food Sciences, King Faisal University, Saudi Arabia

8. References

Ade-Omowaye, B. I. 0., Eshtiaghi, N., and Knorr, D., (2000), Impact of high intensity electric field pulses on cell permeabilisation and as pre-processing step in coconut processing, *lnnov. Food Sci. Emerg. Techno!.* 1(3): 203-209.

Ade-Omowaye, B. I. 0., Taiwo, K, and Knorr, D., (2002), Use of pulsed electric field pretreatment to improve dehydration characteristics of plant based foods, *Trends Food Sci. Technol.* 12: 285-295.

Alvarez, 1., Pagan, R., Condon, S., and Raso, J., 2003b, The influence of process parameters for the inactivation of Listeria monocytogenes by pulsed electric fields, Int. J. Food Microbiol. 87: 87-95.

Alvarez, 1.,Raso, 1., Sala, F. 1., and Condon, S., 2003c, Inactivation of Yersinia enterocolitica by pulsed electric fields, Food Microbiol. 20: 691-700.

Amiali, M., Ngadi, M. O., Raghavan, G. S. V., & Smith, J. P. (2006). Inactivation of *Escherichia coli* O157:H7 and *Salmonella enteritidis* in liquid egg white using pulsed electric field. *Journal of Food Science*, 71, 88–94.

Angersbach, A., and Knorr, D., (1997), High intensity electric field pulses as pretreatment for affecting dehydration characteristics and rehydration properties of potato cubes, *Nahrung* 55: 143-146.

Bajgai, T. R., and Hashinaga,E, 2001, High electric field drying of Japanese radish, Drying Technol. 19: 2291-2302.

Barbosa-Cánovas, G. v., G6ngora-Nieto, M. M., Pothakamury, U. R., and Swanson,B. G., 1998,Preservation of Foods with Pulsed Electric Fields,Academic Press, London.

Barbosa-Cánovas, G. V., Gongora-Nieto, M. M., Pothakamury, U. R., Swanson, B. G. (1999). Fundamentals of high-intensity pulsed electric fields (PEF). In "Preservation of foods with pulsed electric fields". G. V. Barbosa-Cánovas et al. edited. Academic Press. San Diego, CA, USA. p. 1-19, 76-107, 108-155.

Barbosa-Cánovas, G. v., Pierson, M. D., Zhang, Q. H., and Schaffner, D. W, (2000), Pulsed electric fields, 1. Food Sci. 65(8): 65-79.

Barsotti, L. & Cheftel, J. C. (1999). Food processing by pulsed electric fields. II. Biological aspects. Food Review International. 15(2):181-213.

Barsotti, L., Merle, P. P., and Cheftel, J., (1999), Food processing by pulsed electric fields, I. Physical aspects, *Food Rev. Int.* 15(2): 163-180.

Barsotti,L., and Cheftel, 1. C., (1998),Traitementdes alimentspar champs electriques pulses. Sci. Aliment. 18: 584-601.

Bartos, F.J. (2000). Medium voltage AC drives. Control Engineering Online, No. 20. http://www.controleng.com/ archives /2000/ctl0201.00/000203.htm.

Bazhal, M. I., (2001), Etude du mecanisme d'electropermeabilisation des tissus vegetux. Application ii l'extraction du jus des pommes, These de Doctorat, Universite de Technologie de Compiegne, France.

Bendicho S, Barbosa-Cánovas GV, Martín O (2003) Reduction of protease activity in simulated milk ultrafiltrate by continuous flow high intensity pulsed electric field treatments.J Food Sci 68(3):952–957

Bendicho, S., Barbosa-Cánovas, G. v., and Martin, 0.,(2002), Milk processing by high intensity pulsed electric fields, *Trends Food* Sci. *Technol. 13(617): 195-204.*

Bermúdez-Aguirre, D., Mawson, R., Versteeg, K., and Barbosa-Cánovas, G.V. (2009). Composition parameters, physical chemical characteristics and shelf-life of whole milk after thermal and thermo-sonication treatments. Journal of Food Quality 32:283–302.

Blackburn C, and McClure, P., (2002), Introduction, In: Foodbome Pathogens: Hazards, Risk Analysis and Control (C Blackburn and P. McClure, eds.), Woodhead, Cambridge, England, pp. 3-10.

Blatt, F., (1989), Principles of physics, 3rd ed., Allyn and Bacon, Needham heights, MA.

Bluhm, H., (2006). Pulsed power systems: principles and applications: Berlin: Springe.

Bouzrara H; Vorobiev E., (2001). Non-thermal pressing and washing of fresh sugarbeet cossettes combined with a pulsed electrical field. Zuckerindustrie, 126(6), 463–466.

Butz, P., Tauscher, B., (2002). Emerging technologies: chemical aspects. Food Research International 35, 279–284.

Chang D.C., Saunders J.A., Chassy B.M. and Sowers A.E. (1992). Overview of electroporation and electrofusion. In: Chang, D.C., Chassy, B.M., Saunders J.A. and Sowers A.E (eds), Guide to Electroporation and Electrofusion. San Diego, CA: Academic Press, Inc. pp. 1–6.

Cserhalmi, Z., Vidacs, I. Beczner, J, and Czukor, B., (2002). Inactivation of Saccharomyces cerevisiae and Bacillus cereus by pulsed electric fields, Innov Food Sci Emerg Technol, 3(1): 41-45.

Dunn, J. E., and Pearlman, J. S. (1987). Methods and apparatus for extending the shelf life of fluid food products. U. S. Patent 4,695,472.

Dunn, J., (2001), Pulsed electric field processing: An overview, In: *Pulsed Electric Fields* in *Food Processing, Fundamental Aspects and Applications* (G. Barbosa-Cánovas and Q. H. Zhang, eds.), Technomic Press, Lancaster, PA, pp. 1-30.

Elez-Martínez P, Aguiló-Aguayo I, Martín-Belloso O (2006) Inactivation of orange juice peroxidase by high-intensity pulsed electric fields as influenced by processing parameters. J Sci Food Agric 86:71–81

EPR! and Army, (1997), *Pulsed Electric Field Workshop //: Minutes,* Industrial and Agricultural Technologies and services, Palo Alto, CA.

Estiaghi,M. N., and Knorr,D., (1999), Method for treating sugar beet, InternationalPatent Nr WO 99/6434.

Evans, G. and Cox, D.N. (2006). Australian consumers' antecedents of attitudes towards foods produced by novel technologies. British Food Journal 108(11):916–930.

Evrendilek, G. A., Jin, Z. T., Ruhlman, K. T., Qui, X., Zhang, Q. H., and Ritcher, E. R., (2000), Microbial safety and shelf life

Fernandez-Díaz MD, Barsotti L, Dumacy E, Chefter JC (2000) Effects of pulsed electric fields on ovalbumin solutions and dialyzed egg white. J Agric Food Chem 48:2332–2339

Floury, J., Grosset, N., Leconte, N., Pasco, M., Madec, M., & Jeantet, R. (2005). Continuous raw skim milk processing by pulsed electric field at non-lethal temperature: effect on microbial inactivation and functional properties. Lait, 86, 43–57.

Gongora-Nieto, M. M., Sepulveda, D., Pedrow, P., Barbosa-Cánovas, G., and Swanson, B., (2002), Food processing by pulsed electric fields: Treatment delivery, inactivation level, and regulatory aspects, *Lebensm. Wiss. Techno!.* 35(5): 375-388.

Grahl, T. and Markl, H. (1996). Killing of microorganisms by pulsed electric fields. *Applied Microbiology and Biotechnology,* 45(1/2), 148-157.

Gulyi,1. S.,Lebovka, N.I.,Mank, V., Kupchik,M. P.,Bazhal,M. I., Matvienko,A B., and Papchenko,A Y., (1994),Scientific and Practical Principles of Electrical Treatment of Foo dProductsand Materials, Ukr INTEI, Kiev (in Russian).

Hendrix, C. M., and Red, J. B., (1995), Chemistry and technology of citrus juices and by-products, In: *Production and Packaging Of Non-Carbonated Fruit Juices and Fruit Beverages*(P.R. Ashurst, ed.), Blackie Academic and Professional, Glasgow, UK, pp. 53-87.

Ho, S. Y, and Mittal, G. S., (1996), Electroporation of cell membranes: A review, *Crit. Rev. Biotechnol.* 16(4): 349-362.

Ho, S. Y; Mittal G S (2000). High voltage pulsed electrical field for liquid food pasteurization. Food Review International, 16, 395–434

Ho, S.Y., Mittal, G.S., Cross, J.D., and Griffiths, M.W. (1995). Inactivation of Pseudomonas fluorescens by High Voltage Electric Pulses. J. Food Sci., 60:1337-1340,1343.

Hodgins, A., Mittal, G., and Griffiths, M., (2002), Pasteurization offresh orange juice using low-energy pulsed electrical field, *J. FoodSci.* 67(6): 2294-2299.

Hülsheger, H., Potel, J., and Niemann, E. G., 1981, Killing of bacteria with electric pulses of high field strength, Radiat. Environ. Biophys. 20: 53-65.

Hülsheger, H., Potel, J., and Niemann, E.G. (1983). Electric field effects on bacteria and yeast cells. *Radiation and Environmental Biophysics,* 22, 149-162.

Jay, J. M., (1992), High temperature food preservation and characteristics of thermophilic microorganisms, In: *Modem Food Microbiology,* Van Nostrand Reinhold, New York, pp. 335.

Jayaram, S., Castle, G. S. P., and Margaritis, A., (1991). Effects of high electric field pulses on Lactobacillus brevis at elevated temperatures, IEEE Ind. Appl. Soc. Meet. 5: 674-681.

Jayaram, S., Castle, G. S. P., and Margaritis, A., (1993), The effect of high field DC pulse and liquid medium conductivity on survivability of *Lactobacillusbrevis, Biotechnol. Bioeng.* 40: 117-122.

Jeantet, R., Baron, F., Nau, F., Roignant, M., and Bruhl, G., (1999), High intensity pulsed electric fields applied to egg white: Effect on Salmonella enteritidis inactivation and protein denaturation, 1. Food Prot. 62: 1381-1386.

Jemai A.B., E. Vorobiev (2007). Pulsed Electric Field Assisted Pressing of Sugar Beet Slices: towards a Novel Process of Cold Juice Extraction. Biosystems Engineering (2006) 93 (1), 57–68.

Jeyamkondan, S., Jayas, D. S., and Holley, R. A., (1999), Pulsed electric field processing of foods: A review, J. Food Prot. 62(9): 1088-1096.

Knorr, D., Ade-Omowaye, B.I.O., and Heinz, V. (2002). Nutritional improvement of plant foods by non-thermal processing. Proceedings of the Nutrition Society 61:311–318.

Lado, B. H., and Yousef, A. E., 2002, Alternative food preservation technologies: Efficacy and mechanism, *Microbes Infect.* 4: 433-440.

Loeffler M. J., (2006). Generation and application of high intensity pulsed electric fields. pulsed electric fields technology for the food industry fundamentals and applications. isbn-13: 978-0387-31053-4.p27:72

Marquez, V.0., Mittal, G. S., and Griffiths, M. w., (1997), Destruction and inhibition of bacterial spores by high voltage pulsed electric field, J. FoodSci. 62(2): 399-401,409.

Martı´n-Belloso O., Vega-Mercado H., Qin B.L., Chang F.J., Barbosa-Ca´ novas G.V. and Swanson B.G. (1997). Inactivation of Escherichia coli suspended in liquid egg using pulsed electric fields. Journal of Food Processing and Preservation 21:193–208.

Martin, 0., Qin, B., Chang, E, Barbosa-Cánovas, G., and Swanson, B., (1997), Inactivation of *Escherichiacoli* in skim milk by high intensity pulsed electric fields, *J. FoodProc. Eng.* 20: 317-336.

Mertens, B., and Knorr, D., (1992), Developments of nonthermal processes for food preservation, *FoodTechno!'* May: 124133.

Min S., Evrendilek GA, Zhang Q H., (2007) Pulsed electric fields: processing system, microbial and enzyme inhibition, and shelf-life extension of foods. IEEE T Plasma Sci 35:59–73

Min S., Jin Z.T. and Zhang Q.H., (2003). Commercial scale pulsed electric field processing of tomato juice. Journal of Agricultural and Food Chemistry 51(11): 3338–3344.

Ortega-Rivas, E., Zárate-Rodríguez, E., & Barbosa-Cánovas, G. V., (1998). Apple juice pasteurization using ultrafiltration and pulsed electric fields. Food and Bioproducts Processing, 76C, 193–198.

Pagan, R., Esplugas, S., Gongora-Nieto, M. M., Barbosa-Cánovas, G. v., and Swanson, B. G., (1998), Inactivation of *Bacillus subtilis* spores using high intensity pulsed electric fields in combination with other food conservation technologies, *FoodSci. Techno!' Int.* 4(1): 33-44.

Palaniappan, S., and Sastry, S. K., (1990), Effects of electricity on microorganisms: A review, J. Food Process Preserv. 14:393-414.

Peleg, M., (1995). A model of microbial survival after exposure to pulsed electric fields. Journal of the Food Science & Agriculture. 67:93-99.

Pothakamury, U.R., Monsalve-González, A., Barbosa-Cánovas, G.V. And Swanson, B.G. (1995). Inactivation of *Escherichia coli* and *Staphylococcus aureus* in model foods by pulsed electric field technology. *Food Research International,* 28(2), 167-171.

Prassanna, G. 1., and Panda, T., (1997), Electroporation: Basic principles, practical considerations and applications in molecular biology, Bioprocess Eng. 16: 261-264.

Puc, M., Čorovič,S., Flisar, K., Petkovsek,M., Nastran,J., and MiklavčičD., (2004), Techniquesof signal generation required for electropenneabilization.Survey of electropenneabilization devices, Bioelectrochemistry 64: 113-124.

Qin, B. L., Barbosa-Cánovas,G. V.,Swanson, B., Pedrow,P. D., and Olsen, R. G., 1998, Inactivating microorganisms using a pulsed electric field continuous treatments system, *IEEE Trans. Ind. Appl.* 34: 43-50.

Qin, B. L., Chang, F., Barbosa-Cfinovas, G. V., and Swanson, B. G. (1995). Nonthermal inactivation of Saccharomyces cerevisiae in apple juice using pulsed electric fields. Lebensm.-Wiss. Technol. 28, 564-568.

Qin, B. L., Pothakamury,U. R, Barbosa-Cánovas,G. *v.*, and Swanson, B., (1996), Nonthennal pasteurization of liquid foods using high intensity pulsed electric fields, *Crit. Rev. Food Sci. Nutr.* 36(6): 603-627.

Qin, B. L., Zhang, Q., Barbosa-Cánovas, G., Swanson, B., and Pedrow, P. D., (1995), Pulsed electric field treatment chamber design for liquid food pasteurization using finite element method, Transactions of the ASAE, 38 (2): 557-565.

Quass, D. W. (1997). Pulsed electric field processing in the food industry. A status report on pulsed electric field. Palo Alto, CA. Electric Power Research Institute. CR- 109742. p.23-35.

Raso, J., Alvarez, I., Condon, S., and Sala, E J., (2000), Predicting inactivation of Salmonella senftenbergby pulsed electric fields, Innov. FoodSci. Emerg. Technol. 1: 21-30.

Raso,J., Calderon,M.L., Gongora,M., Barbosa-Cánovas,G. v., andSwanson,B.G., (1998), *InactivationofZygosaccharomyces bailii* in fruit juices by heat, high hydrostatic pressure and pulsed electric fields, *J. Food Sci.* 63(6): 1042-1044.

Rastogi, N., Eshtiaghi, M., and Knorr, D., (1999), Accelerated mass transfer during osmotic dehydration of high intensity electric field pulse pre-treated carrots, 1. *Food Sci.* 64: 1020-1023.

Resurreccion, A.V.A., Galvez, F.C.F., Fletcher, S.M., Misra, S.K., (1995). Consumer attitudes toward irradiated food: results of a new study. Journal of Food Protection 58, 193–196.

Ross, A.I.V., Griffiths, M.W., Mittal, G.S., and Deeth, H.C.(2003). Combining nonthermal technologies to control food borne microorganisms. International Journal of Food Microbiology 89:125–138.

Sale, A. J. H., and Hamilton, W. A., (1967), Effect of high electric fields on microorganisms. I. Killing of bacteria and yeast, Biochim. Biophys.Acta 148: 781-788.

Sampedro F., A. Rivas, D. Rodrigo, A. Martınez., and Rodrigo M., (2007). Pulsed electric fields inactivation of Lactobacillus plantarum in an orange juice–milk based beverage: Effect of process parameters Journal of Food Engineering 80 : 931–938.

Sepulveda, D. R. & Barbosa-Cánovas, G. V. (2005). Present status and the future of PEF technology. In "Novel food process technologies". G. V. Barbosa-Cánovas, M. S. Tapia & M. P. Cano edited. CRC Press, Boca Raton, FL. p. 1-45.

Taiwo,K. A., Angersbach,A., and Knorr, D., (2002), Influenceof high intensity electric field pulses and osmotic dehydration on the rehydrationcharacteristicsof apple slices at different temperatures.J. Food Eng. 52: 185-192.

Tiwari, B.K., O'Donnell, C.P., and Cullen, P.J. (2009). Effect of nonthermal processing technologies on the anthocyanin content of fruit juices. Trends in Food Science and Technology 20:137–145.

Vega-Mercado H, Martín-Belloso O, Qin BL, Chang FJ, Góngora-Nieto MM, Barbosa-Cánovas GV, Swanson BG (1997) Non-thermal food preservation: pulsed electric fields. Trends Food Sci Technol 8(5):151–157

Vega-Mercado H., Pothakamury U.R., Chang F.J., Barbosa-Cánovas G.V. and Swanson B.G. (1996). Inactivation of Escherichia coli by combining pH, ionic strength and pulsed electric fields hurdles. Food Research International 29:117–121.

Vorobiev E., N. Lebovka (2008). Electrotechnologies for Extraction from Food Plants and Biomaterials, DOI: 10.1007/978-0-387-79374-0 2, C-Springer Science+Business Media, LLC 2008 Pulsed-Electric-Fields-Induced Effects in Plant Tissues: Fundamental Aspects and Perspectives of Applications.

Vorobiev, E., Andre, A., Bouzrara, H., and Bazhal, M., (2000). Procede d'extraction de liquide d'un materiau cellulaire, et dispositifsde mise en ouvre du dit procede, Demande de brevet en France, N° 0002159 du 22.02.00.

Vorobiev, E., Jemai, A. B., Bouzrara, H., Lebovka, N. I., and Bazhal, M. I., (2004). Pulsed electric field assisted extraction of juice from food plants, In: Novel Food Processing Technologies (G. V. Barbosa-Cánovas and M. P. Cano, eds.), Marcel Dekker,NewYork, pp. 105-130.

Weise,Th. H. G. G., and Loeffler, M. J., (2001). Overviewon pulsed powerapplications, In: Proceedings of the International Conference on Pulsed Power Applications, Gelsenkirchen,Germany,pp. A.0I/I-A.01/8.

Wirjantoro, T. I., and Lewis,M . J., (1997). Effect of nisin and high temperature pasteurization of shelf life of whole milk, *J. Soc. Dairy Technol.* 49(4): 99-1 02.

Wouters, P. C, Dutreux, N., Smelt, J. P. P. M. & Lelieveld, H. L. M., (1999). Effects of pulsed electric fields on inactivation kinetics of *Listeria innocua.* Applied Environmental Microbiology. 65:5364-5371.

Wouters, P., Alvarez, I., and Raso, I., (2001). Critical factors determining inactivation kinetics by pulsed electric field food processing, Trends FoodSci. Technol. 12: 112-121.

Yeorn, H., Streaker, C., Zhang, Q., and Min, D., (2000). Effects of pulsed electric fields on the quality of orange juice and comparison with heat pasteurization, *J. Agric. Food Chem.* 48: 4597-4605.

Yeorn, H., Zhang, Q., and Dunne, C; (1999). Inactivation of papain by pulsed electric fields in a continuous system, *Food Chem.* 67: 53-59.

Zhang, Q. H., Barbosa-Cánovas, G. V., & Swanson, B. G. (1995). Engineering aspects of pulsed electric field pasteurization. Journal of Food Engineering, 25, 261–281.

Zhang, Q. H., Chang, E-J., and Barbosa-Cánovas, G. V., (1994), Inactivation of microorganisms in a semisolid model food using high voltage pulsed electric fields, *Lebensm Wiss. Technol*. 27(6): 538-543.

Zhang, Q. H., Qin, B. L., Barbosa-Cánovas, G. V., Swanson, B. G., and Pedrow, P. D., (1996), Batch mode for treatment using pulsed electric fields, US Patent, 5,549,041.

Zhang, Q. H., Qiu, x., and Sharma, S. K. (1997). Recent developments in pulsed electric processing. In "New Technologies Yearbook" (D. I. Chandrana, ed.), pp. 31-42. National Food Processors Association, Washington, D.C.

Zimmermann U, and Benz R, (1980), 'Dependence of the electrical breakdown voltage on the charging time in Valonia utricularis', J Membr Biol, 53, 33-43.

Zimmermann, U., 1986, Electric breakdown, electropermeab ilization and electrofusion, Rev. Phys. Biochem. Pharmacol.105: 196-256.,176-257.

Chapter 11

Understanding Color Image Processing by Machine Vision for Biological Materials

Ayman H. Amer Eissa and Ayman A. Abdel Khalik

Additional information is available at the end of the chapter

1. Introduction

Handling (Post harvest) process of fruits is completed in several steps: washing, sorting, grading, packing, transporting and storage. The fruits sorting and grading are considered the most important steps of handling. Product quality and quality evaluation are important aspects of fruit and vegetable production. Sorting and grading are major processing tasks associated with the production of fresh-market fruit types. Considerable effort and time have been invested in the area of automation.

Suitable handling (Post harvest) process of fruits and vegetables is considered the most important process that leads to conserve the fruits quality until reach to the consumers, improve the quality of industry food products and decrease the losses of fruits that estimated as 30% of crops in Egypt (Reyad, 1999).

Sorting is a separation based on a single measurable property of raw material units, while grading is "the assessment of the overall quality of a food using a number of attributes". Grading of fresh product may also be defined as 'sorting according to quality', as sorting usually upgrades the product (Brennan, 2006).

Sorting of agricultural products is accomplished based on appearance (color and absence defects), texture, shape and sizes. Manual sorting is based on traditional visual quality inspection performed by human operators, which is tedious, time-consuming, slow and non-consistent. It has become increasingly difficult to hire personnel who are adequately trained and willing to undertake the tedious task of inspection. A cost effective, consistent, superior speed and accurate sorting can be achieved with machine vision assisted sorting.

In recent ten years, operations in grading systems for fruits and vegetables became highly automated with mechatronics, and robotics technologies. Machine vision systems and near infrared inspection systems have been introduced to many grading facilities with mechanisms for inspecting all sides of fruits and vegetables (Kondo, 2009).

Machine vision and image processing techniques have been found increasingly useful in the fruit industry, especially for applications in quality inspection and defect sorting applications. Research in this area indicates the feasibility of using machine vision systems to improve product quality while freeing people from the traditional hand-sorting of agricultural materials.

The use of machine vision for the inspection of fruits and vegetables has increased during recent years. Nowadays, several manufacturers around the world produce sorting machines capable of pre-grading fruits by size, color and weight. Nevertheless, the market constantly requires higher quality products and consequently, additional features have been developed to enhance machine vision inspection systems (e.g. to locate stems, to determine the main and secondary color of the skin, to detect blemishes).

Automated sorting had undergone substantial growth in the food industries in the developed and developing nations because of availability of infrastructures. Computer application in agriculture and food industries have been applied in the areas of sorting, grading of fresh products, detection of defects such as cracks, dark spots and bruises on fresh fruits and seeds. The new technologies of image analysis and machine vision have not been fully explored in the development of automated machine in agricultural and food industries (Locht et al, 1997).

Rapid advances in artificial intelligent automated inspection of orange and tomato fruits by computer vision feasible. An intelligent vision system to evaluate fruit quality (size, color, shape, extent of blemishes, and maturity) and assign a grade would significantly improve the economic benefits to the orange and tomato fruits industries. It would potentially increase the consumer confidence in the quality of fruit.

Research efforts have concentrated on the implementation of machine vision to replace manual sorters.

The aim of this study is to develop machine vision techniques based on image processing techniques for estimation the quality of orange and tomato fruits and to evaluate the efficiency of these techniques regarding the following quality attributes: size, color, texture and detection of the external blemishes.

The specific objectives are to quantify the following attributes for inspection of orange and tomato fruits:

1. Color,
2. Texture (homogeneity or non-homogeneity),
3. Size (projected area),
4. External blemishes (detect defects).

5. Develop image processing techniques to sorting orange and tomato fruits into quality classes based on size, color and texture analysis,
6. Evaluate the performance of the system using some orange and tomato fruits, and
7. Evaluate the accuracy of the techniques by comparison with manual inspection.

2. Sorting and grading of fruits and vegetables

Handling (Post harvest) process of fruits is completed in several steps: washing, sorting, grading, packing, transporting and storage. The fruits sorting are considered one of the most important steps of handling.

Product quality and quality evaluation are important aspects of fruit and vegetable production. Sorting and grading are major processing tasks associated with the production of fresh-market fruit types. Considerable effort and time have been invested in the area of automation, but the complexity of fruit sorting and required sorting rates have forced the sorting of most fruit types to be performed manually. Although they currently achieve the best performance, human graders are inconsistent and represent large labor costs.

Machine vision is the study of the principles underlying human visual perception, and it attempts to provide the computer-camera system the visual capabilities easily accomplished by humans. In the human eye-brain system the human eye receives light from an object and then converts the light into electric signals. It does not interpret these signals nor make decision based upon the nature of the image. Image interpretation and decision-making are performed by the brain. Similarly, a machine vision system has an eye, which may be a camera or a sensor. Image interpretation and decision-making are done by appropriate software and hardware. Machine vision, often referred to as computer vision, can be defined as a process of producing description of an object from its image.

In manual inspection, a human inspector evaluates individual fruit in order to assign a grade. This process is tedious, labor intensive, and subjective. It has become increasingly difficult to hire personnel who are adequately trained and willing to undertake the tedious task of inspection (Morrow et al., 1990).

Mcrae (1985), mentioned that, the term "grading" can be applied to two distinct operations which are: (1) sizing, in which the grades are segregated according to their dimensions and (2) inspection, in which grades are based on the proportion of undesirable characteristics such as greening, cuts or other blemishes which are allowed to remain with the sound tubers and involves the elimination of unwanted material.

Leemans and Destain (2004) mentioned that, fresh market fruits like apples are graded into quality categories according to their size, color and shape and to the presence of defects. The two first quality criteria are actually automated on industrial graders, but fruits grading according to the presence of defects is not yet efficient and consequently remains a manual operation, repetitive, expensive and not reliable.

Brennan (2006) stated that, sorting and grading are terms which frequently used interchangeably in the food processing industry. Sorting is a separation based on a single measurable property of raw material units, while grading is "the assessment of the overall quality of a food using a number of attributes". Grading of fresh product may also be defined as 'sorting according to quality', as sorting usually upgrades the product.

Kondo (2009) reported that, in recent ten years, operations in grading systems for fruits and vegetables became highly automated with mechatronics, and robotics technologies. Machine vision systems and near infrared inspection systems have been introduced to many grading facilities with mechanisms for inspecting all sides of fruits and vegetables.

Sorting of agricultural products is accomplished based on appearance, texture, shape and sizes. Manual sorting is based on traditional visual quality inspection performed by human operators, which is tedious, time-consuming, slow and non-consistent. A cost effective, consistent, superior speed and accurate sorting can be achieved with machine vision assisted sorting.

Automated sorting had undergone substantial growth in the food industries in the developed and developing nations because of availability of infrastructures. Computer application in agriculture and food industries have been applied in the areas of sorting, grading of fresh products, detection of defects such as cracks, dark spots and bruises on fresh fruits and seeds. The new technologies of image analysis and machine vision have not been fully explored in the development of automated machine in agricultural and food industries. There is increasing evidence that machine vision is being adopted at commercial level but the slow pace of technological development in Egypt which are not available are among the factors that will limit the processes that requires computer vision and image analysis (Locht et al, 1997).

3. Manual inspection

The method used by the farmers and distributors to sort agricultural products is through traditional quality inspection and handpicking which is time-consuming, laborious and less efficient.

The maximum manual sorting rate is dependent on numerous factors, including the workers experience and training, the duration of tasks, and the work environment (temperature, humidity, noise levels, and ergonomics of the work station). More fundamentally, viewing conditions (illumination, defect contrast, and viewing distance) must be optimal to achieve maximum sorting rates.

Attempts to develop automatic produce sorters have been justified mostly by the inadequacies of manual sorters, but few authors provide results that demonstrate the degree of manual sorting inefficiencies. Flaws were more accurately identified when the inspector knew that only one type of flaw was present in the sample. The detectability of each flaw decreased when the sample contained more than one type of flaw. The authors indicated that different flaws must be mentally processed separately in a limited amount of time, and that these separate decisions may interfere with each other when more than one flaw is present in the sample. It was also proposed that a speed-accuracy relationship existed.

Geyer and Perry (1982) showed that samples with more than one flaw required a longer inspection time to achieve similar accuracy than a sample with only one flaw type. It was thought that inspector would have to search differently types of flaws, and this may have contributed to the longer inspection time. The increased inspection time improved correct rejection. The rejection of sound items was blamed on the increased false alarm rate due to more decision cycles.

More than the ability to discern a defect is required for optimal defect detection. Meyers et al. (1990) indicated that inspection tasks were complicated by the fact that acceptable defect limits periodically change. Also, individuals must apply absolute limits to continuous variables, such as color. In addition to the interpretation of the allowable limits, inspector must be able to see the defect if they are to reject the produce using a standard peach grading line with uniform spherical balls, theoretically only 88.7% of the surface area was presented to the inspector when standing at the side of the conveyor . Actual tests showed that only 82% of the defects on the balls were made visible to the inspector. The amount of surface area inspected is increased by placing multiple manual graders at both sides of a conveyor.

Many of the decision that are made during manual inspection are based on qualitative measurements, and Muir et al. (1989) illustrated individual "human sensors" are quite variable and difficult to calibrate. When qualified inspector were asked to quantify the amount of surface defect on a potato (in percentage of the total tuber surface), the values for a single sample ranged from 10 to 70%. The repeatability of individual inspectors was also very poor. Differences between two consecutive readings were as high as 40 percentage points in some cases. Appropriate imagines sensors are more accurate, with a maximum variation of 15 percentage points.

Rehkugelr and Throop (1976) indicate that a manual sorter was able to remove bruised apples from sound fruit with acceptable sorting efficiencies at a rate of approximately 1fruit/s. Similarly, Stephenson (1976) showed that rates for sorting tomatoes into immature and mature lots should not exceed 1fruit/s per inspector. A slightly faster rate, 1.2 fruit/s, was identified as the maximum rate for an inspector to reject 72% of serious defects in oranges. These results demonstrated the shortfalls of manual inspection and re-enforced the need for a more consistent grading system. Implementation of automated sorting machines may improve accuracy, decrease labor costs, and result in a final product free of defects.

The method used by the farmers and distributors to sort agricultural products is through traditional quality inspection and handpicking which is time-consuming, laborious and less efficient. Sun et al (2003) observed that the basis of quality assessment is often subjective with attributes such as appearance, smell, texture and flavour frequently examined by human inspectors. Francis (1980) found that human perception could easily be fooled. It is pertinent to explore the possibilities of adopting faster systems which will save time and more accurate in sorting of crops. One of such reliable method is the automated computer vision sorting system.

4. Machine vision applications

Machine vision technology uses a computer to analyze an image and to make decisions based on that analysis. There are two basic types of machine vision applications – inspection and control. In inspection applications, the machine vision optics and imaging system enable the processor to "see" objects precisely and thus make valid decisions about which parts pass and which parts must be scrapped. In control applications, sophisticated optics and software are used to direct the manufacturing process. Machine-vision guided assembly can eliminate any operator error that might result from doing difficult, tedious, or boring tasks; can allow process equipment to be utilized 24 hours a day; and can improve the overall level of quality.

The following process steps are common to all machine vision applications:

- **Image acquisition**

An optical system gathers an image, which is then converted to a digital format and placed into computer memory.

- **Image processing**

A computer processor uses various algorithms to enhance elements of the image that are of specific importance to the process.

- **Feature extraction**

The processor identifies and quantifies critical features in the image (e.g., the position of holes on a printed circuit board, the number of pins in a connector, the orientation of a component on a conveyor) and sends the data to a control program.

- **Decision and control**

The processor's control program makes decisions based upon the data. Are the holes within specification? Is a pin missing? How must a robot move to pick up the component? Machine vision technology is used extensively in the automotive, agricultural, consumer product, semiconductor, pharmaceutical, and packaging industries, to name but a few. Some of the hundreds of applications include vision-guided circuit-board assembly, and gauging of components, razor blades, bottles and cans, and pharmaceuticals.

4.1. Use of machine vision to classify agricultural products

Machine vision is the use of a computer to analyze a picture in order to extract meaningful information out of the picture. Using this powerful tool, accurate information, such as the images shape, size or appearance, can be obtained from an object that could not be easily obtained by human observation. To better classify the shape and appearance of agricultural products several studies have looked into using machine vision to classify various agricultural products. These include studies by Nielsen et al 1998, Paulus et al 1997, and Heinemann et al 1994.

5. Machine vision system

Grading and sorting machine vision system consist of feeding unite, a belt conveyor to convey the fruit, a color CCD camera located in an image acquisition chamber with lighting system for image capturing, control unite for open and close gates according to signals from computer unite and a computer with an image frame grabber to process the captured image.

The acquisition of an image that is both focused and illuminated is one of the most important parts of any machine vision system. **Figure (1)** shows the general steps required in obtaining results from an image of an object Sun et al (2003).

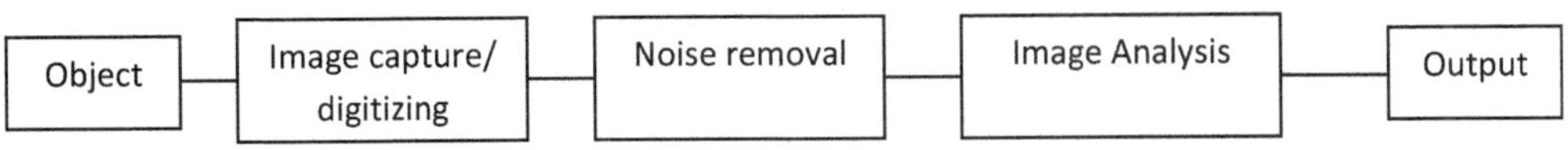

Figure 1. Imaging flowchart

Originally the image capture and digitizing of the image was accomplished by using a combination of a video camera and a frame grabber program. This method has almost been entirely replaced with CCD (charge-coupled device) and CMOS (complementary metal oxide semiconductor) chips. These chips use electrical circuits to directly convert light intensities into a digital image. They combine the video camera and frame grabber into one tool that can operate faster and with less distortion of the image. These chips also have the advantage that they can produce images at a much higher resolution then the frame grabber method (Mummert, 2004).

Noise is the incorrect representation of a pixel inside an image. It is best observed in variations in the color of a uniformly colored surface. Noise can be caused by numerous electrical sources and its removal is important, since the noise can cause the features of an image to appear distorted. This distortion can cause the features to be measured and classified incorrectly. While many algorithms have been proven useful to remove noise, the simplest method is to take multiple images of the same object and averaging the images together (Mummert, 2004). Since the noise is not the same in every image, when averaged the noise will blend into its surroundings, making the resulting image much clearer. Preprocessing of an image can include thresholding, cropping, gradient analysis, and many more algorithms. All of the processes permanently change the pixel values inside an image so that it can be analyzed by a computer. For example, in grey scale thresholding, a value for the intensity is selected and any pixel whose intensity value is less then the selected value intensity is set to 0 (black), if greater the value is set to 255 (white). After thresholding, the resulting grey scale image can easily have features classified and measured. The outputs from a machine vision system can be varied, in robotics the output might represent the location of an object to be moved, in inspections the output would be a pass or fall result, or in the case of this study the output is the sweet potatoes size and shape.

Machine vision and image processing techniques have been found increasingly useful in the fruit industry, especially for applications in quality inspection and defect sorting applications.

Research in this area indicates the feasibility of using machine vision systems to improve product quality while freeing people from the traditional hand-sorting of agricultural materials (Tao, 1996 a,b; Heinemann et al.,1995; Crowe and Delwiche, 1996; Throop et al., 1993; Yang, 1993; Upchurch et al., 1991). However, automating fruit defect sorting is still a challenging subject due to the complexity of the process. From fruit industry perspective, the fundamental requirements for an imaging-based fruit sorting system include: (1) 100% total inspection so that each piece of fruit is checked; (2) high-speed on-line and adaptation to existing packing lines; (3) sorting accuracy comparable to human sorters; and (4) the flexibility to adapt to fruits natural variations in shape, size, brightness, and various defect (Tao, 1998; Wen and Tao, 1997; Rigney et al., 1992). Machine-vision systems distinguish between good and defective fruit by contrasting the differences in light reflectance off the fruits surfaces (Miller, 1995; Thai et al., 1992; Guyer et al., 1994). Machine vision is increasingly used for automated inspection of agricultural commodities (Brosnan and Sun, 2004; Chen et al., 2002). Research results suggest that it is feasible to use machine vision systems to inspect fruit for quality related problems (Bennedsen and Peterson, 2005, and Brosnan and Sun, 2004). For fruit such as apples, commercial systems are available that allow sorting based on physical characteristics like weight, size, shape, and color. Automated fruit grading, standards assigned to fruit based on exterior quality, is also possible with machine vision (Leemans et al., 2002). Commercial sorters frequently use a conveyor system with either shallow cups (each cup holding one apple as it is moved) or bicone rollers that allow apples to rotate while moving along the conveyor (Figure 2). To be considered commercially applicable, automated systems must be able to handle fruit at rates of at least 6-10 fruit per second (Throop et al., 2001).

A camera or cameras above the conveyor are commonly used to capture images in these systems, sometimes in conjunction with mirrors below the fruit. The rotation of apples produced by bi-cone rollers allows for the imaging of multiple aspects of each apple's surface by using two or more cameras spaced apart along the conveyer. This approach has not been proven to be viable for defect detection for a number of reasons, including non-uniform rotation due to differences in apple sizes and frequent bouncing due to non-uniform shapes.

Figure 2. A Compac™ apple sorter. Courtesy of Compac, Inc., Visalia, CA.

Currently, there is no imaging process commercially used to detect defects or contamination due to lack of a method for imaging 100% of the entire surface of individual fruit. Thus, manual sorting remains the primary method for removal of apples with defects (Bennedsen & Peterson, 2005).

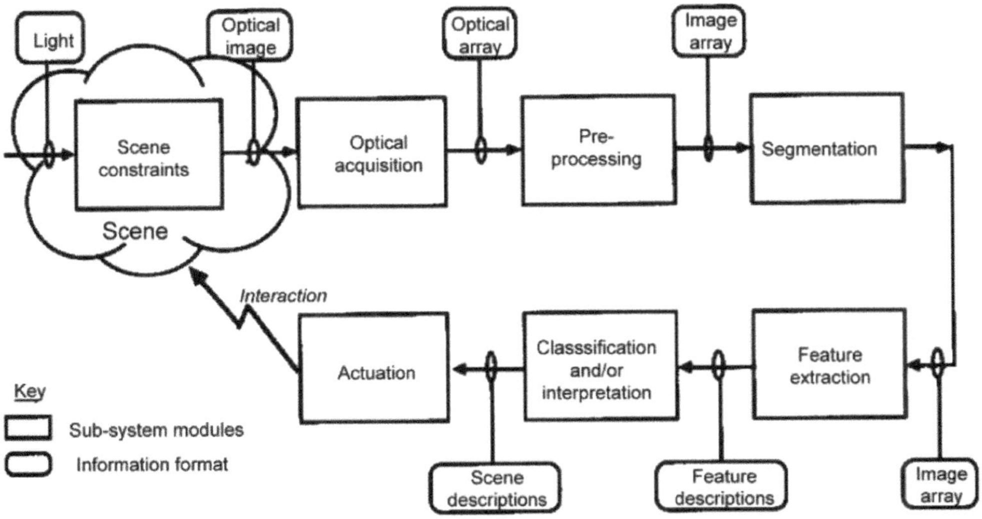

Figure 3. A simple block diagram for a typical vision system operation.

The main components of a typical vision system have been described in this study. Several tasks such as the image acquisition, processing, segmentation, and pattern recognition are conceivable. The role of image-acquisition sub-system in a vision system is to transform the optical image data into an array of numerical data, which may be manipulated by a computer. **Fig. 3** shows a simple block diagram for such a machine vision system. It includes systems and sub-systems for different processes. The big rectangles show the sub-systems while the parts for gathering information are presented as small rectangles in **Fig. 3**. As can be seen in **Fig. 3**, the light from a source illuminates the scene (it can be an industrial environment), and an optical image is generated by image sensors. Image arrays, digital camera, or other means are used to convert optical image into an electrical signal that can be converted to an ultimate digital image.

Typically, cameras incorporating either the line scan or area scan elements are used, which offer significant advantages. The camera system may use either charge coupled device (CCD) sensor or vidicon for the light detection. The preprocessing, segmentation, feature extraction and other tasks can be performed utilizing this digitized image. Classification and interpretation of image can be done at this stage and considering the scene description, the actuation operation can be performed in order to interact with the scene. The actuation sub-system therefore provides an interaction loop with the original scene in order to adjust or modify any given condition for a better image taking. (Golnabia and Asadpour, 2007).

The automated strawberry grading system (Liming and Yanchao, 2010) was developed based on three characteristics: shape, size and color. The automated strawberry grading system **(Fig. 4)** mainly consists of a mechanical part, an image processing part, a detection part and a control part. The mechanical part mainly consists of a conveyer belt, a platform, a leading screw, a gripper and two motors to implement the strawberry transport and gradation. The image processing part consists of camera (WV-CP470, Panasonic), image collecting card (DH-CG300, Daheng company), a closed image box and a computer (PCM9575) to implement image preprocessing, segmentation, extracting grading characteristic and to grade the strawberry by these characteristics.

The detection part consists of two photoelectrical sensors and two limit switches. The photoelectrical sensors are used to detect the strawberry position; the limit switch is used to protect the slider on the leading screw during the detection. The control part adopts the single-chip-microcomputer (SCM) to receive the signals from the photoelectrical sensor, the limit switch and the computer, finally to control the motors.

The results show that the strawberry classification algorithm is designed viable and accurately. Strawberry size error is less than 5%, the color grading accuracy rate is 88.8%,

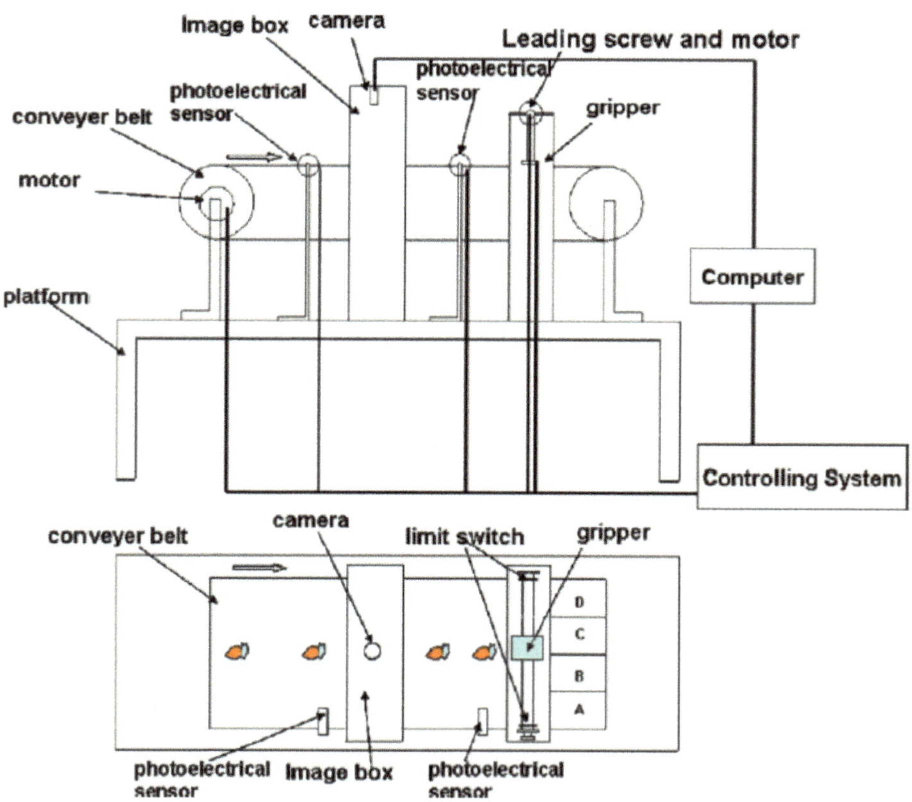

Figure 4. The structure of the strawberry automated grading system.

and the shape classification accuracy rate is over 90%. The average time to grade one strawberry is no more than 3 s.

(Blasco, et al. 2009) developed an engineering solution for the automatic sorting of pomegranate arils. The prototype (**Fig. 5**) basically consisted of three major elements that corresponded to the feeding, inspection and sorting units. These are described below. The prototype used two progressive scan cameras to acquire 512 _ 384 pixel RGB (Red, Green and Blue) images with a resolution of 0.70 mm/pixel. Both cameras were connected to a computer, the so-called "vision computer" (Pentium 4 at 3.0 GHz), by means of a single frame grabber that digitized the images and stored them in the computer's memory.

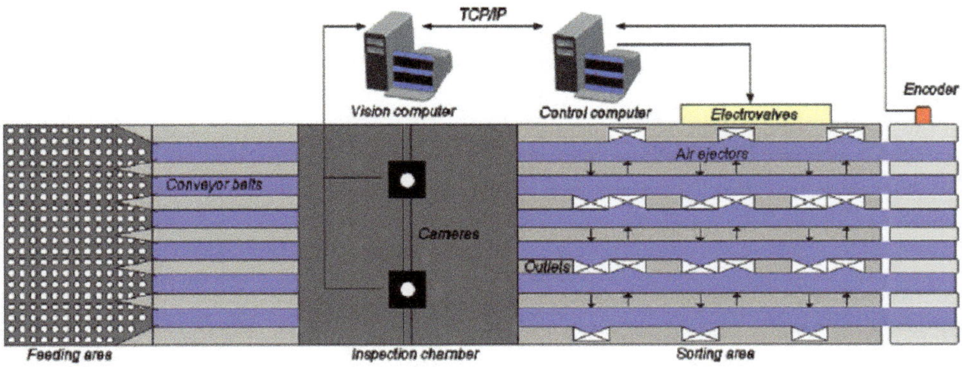

Figure 5. Scheme of the sorting machine.

The illumination system consisted of two 40 w daylight compact fluorescent tubes located on both sides of each conveyor belt. The scene captured by each camera had a length of approximately 360 mm along the direction of the movement of the objects and a width that allowed the system to inspect three conveyor belts at the same time. The entire system was housed in a stainless steel chamber.

The sorting area followed the inspection chamber. Three outlets were placed on one side of each of the conveyor belts. In front of each outlet air ejectors were suitably placed to expel the product. The separation of the arils was monitored by the control computer, in which a board with 32 digital outputs was mounted. This board was used to manage the air ejectors. The computer tracked the movement of the objects on the conveyor belts by reading the signals produced by the optical encoder attached to the shaft of the carrier roller.

Concluded that the prototype for inspecting and sorting the arils was developed and successfully commissioned, which could handle a maximum throughput of 75 kg/h. The inspection unit, which had two cameras connected to a computer vision system, had enough capacity to achieve real-time specifications and enough accuracy to fulfill the commercial requirements. The sorting unit was able to classify the product into four categories.

(Aleixos, et al. 2002). Developed a multispectral inspection of citrus in real-time using machine vision and digital signal processors. Describes a new machine vision system for citrus inspection, including a parallel hardware and software architecture, able to determine the external quality of the fruit in real-time at a speed of 10 fruits/s.

The vision system has been placed on a commercial fruit sorter having four independent inspection lines. As the first step, the sorter singles the fruit before they enter into the inspection site by means of bi-conic rollers. In principle, each individual fruit is located in a space between two rollers (what is called a cup), although sometimes, when there is an excessive loading, two or more fruits are located in the same cup or fruit are located between two filled cups. The inspection site **(Fig. 6)** provides adequate lighting to the scene by fluorescent tubes, incandescent lamps and polarised filters that remove reflections from the surface of the fruit. The scene is composed of three complete fruit, imaged with a multispectral camera that simultaneously captures four bands: the three conventional color bands (R, G and B) and another centred at 750 nm (near infrared, denoted I). The camera **(Fig. 7)** has two CCDs, one a color CCD that provides RGB information and the other monochromatic, to which has been coupled an infrared filter, centred on 750 nm (±10), that provides I information. The light coming from the scene reaches a semi-transparent mirror that refracts 50% of the light towards the infrared (A) CCD and reflects the other 50% to a second mirror (B), which reflects all the light towards the color CCD. The system guarantees at least three whole fruits on each image with a resolution of 0.7 mm/pixel.

The fruit rotates while passing below the camera due to a forced rotation of the rollers. To single the fruits and estimate their size and shape, the system uses only the I information, but for color estimation and defect detection it is necessary to work also with the color bands.

This fact has been used to set up a parallel strategy based on dividing the inspection tasks between two digital signal processors (DSP), so during on-line work, two image analysis procedures are performed by the two DSP running in parallel in a master/slave architecture. The master processor calculates the geometrical and morphological features of the fruit using only the I band, and the slave processor estimates the fruit color and detects the skin defects using the four RGBI bands. After the image processing, the master processor collects the information from the slave and sends the result to a control computer. The system was tested under laboratory conditions at two common sizer speeds: 300 and 600 fruits/min (5–10 fruits/s).

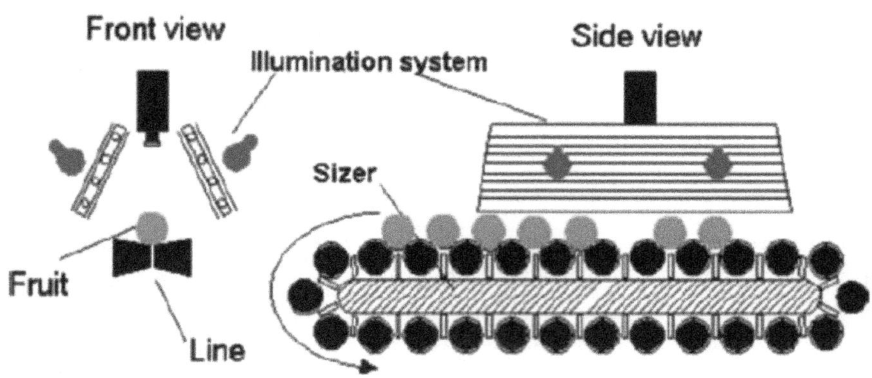

Figure 6. Scheme of the sorter and lighting system.

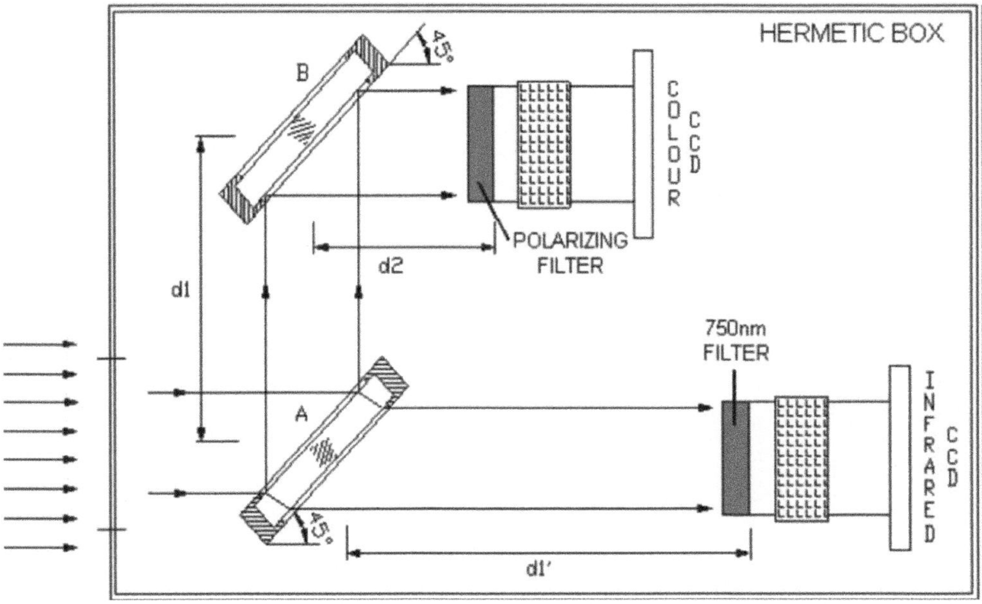

Figure 7. Scheme of the multispectral camera.

An image processing based technique was developed by **(Omid et al. 2010)** to measure volume and mass of citrus fruits such as lemons, limes, oranges, and tangerines. The technique uses two cameras to give perpendicular views of the fruit as shown in **(figure 8)**. An efficient algorithm was designed and implemented in Visual Basic (VB) language. The product volume was calculated by dividing the fruit image into a number of elementary elliptical frustums. The volume is calculated as the sum of the volumes of individual frustums using VB.

The volumes computed showed good agreement with the actual volumes determined by water displacement method. The coefficient of determination (R2) for lemon, lime, orange, and tangerine were 0.962, 0.970, 0.985, and 0.959, respectively. The Bland–Altman 95% limits of agreement for comparison of volumes with the two methods were (_1.62; 1.74), (_7.20; 7.57), (_6.54; 6.84), and (_4.83; 6.15), respectively. The results indicated citrus fruit's size has no effect on the accuracy of computed volume. The characterization results for various citrus fruits showed that the volume and mass are highly correlated. Hence, a simple procedure based on computed volume of assumed ellipsoidal shape was also proposed for estimating mass of citrus fruits. This information can be used to design and develop sizing systems.

Computer vision is the construction of explicit and meaningful descriptions of physical objects from images. States that it encloses the capturing, processing and analysis of two-dimensional images, with others noting that it aims to duplicate the effect of human vision by electronically perceiving and understanding an image. The basic principle of computer vision is described in **Fig. 9**. Image processing and image analysis are the core of computer vision with numerous algorithms and methods available to achieve the required classification and measurements.

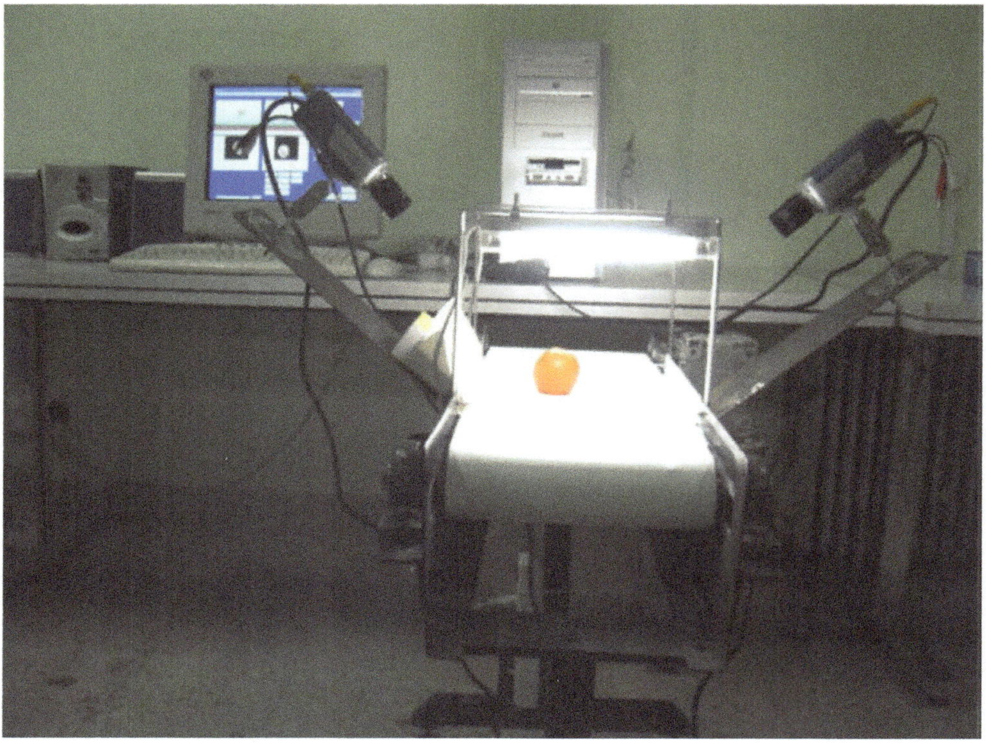

Figure 8. The developed machine vision system.

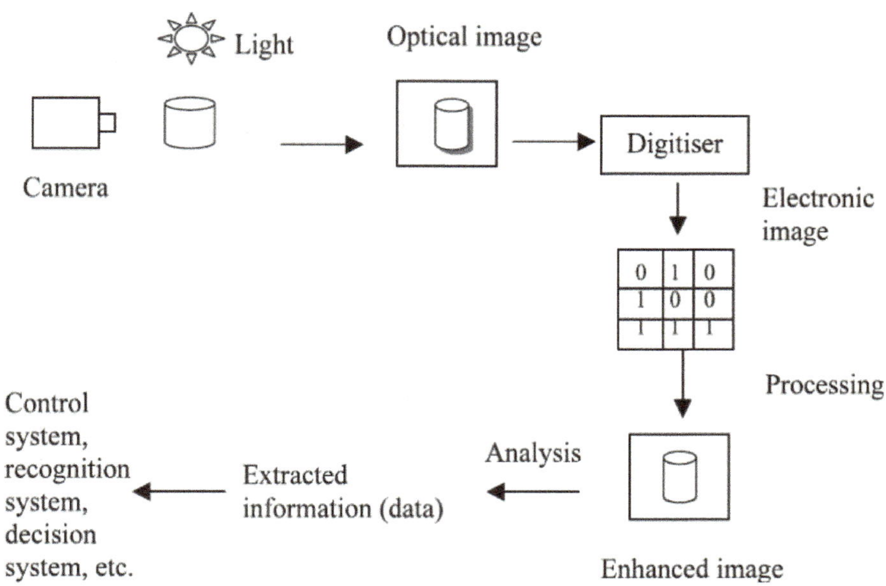

Figure 9. Principle of computer vision system.

Computer vision systems have been used increasingly in the food and agricultural industry for inspection and evaluation purposes as they provide suitably rapid, economic, consistent and objective assessment. They have proved to be successful for the objective measurement and assessment of several agricultural products. Over the past decade advances in hardware and software for digital image processing have motivated several studies on the development of these systems to evaluate the quality of diverse and processed foods. Computer vision has long been recognized as a potential technique for the guidance or control of agricultural and food processes. Therefore, over the past 20 years, extensive studies have been carried out, thus generating many publications.

Computer vision is a rapid, economic, consistent and objective inspection technique, which has expanded into many diverse industries. Its speed and accuracy satisfy ever-increasing production and quality requirements, hence aiding in the development of totally automated processes. This non-destructive method of inspection has found applications in the agricultural and food industry, including the inspection and grading of fruit and vegetable. It has also been used successfully in the analysis of grain characteristics and in the evaluation of foods such as meats, cheese and pizza (Brosnan and Sun, 2002).

(Jarimopas and Jaisin, 2008) develop an efficient machine vision experimental sorting system for sweet tamarind pods based on image processing techniques. Relevant sorting parameters included shape (straight, slightly curved, and curved), size (small, medium, and large), and defects. The variables defining the shape and size of the sweet tamarind pods were shape index and pod length. A pod was said to have defects if it contained cracks.

The sorting system involved the use of a CCD camera which was adapted to work with a TV card, microcontrollers, sensors, and a microcomputer as shown in **figure 10**. Conveyor belt

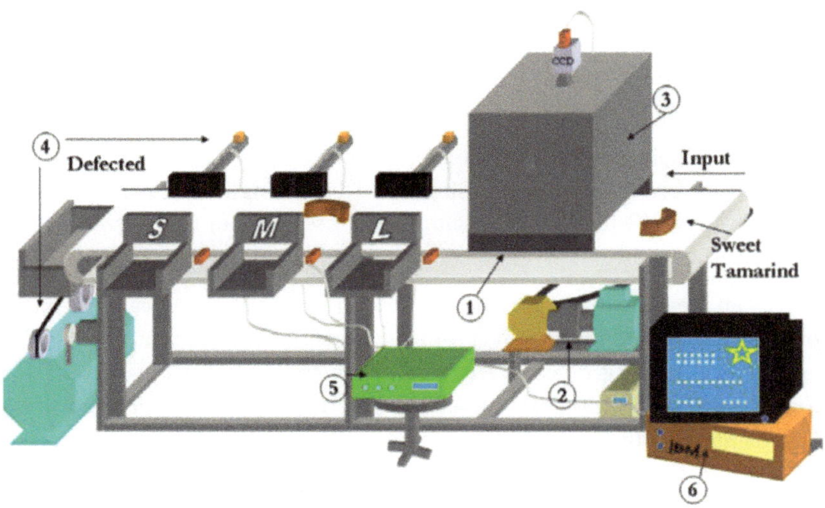

Figure 10. An experimental machine vision system for sorting sweet tamarind pods (1 is conveyor; 2 is power drive; 3 is light source and CCD camera; 4 is pneumatic segregator and compressed air tank; 5 is control unit; and 6 is microcomputer).

30 cm wide and 180 cm long with four receivers for the sorted sweet tamarind. On the right side of the belt was a box with a CCD camera mounted on the top and four 14-watt energy saving lamps at each corner of the box to give uniform light intensity with minimum shadows. The camera, which was mounted about 41 cm above the belt, had a focal length of 38–72 mm and provided a resolution of 520 vertical TV lines. A cylinder of compressed air was used to drive the three pneumatic segregators. The sorting system was so designed as to sort sweet tamarind into three sizes (large, medium, and small).

The defective pods were rejected at the left hand end of the conveyor. The control unit components were assembled in a box and placed under the sorting system.

The results showed that the three control factors did not significantly affect shape, size, and defects at a significance level of 5%. The averaged shape indexes of the straight, slightly curved, and curved pods were 51.1%, 61.6%, and 75.8%, respectively. Pod length was found to be influenced by size and cultivar, with Sitong and Srichompoo pods ranging from 10.0 to 14.0 cm and 8.5 to 12.4 cm, respectively. The vision sorting system could separate Sitong tamarind pods at an average sorting efficiency (E_W) of 89.8%, with a mean contamination ratio (C_R) of 10.2% at a capacity of 1517 pod/h.

Orange grading operations have been mechanized from a couple of decades. At the first stage of the mechanization, plates with holes of orange fruit sizes were used for sorting. Machine vision and near infrared (NIR) technologies have been utilized and improved with engineering design to convey fruits to detect fruit size, shape, color, sugar content and acidity since about ten years ago. The system inspects fruit with color CCD cameras installed at six different positions on a line to provide all side fruit images with lighting devices. The light devices are made by halogen lamps or LEDs fitted with PL (polarizing) filters to eliminated halation on glossy fruit surfaces. The near infrared inspection systems consist of halogen lamps and a spectrophotometer to analyze absorption bands of transmissive light from fruits. Furthermore, an X-ray imaging system is sometimes installed on each line to find internal defects such as rind-puffing.

Fig. 11 shows a whole inspection system on an orange grading line. After dumping containers filled by oranges, fruits are singulated by a singulating conveyor. Singulated fruits are sent to the NIR inspection system (transmissive type) to measure sugar content (brix equivalent) and acidity.

In addition, it can measure the granulation level of the fruit which indicates the inside water content of fruit. The second inspection is X-ray imaging for internal structural quality. Rind-puffing, a biological defect is detected from the image. In the external inspection stage, color images from six machine vision sets under random trigger mode, are copied to the image grabber boards fitted on the image processing computers whenever a trigger occurs.

The four cameras are set for acquiring side images, while the two cameras are from top. The final camera acquires a top image of each fruit after fruit turning over because both top and bottom sides are inspected. All the images are processed using specific algorithms for detecting image features of color, size, shape, and external defect. Output signals from

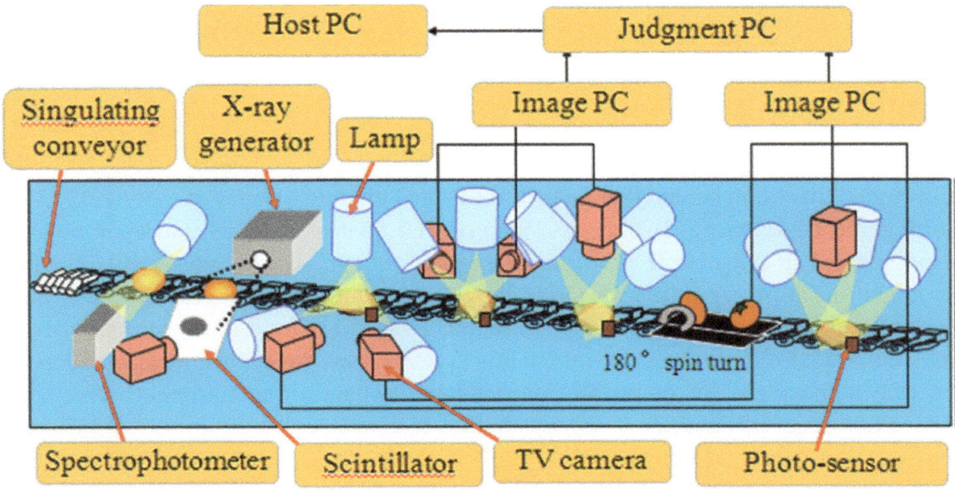

Figure 11. A whole orange fruit grading system on a line manufactured by SI Seiko Co., Ltd., Japan.

image processing are transmitted to the judgment computer where the final grading decision (usually into several grades and several sizes) is made based on fruit features and internal quality measurements.

Fig. 12 shows a fruit grading robot system installed at JA Shimoina, Japan. The robot system consists of two 3 DOF manipulators, in which one is a providing robot, while the other is a grading robot with 12 machine vision systems. After container comes under the providing robot (1), 12 fruits are sucked up by suction pads at a time (2) and are transported to intermediate stage making space toward vertical direction on this page between fruits (3).

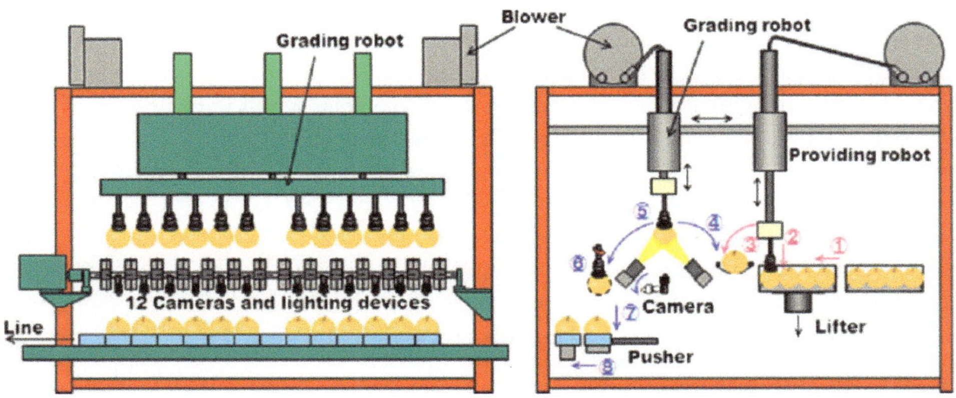

Figure 12. A fruit grading robot system manufactured by SI Seiko Co., Ltd., Japan (Left: front view, Right: side view).

The grading robot picks 12 fruits up again (4) and 12 bottom images of the fruits are acquired during the manipulator moving to trays on a conveyor line (5). Just before releasing the fruits to the trays (7), 4 side images of each fruit are acquired by rotating the suction pads for 270_ (6). The fruits are pushed out onto a line (8) and top images are acquired by another color camera stationed on each line. Software algorithms of machine vision are similar with that of the orange grading system. Fruit color, size, shape, and defect are measured.

Concluded that it can be said that roles of automated grading systems as follows: 1) Efficient sorting and labor saving, 2) Uniformization of fruit quality, 3) Enhancing market value of products, 4) Fair payment to producers based not only on quantity but on quality of each product, 5) Farming guidance from grading results and GIS (Geographical Information System), and 6) Contribution to the traceability system for food safety and security. The most important difference of the automation systems from the conventional machines is to be able to handle a lot of precise information. To handle the comprehensive data on agricultural products and foods, understanding of diversity and complexity of biomaterial properties is required and sensors to collect data should be often designed based on the properties. Through the traceability system in which all the data of producers, distributors, and consumers are linked and opened to them, it is expected that mutual information exchange among them makes more effective procedure at each stage and produces more safety and higher quality products (Kondo, 2010).

Identification of apple stem-ends and calyxes from defects on process grading lines is a challenging task due to the complexity of the process. An in-line detection of the apple defect is developed in this article. Firstly, a computer controlled system using three color cameras is placed on the line. In this system, the apples placed on rollers are rotating while moving, and each camera is capturing three images from each apple. In total nine images are obtained for each apple allowing the total surface to be scanned. Secondly, the apple image is segmented from the black background by multi-threshold methods. The defects, including the stem-ends and calyxes, called regions of interest (ROIs), are segmented and counted in each of the nine images. Thirdly, since a calyx and stem-end cannot appear at the same image, an apple is defective if any one of the nine images has two or more ROIs. There are no complex imaging processes or pattern recognition algorithms in this method, because it is only necessary to know how many ROIs are there in a given apple's image. Good separation between normal and defective apples was obtained. The classification error of unjustified acceptance of blemished apples reduced from 21.8% for a single camera to 4.2% for the three camera system, at the expense of rejecting a higher proportion of good apples. Averaged over false positive and false negative, the classification error reduced from 15 to 11%. The disadvantage of this method is that it could not distinguish different defect types. Defects such as bruising, scab, fungal growth, and disease, are treated as the same.

The lighting and image acquisition system were designed to be adapted on an existing single row grading machine (prototype from Jiangsu Univ., China). Six lighting tubes (18 W, type 33 from Philips, Netherlands) were placed at the inner side of a lighting box while three cameras (color 3CCD uc610 from Uniq, USA), two having their optical axis in a plane perpendicular to the fruit movement and inclined at 60° with respect to the vertical and one above observed the grading line in the box, as shown in **Figs. 13 and 14**. The lighting box is

1000mm in length and 1000mm in width. The distance between apple and camera is 580mm, thus there are three apples in the view field of each camera with a resolution of 0.4456mm per pixel. The images were captured using three Matrox/meteorII digitized frame-grabbers (Matrox, Canada) loaded in three separate computers. The standard image treatment functions were based on the Matrox libraries (Matrox, Canada) with remaining algorithms implemented in C++. A local network was built among the computers in order to communicate results data.

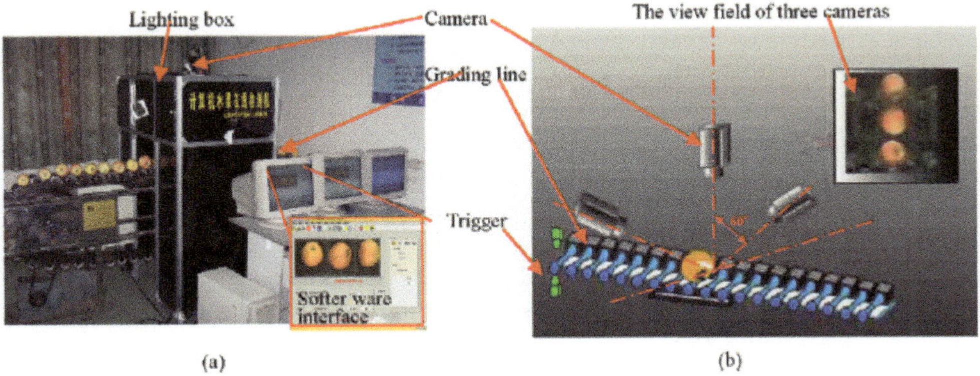

Figure 13. Hardware system of apple in-line detection.

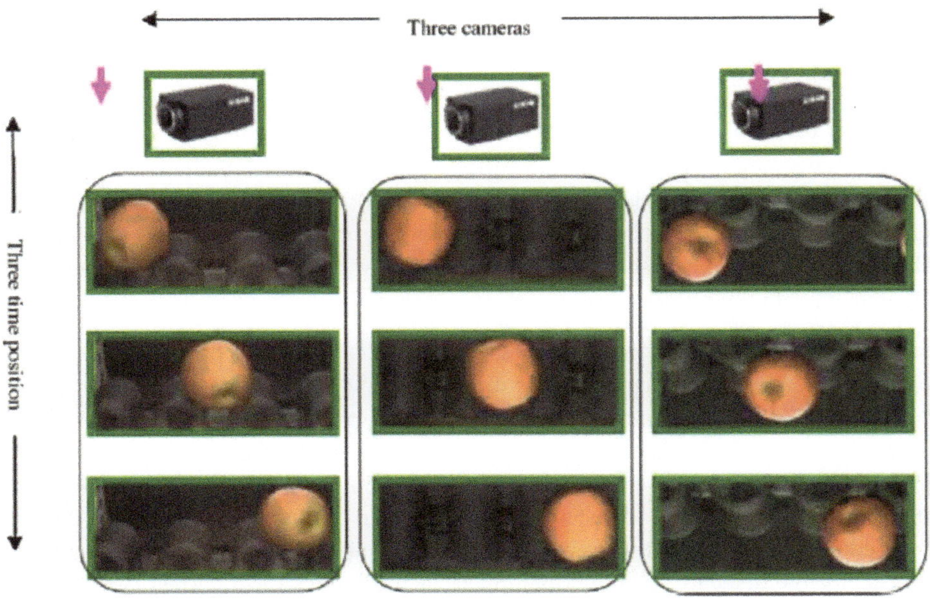

Figure 14. Trigger grab of nine images for an apple by three cameras at three positions.

The central processing unit of each computer was a Pentium 4 (Intel, USA) clocked at 3.0 GHz. The fruits placed on corn-shaped rollers are rotating while moving. The friction between rollers and the belt on the conveyor rack makes the corn-shaped roller rotate while moving through the field-of-view of the cameras. This was adjusted in such a way that a spherical object having a diameter of 80mm made a rotation in exactly three images when passed through the view field of camera. The moving speed in the range 0–15 apples per second could be adjusted by the stepping motor (Xiao-bo, et al, 2010).

One of the main problems in the post-harvest processing of citrus is the detection of visual defects in order to classify the fruit depending on their appearance. Species and cultivars of citrus present a high rate of unpredictability in texture and color that makes it difficult to develop a general, unsupervised method able of perform this task. In this paper we study the use of a general approach that was originally developed for the detection of defects in random color textures. It is based on a Multivariate Image Analysis strategy and uses Principal Component Analysis to extract a reference eigenspace from a matrix built by unfolding color and spatial data from samples of defect-free peel. Test images are also unfolded and projected onto the reference eigenspace and the result is a score matrix which is used to compute defective maps based on the T2 statistic. In addition, a multiresolution scheme is introduced in the original method to speed up the process. Unlike the techniques commonly used for the detection of defects in fruits, this is an unsupervised method that only needs a few samples to be trained. It is also a simple approach that is suitable for real-time compliance. Experimental work was performed on 120 samples of oranges and mandarins from four different cultivars: Clemenules, Marisol, Fortune, and Valencia. The success ratio for the detection of individual defects was 91.5%, while the classification ratio of damaged/sound samples was 94.2%. These results show that the studied method can be suitable for the task of citrus inspection.

The method performs novelty detection, and also is able to identify new unpredictable defects, by using a model of sound color textures and considering those locations that do not fit this model as being defective. It also needs only a few samples to carry out the unsupervised training. For this reason, it is suitable for citrus inspection as these systems need frequent tuning to adjust to the inspection of new cultivars and even the features of each batch of fruit within the same cultivar.

Experimental work was performed using 120 samples (images) of randomly selected oranges and mandarins belonging to four different cultivars: Marisol, Clemenules, Fortune and Valencia. First, a set of experiments were carried out to tune the parameters of the method for each cultivar. These included the number of principal eigenvectors used to define the reference eigenspace, the T2 threshold (percentile in the T2 cumulative histogram) used to determine if locations in test samples were sound or defective, and finally, the set of scales used in the multiresolution framework. Once the parameters were tuned, wecompiled the results for the detection of individual defects achieving 91.5% of correct detections and 3.5% of false detections. By using chromatic and textural features, the main contribution of this method is the capability of detecting external defects in different cultivars of citrus that present different textures carrying out only a single previous unsupervised training. The method achieved a performance rate of 94.2% successful classification of complete samples of fruit as either damaged or sound. These results show that the MIA approach studied here can be adequate for the task of citrus inspection (Fernando. et al, 2010).

Contemporary Vision and Pattern Recognition problems such as face recognition, fingerprinting identification, image categorization, and DNA sequencing often have an arbitrarily large number of classes and properties to consider. To deal with such complex problems using just one feature descriptor is a difficult task and feature fusion may become mandatory. Although normal feature fusion is quite effective for some problems, it can yield unexpected classification results when the different features are not properly normalized and preprocessed. Besides it has the drawback of increasing the dimensionality which might require more training data. To cope with these problems, this paper introduces a unified approach that can combine many features and classifiers that requires less training and is more adequate to some problems than a naïve method, where all features are simply concatenated and fed independently to each classification algorithm. Besides that, the presented technique is amenable to continuous learning, both when refining a learned model and also when adding new classes to be discriminated. The introduced fusion approach is validated using a multi-class fruit-and-vegetable categorization task in a semi-controlled environment, such as a distribution center or the supermarket cashier. The results show that the solution is able to reduce the classification error in up to 15 percentage points with respect to the baseline.

Oftentimes, when tackling complex classification problems, just one feature descriptor is not enough to capture the classes' separability. Therefore, efficient and effective feature fusion policies may become necessary. Although normal feature fusion is quite effective for some problems, it can yield unexpected classification results when not properly normalized and preprocessed. Additionally, it has the drawback of increasing the dimensionality which might require more training data.

This paper approaches the multi-class classification as a set of binary problems in such a way one can assemble together diverse features and classifier approaches custom-tailored to parts of the problem. It presents a unified solution (Section 4) that can combine many features and classifiers. Such technique requires less training and performs better if compared with a naïve method, where all features are simply concatenated and fed independently to each classification algorithm.

The results show that the introduced solution is able to reduce the classification error in up to 15 percentage points with respect to the baseline. A second contribution of this paper is the introduction to the community of a complete and well-documented fruit/vegetables image data set suitable for content-based image retrieval, object recognition, and image categorization tasks. We hope this data set will be used as a common comparison set for researchers workingin this space.

Although we have showed that feature and classifier fusion can be worthwhile, it seems not to be advisable to combine weak features with high classification errors and features with low classification errors. In this case, most likely the system will not take advantage of such combination.

The feature and classifier fusion based on binary base learners presented in this paper represents the basic framework for solving the more complex problem of determining not only the species of a produce but also its variety. Since it requires only partial training for the added features and classifiers, its extension is straightforward. Given that the introduced solution is general enough to be used in other problems, we hope it will endure beyond this paper.

Whether or not more complex approaches such as appearance based descriptors provides good results for the classification is still an open problem. It would be unfair to conclude they do not help (Anderson et al, 2010).

Color is an important quality attribute that dictates the quality and value of many fruit products. Accurately measuring and describing heterogenous fruit color changes during ripening is difficult with the instrumentation available (chromometer and colorimeter) due to the small viewing area of the equipment. Calibrated computer vision systems (CVS) provide another technique that allows capture and quantitative description of whole fruit color characteristics. Published research has demonstrated errors in CVS due to product curvature. In this work, it was confirmed that of the measured a* and b* color values on a curved surface, 55% and 69% of the values were within the range measured for the same flat surface. This deviation of measurement results in description of hue angle and chroma with an average error of 2◦ and 2.5, respectively. The system developed allows capture of hue angle data of whole fruit of heterogeneous colour. The usefulness of the device for capturing descriptive colour data during maturation of fruit is demonstrated with 'B74' mangoes (Kang et al, 2008).

Hyperspectral images of the apples (normal and injured) were acquired using a lab-scale hyperspectral imaging system (**Fig. 15**) that consisted of a charge-coupled device (CCD) camera (PCO-1600, PCO Imaging, Germany) connected to a spectrograph (ImSpector V10E, Optikon Co., Canada) coupled with a standard C-mount zoom lens. The optics of this imaging system allowed studying fruit properties associated with the spectral range of 400–1000 nm.

The camera faced downward at a distance of 400mm from the target. The sample was illuminated through a cubic tent made of white nylon fabric to provide uniform lighting conditions. The light source consisted of two 50Whalogen lamps mounted at a 45◦ angle from horizontal, fixed at 500mm above the sample and spaced 900mm apart on two opposite sides of the sample. The sample was put in a position that corresponded with the center of the field of view of the camera (300mm×300mm), with calyx–stem end perpendicular to the camera lens to avoid any discrepancy between the normal surface and stem or calyxes. The camera spectrograph assembly was provided with a stepper motor to move this unit through the camera's field of view to scan the apple line-by-line.

The spectral images were collected in a dark room where only the halogen light source was used. The exposure time was adjusted to 200ms throughout the test. Each collected spectral image was stored as a three-dimensional image (x, y,). The spatial components (x, y) included 400×400 pixels, and the spectral component () included 826 bands within

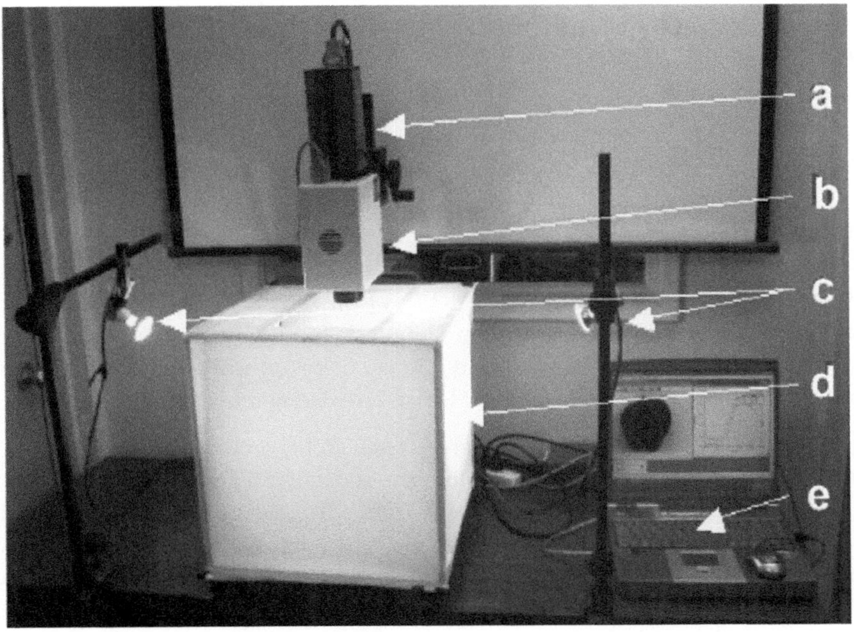

Figure 15. The hyperspectral imaging system: (a) a CCD camera; (b) a spectrograph with a standard C-mount zoom lens; (c) an illumination unit; (d) a light tent; and (e) a PC supported with the image acquisition software.

400–1000nm range. The hyperspectral imaging system was controlled by a laptop Pentium M computer (processor speed: 2.0 GHz; RAM: 2.0 GB) preloaded and configured with the Hypervisual Image Analyzer® software program (ProVision Technologies, Stennis Space Center, MO, USA). All spectral images acquired were processed and analyzed using the Environment for Visualizing Images software program (ENVI 4.2, Research Systems Inc., Boulder, CO, USA).

The hyperspectral images were calibrated with a white and a dark references. The dark reference was used to remove the dark current effect of the CCD detectors, which are thermally sensitive.

Hyperspectral imaging (400–1000 nm) and artificial neural network (ANN) techniques were investigated for the detection of chilling injury in Red Delicious apples. Ahyperspectral imaging system was established to acquire and pre-process apple images, as well as to extract apple spectral properties. Feed-forward back propagation ANN models were developed to select the optimal wavelength(s), classify the apples, and detect firmness changes due to chilling injury. The five optimal wavelengths selected by ANN were 717, 751, 875, 960 and 980 nm. The ANN models were trained, tested, and validated using different groups of fruit in order to evaluate the robustness of the models. With the spectral and spatial responses at the selected five optimal wave lengths, an average classification accuracy of 98.4%was achieved for distinguishing between normal and injured fruit. The correlation coefficients between measured and predicted firmness values were 0.93, 0.91 and 0.92 for the training, testing, and validation sets, respectively (Elmasry et al, 2009).

Naoshi et al (2008). Mentioned that, there are many types of citrus fruit grading machine with machine vision capability. While most of them sort fruit by size, shape, and color, detection of rotten fruit remains challenging because their appearances are similar to normal parts. Objectives of this research were to investigate if fluorescence would be a good indicator of the fruit rot, and to develop an economical solution to add the rot inspection capability to an existing machine vision fruit inspection station. A machine vision system consisting of a pair of white and ultra violet (UV) LED lighting devices and a color CCD camera was proposed for the citrus fruit grading task. Since the time lag between the color and fluorescence image captures was short (14ms), it was possible to inspect color, shape, size, and rot of a fruit on the move before it leaves an existing industrial inspection chamber.

Cheng et al (2003). Mentioned that, a near–infrared (NIR) and mid–infrared (MIR) dual–camera imaging approach for online apple stem–end/calyx detection is presented in this article. How to distinguish the stem–end/calyx from a true defect is a persistent problem in apple defect sorting systems. In a single–camera NIR approach, the stem–end/calyx of an apple is usually confused with true defects and is often mistakenly sorted. In order to solve this problem, a dual–camera NIR/MIR imaging method was developed. The MIR camera can identify only the stem–end/calyx parts of the fruit, while the NIR camera can identify both the stem–end/calyx portions and the true defects on the apple. A fast algorithm has been developed to process the NIR and MIR images. Online test results show that a 100% recognition rate for good apples and a 92% recognition rate for defective apples were achieved using this method. The dual–camera imaging system has great potential for reliable online sorting of apples for defects.

Sunil et al (2009). Identification of the insect damage is critical in the pecan processing. The insect damage is positively linked to the production of carcinogenic toxins in many food products. Previously, X-ray images were used for pecan defect identification, but the feature extraction was done manually. The objective of this article was to automate the feature extraction. Three energy levels (30 kV and 1 mA, 35 kV and 0.5 mA, and 40 kV and 0.75 mA) were used to acquire the images of the good pecans, pecans with insect exit holes, and nutmeat eaten pecans. After thresholding, three features were extracted. The features used were area ratio (ratio of area of the nutmeat and shell to the area of the total nut), mean local intensity variation, and average pixel intensity. The local adaptive methods performed well for the selected energy levels. The results indicate that it is feasible to distinguish between the good pecans and eaten nutmeat pecans. However, the selected features were not able to distinguish between the good pecans and the pecans with one or two insect exit holes.

Jun et al (2004). In this study, a mobile fruit grading robot for information-added product in precision agriculture was developed. The prototype robot, which consisted of a manipulator, an endeffector, a machine vision system, and a mobile mechanism, was made. The robot could acquire five fruit images from four sides and the top while its manipulator transported the fruit received from the operator. A preliminary experiment was conducted with 372 samples of sweet pepper in variety of "TosahikariD" in laboratory. A fruit mass prediction method was developed by use of the five images.

A high spatial resolution (0.5–1.0 mm) hyperspectral imaging system is presented as a tool for selecting better multispectral methods to detect defective and contaminated foods and agricultural products. Examples of direct linear or non-linear analysis of the spectral bands of hyperspectral images that resulted in more efficient multispectral imaging techniques are given. Various image analysis methods for the detection of defects and/or contaminations on the surfaces of Red Delicious, Golden Delicious, Gala, and Fuji apples are compared. Surface defects/contaminations studied include side rots, bruises, flyspecks, scabs and molds, fungal diseases (such as black pox), and soil contaminations. Differences in spectral responses within the 430–900 nm spectral range are analyzed using monochromatic images and second difference analysis methods for sorting wholesome and contaminated apples. An asymmetric second difference method using a chlorophyll absorption waveband at 685 nm and two bands in the near-infrared region is shown to provide excellent detection of the defective/contaminated portions of apples, independent of the apple color and cultivar. Simple and requiring less computation than other methods such as principal component analysis, the asymmetric second difference method can be easily implemented as a multispectral imaging technique.

Fig. 16 is a schematic diagram of the ISL hyperspectral imaging system. It consists of a charge coupled device (CCD) camera system SpectraVideoe Camera from PixelVision, Inc.

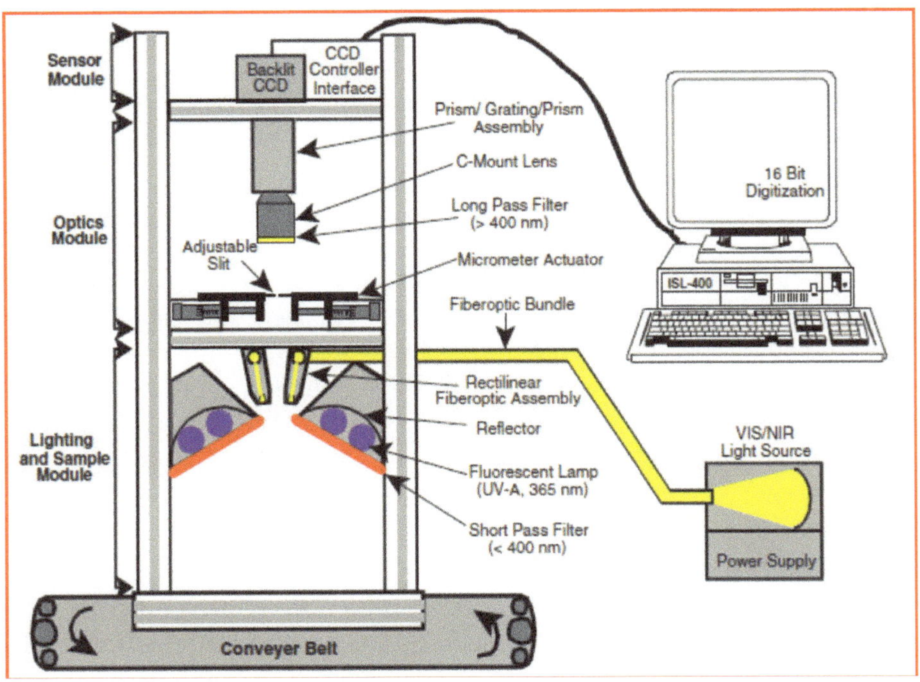

Figure 16. Schematic of the hyperspectral imaging system.

(Tigard, OR, USA) equipped with an imaging spectrograph SPECIM ImSpector version 1.7 from Spectral Imaging Ltd. (Oulu, Finland). The Im-Spector has a fixed-size internal slit to define the field of view for the spatial line and a prism/grating/prism system for the separation of the spectra along the spatial line. To improve the spatial resolution of the hyperspectral images, an external adjustable slit is placed between the sample and the camera optical set. This better defines the field of view and increases the spatial resolution. The image acquisition and recording is performed with a Pentium-based PC using a general purpose imaging software, PixelViewe 3.10 Beta 4.0 from Pixel-Vision, Inc. (Tigard, OR, USA.).

A C-mount set with a focus lens and an aperture diaphragm allows for focusing and aperture adjustments, for which the circular aperture is opened to its maximum and the external slit is adjusted with micrometer actuators to optimize light flow and resolution. The light source consists of two 21 V, 150 W halogen lamps powered with a regulated DC voltage power supply from Fiber-Lite A-240P from Dolan-Jenner Industries, Inc. (Lawrence, MA, USA). The light is transmitted through two optical fibers towards a line light reflector. The sample is placed on a conveyor belt with an adjustable speed AC motor control Speedmaster from Leeson Electric Motors (Denver, CO, USA). The sample is scanned line by line with an adjustable scanning rate, illuminated by the two line sources as it passes through the camera's field of view (Patrick, et al, 2004).

Naoshi et al (2008). Mentioned that, a complete fruit quality inspection system should be able to examine two opposite sides of each fruit. In automating such an inspection system, it is a well-known challenge when there is a need to mechanically manipulate fruits of irregular shapes and sizes. An innovatively designed rotary tray was developed for use in an eggplant fruit grading system. The rotary tray enables the presentation of two opposite sides of each fruit for inspection by machine vision systems. The rotary tray was designed for handling baby eggplants and mainly consisted of two cover plates and six side plates.

It is capable of performing five tasks on a fruit: receiving, presenting, holding, rotating, and releasing. The sequence of stages that a rotary tray goes through while moving along an inspection line are: 1. receiving a fruit, 2. presenting the fruit during the first image acquisition, 3. holding the fruit by closing one cover plate, 4. turn the fruit to its opposite side by rotating the entire tray, 5. opening the other cover plate , 5. presenting the opposite side of the fruit during the second image acquisition, 6. holding the fruit while the decision on its quality is being made by the machine vision algorithms, and 7. releasing the fruit to a particular location according to the inspection result.

The motions of a rotary tray are activated along a grading line by lifting guides, rotary pushers, clicks, and cams. The actions at stages 1 through 6 are performed by mechanical devices strategically placed along a motor driven grading conveyor. The releasing action is triggered by a rotary solenoid when the fruit arrives at a proper location. Six eggplant grading lines, each containing a series of the rotary trays, are being operated at an agricultural cooperative facility in Japan.

Jiangsheng and Yibin (2006). In this research, a novel approach for fruit shape detection was proposed, which based on multi scale level set framework. An image was first decomposed from coarse to fine by wavelet analysis method and a serial of images were formed. Then we use region homogeneity in a level set approach to extract fruit shape boundary at the coarse scale. At the finer scale, these coarse boundaries are used to initialize boundary detection and serve as a priori shape knowledge to guide contour evolution. This new algorithm doesn't need any noise removal preprocessing, and can find accurate shape boundary without any assumption in a noisy image. The proposed method has been applied to fruit shape detection with more promising result than traditional method.

Color is important in evaluating quality and maturity level of many agricultural products. Color grading is an essential step in the processing and inventory control of fruits and vegetables that directly affects profitability. Dates are harvested at different levels of maturity that require different processing before the dates can be packed. Maturity evaluation is crucial to processing control, but conventional methods are slow and labor-intensive. Because date maturity level correlates strongly with color, automated color grading could be used. A novel and robust color space conversion and color index distribution analysis technique for automated date maturity evaluation that is well suited for commercial production is presented in this paper. In contrast with more complex color grading techniques, the proposed method makes it easy for a human operator to specify and adjust color preference settings for different color groups representing distinct maturity levels. The performance of this robust color grading technique is demonstrated using date samples collected from field testing.

Concluded that A new color space conversion method and color index distribution analysis technique specifically for automated date maturity evaluation has been presented. The proposed approach uses a third-order polynomial to convert 3D RGB values into a simple 1D color space. Unlike other color grading techniques, this approach makes the selection and adjustment of color preferences easy and intuitive. Moreover, it allows a more complicated distribution analysis of fruit surface colors. The user can change color and consistency cutoff points in a manner consistent with human color perception, simply sliding a cutoff point to include fruit that is "slightly darker" or "lighter red". Moreover, changes in preferred color ranges can be completed without reference to precise color values. Furthermore, by converting 3D colors to a linear color space, color distribution analysis required for date maturity evaluation is much more straightforward. The implementation of this new color space conversion method and the results presented demonstrate the simplicity and accuracy of the proposed technique. To calibrate the system, an experienced grader specifies a set of colors of interest, each accompanied by a preferred index value on a linear scale. Provided that the selected color samples cover the complete range of expected colors, accurate color grading will result. This new technique can be applied to other color grading applications that require the setting and adjustment of color preferences.

6. Light

6.1. Electromagnetic spectrum

Radiation energy travels in space at the speed of light in the form of sinusoidal waves with known wavelengths. Arranged from shorter to longer wavelengths, the electromagnetic spectrum provides information on the frequency as well as the energy distribution of the electromagnetic radiation.

When electromagnetic radiation strikes an object, the resulting interaction is affected by the properties of an object such as color, physical damage, and presence of foreign material on the surface. Different types of electromagnetic radiation can be used for quality control of foods. For example, near-infrared radiation can be used for measuring moisture content, and internal defects can be detected by X-rays.

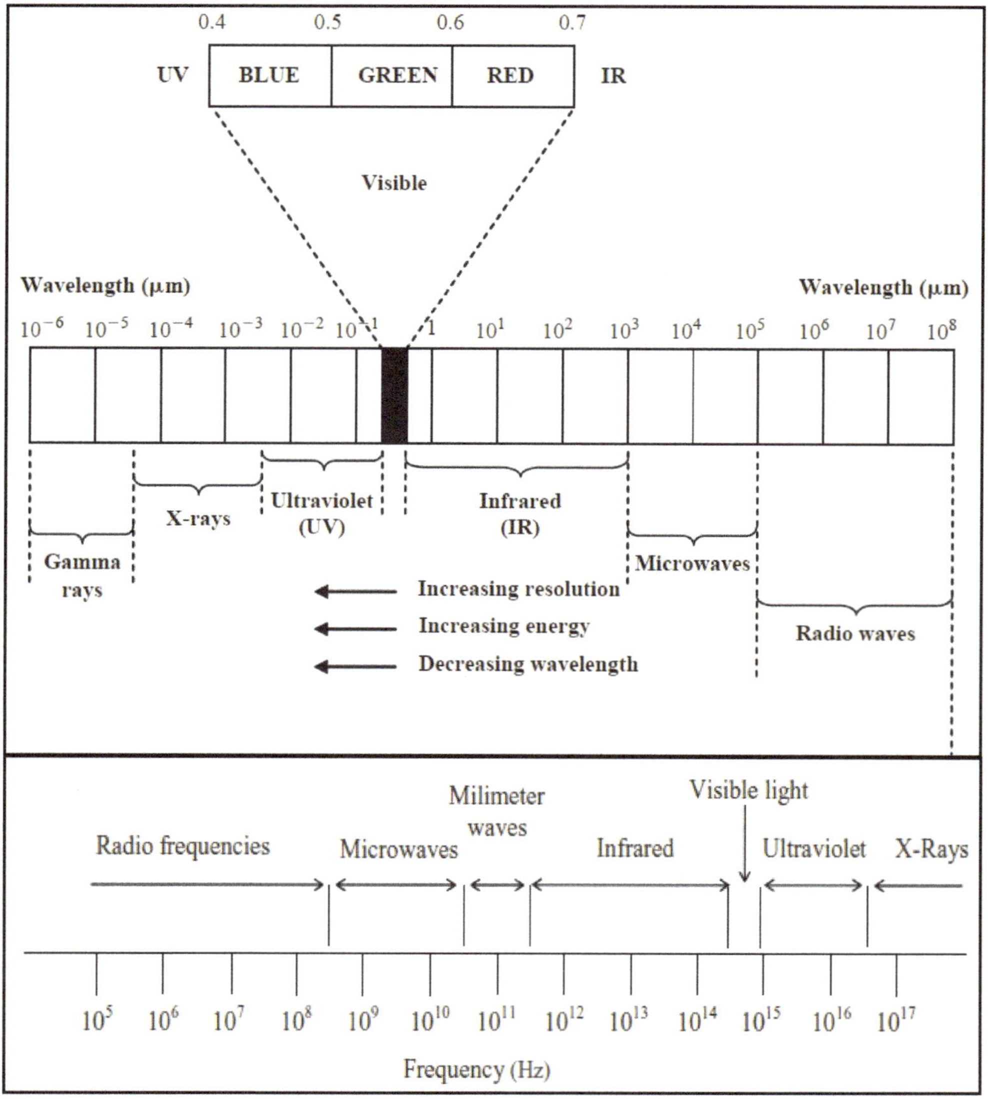

Figure 17. The electromagnetic spectrum comprises the visible and non-visible range.

Electromagnetic radiation is transmitted in the form of waves and it can be classified according to wavelength and frequency. The electromagnetic spectrum is shown in **Fig 17**.

Referring to Figure 17, the gamma rays with wavelengths of less than 0.1 nm constitute the shortest wavelengths of the electromagnetic spectrum. At the other end of the spectrum, the longest waves are radio waves, which have wavelengths of many kilometers. The well-known ground-probing radar (GPR) and other microwave-based imaging modalities operate in this frequency range.

Traditionally, gamma radiation is important for medical and astronomical imaging, leading to the development of various types of anatomical imaging modalities such as computed tomography (CT), magnetic resonance imaging (MRI), nuclear magnetic resonance (NMR), single photon emission computed tomography (SPECT) and positron emission tomography (PET) operate at shorter wavelengths ranging from 10^{-8} m to 10^{-13} m.

Located in the middle of the electromagnetic spectrum is the visible range, consisting of narrow portion of the spectrum with wavelengths ranging from 400 nm (blue) to 700 nm (red). The popular charge-coupled device or CCD camera operates in this range.

Infrared (IR) light lies between the visible and microwave portions of the electromagnetic band. As with visible light, infrared has wavelengths that range from near (shorter) infrared to far (longer) infrared.

Ultraviolet (UV) light is of shorter wavelength than visible light. Similar to IR, the UV part of the spectrum can be divided, this time into three regions: near ultraviolet (NUV) (300 nm) (NUV), far ultraviolet (FUV) (30 nm), and extreme ultraviolet (EUV) (3 nm). NUV is closest to the visible band, while EUV is closest to the X-ray region and therefore is the most energetic of the three types. FUV, meanwhile, lies between the near and extreme ultraviolet regions, and is the least explored of the three.

Electromagnetic waves travel at the speed of light and are characterized by their frequency (f) and wavelength (λ). In a medium, these two properties are related by:

$$c = \lambda f \tag{1}$$

where c is the speed of light in vacuum (3.0×10^8 m/s).

Radiation can exhibit properties of both waves and particles. Visible light acts as if it is carried in discrete units called photons. Each photon has an energy, E, that can be calculated by:

$$E = h\,f \tag{2}$$

where h is Planck's constant (6.626×10^{-34} J·s). (Sahin & Sumnu, 2005; Sun, 2008).

6.2. Illumination

The provision of correct and high-quality illumination, in many vision applications, is absolutely decisive. Engineers and machine vision practitioners have long recognized lighting as being an important piece of the machine vision system. However, choosing the right lighting strategy remains a difficult problem because there is no specific guideline for integrating lighting into machine vision applications.

Therefore, the illuminant is an important factor that must be taken into account when considering machine vision integration. Frequently, knowledgeable selection of an illuminant is necessary for specific vision applications.

For detection of differences in color under diffuse illumination, both natural daylight and artificial simulated daylight are commonly used. A window facing north that is free of direct sunshine is the natural illuminant normally employed for visual color examination. However, natural daylight varies greatly in spectral quality with direction of view, time of day and year, weather, and geographical location. Therefore, simulated daylight is commonly used in industrial testing. Artificial light sources can be standardized and remain stable in quality. The Commission Internationale de l'Eclairage (CIE) (The International Commission on Illumination) recommended three light sources reproducible in the laboratory in 1931. Illuminant A defines light typical of that from an incandescent lamp, illuminant B represents direct sunlight, and illuminant C represents average daylight from the total sky. Based on measurements of daylight, CIE recommended a series of illuminants D in 1966 to represent daylight. These illuminants represent daylight more completely and accurately than illuminants B and C do. In addition, they are defined for complete series of yellow to blue color temperatures. The D illuminants are usually identified by the first two digits of their color temperature Sahin & Sumnu, 2005; Sun, 2008).

Traditionally, the two most common illuminants are fluorescent and incandescent bulbs, even though other light sources (such as light-emitting diodes (LEDs) and electroluminescent sources) are also useful.

Computer Vision Systems are affected by the level and quality of illumination as with the human eye. The performance of the illumination system greatly influences the quality of image and plays an important role in the overall efficiency and accuracy of the system. Illumination systems are the light sources. The light focuses on the materials (especially when used). Lighting type, location and color quality play an important role in bringing out a clear image of the object. Lighting arrangements are grouped into front- or back-lighting. Front lighting serve as illumination focusing on the object for better detection of external surface features of the product while back-lighting is used for enhancing the background of the object. Light sources used include incandescent lamps, fluorescent lamps, lasers, X-ray tubes and infra-red lamps (Narendra and Hareesh, 2010).

7. Color

Color is one of the important quality attributes in foods. Although it does not necessarily reflect nutritional, flavor, or functional values, it determines the acceptability of a product by consumers. Sometimes, instead of chemical analysis, color measurement may be used if a correlation is present between the presence of the colored component and the chemical in the food since color measurement is simpler and quicker than chemical analysis.

It may be desirable to follow the changes in color of a product during storage, maturation, processing, and so forth. Color is often used to determine the ripeness of fruits. Color of potato chips is largely controlled by the reducing sugar content, storage conditions of the potatoes, and subsequent processing. Color of flour reflects the amount of bran. In addition, freshly milled flour is yellow because of the presence of xanthophylls.

Color is a perceptual phenomenon that depends on the observer and the conditions in which the color is observed. It is a characteristic of light, which is measurable in terms of intensity and wavelength. Color of a material becomes visible only when light from a luminous object or source illuminates or strikes the surface.

Light is defined as visually evaluated radiant energy having a frequency from about 3.9×10^{14} Hz to 7.9×10^{14} Hz in the electromagnetic spectrum. Light of different wavelengths is perceived as having different colors. Many light sources emit electromagnetic radiation that is relatively balanced in all of the wavelengths contained in the visible region. Therefore, light appears white to the human eye. However, when light interacts with matter, only certain wavelengths within the visible region may be transmitted or reflected. The resulting radiations at different wavelengths are perceived by the human eye as different colors, and some wavelengths are visibly more intense than others. That is, the color arises from the presence of light in greater intensities at some wavelengths than at the others.

The selective absorption of different amounts of the wavelengths within the visible region determines the color of the object. Wavelengths not absorbed but reflected by or transmitted through an object are visible to observers.

Physically, the color of an object is measured and represented by spectrophotometric curves, which are plots of fractions of incident light (reflected or transmitted) as a function of wavelength throughout the visible spectrum (**Figure 18**).

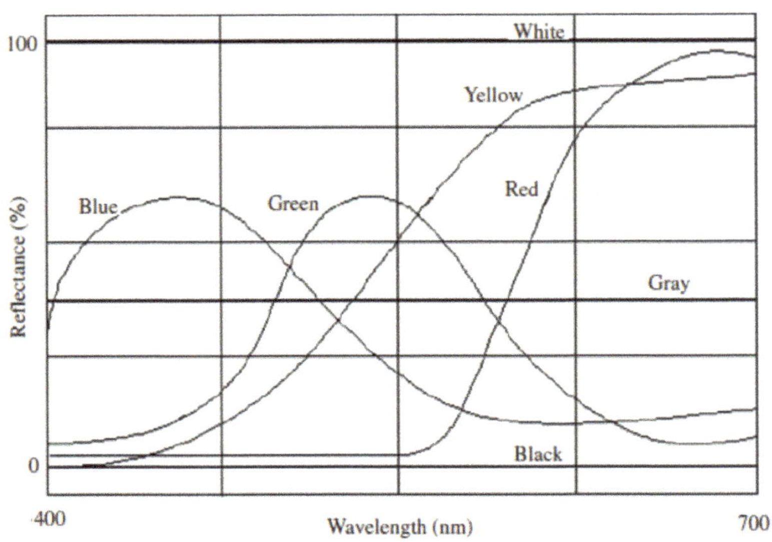

Figure 18. Shows spectrophotometric curves

7.1. Color fundamentals

The different colors we perceive are determined by two factors: the nature of the light reflected from the object and the source of the light. The reason tomatoes look red is that they absorb most of the violet, blue, green, and yellow components of the daylight, and reflect mainly the red components.

Leaves look green because they only reflect the green colors and absorb the red and blue colors. The source of light determines what colors can be reflected. Sunlight combines all lights of wavelengths, so objects appear colored in daylight. If the light source has a single wavelength, then objects just reflect this wavelength light and no other lights.

7.2. Trichromatic theory

The presence of three types of color receptors in the retinal layer confirmed the ideas that had been proposed in the trichromatic theory of human color vision.

This states that the magnitudes of three stimuli determine the perception of a color and not the detailed distribution of light energy across the visible spectrum. The concept is illustrated in **Figure 19.** If these stimuli are the same for two different light distributions, then the color appearance of the lights will be the same, irrespective of their spectrum. The trichromatic theory is important since it forms the basis of most methods of expressing color in terms of numbers and of the methods of reproduction of colored images.

The idea that three different types of photoreceptors participate in a population code for color is often referred to as the "trichromatic theory" of color vision.

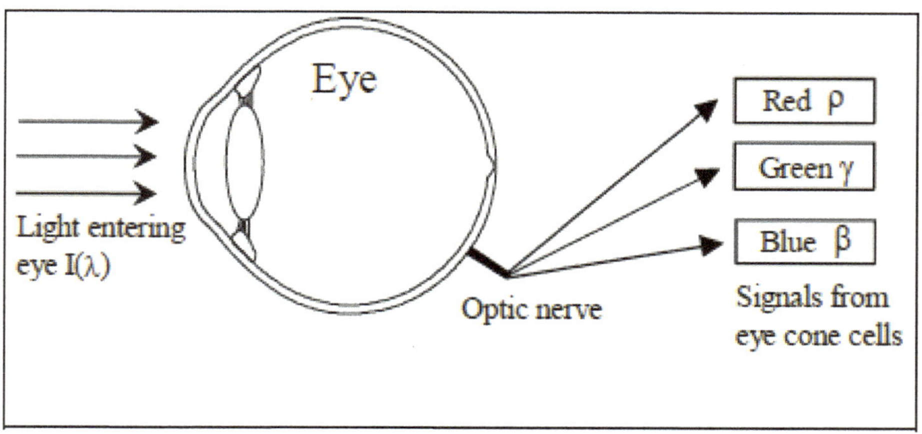

Figure 19. Show the signals from the eye cone cells

Therefore any light can be matched with a combination of any three others. Three receptors are types of cones:

- S (Short): most receptive at 419nm

- M (Medium): most receptive at 531nm
- L (Long): most receptive at 558nm as shown in **Figure 20**.

Red and green are not only unique hues but are also psychologically opponent color sensations. A color will never be described as having both the properties of redness and greenness at the same time; there is no such color as a reddish green. In the same way, yellow and blue are an opponent pair of color perceptions.

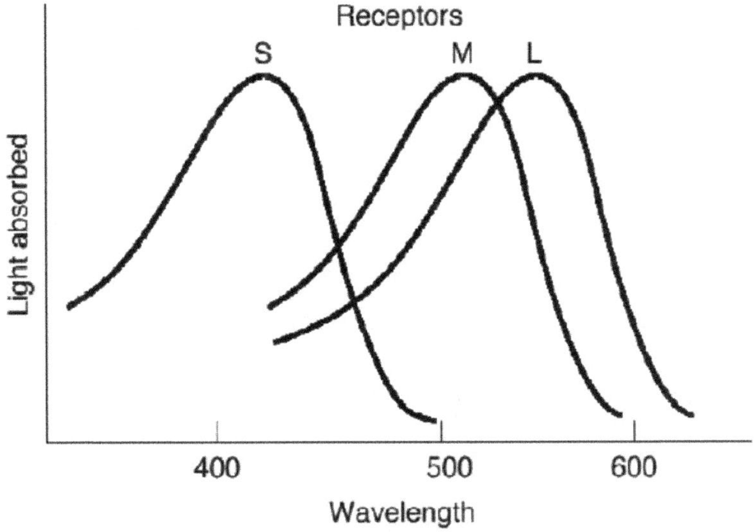

Figure 20. Show the cone absorption spectra

The six properties can be grouped into two opponent pairs, red/green and yellow/blue and the luminance property of white/black. The second stage of color vision is thought to arise from the action of neurons and in particular by inhibitory synapses. **Figure 21** illustrates the signal pathways and the processing required accounting for the properties described in the opponent theory. The human eye has receptors for short (S), middle (M), and long (L) wavelengths, also known as blue, green, and red receptors.

Three cone types are combined to form three opponent process channels:-

- S vs (M + L) = Blue/Yellow
- (L+S) vs M = Red/Green
- M + L = Black/White

In addition to the existence of the three different classes of cone photo pigments, considerable support for the dichromatic theory comes from observations of human color perception. For example, experiments in which subjects are shown different colors and asked to match them by mixing only three pure wavelengths of light in various proportions show that humans can, indeed, match any color using only three wavelengths of light - red, green and blue (Colour4Free, 2010).

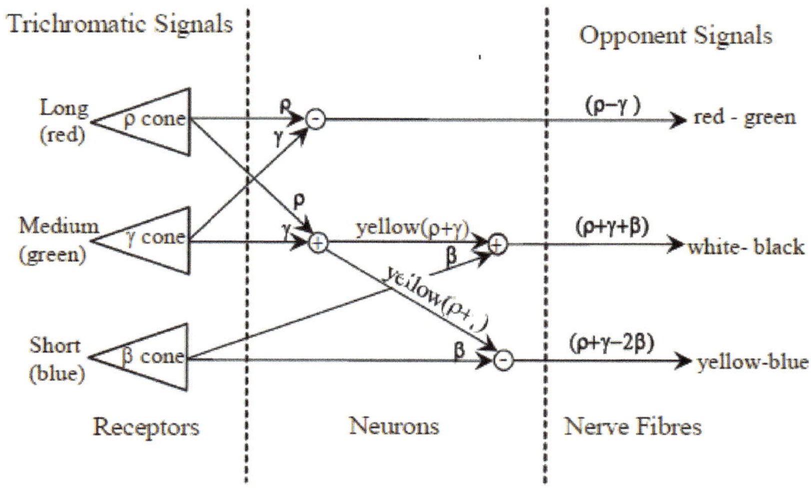

Figure 21. Show a set of signal paths consistent with the two stages of color vision.

7.3. The CIE chromaticity system

In 1931, the International Commission on Illumination, CIE (Commission Internationale del'Eclairage), defined three standard primary colors to be combined to produce all possible perceivable colors. The three standard primaries of the 1931 CIE, called X, Y, and Z, are imaginary colors.

The three dimensional color space CIE XYZ is the basis for all color management systems. This color space contains all perceivable colors - the human gamut. The two dimensional CIE chromaticity diagram xyY (below) shows a special projection of the three dimensional CIE color space XYZ. Some interpretations are possible in xyY, others require the three dimensional space XYZ or the related three dimensional space CIELab.

The new color-matching functions $x(\lambda)$, $y(\lambda)$,$z(\lambda)$ have non-negative values, as expected. The functions $x(\lambda)$, $y(\lambda)$,$z(\lambda)$ can be understood as weight factors. For a spectral pure color C with a fixed wavelength λ read in the diagram the three values as shown in **figure 23.** Then the color can be mixed by the three Standard Primaries:

$$C = x(\lambda)\,X + y(\lambda)\,Y + z(\lambda)\,Z \tag{3}$$

Generally we write

$$C = X\,X + Y\,Y + Z\,Z \tag{4}$$

and a given spectral color distribution $P(\lambda)$ delivers the three coordinates XYZ by these integrals in the range from 380nm to 700nm or 800nm:

$$X = k\int P(\lambda)\bar{x}(\lambda)d(\lambda) \tag{5}$$

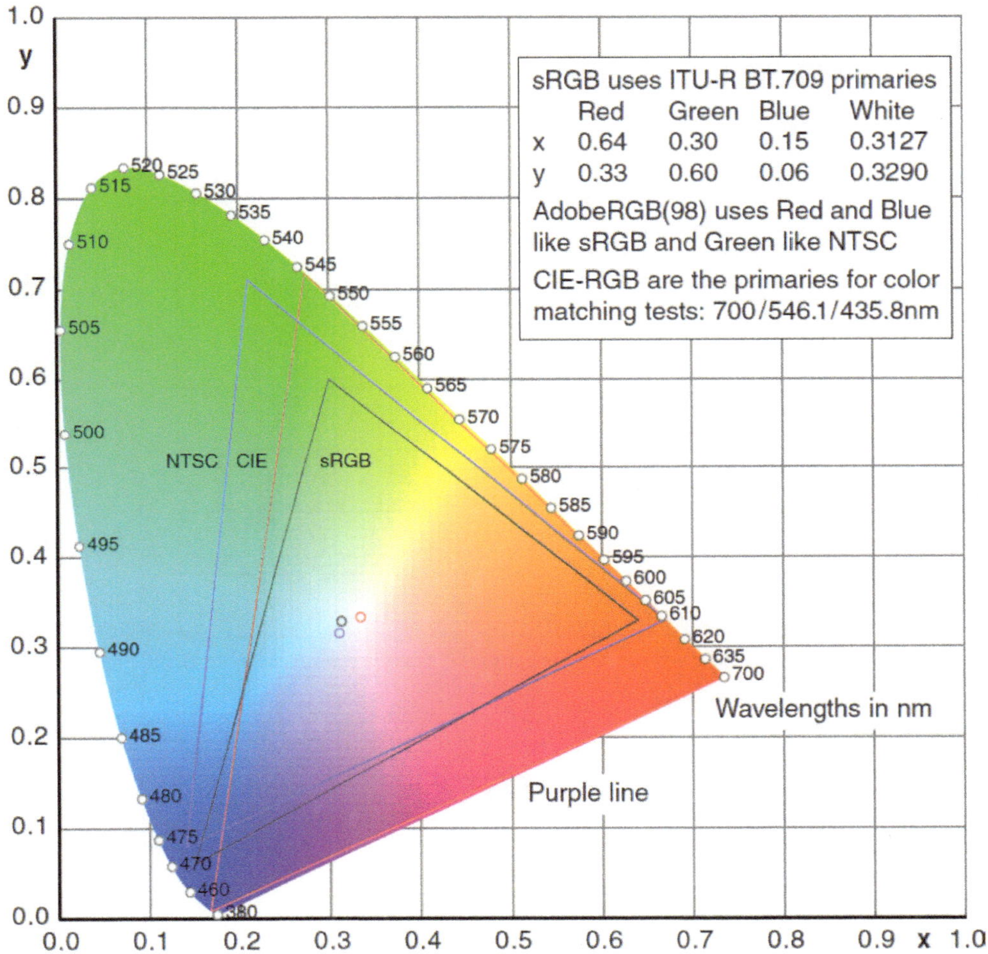

Figure 22. Show the CIE chromaticity diagram

$$Y = k\int P(\lambda)\bar{y}(\lambda)d(\lambda) \tag{6}$$

$$Z = k\int P(\lambda)\bar{z}(\lambda)d(\lambda) \tag{7}$$

where, k is a constant; it is 680 lumens/watt for a CRT; the λx, λy, and λz are color-matching functions.

The chromaticity values x,y,z depend only on the hue or dominant wavelength and the saturation. They are independent of the luminance:

$$x = \frac{X}{X+Y+Z} \tag{8}$$

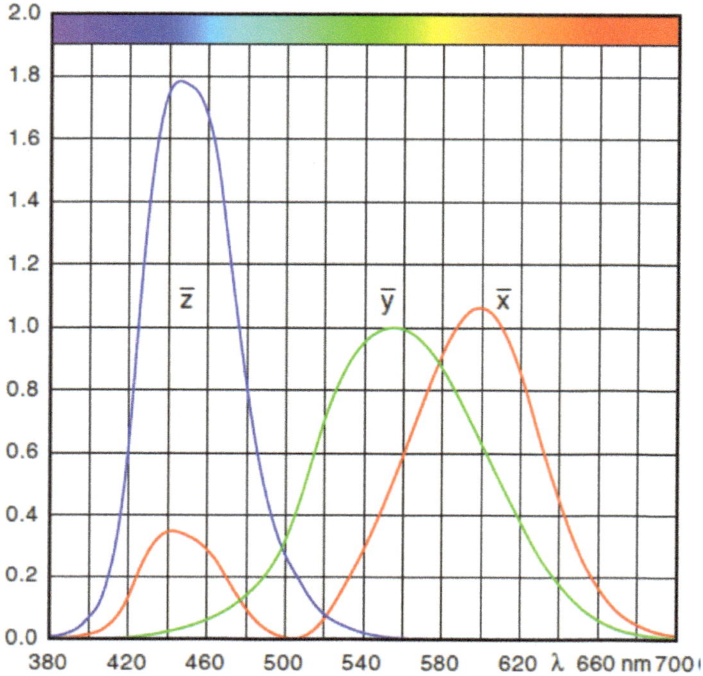

Figure 23. Show the XYZ Color-matching functions.

$$y = \frac{Y}{X+Y+Z} \tag{9}$$

$$z = \frac{Z}{X+Y+Z} \tag{10}$$

Obviously we have x + y + z = 1. All the values are on the triangle plane, projected by a line through the arbitrary color XYZ and the origin, if we draw XYZ and xyz in one diagram. This is a planar projection. The center of projection is in the origin as shown in **figure 24.**

The vertical projection onto the xy-plane is the chromaticity diagram xyY (view direction). To reconstruct a color triple XYZ from the chromaticity values xy we need an additional information, the luminance Y.

$$z = 1 - x - y \tag{11}$$

$$X = \frac{x}{y}Y \tag{12}$$

$$Z = \frac{z}{y}Y \tag{13}$$

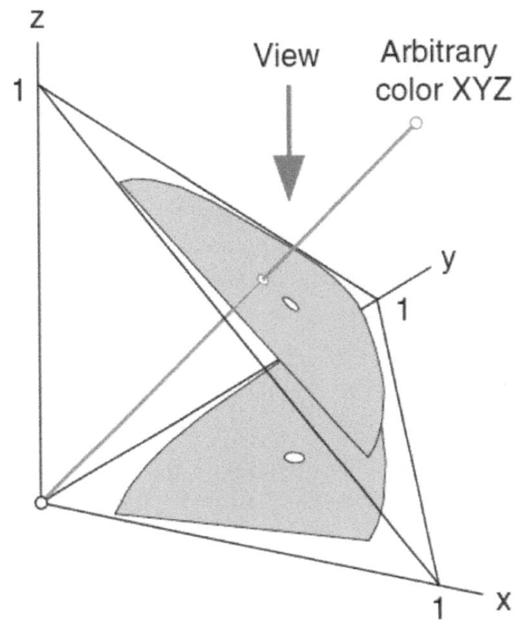

Figure 24. Show projection and chromaticity plane.

The interior and boundary of the diagram represent all visible chromaticity values. The boundary of the diagram represents the 100 percent pure colors of the spectrum. The line joining the red and violet spectral points, called the purple line, is not part of the spectrum. The center point E of the diagram represents a standard white light, which approximates sunlight. Luminance values are not available in the chromaticity diagram because of normalization. Colors with different luminance but the same chromaticity have the same point. The chromaticity diagram is useful for the following:-

- Comparing color gamut for different sets of primaries.
- Identifying complementary colors.
- Determining the dominant wavelength and purity of a given color. (Hoffmann, 2000).

7.4. Color gamut

Color gamuts are represented on the chromaticity diagram as straight-line segments or as polygons.

Each color model uses a different color representation. The term color gamut is used to denote the universe of colors that can be created or displayed by a given color system or technology. The colors that are perceivable by the human visual system fall within the boundaries of the horse-shoe shape derived from the CIE-XYZ color space diagram, while the RGB colors (that can be displayed on an RGB monitor) fall within the red triangle that connects the RGB primary dots.

It is obvious that, the full range of perceptible color by humans is not available by the RGB color model and the transformations from one space to another may create colors outside the color gamut.

7.5. Color models

A color model is a method by which humans can specify, create and visualize color. A color model is a specification of a 3D color coordinate system and a visible subset in the coordinate system within which all colors in a particular color gamut lie. For example, the RGB color model is the unit cube subset of the 3D Cartesian coordinate system. There is more than one color model. The purpose of a color model is to allow convenient specification of colors within some color gamut. However, no color model can be used to specify all visible colors.

The choice of a color model is based on the application. Some equipment has limiting factors that dictate the size and type of color model that can be used; for example, the RGB color model is used with color CRT monitors, the YIQ color model is used with the broadcast TV color system, and the CMY color model is used with some color-printing devices. Unfortunately, none of these models are particularly easy to use comparing with human perception. According to human intuitive color concepts, it is easy to describe the color in terms of shade, tint, and tone, or hue, saturation, and brightness. Color models which attempt to describe colors in this way include HSV, HLS, CIEL*a*b*, CIEL*C*H*, CIEL*u*v*. (Shen, 2003) (Fairchild, 1997) (Findling, 1996).

7.5.1. RGB color model

Based on the tri-stimulus theory of the vision of human eyes, the RGB (short for "red, green, and blue") color model describes colors as positive combinations of three appropriately defined red, green, and blue primaries in a Cartesian coordinate system; this is an example of an additive color model.

The RGB color space can be defined by mapping the red, green, and blue intensity components into the Cartesian coordinate system. The dynamic range of the intensity values is scaled from 0 to 255 counts, and each primary color is represented by eight bits. The RGB color space shown in **(Figure 25)** displays 16.77 million discrete colors. The red, green, and blue corners of the cube indicate 100 percent color saturation.

An imaginary line can be drawn from the origin of the cube to the furthest opposite corner. Along this line are 256 achromatic colors representing possible shades of gray. Black resides at the origin of the color cube, and white is at the opposite corner. The RGB system enables the reproduction of any color within the color space by using an additive mixture of the primary colors. For an example, White is the sum of 255 counts of red, green, and blue, and the function is usually expressed by RGB (255, 255, 255).

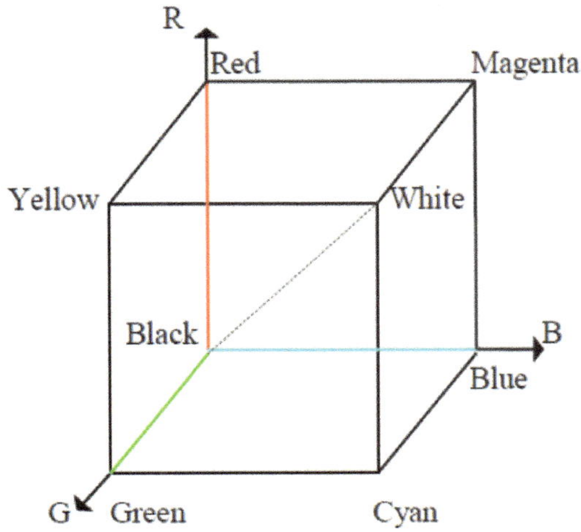

Figure 25. Show the RGB color model

7.5.2. The CMY & CMYK color models

Like the RGB color model, CMY color space is a subspace of standard three-dimensional Cartesian space, taking the shape of a unit cube. Each axis represents the basic secondary colors cyan, magenta, and yellow. Unlike RGB, however, CMY is a subtractive color model, meaning that where in RGB the origin represents pure black, the origin in CMY represents pure white. In other words, increasing values of the CMY coordinates move towards darker colors where increasing values of the RGB coordinates move towards lighter colors see **Figure 26**. Conversion from RGB to CMY can be done using the simple formula where it has been assumed that all color where it has been assumed that all color values have been normalized

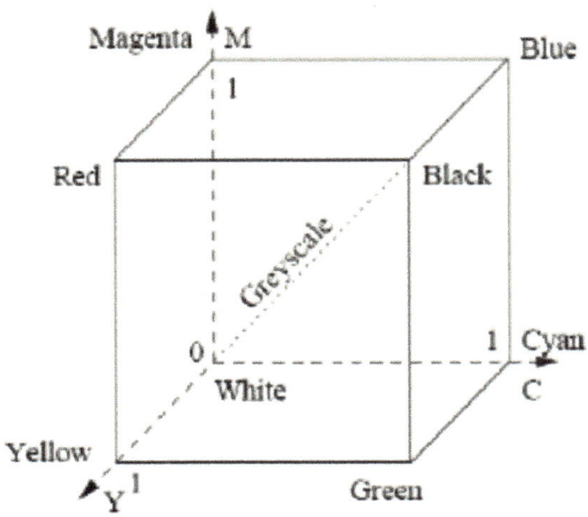

Figure 26. Show the CMY color model

$$\begin{bmatrix} C \\ M \\ Y \end{bmatrix} = \begin{bmatrix} 1 \\ 1 \\ 1 \end{bmatrix} - \begin{bmatrix} R \\ G \\ B \end{bmatrix} \quad (14)$$

to the range [0, 1]. This equation reiterates the subtractive nature of the CMY model. Although equal parts of cyan, magenta, and yellow should produce black, it has been found that in printing applications this leads to muddy results.

Thus in printing applications a fourth component of true black is added to create the CMYK color model. Four-color printing refers to using this CMYK model. As with the RGB model, point distances in the CMY space do not truly correspond to perceptual color differences.

7.5.3. YIQ color model

Developed by and for the television industry, the YIQ color system arose from a need to compress broadcasted digital imagery with as little visual degradation as possible.

The YIQ model is used in U.S.A. commercial color television broadcasting and is closely related to color raster graphics, which is suited to monochrome as well as color CRT display historically. The parameter Y is luminance, which is the same as in the XYZ model. Parameters I and Q are chromaticity, with I containing orange-cyan hue information, and Q containing green-magenta hue information. There are two peculiarities with the YIQ color model. The first is that this system is more sensitive to changes in luminance than to changes in chromaticity; the second is that color gamut is quite small, it can be specified adequately with one rather than two color dimensions. These properties are very convenient for the transfer of TV signals.An approximate linear transformation from a given set of RGB coordinates to the YIQ space is given by the following formula:

$$\begin{bmatrix} Y \\ I \\ Q \end{bmatrix} = \begin{bmatrix} 0.30 & 0.59 & 0.11 \\ 0.60 & -0.28 & -0.32 \\ 0.21 & -0.52 & 0.31 \end{bmatrix} \begin{bmatrix} R \\ G \\ B \end{bmatrix} \quad (15)$$

7.5.4. HSV & HSL color models

The RGB, CMY, and YIQ color models are hardware-oriented. These do not provide an intuitive method to reproduce the colors according to human vision. For a specified color, people prefer to use tint, shade, and tone to describe a color. Thehsv (hue, saturation, and value) and HSL (hue, saturation, lightness) color models are very different from the previously explored RGB and CMY/K and YIQ color models in that both systems separate the overall intensity value of a point from its chromaticity. The HSV and HSL models can be visualized in three dimensions as a downward pointing hexcone. The HSV color model is a color model defined to describe the colors similarly to human vision. The HSV color model can be derived from the RGB cube. By looking along the diagonal of the RGB cube, which is from origin to (1,1,1), a hexagonal cone is seen from the outline of the cube as shown in **(Figure 27).**

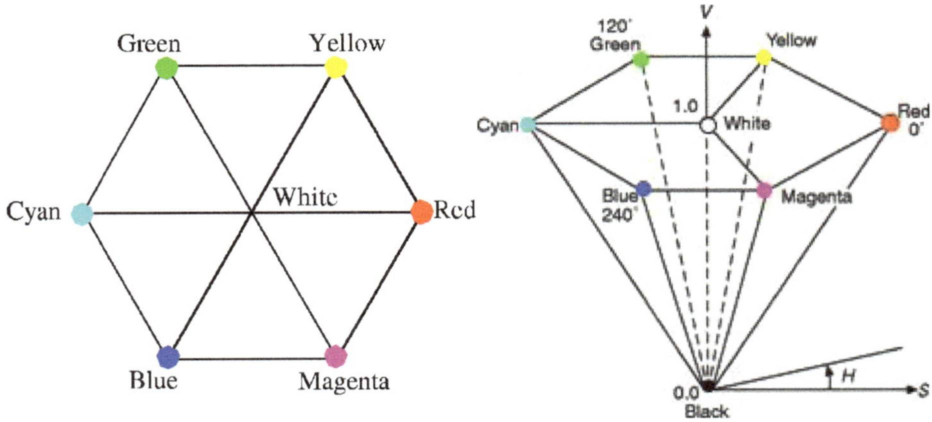

Figure 27. Color hexacone for HSV representation

The boundary of the hexcone represents the various hues, the saturation is measured along a horizontal axis, and value is along a vertical axis through the centre of the hexcone. The color wheel is varied same as the human perception.

Hue is represented by the angle around the vertical axis, with starting red at 0°, then yellow, green, cyan, blue, and magenta respectively, each interval is 60°. Any two colors with 180° difference are complementary colors. Saturation (S) varies from 0 to 1. It is the fraction of distance from center to edge of hexcone. At the S = 0, it is the grey scale. Value (V) varies from 0 to 1 at the top. At the origin, it represents black; and at the top of the hexcone, colors have their maximum intensity. As S =1, the colors have the pure hues.

The HSL color model is very much similar to the HSV system. A double hexcone, with two apexes at both pure white and pure black rather than just one at pure black, is used to visualize the subspace in three-dimensions as shown in **(figure 28).**

In HSL, the saturation component always goes from a fully saturated color to the corresponding gray value; whereas in HSV, with V at its maximum, saturation goes from a fully saturated color to white, which may not be considered intuitive to some. Additionally, in HSL the intensity component always spans the entire range from black through the chosen hue to white. In HSV, the intensity component only goes from black to the chosen hue. Because of the separation of chromaticity from intensity in both the HSV and HSL color spaces, it is possible to process images based on intensity only, leaving the original color information untouched. Because of this, HSV and HSL have found wide spread use in computer vision research.

*7.5.5. CIEL*a*b* color model*

CIEL*a*b* (or CIELAB) is another color model that separates the color information in ways that correspond to the human visual system. It is based on the CIEXYZ color model and was adopted by CIE in 1976. CIEL*a*b* is an opponent color system (no color can involve the opponent colors at same time) based on the earlier (1942) system of Richard Hunter called L, a, b.

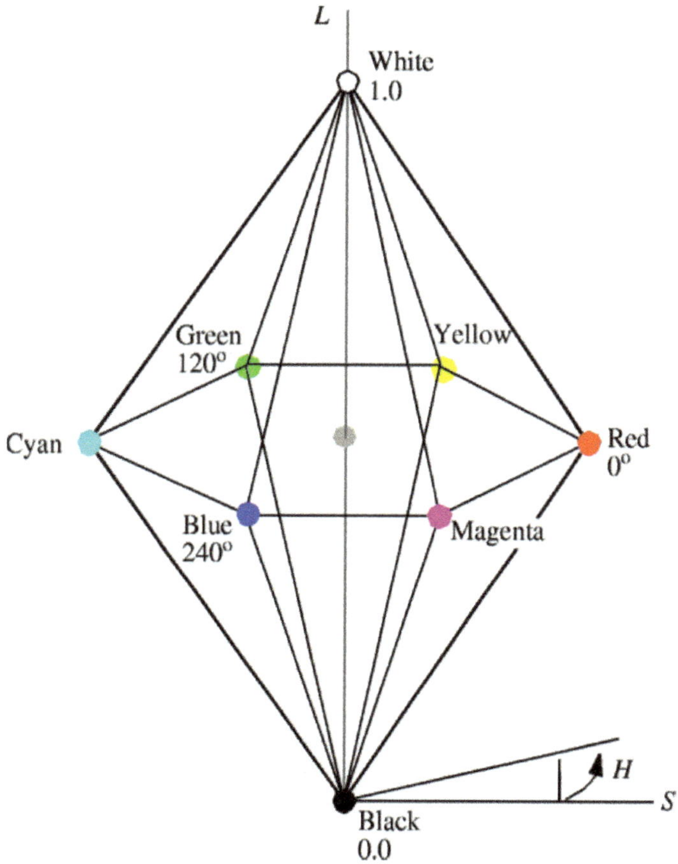

Figure 28. Show the HSL color model

The CIELAB color measurement method was developed in 1976 and offers more advantages over the system developed in 1931. It is more uniform and based on more useful and accepted colors describing a theory of opposing colors.

CIEL*a*b* defines L* as lightness; a* and b* are defined as the color axes to describe the hue and saturation. The color axes are based on the fact that a color can't be red and green, or both blue and yellow, because these colors oppose each other. The a* axis runs from red (+ a) to green (- a) and the b* axis from yellow (+ b) to blue (- b) as shown in **Figure 29.** Hue values do not have the same angular distribution in CIEL*a*b* color model as the hue value in HSV. In fact, CIEL*a*b* is intended to mimic the logarithmic response of the human eye. [For98] The CIEL*a*b* overcomes the limitations of color gamut in the CIE chromaticity diagrams. However, in order to convert to other color models, L* is defined form 0 (black) to 100 (white), a* is from –100 (green) to 100 (red), b* is from –100 (blue) to 100 (yellow).

CIEL*C*H* has the same definition with the CIEL*a*b* except that its values are defined in a polar coordinate system. Thus in CIEL*C*H*, L* measures brightness, C* measures saturation and H* measures hue. We will use this model instead of HSV, as CIEL*C*H* is based on CIEL*a*b* and not on RGB, and hence is device-independent.

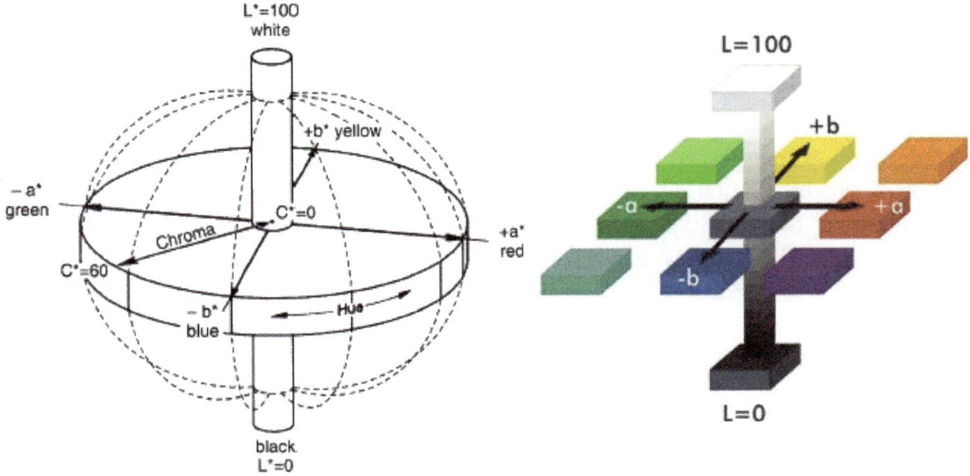

Figure 29. Show the CIEL*a*b* color model

The color models which are used in computer graphics have been traditionally designed for specific devices, such as RGB color model for CRT displays and CMY color model for printers. They are device dependent. Therefore, it becomes meaningless to compare the colors with different devices or the same device under different conditions.

CIEL*a*b* is a device independent color model, and is used for color management as the device independent model of the ICC (International Color Consortium) device profiles (Shen, 2003) (Fairchild, 1997) (CIE, 1999) (CIE, 1998) (Snead, 2005) (Findling, 1996) (Braun, et al. 1998).

7.5.6. sRGB color model

In order to avoid the color difference with different display systems, the IEC (International Electrotechnical Commission) introduced sRGB (IEC 61966-2-1) as a standard color model solution for office, home and web markets. The sRGB model serves the needs of PC and Web based color imaging systems and is based on the average performance of CRT displays. The sRGB solution is supported by the following observations:

- Most computer displays are similar in their phosphor chromaticities (primaries) and transfer function.
- The RGB color model is native to CRT displays, scanners and digital cameras, which are the devices with the highest performance constraints.
- The RGB color model can be made device independent in a straightforward way. It is also possible to describe color gamuts that are large enough for all but a small number of applications.

The accurate handling of color characteristics of digital images is a non-trivial task because RGB signals generated by digital cameras are 'device-dependent', i.e. different cameras produce different RGB signals for the same scene. In addition, these signals will change over time as they are dependent on the camera settings and some of these may be scene dependent, such as the shutter speed and aperture diameter. In other words, each camera defines a custom device-dependent RGB color space for each picture taken. As a consequence, the term RGB (as in RGB-image) is clearly ill-defined and meaningless for anything other than trivial purposes. As measurements of colors and color differences in this paper are based on a standard colorimetric observer as defined by the CIE (Commission Internationale de l'Eclairage), the international standardizing body in the field of color science, it is not possible to make such measurements on RGB images if the relationship between the varying camera RGB color spaces and the colorimetric color spaces (color spaces based on said human observer) is not determined. However, there is a standard RGB color space (sRGB) that is fixed (device independent) and has a known relationship with the CIE colorimetric color spaces. Furthermore, sRGB should more or less display realistically on most modern display devices without extra manipulation or calibration.

The sRGB tristimulus values are linear combinations of the 1931 CIE XYZ values as measured on the faceplate of the display, which assumes the absence of any significant veiling glare. A linear portion of the transfer function of the dark end signal is integrated into the encoding specification to optimise encoding implementations.

A calibrated, nonlinear standard RGB color space called sRGB has been proposed by Microsoft and Hewlett-Packard. Benefits of sRGB are easier portability of RGB color images (especially on the Internet) and faster computational performance than in the uniform CIE spaces. The white point of sRGB is D65 as in the ITU-R BT.709 standard. The phosphor chromaticities are also from BT.709. The sRGB color space is large enough to fit in most device RGB spaces.

The suggested CRT gamma is 2.2 which complies with most monitors. The sRGB color space is computationally fast enough for interactive video and is becoming the future de facto Internet standard (Shen, 2003) (CIE, 1999) (CIE, 1998).

Author details

Ayman H. Amer Eissa*

Department of Agriculture Engineering, Faculty of Agricultural, Minoufiya University, Egypt
Department of Agriculture Systems Engineering, College of Agricultural and Food Sciences, King Faisal University, Saudi Arabia

Ayman A. Abdel Khalik

Agriculture Engineering Research Institute (AEnRI), Giza, Egypt

* Corresponding Author

8. References

Alonge, A. F., & Adigun, Y. J. (1999). Some physical and aerodynamic properties of sorghum as related to cleaning. In Paper presented at the 21st Annual Conference of the Nigerian Society of Agricultural Engineers (NSAE) at Federal Polytechnic, Bauchi, Nigeria.

Anderson, R., Daniel, C, H., Jacques, W., and Siome, G. (2010). Automatic fruit and vegetable classification from images. Computers and Electronics in Agriculture 70. pp, 96–104.

Bassiou, N., C, Kotropoulos. 2007. Color image histogram equalization by absolute discounting back-off. Computer Vision and Image Understanding.V.107 PP: 108–122.

Bennedsen, B.S., D.L., Peterson, and A.Tabb, (2005). Identifying defects in images of rotating apples. Comput. Electron. Agr. 48(2): 92-102.

Bennedsen, B.S. and D.L. Peterson, (2005). Performance of a system for apple surface defect identification in near-infrared images. Biosyst. Eng. 90(4): 419-431.

Bennedsen, B.S. and D.L. Peterson, (2004). Identification of apple stem and calyx using unsupervised feature extraction. Trans. ASAE 47(3): 889-894.

Braun, G. J., F. Ebner, and M. D. Fairchild.1998. Color Gamut Mapping in a Hue-Linearized CIELAB Color Space, IS&T/SID 6th Color Imaging Conference, Scottsdale, 163-168.

Brennan J. G., (2006). Food Processing Handbook. Chapter(1) Postharvest Handling and Preparation of Foods for Processing. WILEY-VCH Verlag GmbH & Co. KGaA, Weinheim

Brosnan, T., and D.W. Sun, (2004). Improving quality inspection of food products by computer vision - a review. J. Food Engin. 61(1): 3-16.

Castleman, K. (1996). Digital image processing. Englewood Cliffs, NJ: Prentice-Hall, 667p.

Chakespari, A. G; Rajabipour, A and H. Mobli (2010). Post Harvest Physical and Nutritional Properties of Two Apple Varieties. Journal of Agricultural Science Vol. 2, No. 3; (61-68).

Chen, Y.R., K. Chao, and M.S. Kim, (2002). Machine vision technology for agricultural applications. Comput. Electron. Agr. 33(2/3): 173-191.

Cheng, X., Y. Tao, Y. R. Chen, and Y. Luo. (2003). Nir/MIR dual–sensor machine vision system for online apple stem–end/calyx recognition. Transactions of the ASAE. Vol. 46(2): 551–558.

CIE. 1998. Color Measurement and Management in Multimedia Systems and Equipment. Part 2-1: Default RGB Color Space – sRGB. http://www.colour.org/tc8-05/Docs/colorspace/.

CIE. 1999. Multimedia Systems and Equipment Color Measurement and Management. Part 2-2: Color management Extended RGB Color Space – sRGB64. http://www.colour.org/tc8-05/Docs/colorspace/.

Colour4Free. 2010. H14 color vision theory V01.doc page 1 of 4. http://colour4free.org.uk/Books/H14ColourVisionTheoryV01.pdf.

Crowe, T.G., and M.J. Delwiche. (1996). Real-time defect detection in fruit-part II: An algorithm and performance of a prototype system. Transactions of the ASAE 39(6): 2309-2317.

Dah, J. L., S. Robert., A. James and M.C. Steve. (2008). Development of a machine vision system for automatic date grading using digital reflective near-infrared imaging. Journal of Food Engineering 86 p.p 388–398.

ElMasrya, G., Wangb, N.,and Vigneault, C.(2009). Detecting chilling injury in Red Delicious apple using hyperspectral imaging and neural networks. Postharvest Biology and Technology 52. pp, 1–8.

Fairchild, M.D. 1997. Status of CIE Color Appearance Models. http://www.colour.org/tc8-01/MDF_AICpaper.pdf.

Fernando, L, G., Gabriela, A, G., José. B., Nuria, A., and José, M, V. (2010). Automatic detection of skin defects in citrus fruits using a multivariate image analysis approach. Computers and Electronics in Agriculture 71. pp, 189–197.

Findling, H. 1996. Fuzzy Algorithm for the Enhancement of Noise degraded Images. Master thesis of science. School of Florida Institute of Technology. Melbourne, Florida.

Francis, F.J. 1980. Colour quality evaluation of horticultural crops. Hort Science, 15(1): 14-15.

Geyer, Lewis H and Ranald F. Perry. (1982). Variation in detectability of multiple flaws with allowed inspection time. Human Factors 24(3): 361-365.

Guyer, D.G. Brown, E. Timm, R. Brook, and E. Marshall.(1994). Lighting systems for fruit and vegetable sorting. Extension Bulletin E-2559.E.Lansing, Mich.: Cooperative Extension Service, Michigan State University.

Heinemann P.H., Hughes R., C.T.Morrow, H.J. Sommer, R.B. Beelman and P.J. Wuest (1994). Grading of mushrooms using a machine vision system. Transactions of the ASAE 37 (5): 1671–1677.

Heinemann, P.H., Z.A. Varghese, C.T. Morrow, H.J. Sommer III, and R.M. Crassweller. (1995). Machine vision inspection of "Golden Delicious" apples. Applied Engineering in Agriculture 11 (6): 901-906.

Hoffmann, G. 2000. CIE color space. http://www.fho-emden.de/~hoffmann/ciexyz29082000.pdf

Jiangsheng, G., and Y., Yibin. (2006). Fruit Shape Detection Based on Multi-scale Level Set Framework. ASABE Paper No 063088.

Jun, Q., S. Akira, S. Sakae, and K. Naoshi. (2004). Mobile Fruit Grading Robot -Concept and prototype. American Society of Agricultural and Biological Engineers, St. Joseph, Michigan www.asabe.org.

Kang, S.P., East, A.R., and Trujillo, F.J. (2008). Colour vision system evaluation of bicolour fruit: A case study with 'B74' mango. Postharvest Biology and Technology 49. pp, 77–85.

Kheiralipour K., Tabatabaeefar, H., Mobli, A., Rafiee, S., Sharifi, M., Jafari, A. and Rajabipour, A. (2008). Some physical and hydrodynamic properties of two varieties of apple (Malus domestica Borkh L.).Int. Agrophysics, vol.22, pp, 225-229.

Kondo, N. (2009). Automation on fruit and vegetable grading system and food traceability", Trends in Food Science & Technology, doi: 10.1016/j.tifs.

Kondo,N. (2010). Automation on fruit and vegetable grading system and food traceability. Trends in Food Science & Technology 21. pp, 145-152.

Leemans, V., Magein, H., and Destain, M.F. (2002). On-line fruit grading according to their external quality using machine vision. Biosyst. Eng. 83(4): 397-404.

Mathworks (2005). Image processing toolbox for use with Matlab. Natick, USA: The Mathworks Inc.

Mendoza, F., & Aguilera, J. M. (2004). Application of image analysis for classification of ripening bananas. Journal of Food Science, 69(9), E471–E477.

Mendoza, F., Dejmek, P., & Aguilera, J. M. (2006). Calibrated color measurements of agricultural foods using image analysis. Postharvest Biology and Technology, 41,285–295.

Mery, D., & Pedreschi, F. (2005). Segmentation of color food images using a robust algorithm. Journal of Food Engineering, 66, 353–360.

Meyers, J.B., S.E. Prussia, C.N. Thai, T.L. Sadosky, and D.T. Campbell. (1990). Visual inspection of agricultural products moving along sorting conveyors. Transactions of the ASAE 33 (2): 367-372.

Miller, W.M. 1995. Optical defect analysis of Florida citrus. Applied Engineering in Agriculture 11 (6): 855-860.

Mirasheh, R. (2006). Designing and making procedure for a machine determining olive image dimensions. Master of Science Thesis, Tehran University.

Mohsenin, N.N. (1970). Physical properties of plant and animal materials. Vol.1. Structure physical characteristics and mechanical properties. Gordon and Breach Science Publications, New York.

Mohsenin, N. N. (1986). Physical properties of plant and animal materials. Gordon of Breach science publishers, New York.

Muir, A.Y., I.D.G. Shirlaw, and D.C. Mckae. (1989). Machine vision using spectral imaging techniques. Agricultural Engineering 44(3): 79-81.

Mummert, C.N. (2004). The development of a machine vision system to measure the shape of a sweetpotato root. Master of thesis of science. Fac, of Agric. North Carolina State University.

Naoshi, K., N. Takahisa, M. Yoshihid, L. Peter, K. Mitsutaka, K. Makoto, D.F. Paolo, and O. Yuichi. (2008). A double image acquisition system with visible and UV LEDs for citrus fruit. American Society of Agricultural and Biological Engineers, ASABE.

Narendra, V.G. and Hareesh, K.S. 2010. Prospects of Computer Vision Automated Grading and Sorting Systems in Agricultural and Food Products for Quality Evaluation. International Journal of Computer Applications. V (1), No. 4.

Nielsen, H.M. and W. Paul, (1998). Modeling image processing parameters and consumer aspects for tomato quality grading. In: Mathematical and Control Application in Agriculture and Horticulture. Proceedings of the Third IFAC Workshop, Pergamon/Elsevier, Oxford, UK.

Onder K.; Aziz, O. and Ibrahim, A. (2006). Physical properties of cactus pear (Opuntia ficus india L.) grown wild in Turkey. Journal of Food Engineering (73): 198–202.

Patrick, M, M., Yud, R, C., Moon, S, K., and Diane, E, C. (2004). Development of hyperspectral imaging technique for the detection of apple surface defects and contaminations. Journal of Food Engineering 61. pp, 67–81.

Paulus, I., R. De Busscher, and E. Schrevens, (1997). Use of image analysis to investigate human quality classification of apples. Journal of Agricultural Engineering Research 68, pp. 341-353.

Peschel S., Franke R., Schreiber L., and Knoche M., 2007. Composition of cuticle of developing sweet cherry fruit. Phytochemistry, 68, 1017-1025.

Rehkugler, G.E. and J.A. Throop. (1976). Optical-mechanical bruised apple sorters. PP. 185-188, 192. In: J.J. Gaffney(ed). Quality Detection in Foods. ASAE. St. Josph, MI.

Rigney M.P., G.H. Brusewitz, and G.A. Kranzler. (1992). Asparagus defect inspection with machine vision. Transactions of the ASAE 35(6): 1873-1878.

Sahin, S. and G. S. Sumnu. 2006. Physical Properties of Foods. Library of Congress Control Number: 2005937128. Springer Science+Business Media, LLC.

Shen, Z. 2003. Color differentiation in digital images. Master thesis of science. School of Computer Science and Mathematics. Victoria University of Technology, Australia.

Snead, M.C.2005. A method of content-based image retrieval for the generation of image mosaics. Master thesis of science. School of Electrical Engineering and Computer Science. University of Central Florida .

Stephenson, K.Q. (1976). Color sorting system for tomatoes. ASAE. PP. 199-201.

Sun, D.W. 2008. Computer Vision Technology for Food Quality Evaluation.Food Science and Technology, International Series. ISBN: 978-0-12-373642-0.

Sun, D.W., and T. Brosnan. (2003). Improving quality inspection of food products by computer vision: A Review. Journal of Food Engineering, 61: 3-16.

Sunil K. M., R. W. Paul, W. Ning, B. Timothy, and O. M. Niels. (2009). Adaptive Thresholding of Pecan X-ray Images using Water Flow Models and Feature Extraction. ASABE.

Tao, Y. (1996a). Methods and apparatus for sorting objects including stable color transformayion. U.S. Patent No. 5,533,628.

Tao, Y. (1996b). Spherical transform of fruit images for on-line defect extraction of mass objects. Opt. Engng. 35(2):344-350.

Tao, Y. (1998). Defective object inspection and separation system. U.s. patent no. 5,732.147.

Throop., J.A., Aneshansley, D.J., Upchurch, B.L., and Anger, B. (2001). Apple orientation on two conveyors: performance and predictability based on fruit shape characteristics. Trans. ASABE. 44(1): 99-109.

Throop, J.A., D.J. Aneshansley, and B.L. Upchurch. (1993). Investigation of texture analysis features to apple bruises. ASAE Paper No. 93-3527.

Upchurch, B.L, H.A. Affeldt, W.R. Hruschka, and J.a. Throop. (1991). Optical detection of bruises and early frost damage on apples. Transaction of the ASAE 34(3): 1004-1009.

Wen, Z., and Y. Tao. (1997). adaptive spherical transform of fruit images for high-speed defect detection. ASAE paper No. 97-3076.

Xiao-bo, Z., Jie-wen, Z., Yanxiao, L., and Holmes, M. (2010). In-line detection of apple defects using three color cameras system. Computers and Electronics in Agriculture 70.pp, 129–134.

Yang, Q. (1993). Finding stalk and calyx of apples using structured lighting. Comput. Electron. Agric. 8: 31-42.

Chapter 12

Public Health Policies and Functional Property Claims for Food in Brazil

Paulo César Stringueta,
Maria da Penha Henriques do Amaral, Larissa Pereira Brumano,
Mônica Cecília Santana Pereira and Miriam Aparecida de Oliveira Pinto

Additional information is available at the end of the chapter

1. Introduction

Specific regulations for functional foods began in Japan in the middle of the 80´s. Known as food for specific health use (FOSHU: *foods for specified health use*), these foods carry a label of approval from the Japanese Ministry of Health and Welfare [1]. The goal was to develop foods that would enable the reduction of expenses on public health, considering the high life expectancy in that country [2]. The term "functional foods" refers to processed foods, similar in appearance to conventional foods, consumed as part of a normal diet, but demonstrating physiological benefits and/or reduction in chronic diseases risk, in addition to its basic nutritional functions [1]. The principle was soon adopted worldwide [2]. In Brazil, functional foods were officially recognized in 1999, with the approval of specific legislation [3]. The designations of the claims, as well as the criteria for their approval vary according to the regulations of each country or economic block. Therefore, it is noteworthy considering the regulations of the intended market for the product.

The tendency of the Codex Alimentarius[1] and of several countries was disciplining these claims on the functional properties of foods or its components, as well as on the security scientific evidence-based, in order to avoid confusion and mistakes to the consumer with nomenclature and claims about properties not scientifically proved [4-7].

Functional foods are an important part of wellness, which also includes a balanced diet and physical activity [8]. The Food Guide for the Brazilian Population of the Ministry of Health

[1] The combined program of the United Nations for Food and Agriculture Organization (FAO) and the World Health Organization (WHO) is an international forum for standardization on food, created in 1962, and its rules aim to protect the health of the population, ensuring fair practices in international and regional trade of food, creating international mechanisms aimed at removing tariff barriers, promoting and coordinating all works held in standardization.

recommends stimuli to physical activity practice, adoption of a varied diet and warn not to mystify food functional components [9].

Brazilian legislation does not define functional food. It defines functional property claim and health property claim. It establishes the basic guidelines for risk assessment and safety, analysis and proof of functional and/or health properties claimed in labeling, as well as the conditions of registration for foods with claims of functional and/or health properties [5,6,10].

Among the guidelines for this type of food it is allowed functional claims related to the physiological role in the growth, development and normal functions of the body and/or also claims about the general maintenance of health and disease risk reduction, as optional. It is not allowed claims which make reference to the cure or prevention of disease. The food or ingredient that claim functional and/or health properties can, in addition to basic functions, when it is a nutrient, produce metabolic and/or physiologic effects and/or beneficial effects to health, and it must be safe for consumption without medical supervision. To submit claims of functional property and/or health, food, as well as bioactive substances and isolated probiotics must be, obligatorily, registered in the competent authority. The advertising content of these products can not be different in their meaning, from that approved for the labeling. The claims should also be in accordance with the guidelines of the public health policies [4-6,11].

The public health policy in Brazil for the specific area of food and nutrition is defined by the Food and Nutrition National Policy (PNAN in Portuguese for *Politica Nacional de Alimentação e Nutrição*) [12]. It presents an interface with the National Policy for Health Promotion (PNPS in Portuguese for *Política Nacional de Promoção da Saúde*) [13]. The Food Guide for Brazilian People[10] constitutes compliance with one of the PNAN guidelines. This guide has incorporated the recommendations of the global strategy[14] and it establishes guidelines for healthy eating and physical activity. The guidelines of these policies are used as criteria for evaluating claims of functional and/or health properties in food.

The outlook reported in 2006 regarding the processes approvals flow in requests for registration of foods claiming functional properties, in the period from 1999 to 2004, evidenced a high proportion of refused cases [15]. There is the need of greater understanding by the industry, regarding the criteria used for evaluating the processes with regard to the terms employed in the law [6] "in accordance with the guidelines of the public health policy", "in light of current scientific knowledge" and "food of occasional consumption". For this understanding it is crucial to understand the meaning of terms employed in the regulations and in the public policies.

In March 2011, the National Agency of Sanitary Surveillance (ANVISA in Portuguese *Agência Nacional de Vigilância Sanitária*) presented in the Management Report for the years 2005 to 2010 the main achievements of the Agency. From the perspective of modernization of the management, from 2006 on, ANVISA has promoted a reform in the food control model, with emphasis on the reduction in bureaucracy of the sanitary registration and on the strengthening of the post-market control. The reduction on the sanitary registration bureaucracy was started in 2000, and broadened in 2005 and 2010.

Currently, the sanitary registration is required only for six categories of food, allowing the administrative machine to look over the control of the product directly offered to the consumer, an international trend in the food regulation [16].

It remains with mandatory registration foods claiming functional and/or health properties, infant feeding, food for enteral nutrition, new foods, new ingredients and bioactive substances and isolated probiotics claiming for functional or health properties. For its innovative character or consumption by specific population groups, these foods require the evaluation of safety and efficacy, reasons for maintaining the registration [16].

2. Brazilian legislation

Responsibilities of the National Health Surveillance Agency (ANVISA):

- To establish standards, monitor and execute policies, guidelines and actions of the health surveillance.
- To grant products registrations, according to the guidelines of its area of action.
- To control, monitor and track, under the prism of health legislation, advertising and publicity of products covered by the health surveillance scheme [17].

In Brazil, since 1990 there had been applications for registration of various products not recognized as food yet, within the traditional concept of food. Given the complexity of the issue and the change in the focus of food analysis, which now considers the risk criterion, ANVISA created the Advisory Technoscientific Commission on Functional Foods and Novel Foods (CTCAF in Portuguese for *Comissão de Assessoramento Tecnocientífico em Alimentos Funcionais e Novos Alimentos*), consisting of professors and outstanding researchers working in universities and research institutions, with the purpose of assisting the Board of Food and Toxicology in decisions related to this issue. The former term CTCAF was further changed to the Advisory Technoscientific Commission on Foods with Claims of Functional and/or Health Property and Novel Foods [18].

After extensive discussion in 1999, ANVISA has approved regulations that deal with basic guidelines:

- **Resolution nº 16/99**: procedures for registration of food and/or new ingredients [19].
- **Resolution nº 17/99**: risk assessment and food safety [20].
- **Ministerial Order 398/99 and Resolution nº 18/99**: basic guidelines for analysis and approval of claims for functional and/or health property mentioned on the labeling of the foods [5,6].
- **Resolution nº 19/99**: procedures for registration of foods claiming functional and/or health properties [10].

These categories of food must also comply with the legislation for food in general, and in no case it is allowed to claim medicinal or therapeutic properties. Other regulations such as:

- Resolution nº 22/2000: Technical Regulation on the basic procedures for registration and for exemption from obligatory registration of imported products relevant to food field, contained in the annex to this resolution [21].

- Resolution nº 23/2000: Demonstrates about the Manual of Basic Procedures for the Registration and for Exemption from Obligatory Registration of Products Pertinent to the Food Area [22].
- Resolution RDC nº 2/2002: Technical Regulation of bioactive substances and isolated probiotics with claim for functional and/or health properties, annexed to that resolution [23].
- Technical Report nº 9/2004: Guidance for use, on food labels, the claims of functional properties of nutrients which function is fully recognized by the scientific community (Item 3.3 of Resolution nº. 18/99) [4].
- Technical Report nº 19/2006: Procedures for adding edible mushrooms in capsules, tablets and pills in the food area [24].
- Resolution RDC nº 27/2010: Demonstrates about the categories of food and packaging which are excepted or compulsory to sanitary registration which apply to all types of food [25].
- Decree Law nº 986/1969: Instructs food basics rules [11].

3. Registration

The registration at ANVISA is compulsory both for bioactive substances and isolated probiotics, as well as for food claiming functional and/or health properties and for new foods and new ingredients, produced in Brazil or imported [21-23]. For this, it is necessary to demonstrate the efficacy and safety of the food consumption. Even for products of animal origin, such as dairy products, the competence of the Ministry of Agriculture Livestock and Supply (MAPA in Portuguese for *Ministério da Agricultura Pecuária e Abastecimento*), the proceedings on the confirmation of the claims must be submitted to ANVISA for analysis.

In the case of foods regulated by the MAPA, such as the dairy products, companies must first fill a petition in ANVISA, referring to the request "Evaluation of Foods with Claims of Functional and/or Health Properties." ANVISA shall send response about the assessment for the company, with a copy to the responsible area of MAPA [26]. For the assessment, it is necessary to include the information and documentation required in the Resolution nº18/1999 [6].

4. New food

According to ANVISA [19], "new food and/or new ingredients" are foods or substances with no history of consumption in this country, or foods with substances already consumed, but with additions or employed at levels much higher than those currently observed in the regular diet. For example: eggplant in capsules, *Agaricus blazei* mushroom in capsules, guarana in capsules.

Foods that may be consumed in the form of capsules, tablets or other dosage forms and which do not claim functional or health properties that are scientifically proven must bring on the label the words "The Ministry of Health warns: there is no proved scientific evidence that this food will prevent, treat or cure diseases". For this category of foods it is not allowed to claim functional and/or health properties [19,24].

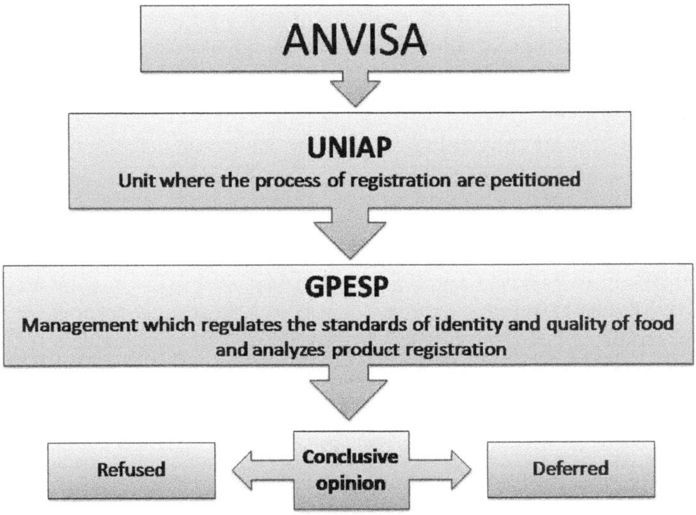

ANVISA - National Agency of Sanitary Surveillance (in Portuguese *Agência Nacional de Vigilância Sanitária*);
UNIAP - Public Unit Service (in Portuguese *Unidade de Atendimento ao Público);*
GPESP - Management of Special Products (in Portuguese *Gerência de Produtos Especiais);*

Figure 1. Steps to registration of food in Brazil

5. Bioactive substances and isolated probiotics

RDC nº 2/2002 [23] is applied to the guidelines to be adopted for the safety assessment, registration and commercialization of bioactive substances and isolated probiotics claiming functional and/or health properties, presented as dosage forms (capsules, tablets, pills, powders, granules, pastilles, suspensions and solutions). The products are classified as: carotenoids, phytosterols, flavonoids, phospholipids, organosulfur compounds, polyphenols and probiotics. Once approved, the claims proposed by the manufacturer are mandatory, and they must be presented in the manner and wording approved by ANVISA [23]. An example of a product registered in this category is lycopene in capsules.

6. Foods with claims of functional and/or health properties

To obtain the registration of food with claims, the company must present combined with the other documentation contained in the legislation for food in general, the scientific-technical report containing the following information [5,6,10]:

- Text and copy of the wording layout of the labeling.
- Description of the product
- Consumption that is foreseen or recommended by the manufacturer;

- Description of the analytical methodology for evaluation of the components which are object of claim;
- Chemical composition with molecular characterization, when appropriate and product formulation;
- Purpose, usage conditions and nutritional value, when appropriate;
- Scientific evidences applicable, when appropriate, to attest the efficacy when calaiming functional and/or health properties:
    - nutritional and/or physiological and/or toxicological tests in experimentation animals;
    - biochemical tests;
    - scientific description of the ingredients of the product, according to species of botanical, animal or mineral origin, when appropriate;
    - epidemiological studies;
    - clinical trials;
    - evidence of traditional use, observed in the population, without harm to health;
    - comprehensive evidence of the scientific literature, international health agencies and internationally recognized legislation on the properties and characteristics of the product;
    - documented information about approval of use of the food or ingredient in other countries, economic blocks, Codex Alimentarius and other internationally recognized authorities.

It shall be considered as scientific evidences, the copies of original papers published in journals of recognized scientific imprint. Therefore, book chapters, non scientific weekly journal articles, among others are not valid as scientific evidence.

The scientific papers in English or Spanish do not need translation. Copies of other works in foreign languages must be accompanied by translation, not necessary sworn.

It is the responsibility of the company submitting the copy of the scientific papers referenced in the Technical Scientific Report. It will not be considered as valid references the abstracts of papers and bibliographic citations.

The Technical Report nº 9, may 2004 [4] was established on the basis that the application of the item 3.3 of the resolution nº 18/99 "for the nutrients with function fully recognized by the scientific community, it is not required to demonstrate efficacy or to analyze this nutrient in order to have the functional claim in the labeling" [6] allowed situations that contradicted the guidelines of public health policies, as well as it was observed an increased use of claims on labels of products exempted from the mandatory register in the trade [4]. According to that report, the following criteria must be met for approval of claims for nutrients with function fully recognized by the scientific community:

- They must be related to nutrients intrinsic to the product, which must be present at least in the amount set for the attribute "source" according to regulations on supplementary nutrition (Ministerial Order nº 27/98) [27].
- They must be specific to the role of the nutrient which is claimed.

- They must be linked to the normal food intake of the population, which should not be of occasional consumption and shall not be present as capsules, tablets, pills, or other dosage forms.

The meeting to the criteria established for the use of the claims set forth in item 3.3 of the Resolution nº 18/99, which are of the company responsibility, dispense it to send the documentation for the technical assessment, noting that the claims can not refer to prevention, treatment and cure of diseases [6]. However, the exemption refers only to the necessity of proving the claims - Item 4.1.1.9 of the RDC 18/99 and not to registration and other items of the technicalscientitific report [6]. Foods added with essential nutrients, which claim for functional properties should be referred for evaluation of each case.

Regarding the expression "it should not be of occasional consuption" [4], according to RDC nº 359/2003, the following products are considered as occasional consumption: whole fruit preserved for adornment (maraschino cherry, raspberry), candies, lollipops and pastilles, chewing gum, chocolates, sweets and similar products; chocolate confectionery *dragees* in general; ice cream, individual units of ice cream; cereal bar with more than 10% fat, nougat, soft drinks, carbonated or not (tea, soy-based beverages and soda); powder for preparing refreshments, sweet biscuits, with or without filling, *brownies* and *alfajors*, candied fruits, *panettone*, fruit cake, cakes and similar with filling and cover; *snacks* prepared from cereals and flour for snack consumption, mix for preparation of sweets, topping for cakes, pies and ice cream [28].

In 2005 the products with claims of functional and/or health properties approved since 1999 were re-evaluated. It was used as basis the current scientific knowledge, as well as reports and studies that demonstrated the difficulties encountered by consumers in understanding the true meaning of the features announced for certain products containing claims. The review considered as assumptions, the need of the claims to be in accordance with the policies of the Ministry of Health and be easily understood by consumers, in addition to complying with the provisions of the resolutions nº 17/99, 18/99 and 19/99. The following products had their claims changed, in order to improve the consumer understanding about the properties of these foods: fatty acids of omega-3 family, carotenoids (lycopene and lutein), dietary fiber (fiber, beta-glucane, fructooligosaccharides, inulin, lactulose, *Psillinum* ou *Psillium*, chitosan), phytosterols, probiotics (*Lactobacillus acidophilus, L. casei shirota, L. casei var. rhammosus, L. casei var. defensis, L. delbrueckii subspécie bulgaricus, Bifidobacterium bifidum, B. lactis, B. longum, Streptococcus salivarius, thermophillus* subspecies) and soy protein. The claims previously approved, regarding to caffeine, sorbitol, xylitol, mannitol, sodium stearate, sodium bicarbonate, omega-6 fatty acids, polyunsaturated and monounsaturated fat acids (in vegetable oils), and the liquid compound ready for consumption, were no longer allowed. Companies should adapt the wording on the label, following this new format of the claims by January 30, 2006 or within the deadline negotiated with regional surveillances for the depletion of the products package [18]. The list of approved claims was updated in July 2008. The list (Annex I) included the claims related to beta-glucane, resistant dextrin in powder, partially hydrolyzed guar gum and polydextrose [29]. The Figure 1 shows the steps to registration of food in Brazil.

The situation in Brazil in 2006, after the review of the claims in 2004, was as follows: no health claim was approved. It was approved 14 claims of generic functional properties with standardized language and 25 kinds of substances or microorganisms with functional claim [15]. The claims about nutrients with fully established function were under discussion [15, 17, 30].

7. Advertising

The advertising and publicity of these products are inspected by ANVISA, through the General Management of Inspection, Quality Monitoring, Control and Surveillance of Imputs, Drugs and Products, Advertising and Publicity (GGIMP in Portuguese for *Gerência Geral de Inspeção, Monitoramento da Qualidade, Controle e Fiscalização de Insumos, Medicamentos e Produtos, Propaganda e Publicidade*) which incorporated in February 2012 the General Managing of Surveillance and Monitoring of Advertising, Publicity, Promotion and Information about Products Subject to Sanitary Surveillance (GGPRO in Portuguese for *Gerência Geral de Monitoramento e Fiscalização da Propaganda, de Publicidade, de Promoção e de Informação de Produtos Sujeitos à Vigilância Sanitária*) [31]. Any consumer information booklet, which compose the product package, or an instrument for its disclosure, can not convey allusive information to their properties other than those approved by the competent authority of ANVISA to be present on its label [11, 18].

In studies performed for Pinto, in reference [32], the advertisements of foods *folder* with claim of functional and/or health properties, of bioactive substances and isolated probiotics, as well as of new foods were assessed regarding the current legislation, and a high proportion of samples did not attend to more than one requirement. The claim for medicinal or therapeutic properties, the omission of mandatory warning statements and the recommended forms of consumption of the products may lead to inadequate intake and cause undesirable reactions in specific groups of consumers, especially people with allergies, children, pregnant women, nursing mothers and those with celiac disease. Moreover, it can slow the search for appropriate treatment, and may worsen diseases. This deterioration increases public spending on curative health.

In Brazil there has not been regulation specific to food advertising yet, which complicates the understanding of the industries and inspection by the health authorities. There are requirements of the relevant legislation on foods that are applied to food advertising which claim functional and/or health properties, bioactive substances and isolated probiotics and new foods, in the same way it is upon approved for labeling [32].

8. Criteria for evaluating the scientific basis for claims

World Health Organization (WHO), combined with the United Nations for Food and Agriculture Organization (FAO/UN) published some recommendations on lifestyle, diet and food intake, suggesting levels of scientific evidence for the risk of development of chronic non-communicable diseases (CNCD) [33, 34]. The classification the strength of evidence of FAO/UN for recommendations in clinical practice for disease prevention, according to the quality, quantity and results of available studies is established as evidence convincing, probable, possible and insufficient [33]. The strength of evidence that relates dietary factors and lifestyle with the risk of developing obesity, type 2 diabetes, cardiovascular disease (CVD) and cancer, classified according to the categories mentioned above are summarized in Table 1.

	Obesity	Type 2 Diabetes	CVD	Cancer
High intake of highly energy food	C↑			
Trans fatty acids			C↑	
Fish and fish oil (EPA and DHA)			C↓	
High intake of dietary fiber (NSP)	C↓	P↓	P↓	
High sodium intake			C↑	
Fruits (including red fruits) and vegetables	C↓[k]	P↓[k]	C↓	P↓[l]
Overweight and obesity		C↑	C↑	C↑[s]
Regular physical activity	C↓	C↓	C↓	C↓

Note: Only convincing evidence (C) and probable evidence (P) are included in this summary table.
C↑: convincing increased risk; C↓: convincing decreased risk; P↑: probable increased risk; P↓: probable decreased risk; EPA: eicosapentaenoic acid; DHA: docosapentaenoic acid; NSP: non-starch polysaccharides. k: based on contributions from fruits and vegetables with non-starch polysaccharides (dietary fibers). l: cancer of the oral cavity, esophagus and colorectal cancer. s: for esophageal cancer, colorectal, breast (in postmenopausal women), endometrium and kidney.

Table 1. Summary of strength of evidence for obesity, type 2 diabetes, cardiovascular disease and cancer [33].

9. General policies: Food and health promotion

There is a strong global interest to improve the quality of nutrition and reduce health care costs through prevention of chronic diseases, improving the quality and active life expectancy. The Brazilian health policies related to food and nutrition exhibit this trend and follow the recommendations of the global strategy on diet, physical activity and health, published by the World health [14]. The strategy recommends that member countries adopt policies that encourage healthy eating and physical activity, as a way to reduce the incidence of CNCD caused by unhealthy diet and sedentary lifestyle [35]. The policies that influence the approval of claims for foods are set out below.

10. National policy for Food and Nutrition

The National Food and Nutrition (PNAN), adopted in 1999, integrates the efforts of the Brazilian State that through a set of proposed policies is supposed to respect, protect, promote and provide human rights to health and nutrition [12].

PNAN, updated in November 2011, aims the improvement in food, nutrition and health conditions of the Brazilian population, by promoting healthy and adequate eating habits, food and nutrition surveillance and the comprehensive prevention of the problems related to food and nutrição [34]. It integrates the National Health Policy (PNS) and presents an interface with the National Policy of Health Promotion (PNPS in Portuguese for *Política Nacional de Promoção da Saúde*), inserting itself in the context of Food and Nutritional Security. Figure 2 shows the interfaces of PNAN and PNPS with the National Health Policy [35].

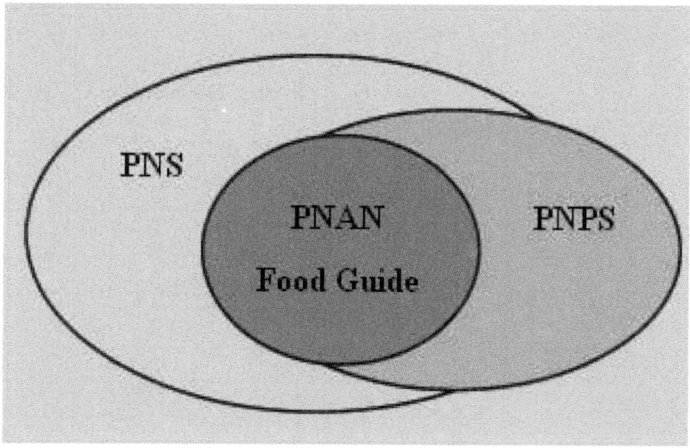

Figure 2. Interfaces of the National Brazilian Health Policies (adopted from Oliveira, 2006) [36].

Some considerations of the National Food and Nutrition are highlighted on the national reality [12]:

- The adoption of the concept of Food and Nutritional Security worldwide and in Brazil facilitated the understanding of the role of the health sector with regard to food and nutrition, recognized as essential to the promotion, protection and restoration of health.
- Obesity in the Brazilian population is becoming much more frequent than the child malnutrition, signaling an epidemiological transition process that should be properly valued in terms of health. Cardiovascular diseases, which are the leading cause of death and disability in adulthood and in old age and are responsible, in Brazil, for approximately 34% of all causes of death, and are related in large part, to obesity and inappropriate eating habits and lifestyles.
- Inappropriate eating habits are a major challenge. Popular culture preserves traditions and erroneous feeding practices on the nutritional value, healing properties, indication or prohibition of food. On the other hand, it emphasizes the proliferation of fast food trade and the increasing use of pre-cooked or quick-cooking food.
- The situation of food and nutrition in Brazil is very complex, where there are typical problems of underdeveloped societies and developed countries.

PNAN present nine guidelines that indicate the lines of action to achieve its purpose, capable of modifying the determinants of health and to promote health. It is consolidated in: 1.Organização of Nutritional Care; 2. Promotion of Healthy and Suitable Food; and 3. Food and Nutrition Surveillance; 4. Management Actions of the Food and Nutrition; 5. Participation and Social Control; 6. Qualification of the Labor Force; 7. Food Control and

Regulation; 8. Research, Innovation and Knowledge on Food and Nutrition; 9. Cooperation and coordination for Food and Nutrition Safety [12].

Some relevant definitions according to PNAN [12]:

- **Suitable and health nutrition:** suitable eating habits to the biological and socio-cultural aspects of individuals, as well as the sustainable use of the environment. They must comply with the requirements of each phase of the life course and with the special dietary needs; be referenced by the food culture and the dimensions of gender, race and ethnicity; be accessible from the physical and financial standpoint, harmonic in quantity and quality, based on adequate and sustainable production practices; with minimal amounts of physical, chemical and biological contaminants.
- **Food and nutritional safety:** consists in the achieving the right of everyone to the regular and permanent access to quality food in sufficient quantity, without compromising access to other essential needs, based on health promoting food practices that respect cultural diversity and that are environmentally, culturally, economically and socially sustainable.
- **Healthy eating practices:** uses, habits and customs that define patterns of food consumption in accordance with scientific knowledge and techniques of good nutrition.
- **Food safety and quality:** deals, in health surveillance, of attributes related to food safety and nutritional value. See also healthy eating.

11. WHO – Global strategy for diet, physical activity and health [14]

The global strategy for diet, physical activity and health approved by the World Health Organization, starting from the recognition that DCNT, such as cardiovascular disease, type 2 diabetes and certain cancers, imposes a significant economic burden on health systems and violates existing high costs on society. The overall strategy is aimed at two main risks of DCNT: diet and physical activity [14].

The overall objective of the global strategy on diet, physical activity and health is to promote and protect health by guiding the development of a qualification of the environment, to support actions at individual, community, national and global levels that, when taken together, lead to a reduction in rates of illness and deaths related to unhealthy diets and physical inactivity. The main specific objectives are:

- To reduce the risk factors for DCNT through the essential action in public health, health promotion and preventive measures.
- To increase the attention and knowledge about diet and physical activity.
- To encourage the development, strengthening and implementation of policies and plans at global, regional, national and community actions that are sustainable, including civil society, private sector and the media.

In relation to diet, it is recommended for individuals and populations:

- To keep the energetic balance and healthy weight.

- To limit energy intake from fats; replace saturated fats for unsaturated fats and eliminate trans fats (hydrogenated).
- To increase the consumption of fruits, vegetables, whole grains and nuts.
- To limit the intake of free sugar.
- To limit the intake of salt (sodium) from every origin and consume iodized salt.

Regarding to physical activity, the overall strategy recommends at least 30 minutes of regular, severe or moderate physical activity, almost every day, to reduce the risk of cardiovascular disease, diabetes, colon cancer and breast cancer and to improve functional status in different stages of life, especially in adulthood and in elderly [14].

12. Responsibilities

To achieve changes in eating habits and physical activity patterns is required the combined effort of many public and private actors (WHO, governments, international partners, civil society and nongovernmental organizations, private sector) for several decades. Among the responsibilities and recommendations, stand out:

- **For Member States:** governments should consider actions that will result in providing accurate and balanced information to consumers to enable them to make healthy choices. Information for consumers should be appropriate to their levels of literacy, communication barriers, local cultures and they should be understood by all segments of the population [14].
- **Marketing, advertising, sponsorship and promotion:** food advertising affects food choices and influences eating habits. The advertising of foods and beverages should not exploit children's credulity and inexperience. Messages that encourage unhealthy dietary practices or physical inactivity should be discouraged, while positive and healthy messages should be encouraged.
- **Labeling:** consumers need accurate, standardized and comprehensible information about foods contents to make healthy choices.
- **Health claims:** As interest in consumer health is increasing, the use of health-related messages by manufacturers is growing. These messages should not confuse the public about nutritional benefits or risks.
- **Promotion of food products consistent with a healthy diet:** Governments could consider measures to encourage the reduction, the salt content of processed foods, the use of hydrogenated fats and sugar content in drinks and sweets.
- **For the private sector:** the food industry, as one of the representatives of the private sector, can be a significant player in promoting healthy diets and physical activity. Initiatives by the food industry to reduce fat, sugar and salt content in processed foods and portion sizes, to increase the introduction of nutritious, healthy and innovative choices, and to review the current marketing practices, could accelerate health gains worldwide.
- Specific recommendations for the food industry
- To promote healthy diet and physical activity in compliance with the guidelines and the national and international standards and with the overall objectives of the global

strategy; limiting the levels of saturated fats, trans fatty acids, free sugars and salt in existing products.

- To continue to develop and provide nutritious, healthy and affordable choices for consumers.
- To consider the introduction of products with better nutritional value.
- To provide consumers with adequate and understandable information about nutrition and products.
- To practice responsible *marketing* that supports the strategy, particularly with respect to promotion and "marketing" of foods with high content of saturated fats, trans fatty acids, free sugars, or salt, especially for children.
- To issue simple, clear and consistent labels and health claims based on evidence, that help consumers make healthy choices regarding the nutritional value of foods.

The implementation of this strategy by all stakeholders contribute to the sustainable improvement of human health [14].

13. Food guide for the Brazilian population [9]

The editing of the first official dietary guidelines of Brazil was part of the implementation strategy of the National Food and Nutrition, a member of the National Health Policy. It consolidates itself as concrete evidence for implementation of recommendations issued by the World Health Organization [14]. The Guide aims to contribute to the direction of feeding practices aimed at health promotion and prevention of diseases related to food [9]:

- Malnutrition and micronutrient deficiencies, such as iron deficiency anemia, vitamin A deficiency and iodine deficiency disorders, which are still public health problems in the country.
- Chronic non-communicable diseases (DCNT): diabetes, obesity, hypertension, cardiovascular disease and some cancers.

Next, it is highlighted some considerations contained in the Food Guide for Brazilian People [9] and some aspects of Guideline 1 - Healthy foods and meals, and the Special Guideline 1 - Physical activity.

In the last two or three generations, Brazilian society has become predominantly urban. Chronic diseases of the total population in the country increased from 34.4% in 1979 to 48.3% in 2003. Services and public policies need to respond to these changes and the complexity of its manifestations in health [9].

14. The epidemiological transition in Brazil

With the urbanization of the population, the patterns of work and leisure shifted to lower energy consumption. On the other hand, the increased consumption of processed foods, high intake in fat, sugar and salt, associated with lower daily energy use due to reduced physical activity, explain the rising trends of overweight and obesity in the Brazilian population, and DCNTs associated [9].

The evolution of DCNT is an additional challenge to the food and nutrition security, which must be combined with efforts to reverse the prevalence of child malnutrition and the control and prevention of micronutrient deficiencies.

15. Healthy lifestyles

Recent scientific evidence shows that health may be more related to how people live than to their biological and genetic determination. The approach in promoting healthy lifestyles, it is identified two dimensions: one that aims to stimulate and encourage healthy behaviors, healthy eating and regular physical activity, and another, aims to inhibit habits and practices harmful to health as consumption of tobacco and alcohol. Healthy eating begins with the practice of breastfeeding, and extends to life by adopting good eating habits. Thus, the suitable nutrition of pregnant women and children must be understood and emphasized as a strategic action part, in order to promote health in adulthood.

16. Healthy eating: some considerations

In general, the food choices are not determined for such preference and habits, but rather for the system of production and food supply [9]. The State, through its public policies, has the responsibility to foster socio-environmental changes at the collective level, to promote healthy choices at individuals or families. Thus, the assumption is promoting healthy eating to expand and foster decision-making autonomy of individuals and groups, through access to information to the choice and adoption of healthy feeding (and life) practices [9].

An alternative action to promote healthy eating should be, for example, the shift in the consumption of unhealthy foods for healthier foods. Overvalue or mythologize certain foods because of their nutritional or functional characteristics should not constitute the practice of promoting healthy eating. Nutritionally rich foods should be valued and will come naturally in the adopted diet, without needing to mythologize one or more of its characteristics, a trend much exploited by advertising and marketing of functional foods and nutritional supplements.

According to the principles of healthy eating, all food groups should make up the daily diet. Some guidelines of the Food Guide for the Brazilian population are highlighted here as guidelines for actions of the government and of the productive sector [9]:

- Guideline 1 – healthy foods and meals
    - Ensure food quality – *in nature* and processed – placed on the market for consumption of the population.
    - Ensure the enforcement of legislation that promotes breastfeeding as the child's right to suitable diet.
    - Regulate food *marketing* strategies across all forms of media, particularly those directed towards children and adolescents.

    - In relation to bioactive compounds present in vegetables, fruits and herbs native to Brazil, the guide points out that on the basis of updated knowledge, the orientation remains the same: "a meal rich in fruits, vegetables, natural sources of vitamins and minerals and bioactive compounds, is fundamental to the maintenance of health".
    - The labels carry wealth essential information to the consumer. An example of a functional property claim that may appear on food labels, since previously evaluated and approved by ANVISA is "Dietary fiber helps the intestinal functioning. Its consumption should be associated with a balanced diet and healthy lifestyle". It is clear that the industry will give greater emphasis to the positive features of its product. Thus, it is important to analyze more than one piece of information. For example, a product with high fiber content - which is a positive feature - may, however, have a high fat, sugar or sodium content. Another product with a high calcium content may have a high concentration of saturated fat. Increasingly, it is important that the consumer has access to information, strengthening the capacity of analysis and decision to choose one product or another, against the indiscriminate amount of information available in various media outlets and advertising.
- Special guideline 1 – physical activity
    - The basic principle to maintain an energy balance is the balance between intake and energy expenditure. One of the guidelines on the recommended actions to government and business sector to develop disclosure forms and media to report and value the adoption of healthy lifestyles, combining the promotion of healthy eating and regular physical activity.

17. National policy for health promotion [13]

PNPS has as general objective "to promote quality of life and to reduce vulnerability and health risks related to its determinants and constraints - ways of living, working conditions, housing, environment, education, leisure, culture, access to essential goods and services".

The strategy for implementing the PNPS related to diet is stimulating the inclusion of health promotion actions at all levels of care, with emphasis on primary care, focused on the actions of body care and health, healthy eating and prevention and tobacco control.

18. Final considerations

The legislation aims to promote and protect the health of consumers through the registration and inspection of food products with claims of functional properties. The knowledge of the legislation and its updates is critical to the successful development and registration of products. It is noted the convergence of the presented policies to ensure reliable information to consumers and to strengthen their capacity to understand this information, so that they can make healthier food choices. It also recognizes that a healthy diet, as a measure of health promotion, can not be dissociated from the adoption of healthy lifestyles, especially physical activity. Therefore, the messages of the claims have been developed and standardized by ANVISA, conditioning the benefit claimed by the consumption of food to the adoption of a balanced diet and healthy lifestyle. It was also established the observations and warning statements to be included in the labeling and advertising of the products. According to these policies, foods that support claims can not contribute to the increased incidence of overweight, obesity and other DCNT. Thus, it must not be approved claims for foods that may discourage breastfeeding, or containing high load of energy, high content of sugars, salt, saturated fats and trans fats. Moreover, the claims should have a solid scientific evidence.

Annex I

List of Claims Approved and Updated in July 2009

1. The horizontal claims, presented below, are part of a continuous and dynamic process of reassessment of approved claims based on scientific evidence, using the principles described in item III. In addition, it aimed the standardization of the claims in order to improve consumer understanding about the information and properties conveyed on labels of these foods.
2. In 2005, the claims, previously approved, were reevaluated, in order to standardize them to improve consumer understanding of the information and properties conveyed on food labels. With this reevaluation, the claims related to the caffeine, sorbitol, xylitol, mannitol, sodium stearate, sodium bicarbonate, omega 6 fatty acids, polyunsaturated and monounsaturated fatty acids (vegetable oils), and the liquid ready for consumption were no longer permitted.
3. Claims for chewing gum related to sorbitol, xylitol and mannitol were reassessed in 2007 based on new scientific evidence and it was approved the claim set out in the table below.
4. Foods that present in their label wording and/or in their advertising material, the claims listed below, must be registered in the category of "Foods with Claims of Functional and/or Health Property". Thus, they must be registered prior to marketing, as Annex II of the RDC nº 27/2010. The registration of food with claims and the evaluation of new claims will be made by scientific evidence of effectiveness of these, given the criteria present in Resolution nº 18/99 and 19/99.
5. The approved claims that relate the functional and/or health property to a nutrient or non-nutrient food, according to item 3.3 of Resolution 18/99. However, the effectiveness of the food claim must be evaluated case by case, considering that variations may occur in the action of the nutrient or non-nutrients, according to the matrix or formulation of product.
6. The portions of food shall be those provided in Resolution RDC nº 359/03 calculated based on food groups referred in that resolution. For further information consult the Guidance Manual for Industries.

7. In the case of combination of nutrients or non-nutrients in one product, the effectiveness of the claim must be proven in the product, with the concomitant use of non-nutrient or nutrients.
8. In the case of foods regulated by the Ministry of Agriculture, Livestock and Supply (MAPA), the companies should initially fill the 403 petition in ANVISA, referring to the request for the Evaluation of Foods with Claims of Functional and/or Health Properties. ANVISA shall send response assessment for the company, with a copy to the authority of the MAPA area.
9. The claims approved relate the functional and/or health properties of a nutrient or non-nutrient of the food, according to item 3.3 of Resolution nº 18/99. However, the proof of the effectiveness of claim must be performed in each case, considering the formulation and features of the food. Therefore, the use of the claims listed below, in any food, will only be allowed after approval by ANVISA.

Anexx II - List of approved claims

OMEGA-3
Claim
"The consumption of Omega-3 fatty acids helps in maintaining healthy levels of triglycerides, since combined to a balanced diet and healthy lifestyle habits".
Specific requirements
This claim should only be used for Omega-3 long chain fatty acids from fish oils (EPA - eicosapentaenoic acid and DHA - docosahexaenoic acid). The product must have a minimum of 0.1g of EPA and/or DHA in the portion or in 100g or in 100mL of the product ready for consumption, once the portion is bigger than 100g or 100mL. The processes must present analysis report, employing recognized methodology, the content of inorganic contaminants in ppm: Mercury, Lead, Cadmium and Arsenic. Using as reference the Decree nº 55871/65, category of other foods. For products in the form of capsules, pills, tablets and other similar forms, the above requirements must be met in the daily intake recommendation of the product ready for consumption, according to the manufacturers. The nutrition facts table shall contain the three types of fats: saturated, monounsaturated and polyunsaturated fats, describing below the content of omega-3 (EPA and DHA). The product label must include the warning highlighted in bold: "People who have diseases or physiological changes, pregnant or breastfeeding (nursing mothers) should consult their doctor before using the product".

Table 2. FATTY ACIDS

LYCOPENE
Claim
"Lycopene has antioxidant action that protects cells against free radicals. Its consumption must be combined with a balanced diet and healthy lifestyle habits".
Specific requirements
The amount of lycopene contained in the portion of the product ready for consumption, must be declared on the label next to the claim. For products in the form of capsules, pills, tablets and other similar forms, it must be declared the lycopene amount in the daily intake recommendation of the product ready for consumption, according to the manufacturers. The detailed process of the substance obtaining and standardization must be present, including solvents and other compounds used. Present report of the content of residue(s) of solvent(s) used. Present report with the purity of the product.

LUTEIN
Claim
"Lutein has antioxidant action that protects cells against free radicals. Its consumption should be associated with a balanced diet and healthy lifestyle".
Specific requirements
The amount of lutein contained in the portion of the product ready for consumption, must be declared on the label next to the claim. For products in the form of capsules, pills, tablets and other similar forms, it must be declared the lutein amount in the daily intake recommendation of the product ready for consumption, according to the manufacturers. The detailed process of the substance obtaining and standardization must be present, including solvents and other compounds used. Present report of the content of residue(s) of solvent(s) used. Present report with the purity of the product.

ZEAXANTHIN
Claim
"Zeaxanthin has antioxidant properties that protects cells against free radicals. Its consumption should be associated with a balanced diet and healthy lifestyle".
Specific requirements
The amount of zeaxanthin, contained in the portion of the product ready for consumption, must be declared on the label next to the claim. For products in the form of capsules, pills, tablets and other similar forms, it must be declared the zeaxanthin amount in the daily intake recommendation of the product ready for consumption, according to the manufacturers. The detailed process of the substance obtaining and standardization must be present,

including solvents and other compounds used. Present report of the content of residue(s) of solvent(s) used. Present report with the purity of the product.

Table 3. CAROTENOIDS

DIETARY FIBERS
Claim
"The dietary fibers assist in the functioning of the intestine. Its consumption should be associated with a balanced diet and healthy lifestyle".
Specific requirements
This claim may be used provided that the portion of the product ready for consumption provide at least 3g of fiber if the food is solid or 1.5g fiber if the food is liquid. In the nutrition facts table shall be declared the amount of dietary fiber. For products in the form of capsules, tablets, pills and other similar forms, the above requirements must be met in the daily intake recommendation of the product ready for consumption, according to the manufacturers. When presented single in capsules, tablets, pills, powders and other similar forms, the following information, in highlighted bold type, shall appear on the product label: "Consumption of this product should be accompanied by fluid intake".

BETA-GLUCAN
Claim
"The beta-glucan (dietary fiber) assists in reducing cholesterol absorption. Its consumption should be associated with a balanced diet and healthy lifestyle".
Specific requirements
This claim may be used provided that the portion of the product ready for consumption provide at least 3g of beta-glucan if the food is solid or 1.5g beta-glucan if the food is liquid. This claim is approved for beta-glucan present in oats. In the nutrition facts table it shall be declared the amount of **beta glucan** below the dietary fiber amount. When presented single in capsules, tablets, pills, powders and other similar forms, the following information, in highlighted bold type, shall appear on the product label: "Consumption of this product should be accompanied by fluid intake".
RESISTANT DEXTRIN
Claim
"The dietary fibers assist in the functioning of the intestine. Its consumption should be associated with a balanced diet and healthy lifestyle".

Specific requirements
This claim may be used provided that the portion of the product ready for consumption provide at least 3g of resistant dextrin if the food is solid or 1.5g of resistant dextrin if the food is liquid. For products in the form of capsules, tablets, pills and other similar forms, the above requirements must be met in the daily intake recommendation of the product ready for consumption, according to the manufacturers. The use of the ingredient shall not exceed the 30g recommended daily intake of the product ready for consumption, as indicated by the manufacturer In the nutrition facts table it shall be declared the amount of resistant dextrin below the dietary fiber amount. When presented single in capsules, tablets, pills, powders and other similar forms, the following information, in highlighted bold type, shall appear on the product label: "Consumption of this product should be accompanied by fluid intake".

FRUCTOOLIGOSACCHARIDES – FOS
Claim
"The fructooligosaccharides - FOS contribute to the balance of intestinal flora. Its consumption should be associated with a balanced diet and healthy lifestyle ".
Specific requirements
This claim may be used provided that the portion of the product ready for consumption provide at least 3g of FOS if the food is solid or 1.5g of FOS if the food is liquid. For products in the form of capsules, tablets, pills and other similar forms, the above requirements must be met in the daily intake recommendation of the product ready for consumption, according to the manufacturers. In the nutrition facts table it shall be declared the amount of **fructooligosaccharides** below the dietary fiber amount. The consumption of the ingredient shall not exceed the 30g recommended daily intake of the product ready for consumption, as indicated by the manufacturer. When presented single in capsules, tablets, pills, powders and other similar forms, the following information, in highlighted bold type, shall appear on the product label: "Consumption of this product should be accompanied by fluid intake".

PARTIALLY HYDROLYZED GUAR GUM
Claim
"The dietary fibers assist in the functioning of the intestine. Its consumption should be associated with a balanced diet and healthy lifestyle".
Specific requirements
This claim may be used provided that the portion of the product ready for consumption provide at least 3g of guar gum if the food is solid or 1.5g of guar gum if the food is liquid. For products in the form of capsules, tablets, pills and other similar forms, the above

requirements must be met in the daily intake recommendation of the product ready for consumption, according to the manufacturers. This claim is approved for the partially hydrolyzed guar gum obtained from the plant species. In the nutrition facts table it shall be declared the amount of partially hydrolyzed guar gum below the dietary fiber amount. If the product is in isolated form, in sachets or powder, for example, the company must inform in the label, the amount of liquid in which the product must be dissolved. When presented single in capsules, tablets, pills, powders and other similar forms, the following information, in highlighted bold type, shall appear on the product label: "Consumption of this product should be accompanied by fluid intake".

INULIN
Claim
"Inulin contributes to the balance of intestinal flora. Its consumption should be associated with a balanced diet and healthy lifestyle".
Specific requirements
This claim may be used provided that the portion of the product ready for consumption provide at least 3g of inulin if the food is solid or 1.5g inulin if the food is liquid. For products in the form of capsules, tablets, pills and other similar forms, the above requirements must be met in the daily intake recommendation of the product ready for consumption, according to the manufacturers. In the nutrition facts table it shall be declared the amount of **inulin** below the dietary fiber amount. The consumption of the ingredient shall not exceed the 30g recommended daily intake of the product ready for consumption, as indicated by the manufacturer When presented single in capsules, tablets, pills, powders and other similar forms, the following information, in highlighted bold type, shall appear on the product label: "Consumption of this product should be accompanied by fluid intake".

LACTULOSE
Claim
"Lactulose assists in the functioning of the intestine. Its consumption should be associated with a balanced diet and healthy lifestyle".
Specific requirements
This claim may be used provided that the portion of the product ready for consumption provide at least 3g of lactulose if the food is solid or 1.5g of lactulose if the food is liquid. For products in the form of capsules, tablets, pills and other similar forms, the above requirements must be met in the daily intake recommendation of the product ready for consumption, according to the manufacturers. In the nutrition facts table it shall be declared the amount of **lactulose** below the dietary fiber amount.

When presented single in capsules, tablets, pills, powders and other similar forms, the following information, in highlighted bold type, shall appear on the product label: "Consumption of this product should be accompanied by fluid intake".

POLYDEXTROSE
Claim
"Polydextrose assists in the functioning of the intestine. Its consumption should be associated with a balanced diet and healthy lifestyle".
Specific requirements
This claim may be used provided that the portion of the product ready for consumption provide at least 3g of Polydextrose if the food is solid or 1.5g of Polydextrose if the food is liquid. For products in the form of capsules, tablets, pills and other similar forms, the above requirements must be met in the daily intake recommendation of the product ready for consumption, according to the manufacturers. In the nutrition facts table it shall be declared the amount of **polydextrose** below the dietary fiber amount. When presented single in capsules, tablets, pills, powders and other similar forms, the following information, in highlighted bold type, shall appear on the product label: "Consumption of this product should be accompanied by fluid intake".

PSILLIUM OR PSYLLIUM
Claim
"*psillium* (dietary fiber) assist in the reduction of fat absorption. Its consumption should be associated with a balanced diet and healthy lifestyle".
Specific requirements
This claim may be used provided that the portion of the product ready for consumption provide at least 3g of *psillium* if the food is solid or 1.5 g if the food is liquid. For products in the form of capsules, tablets, pills and other similar forms, the above requirements must be met in the daily intake recommendation of the product ready for consumption, according to the manufacturers. The only species to be evaluated was *Plantago ovata*. Any other species should be evaluated for safety in use. In the nutrition facts table it shall be declared the amount of *Psillium* below the dietary fiber amount. When presented single in capsules, tablets, pills, powders and other similar forms, the following information, in highlighted bold type, shall appear on the product label: "Consumption of this product should be accompanied by fluid intake".

CHITOSAN
Claim
"*Chitosan* assists in the reduction of fat and cholesterol absorption. Its consumption should be associated with a balanced diet and healthy lifestyle".
Specific requirements
This claim may be used provided that the portion of the product ready for consumption provide at least 3g of chitosan if the food is solid or 1.5g of if the food is liquid. For products in the form of capsules, tablets, pills and other similar forms, the above requirements must be met in the daily intake recommendation of the product ready for consumption, according to the manufacturers. The processes must present analysis report, using recognized methodology, presenting the content of inorganic contaminants in ppm: Mercury, Lead, Cadmium and Arsenic. Use as a reference the Decree 55871/65, category of other foods. It must be submitted analysis report with physical chemical composition, including the fiber and ash amount. In the nutrition facts table it shall be declared the amount of *chitosan* below dietary fiber amount. In the label it must contain the warning phrase highlighted and in bold: "People allergic to fish and shellfish should avoid consumption of this product ". When presented single in capsules, tablets, pills, powders and other similar forms, the following information, in highlighted bold type, shall appear on the product label: "Consumption of this product should be accompanied by fluid intake".

Table 4. DIETARY FIBERS

PHYTOSTEROLS
Claim
"Phytosterols assist in reducing cholesterol absorption. Their consumption should be associated with a balanced diet and healthy lifestyle".
Specific requirements
The portion of the product ready for consumption should provide at least 0.8g of free phytosterols. Lower amounts may be used provided that proved in the food matrix. The daily intake recommendation of the product, which should be between 1-3 portions/day must ensure the ingestion of 1-3 grams of free phytosterols per day. In the description of the product information should be included "... with phytosterols". The amount of phytosterols contained in the portion of the product ready for consumption, must be declared on the label next to the claim. Phytosterols refer to sterols both as stanols, as well as to esterified. It must present the detailed process of obtaining and standardization of substance, including solvents and other compounds used. Present report with the content(s) of residue(s) of employed solvent(s).

Present report with the product purity and the characterization of the phytosterols/phytostanols present. The label must contain the following warning phrases highlighted in bold: **"People with high cholesterol level should seek for medical advice."** "Phytosterols do not provide additional benefits when consumed over 3 g/day." "The product is not suitable for children under five, pregnant women and milk feeding babies."

Table 5. PHYTOSTEROLS

Mannitol/Xylitol/Sorbitol
Claim
"Mannitol/Xylitol/Sorbitol does not produce acids that damage teeth. The consumption of the product does not replace proper oral hygiene habits and feeding"
Specific requirements
Claims approved only for sugar-free chewing gum.

Table 6. POLYOLS

Lactobacillus acidophilus *Lactobacillus casei shirota* *Lactobacillus casei variedade rhamnosus* *Lactobacillus casei variedade defensis* *Lactobacillus paracasei* *Lactococcus lactis* *Bifidobacterium bifidum* *Bifidobacterium animallis* (including *B. lactis subspecies)* *Bifidobacterium longum* *Enterococcus faecium*
Claim
"The (indicates the microorganism species) (probiotic) contributes to the balance of intestinal flora. Its consumption should be associated with a balanced diet and healthy lifestyle".
Specific requirements
The minimum viable amount of probiotics must be in the range of $1x10^8$ to $1x10^9$ colony forming units (CFU) in the daily recommendation of the product ready for consumption, as indicated by the manufacturer. Smaller values may be accepted, since the company proves its effectiveness. The documentation relating to proof of efficacy should include: Report of the product analysis that prove the amount of the minimum viable microorganism until the end of shelf life.

Resistance test of the culture employed in product against the gastric acidity and bile salts. The amount of probiotics in CFU, contained in the daily recommendation intake in the product ready for consumption, must be declared on the label next to the claim. The microorganisms Lactobacillus delbrueckii (bulgaricus subspecies) and Streptococcus salivarius (thermophillus subspecies) were removed from the list, given that in addition of being species necessary for yoghurt production, they do not have probiotic scientifically proven effect.

Table 7. PROBIOTICS

SOY PROTEIN
Claim
"Daily consumption of at least 25 of soy protein can help to lower cholesterol. Its consumption should be associated with a balanced diet and healthy lifestyle ".
Specific requirements
The amount of soy protein contained in the portion of the product ready for consumption, must be declared on the label next to the claim. In the case of products in the form of capsules, tablets, pills and the similar forms, it should be declared the amount of soy protein in the daily intake recommendation of the product Ready for consumption, as indicated by the manufacturer. **"The wording on the labeling and on the advertising of products based on soy can not present any claim on the basis of isoflavones, regarding the amount ("contains"), functional, health and therapeutic properties (prevention, treatment and cure of diseases)"**

Table 8. SOY PROTEIN

Author details

Paulo César Stringueta
Department of Food Science and Technology – Federal University of Viçosa, Brazil

Maria da Penha Henriques do Amaral and Miriam Aparecida de Oliveira Pinto
Department of Pharmaceutical Sciences – Federal University of Juiz de Fora, Brazil

Larissa Pereira Brumano
Federal University of Juiz de Fora, Brazil

Mônica Cecília Santana Pereira
Sanitary Surveillance Department of the Municipality of Juiz de Fora, Brazil

#. References

[1] ______. Functional Foods: their role in disease prevention and health promotion. Food Technology 1998; 52(11):63-70.

[2] Hasler CM. Functional Foods: benefits, concerns and challenges – A Position Paper from the American Council on Science and Health. The Journal of Nutrition 2002; 132(12):3772-81.

[3] Bianco AL. A Construção das Alegações de Saúde para Alimentos Funcionais. 1.ed. Brasília: Embrapa Informação Tecnológica, 2008. p. 113.

[4] Brasil. Ministério da Saúde. Agência Nacional de Vigilância Sanitária. Informe Técnico nº 9, de 21 de maio de 2004, Orientação para utilização, em rótulos de alimentos, de alegação de propriedades funcionais de nutrientes com funções plenamente reconhecidas pela comunidade científica (item 3.3 da Resolução ANVISA/MS nºs18/99). Disponível em: http://www.anvisa.gov.br/alimentos/informes/09_210504.htm. Acesso em: 6 abr. 2005. 2004b.

[5] Brasil. Ministério da Saúde. Agência Nacional de Vigilância Sanitária. Portaria nº 398, de 30 de abril de 1999. Regulamento Técnico que Estabelece as Diretrizes Básicas para Análise e Comprovação de Propriedades Funcionais e ou de Saúde Alegadas em Rotulagem de Alimentos. Diário Oficial da República Federativa do Brasil. Brasília, 03 maio. 1999. Disponível em: http://e-legis.bvs.br/leisref/public/showAct.php?id=11297&mode=PRINT_ VERSION. Acesso em: 24 out.2005. 1999b.

[6] Brasil. Ministério da Saúde. Agência Nacional de Vigilância Sanitária. Resolução nº 18, de 30 de abril de 1999. Aprova o Regulamento Técnico que Estabelece as Diretrizes Básicas para Análise e Comprovação de Propriedades Funcionais e ou de Saúde Alegadas em Rotulagem de Alimentos. Diário Oficial da República Federativa do Brasil. Brasília, 03 nov. 1999. Disponível em: http://e-legis.bvs.br/leisref/public/showAct.php?id=109. Acesso em: 6 abr. 2005. 1999e.

[7] Lajolo FM. Functional foods: latin american perspectives. British Journal of Nutrition, 2002; 88(Suppl 2):S145-S50.

[8] International Food Information Council Foundation. IFIC. Alimentos Funcionales. 2004. Disponível em: http://www.ific.org/sp/nutrition/functional/index.cfm?render forprint=1. Acesso em: 21 out. 2005.

[9] Brasil. Ministério da Saúde. Secretaria de Atenção à Saúde. Coordenação-Geral da Política de Alimentação. Guia alimentar para a população brasileira. Brasília: Ministério da Saúde. Série A. Normas e Manuais Técnicos. 2008. Disponível em: http://bvsms.saude.gov.br/bvs/publicacoes/guia_alimentar_populacao_brasileira.pdf. Acesso em: 03 maio. 2012.

[10] Brasil. Ministério da Saúde. Agência Nacional de Vigilância Sanitária. Resolução nº 19, de 30 de abril de 1999. Aprova o Regulamento Técnico de Procedimentos para Registro de Alimento com Alegação de Propriedades Funcionais e ou de Saúde em sua Rotulagem. Diário Oficial da República Federativa do Brasil. Brasília, 10 dez. 1999. Disponível em:

http://e-legis.bvs.br/leisref/public/showAct.php?id=110. Acesso em: 30 ago. 2005. 1999f.

[11] Brasil. Decreto-Lei nº 986, de 12 de outubro de 1969. Institui normas básicas sobre alimentos. Diário Oficial da União (D.O.U.), Brasília, 21 de outubro de 1969. Diário Oficial da República Federativa do Brasil. 1969. Brasília, Disponível em: http://e-legis.bvs.br/leisref/ public/show Act.php. Acesso em: 6 abr. 2005.

[12] Brasil. Ministério da Saúde. Secretaria de Atenção à Saúde. Departamento de Atenção Básica. Política Nacional de Alimentação e Nutrição. Série B. Textos Básicos de Saúde. 1. ed. rev. Brasília: Ministério da Saúde, 2012. p. 82.

[13] Brasil. Ministério da Saúde. Política Nacional de Promoção da Saúde. Série B. Textos Básicos de Saúde. Série Pactos pela Saúde, 3.ed , Brasília: Ministério da Saude, 2010. p. 60. Disponível em: http://bvsms.saude.gov.br/bvs/publicacoes/politica_nacional_promocao_saude_3ed.pdf.A cesso em: 03 maio. 2012.

[14] World Health Organization (WHO). Global strategy on diet, physical activity and health: fifty-seventh World Health Assembly Wha 57.17. 22 May 2004. Disponível em: <www.who.int>. Acesso em 10 jul. 2005.

[15] Santos CF. Panorama Brasil – fluxo de aprovação das alegações. In: I Congresso Internacional sobre Alimentos Funcionais – Ciência, Inovação e Regulamentação. FIESP. 24 de abril. São Paulo. 2006. Disponível em: http://www.fiesp.com.br/alimentosfuncionais. Acesso em: 05 maio. 2006.

[16] Brasil. Ministério da Saúde. Agência Nacional de Vigilância Sanitária. Gestão 2005-2010 Principais Realizações, de 30 de março de 2011. Disponível em: http://portal.anvisa.gov.br/wps/wcm/connect/7122b3804745869c8feadf3fbc4c6735/Relat orio_5_anos_DEFINITIVO.pdf?MOD=AJPERES. Acesso em: 02 maio. 2012.

[17] Brasil. Decreto nº 3571, de 21 de agosto de 2000. Dá nova redação a dispositivos do Regulamento da Agência Nacional de Vigilância Sanitária, aprovado pelo Decreto n° 3029, de 16 de abril de 1999. Diário Oficial da República Federativa do Brasil. Brasília, 22 ago. 2000. Disponível em: http://e-legis.bvs.br/leisref/public/showAct.php?id+658&word+decreto. Acesso em: 10 nov. 2005.

[18] Brasil. Ministério da Saúde. Agência Nacional de Vigilância Sanitária. Alimentos. Comissões e Grupos de Trabalho. Comissão Tecnocientífica de Assessoramento em Alimentos Funcionais e Novos Alimentos. Alimentos com alegações de propriedades funcionais e ou de saúde, novos alimentos/ingredientes, substâncias bioativas e probióticos. Atualizado em 11 de janeiro de 2005. VIII – Lista das Alegações Aprovadas. Disponível em: http:www.anvisa.gov.br/ alimentos/comissões/tecno.htm. Acesso em: 1 set. 2005a.

[19] Brasil. Ministério da Saúde. Agência Nacional de Vigilância Sanitária. Resolução nº 16, de 30 de abril de 1999. Aprova o Regulamento Técnico de Procedimentos para Registro de Alimentos e ou Novos Ingredientes. Diário Oficial da República Federativa do Brasil. Brasília, 03 dez. 1999. Disponível em:

http:// e-legis.bvs.br/leisref/public/ showAct.php?id=107. Acesso em: 6 abr. 2005. 1999c.

[20] Brasil. Ministério da Saúde. Agência Nacional de Vigilância Sanitária. Resolução nº 17, de 30 de abril de 1999. Aprova o Regulamento Técnico que Estabelece as Diretrizes Básicas para Avaliação de Risco e Segurança dos Alimentos. Diário Oficial da República Federativa do Brasil. Brasília, 03 dez. 1999. Disponível em: http://e-legis.bvs.br/leisref/ public/showAct.php?id=108. Acesso em: 6 abr. 2005. 1999d.

[21] Brasil. Ministério da Saúde. Agência Nacional de Vigilância Sanitária. Resolução nº 22, de 15 de março de 2000. Dispõe sobre os Procedimentos Básicos de Registro e Dispensa de Registro de Produtos Importados Pertinentes à Área de Alimentos. Diário Oficial da República Federativa do Brasil. Brasília, 16 mar. 2000. Disponível em: http://e-legis.bvs.br/leisref/public/showAct.php?id=136. Acesso em: 28 set. 2005. 2000a.

[22] Brasil. Ministério da Saúde. Agência Nacional de Vigilância Sanitária. Resolução nº 23, de 15 de março de 2000. Dispõe sobre o Manual de Procedimentos Básicos para Registro e Dispensa da Obrigatoriedade de Registro de Produtos Pertinentes à Área de Alimentos. Diário Oficial da República Federativa do Brasil. Brasília, 07 fev. 2003. Disponível em: http://www.engetecno.com.br/resolucao_23_15_03_00.htm. Acesso em: 28 nov. 2005. 2000b.

[23] Brasil. Ministério da Saúde. Agência Nacional de Vigilância Sanitária. Resolução RDC nº 2, de 7 de janeiro de 2002. Aprova o Regulamento Técnico de Substâncias Bioativas e Probióticos Isolados com Alegação de Propriedades Funcionais e ou de Saúde. Diário Oficial da República Federativa do Brasil. Brasília, 9 jan. 2002. Disponível em: http://e-legis.bvs.br/leisref/public/showAct.php?id=9059 &Word=. Acesso em: 30 ago. 2005. 2002.

[24] Brasil. Ministério da Saúde. Agência Nacional de Vigilância Sanitária. Informe Técnico nº 19, de 29 de agosto de 2006. Assunto: Procedimentos para o enquadramento dos Cogumelos comestíveis em cápsulas, comprimidos e tabletes na área de alimentos. Disponível em: http://www.anvisa.gov.br/alimentos/informes/19_300806.htm. Acesso em: 5 set. 2006.

[25] Brasil. Ministério da Saúde. Agência Nacional de Vigilância Sanitária. Resolução RDC nº 27, de 06 de agosto de 2010. Dispõe sobre as categorias de alimentos e embalagens isentos e com obrigatoriedade de registro sanitário. Diário Oficial da República Federativa do Brasil. Brasília, 09 ago. 2010. Disponível em: http://bvsms.saude.gov.br/bvs/saudelegis/anvisa/2010/res0027_06_08_2010.html. Acesso em: 01 maio. 2012.

[26] Santos FF, Almeida MA, Brumano LP, Pereira MCS, Stringheta PC, Pinto MAO. Elaboração do Relatório Técnico-Científico para registro de leites fermentados com alegação de propriedade funcional. Revista do Instituto de Laticínios Cândido Tostes 2011; 383(66):40-8.

[27] Brasil. Ministério da Saúde. Secretaria de Vigilância Sanitária. Portaria nº 27, de 13 de janeiro de 1998. Aprova o Regulamento Técnico referente à Informação Nutricional Complementar (declarações relacionadas ao conteúdo de nutrientes), constantes do anexo desta Portaria. Diário Oficial da República Federativa do Brasil. Brasília, 16 jan. 1998. Disponível em: http://e-legis.bvs.br/leisref/public/showAct.php?id+97&word=. Acesso em: 5 nov. 2005. 1998.

[28] Brasil. Ministério da Saúde. Agência Nacional de Vigilância Sanitária. Resolução RDC nº 359, de 23 de dezembro de 2003. Aprova Regulamento Técnico de Porções de Alimentos Embalados para Fins de Rotulagem Nutricional. Diário Oficial da República Federativa do Brasil. Brasília, 26 dez. 2003. Disponível em: http://e-legis.anvisa.gov.br/leisref/public/ showAct.php?id=9058&word=. Acesso em: 5 set. 2006. 2003.

[29] Brasil. Ministério da Saúde. Agência Nacional de Vigilância Sanitária. Alimentos. Comissões e Grupos de Trabalho. Comissão Tecnocientífica de Assessoramento em Alimentos Funcionais e Novos Alimentos. Alimentos com Alegações de Propriedades Funcionais e ou de Saúde, Novos Alimentos/Ingredientes, Substâncias Bioativas e Probióticos. VIII – Lista das Alegações Aprovadas. Disponível em: http://www.anvisa.gov.br/alimentos/comissoes/tecno.htm. Acesso em: 21 jan. 2012.

[30] Federação das Indústrias do Estado de São Paulo. Introdução: alimentos funcionais. In: I Congresso Internacional sobre Alimentos Funcionais – Ciência, Inovação e Regulamentação. FIESP. 24 de abril. São Paulo. 2006. Disponível em: http://www.fiesp.com.br/alimentosfuncionais. Acesso em: 05 maio. 2006.

[31] Brasil. Ministério da Saúde. Agência Nacional de Vigilância Sanitária. Gerência de Monitoração e Fiscalização da Propaganda, de Publicidade, de Promoção e de Informação de Produtos Sujeitos à Vigilância Sanitária. GPROP. Guia do Projeto de Monitoração de Propaganda e Publicidade de Produtos Sujeitos à Vigilância Sanitária. Versão Dez. 2004 Brasília: 2004, 110 p.

[32] Pinto MAO. Aspectos legais e análise de conteúdo de propagandas impressas de alimentos com alegações de propriedades funcionais [tese de doutorado]. Viçosa: Departamento de Tecnologia de Alimentos da UFV; 2008.

[33] World Health Organization (WHO). Diet, nutrition and the prevention of chronic diseases. Report of a joint WHO/FAO expert consultation. (WHO Technical Report Series, 916).Geneva. 2003. p. 149. 2003. Disponível em: http://whqlibdoc.who.int/trs/who_TRS_916.pdf. Acesso em: 31 maio. 2006.

[34] Stringheta PA, Vilela MAP, Oliveira TT, Nagem TJ. Alimentos "funcionais" Conceitos, contextualização e regulamentação. 1. ed. Juiz de Fora:Templo Gráfica e Editora, 2007. p. 246.

[35] Stringheta PA, Oliveira TT, Gomes RC, Amaral MPH, Carvalho AF, Vilela MAP. Politicas de saúde e alegações de propriedades funcionais e de saúde para alimentos no Brasil. Revista Brasileira de Ciências Farmacêuticas 2007;43(2):181-94.

[36] Oliveira TP. A aplicação da Política Nacional de Alimentação e Nutrição. In: I Congresso Internacional Sobre Alimentos Funcionais – Ciência, Inovação e Regulamentação. Apresentação Power Point. São Paulo. abril. 2006. Disponível em: http://www.fiesp.com.br/alimentos funcionais. Acesso em: 05 maio. 2006.

Section 4

Molecular Basis of Physiological Responses

Chapter 13

Nucleic Acid-Based Methods to Identify, Detect and Type Pathogenic Bacteria Occurring in Milk and Dairy Products

Vincenzina Fusco and Grazia Marina Quero

Additional information is available at the end of the chapter

1. Introduction

Foodborne illnesses caused by pathogenic microorganisms, including bacteria, viruses and parasites are among the most serious public health concerns worldwide. A recent document of the Center for Disease Control and Prevention (CDC) estimates that each year 1 out of 6 American (or 48 million people) get sick, 128,000 are hospitalized and 3,000 die due to foodborne diseases, with *Norovirus*, nontyphoidal *Salmonella*, *Clostridium perfringens*, *Campylobacter* spp. and *Staphylococcus* (*S.*) *aureus* being the top five pathogens causing domestically acquired foodborne illnesses [1]. In the European Union, during 2009, 5,550 food-borne outbreaks occurred, mainly due to *Salmonella*, viruses and bacterial toxins, causing 48,964 human cases, 4,356 hospitalisations and 46 deaths [2]. Bacteria such as *Salmonella* spp., *Campylobacter* spp., *Listeria* (*L.*) *monocytogenes*, *Escherichia* (*E.*) *coli* O157:H7 and (*S.) aureus* have generally been identified as etiologic agents of most food-borne illnesses, with milk and its derivatives products among the most frequently involved food matrices. Moreover, although "less hazardous", these pathogens are a constant threat to the agro-food security, since they can be used to contaminate the environment, crops and animals, causing heavy damage to public health, agriculture and environment [3].

Traditionally, cultivation methods, ranging from plate counting to biochemical characterization, have been used to monitor pathogenic microorganisms in foods. However, these methodologies are labour-intensive and time-consuming, requiring from days to weeks to get results, with the consequence that products are often released for sale before the microbiological results become available. Moreover, these traditional methods as well as their advanced (such as cell wall composition analysis, whole-cell protein fingerprinting and fatty acid analysis) and automated (miniaturised kits or devices) applications often lead to uncertain identification or even misidentification, especially in cases of phenotypically closely related species. Failure to detect pathogens can have adverse health effects as well as substantial economic losses and fatalities. New approaches based on the application of molecular methods have being developed in the last years, bringing new insights in the detection of pathogenic bacteria in milk and milk-based products. In this chapter, we will endeavour to touch upon several nucleic acid based methods (such as PCR and its derivatives, real time PCR, REA-PFGE, fAFLP, etc.) and their application in milk and dairy products.

2. Nucleic acid-based detection and identification of milk- and dairy-borne pathogens

Detection and identification methods to detect milk- and dairy- pathogens may be traced back to at least two basic techniques: direct hybridization and *in vitro* amplification. In the following paragraphs, due to their importance in the microbial safety of milk and its derivatives, and since most of the advanced molecular methods derive from these fundamental techniques, we will provide the basics of nucleic acid hybridization and polymerase chain reaction (PCR) as well as an excursus of the most used nucleic acid-based techniques to identify, quantitatively detect and type pathogenic microorganisms occurring in milk and dairy products.

2.1. Nucleic acid hybridization

2.1.1. Basics

DNA hybridization is mainly based on an intrinsic feature of the DNA molecule, such as the high specificity of base pairing (Figure 1) between homologous strands of single-stranded DNA. The deoxyribonucleic acid (DNA) structure consists in a double helix conformation of two polynucleotide strands held together by hydrogen bonds. DNA is composed of four repeating nucleotides: Adenine, Thymine, Cytosine, and Guanine (Figure 1). Each base is linked to a deoxyribose molecule, which is attached to a phosphate moiety. The various nucleotides are linked together via the 5′ carbon of the deoxyribose molecule and the phosphate group attached to the 3′ carbon (Figure 1). Each nucleotide base in the DNA strand will cross-link (via hydrogen bonds) with a nucleotide base in a second strand of DNA forming a structure that resembles a ladder (Figure 1). These bases cross-link in a very specific order: Adenine will only link with Thymine (and vice-versa), and Cytosine will only link with Guanine (and vice-versa) (Figure 1). Two single strands of DNA will bond together only if their base-pairs match up properly or complement one another [4].

The double stranded DNA may be broken by heat or high pH. The reannealing between single stranded DNAs from different sources is called hybridization (Figure 2).

Standard nucleic acid hybridization assays require the use of a labelled nucleic acid probe (a denaturated DNA fragment varying in size from ten basepairs to kilobasepairs) to identify the target homologous DNA or RNA molecules within a complex mixture of unlabeled nucleic acid molecules, with the stability of the hybrid depending on the extent of base pairing that occurs [5]. Experimentally, the probe is usually labelled and the denatured target nucleic acid (DNA or RNA) is immobilized on a membrane or a polymer support or, if size information of the hybridization target is required, the target DNA is first run through agarose gel electrophoresis and then transferred to a membrane. The labelled probe is then added in a solution allowing the hybridization. After a suitable incubation, the membrane is washed in order to remove any non-specifically bound probe, leaving only the probe base-paired with the target DNA. By controlling the stringency of the washing conditions, DNA sequences 100% complementary to the probe or with lower degrees of similarity (i.e. with some mismatching) might be detected. In particular, having a sequence complementary to that of the target DNA, the probe might bind specifically to the target (previously denatured) and form with it a duplex DNA hybrid, recognizable by the labelled probe. The intensity of the spot is proportional to the amount of hybridized probe and therefore is proportional to the amount of target DNA in the sample. The intensity of the spot can be compared visually with the intensity of spots that correspond to a standard curve yielding semi-quantitative results (i.e., visual quantification), or the intensity can be determined using an instrument (e.g., densitometer) to create a quantitative value comparable with values obtained from the standard curve [6]. To overcome biohazards associated with the use of probes labelled with radioactive isotopes (usually P^{32} and S^{35}), biotin, digoxigenin and different fluorochromes have been used for labelling [7,8,9,10].

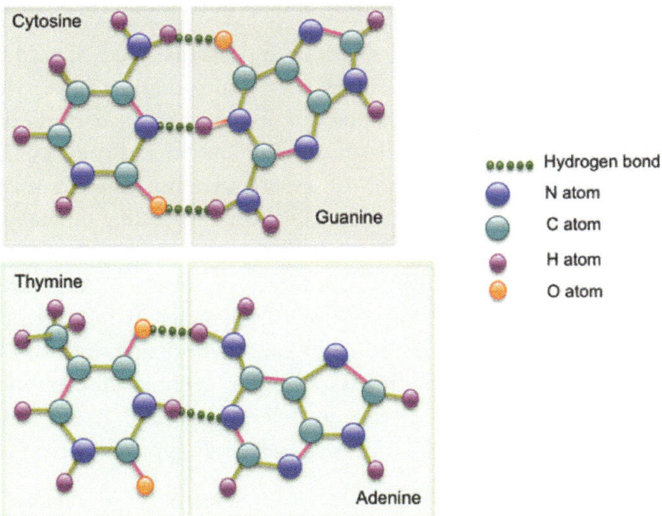

Figure 1. DNA base-pairing.

2.1.2. Application

Several technologies based on the nucleic acid hybridization, such as dot-blot [11], Southern- [12] and Northern-blot [13], colony hybridization [14], colorimetric DNA hybridization [15] etc. have been developed and successfully applied to the pathogen detection [16-41] in milk and dairy products.

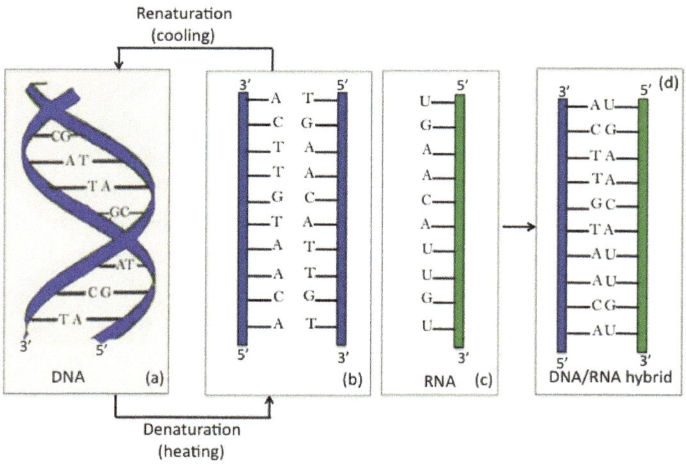

Figure 2. Nucleic acid hybridization.

However, a major drawback of the hybridization assays is their lack of sensitivity, which limits the use of these analyses to populations of cells or genes occurring in relatively high numbers in samples. For this reason, hybridization assays are currently mainly used for culture confirmation rather than direct detection and identification.

Among all the hybridization assays to date available, a particular focus should be given to the fluorescence *in situ* hybridization (FISH). FISH uses fluorescently labeled ribosomal RNA (rRNA) targeted probes and fluorescent microscopy to detect intact bacteria directly in food and clinical specimens, such as blood and tissue, or after enrichment culture [42,43]. Since this technique can visualize the precise location of a particular nucleic acid in the cytoplasm, organelli or nuclei of biological materials, it allows detecting metabollically active microorganisms directly in the environment without cultivation also providing useful information on the spatial distribution of the target organism in the colonised matrix. Experimentally, the procedure consists in preparing the samples, fixing it, preparing a smear or section on a microscope slide, permeabilizing the cells, hybridizing the probe to the DNA or RNA target in the sample and detecting the hybridization event by fluorescence microscopy. FISH assays have been developed and used to detect at family, genus and species level *Staphylococcus* spp., *Listeria* spp., *Campylobacter* spp., *Salmonella* spp. and *E. coli* [44,45,46,47,48].

As the other hybridization assays, FISH suffers from sensitivity. Moreover, FISH may be hindered by microorganism and substrate inherent autofluorescence, insufficient permeability of cell walls, non-specific binding of probes and low ribosome contents.

2.2. Polymerase chain reaction

2.2.1. Basics

The polymerase chain reaction (PCR) succeeded in revolutionizing the analysis of nucleic acids, so much that a Nobel Prize was conferred to Kary Mullis [49]. It is an *in vitro* three-step amplification process first introduced by Saiki and co-workers [50]. In PCR reaction (Figure 3), mixtures of oligonucleotides (primers), properly designed to be complementary to the flanking regions of the target sequence to be amplified, are mixed in molar excess with the DNA template, free deoxyribonucleotides and a DNA polymerase enzyme in an appropriate buffer. Following heating to denature the original strands and cooling to promote primer annealing, the oligonucleotide primers bind to their complementary sequences in the target DNA. Then, the temperature is raised to the optimal temperature of a DNA polymerase, which begins polymerization, adding nucleotides to the 3′ end of each primer attached to a single DNA strand. After one complete cycle, there are two double stranded copies of the target DNA. This process of denaturation, annealing, and polymerase extension repeated cyclically, produces many copies. Theoretically, 30 cycles over a billion copies of the target sequence (2^{30} = 1.07 billion) could be provided.

The availability of both thermostable DNA polymerases, which resist to inactivation at the high temperatures used during the thermal cycling, and thermal cyclers, which could shift their temperatures up and down rapidly, automatically and in a programmed manner, have allowed the PCR to be automated. Amplicons, i.e. PCR products attended, can be visualized through several methods. Apart from DNA hybridization, one of the most used techniques to accomplish amplicon detection is the agarose gel electrophoresis using a buffer stained with a dye (ethidium bromide, SYBR Green etc.) that binds double stranded DNA and fluoresces upon excitation with UV light. By this way, it is possible to observe and photograph the gel by using an apparatus with a UV light source and an appropriate camera [51]. The presence of the target pathogen, regardless of its conditions (live or dead), can be ascertained by the presence in gel of the band relevant to the specific amplicon. Moreover, due to the inverse linear correlation recognised between the $\log_{10}$ size of the DNA fragment (basepair) and the distance migrated by the DNA fragment in the agarose gel during electrophoresis, it is possible to estimate the size of the amplicon from DNA standards loaded in the agarose gel. Other ways to accomplish the detection of PCR products include DNA hybridization and non-gel methods such as the enzyme-linked immunosorbent assay (ELISA) [52].

2.2.2. *Application*

PCR, together with culture and counting methods is been indicated as the most popular method used in pathogen detection [53]. Sequencing analysis of rRNA genes intergenic spacer regions as well as other phylogenetically important genes such as *rpoB, hsp60* etc. and species-specific PCR of DNA fragments that are unique for a given microbial species, have been used to identify, detect and characterize pathogenic microorganisms. In Table 1, a list of PCR based methods developed and used to detect in milk and dairy products the five most concerned pathogens (*Campylobacter* spp., *Salmonella* spp., *E. coli* O157:H7, *L. monocytogenes* and *S. aureus*) is reported.

Among the different PCR variants to date available, multiplex PCR is very useful as it allows the simultaneous detection of several organisms by introducing different primers to amplify DNA regions coding for specific genes of each bacterial strain targeted. Apart from 16S and

23S rRNA and intergenic spacer regions, several genes have been targeted to allow multiplex milk- and dairy-borne pathogens identification, detection and characterization [54-66].

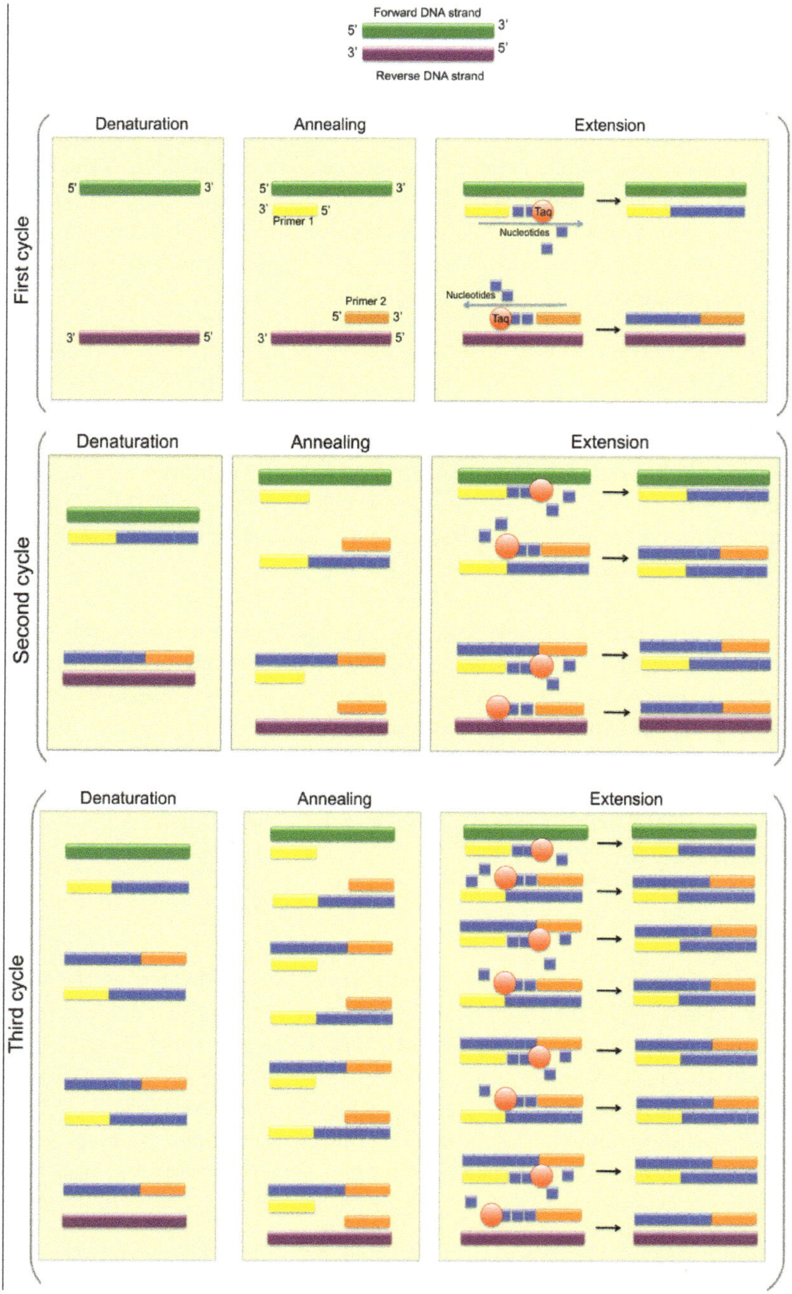

Figure 3. Polymerase chain reaction.

Several targets have been used and several researchers have used the same target to develop different primer pairs to detect the same bacterial species. However, individual assessments of the specificity and sensitivity of a given assay may differ markedly and care should be taken in designing a PCR [67,68]. Moreover, taxonomy of several pathogenic bacteria is being continuously changed in the recent years, thus calling for a periodic re-validation of PCR assays. As an example, the taxonomy of *Campylobacter* has evolved every year since 1988 and to date it contains over 100 *taxa*. Within the EC 6th Framework "MoniQA" Network of Excellence project on harmonisation and standardisation of detection methods for foodborne hazards, we have recently participated, with other 7 laboratories of other 7 countries, in a revalidation study of 31 PCR assays for the identification of *Campylobacter* species, with particular reference to taxa described since 2004, which are closely related to the most concerned pathogenic species *C. jejuni* and *C. coli* [69]. It resulted that i) the sensitivity and specificity of these PCR assays varied considerably, ii) PCR assays recently developed to identify and detect a novel *Campylobacter* species (namely *C. lari*) were no successful in detecting all strains of this species, probably reflecting its complex taxonomy; and iii) several PCR assays gave false positive results for *Campylobacter* species described since PCR tests were reported (including *C. cuniculorum, C. subantarticus, C. peloridis* and *C. volucris*), thus highligthing the need for attention to detail in the design and evaluation of a PCR assay and also for ongoing revalidation of previously already validated PCR assays [69].

Species	Target gene or fragment	Encoded product	Reference	
			a	b
Listeria monocytogenes	16S	16S rRNA	[70]	
	*hly*A	Listeriolysin	[71,72]	[73,74]
Campylobacter spp.	*act*A	Actin assembly-inducing protein	[75]	
	16S	16S rRNA	[76]	
Escherichia coli O157:H7	*eae*	Intimin	[77]	
	stx1 (or VT1)	Shiga-like toxin 1		[78]
	stx2 (or VT2)	Shiga-like toxin 2		[74,79,80]
	hlyA	Haemolysin		[62]
	*rfb*E	Lipopolysaccharide O side chain of *E. coli* O157		[79]
Salmonella spp.	*inv*A	Invasion protein		[81,82]
	stn	Enterotoxin determinant		[74]
Staphylococcus aureus	*nuc*	Thermostable nuclease of *S. aureus*	[83]	[74,84-88]
	sea	Staphylococcal enterotoxin A	[89]	[84,90-96]
	seb	Staphylococcal enterotoxin B	[89,97]	[84,90-96]
	sec	Staphylococcal enterotoxin C	[89]	[84,90-96,98]
	sec1	Staphylococcal enterotoxin C1	[97]	[99]

Species	Target gene or fragment	Encoded product	Reference	
			a	b
	sed	Staphylococcal enterotoxin D	[89]	[84,90-96]
	see	Staphylococcal enterotoxin E	[89]	[84,90-96]
	seg	Staphylococcal enterotoxin G	[89]	[84,91-93, 96,100,101]
	seh	Staphylococcal enterotoxin H	[89]	[91,92,94,96]
	sei	Staphylococcal enterotoxin I	[89]	[91,92,94,96,1 00,101]
	selj	Staphylococcal enterotoxin-like J	[89]	[84,91,94,96]
	selk	Staphylococcal enterotoxin-like K		[94]
	sell	Staphylococcal enterotoxin-like L	[89]	[91,94]
	selm	Staphylococcal enterotoxin-like M	[84,102]	[94,100,101]
	seln	Staphylococcal enterotoxin-like N	[84,102]	[94,100,101]
	selo	Staphylococcal enterotoxin-like O	[84,102]	[94,100,101]
	ser	Staphylococcal enterotoxin R		[92]
	ses	Staphylococcal enterotoxin S		[92]
	set	Staphylococcal enterotoxin T		[92]
	selu	Staphylococcal enterotoxin-like U		[101]
	selv	Staphylococcal enterotoxin-like A		[101]
	ϕent1	Pseudogene *ϕent1*		[100,101]
	ϕent2	Pseudogene *ϕent2*		[100,101]
	tsst	Toxic-shock syndrome toxin		[84,92]
	egc	Enterotoxin gene cluster		[84,92,100,101]
	spa	*Staphylococcus aureus* protein A		[103-106]

Table 1. List of target genes used in conventional PCR protocols for the identification, detection and characterization of some of the most concerned foodborne pathogens. Assays either specifically developed or employing already existing protocols for milk- and dairy- borne pathogens are listed in columns "a" and "b", respectively.

2.3. Real time PCR

Conventional PCR-based detection requires post-amplification confirmative analyses, which, apart from the potential DNA carry-over, are time- and labour-consuming. In real time PCR, fluorescent dyes are used to directly monitor the amplification of the target DNA. Moreover, because fluorescence increases in direct proportion to the amount of specific amplicons, real time PCR can be used for quantification.

SYBR Green is one of the most frequently double-stranded (ds) DNA-specific dyes used in real-time PCR today. It is an asymmetric cyanine dye that can be excited with blue light with a wavelength of 480 nm and having an emission spectrum comparable to that of fluorescein with a maximum at 520 nm [108]. Being a DNA binding dye, SYBR Green allows the detection of any double-stranded DNA during the PCR. Strength and, at the same time, weakness of this system is that, being nonspecific, it can also bind any spurious product (dimers, artefacts etc.). To overcome this problem a melting curve analysis may be carried out at the end of the real time PCR amplification. The strand-specific methods have a higher specificity since they employ fluorophore-coupled nucleic acids to interact with reaction products, probing accumulating PCR products for the presence of the target sequence.

The most commonly used fluorogenic oligoprobes rely upon the fluorescence resonance energy transfer (FRET) between either fluorogenic labels or a fluorophore and a dark or blackhole non fluorescent quencher (NFQ), which disperses energy as heat rather than fluorescence [107]. The FRET spectroscopic process consists in an energy transfer between molecules separated by 10-100 Angstroms, which have overlapping emission and absorption spectra [107].The theory behind this non-radioactive induced dipole interaction process was developed by Förster [109]. The efficiency of this process mainly depends on the distance between fluorophores. Indeed, the sequence-specific signals are generated due to the PCR-product-dependent change in distance between fluorophores [108].

TaqMan probes contain two dyes, a reporter dye (e.g. 6- carboxy-fluorescein; FAM) at the 5′ end and a quencher dye (e.g. 6-carboxy-tetrametyl-rhodamine; TAMRA) at the 3′ end. The proximity of the quencher dye to the reporter in an intact probe allows the quencher to suppress, or "quench", the fluorescence signal of the reporter dye through FRET. If the target of interest is present, these probes specifically anneal between the forward and reverse primer sites. During the real time PCR amplification, the 5′ to 3′ nucleolytic activity of the Taq DNA polymerase cleaves the probe between the reporter and the quencher, only if the probe hybridizes to the target. Thereafter, the quencher is released from the fluorophor, which now fluoresces after excitation.

The signal increases in direct proportion to the amount of PCR product in a reaction [107]. Accumulation of PCR products is detected directly by monitoring the increase in fluorescence of the reporter dye. The average background fluorescence is usually measured for the first 10 cycles of the reaction and subtracted from each fluorescent reading, resulting in a standardized amplification plot of fluorescence intensity against cycle number for each reaction. The cycle threshold (Ct), defined as the first cycle in which there is a significant increase in fluorescence above a specified threshold, is then calculated and the fluorescence against the cycle number is plotted to obtain a curve that represents the accumulation of PCR products in function of time. Running several reactions containing dilutions of known amount of target DNA a standard curve can be created and used to quantify unknown amounts of target DNA [107].

Being time-saving (especially the "fast systems" and requiring reduced handling, avoiding the risk of carryover contaminations, real time PCR is revolutioning the clinical, food and environmental diagnostics. Moreover, depending on the detection platform utilised, it is very highthroughput being possible to process either 96 or 384 samples per run even in a multiplexing format. Simplex and multiplex real time PCR assays have been developed and used to identify and quantitatively detect directly and undirectly in milk and dairy products, *S. aureus, L. monocytogenes, Salmonella* spp., *E. coli* O157:H7 and *Campylobacter* spp. (Table 2). When the culture-independent approach is performed the availability of an appropriate DNA extraction and purification protocol is crucial. Milk and its derivatives, due to their intrinsic complexity (in terms of composition and structure) and the likely presence (in raw milk based products) of abundant background microflora, may significantly affect the efficiency of both the nucleic acid extraction and PCR amplification [66,86, 110-113]. Moreover, in such cases, it is of outmost importance to use standard curves appropriate to the specific case to be analysed. Fusco et al. [101] developed a TaqMan and a SYBR Green rt-PCR based assay targeting the enterotoxin gene cluster of *S. aureus*, regardless of its variants, for the rapid and reliable identification and quantitative detection of *egc*$^{+}$ *S. aureus* strains (i.e. *S. aureus* harbouring the *enterotoxin gene cluster*). In addition, given the well recognised role of *S. aureus* as one of the commonest aetiological agent of clinical and sub-clinical mastitis [114-116] and considering that milk and milk-based products contaminated with this pathogen are some of the food matrices more often involved in staphylococcal food poisoning [117-122], they evaluated the effectiveness of these novel assays in artificially and naturally contaminated raw milk [101]. To achieve these goals, standard curves were constructed using ten-fold dilutions of target pre-purified DNA, or DNA extracted from ten-fold dilutions of a *egc*$^{+}$ *S. aureus* strain in broth and in raw milk [101]. Moreover, in order to determine the diagnostic sensitivity, defined as a measure of the degree to detect the target pathogen in the biological matrix, and to assess the applicability of the assay in simulated staphylococcal food poisoning conditions (i.e enterotoxin positive *S. aureus* concentration above 10^5 cfu mL^{-1}), three standard curves in raw milk were constructed using log phase broth cultures of either a single *egc*$^{+}$ *S. aureus* strain, a mix of *egc*$^{+}$ *S. aureus* strains and a mix of *egc*$^{+}$ and *egc*$^{-}$ *S. aureus* strains harbouring (and not) some of the commonest enterotoxin genes associated to this syndrome [101]. Overall, the TaqMan assay revealed less sensitivity (limit of quantification=10^2 cfu equivalents per reaction mixture of *egc*$^{+}$ *S. aureus* either singly, in mix and in mix with *egc*$^{-}$ *S. aureus* strains) in milk than in DNA (pre-purified and not), thus highlighting the importance of using a standard curve in raw milk to accurately quantify *egc*$^{+}$ *S. aureus* in real raw milk samples [101].

Species	Target gene	Protein	Reference	
			a	b
Listeria monocytogenes	*hly*A	Listeriolysin	[123,124]	[125]
	*ssr*A	tmRNA	[126]	[127]
	*prf*A	Transcriptional regulator PrfA; listeriolysin positive regulatory protein	[128,129]	
	16S	16S rRNA	[130]	

Species	Target gene	Protein	Reference	
			a	b
	16S-23S rRNA IGS	Intergenic region spacer between the 16S and 23S rRNA	[131]	[132]
Staphylococcus aureus	*egc*	Enterotoxin gene cluster	[101]	
	*htr*A	High-temperature-requirement A protein	[133]	
	nuc	Thermostable nuclease	[134]	
Escherichia coli	*stx*1	Shiga-like toxin 1	[135]	
	*stx*2	Shiga-like toxin 2	[135,136]	
	eae	Intimin	[135]	[30]
	stx1, stx2	Shiga-like toxin 1 e 2	[137]	[30]
Campylobacter spp.	VS1	*C. jejuni* specific fragment	[138]	
Salmonella spp.	*ttr*RSBCA	Proteins involved in tetrathionate respiration	[139]	
	*inv*A	Invasion protein	[140]	

Table 2. List of target genes used in simplex and multiplex real-time PCR protocols for the identification, (quantitative) detection and characterization of some of the most concerned foodborne pathogens. Assays either specifically developed or employing already existing protocols for milk- and dairy- borne pathogens are listed in columns "a" and "b", respectively.

2.4. Detecting stressed or injured pathogens: EMA and PMA PCR/real time PCR, reverse transcription PCR/real time PCR and NASBA

The complexity and variability of food composition as well as physical and/or chemical stresses that pathogenic microorganisms encounter in the environment, in foods and food preparation/production/storage processes, if inadequate or sub lethal, may result in incomplete inactivation [141,142]. Such injured or stressed bacteria are a potential risk since they can, under appropriate conditions, recover and regain or even enhance their pathogenicity [141,143-145]. Failure to detect injured pathogens can have adverse health effects as well as fatalities and economic losses. All these findings prompt the need for improved enumeration methods capable of discriminating among viable, dead, and injured microbial cells. Conventional culture based methods do not allow the enumeration of stressed or injured bacteria, as they use selective agents whose injured or stressed pathogens' cells, depending on the site and degree of damaging, are extremely sensitive [142]. The inadequateness of highly selective solid and liquid media remarkably complicates the detection of pathogenic bacteria in foods characterized by a complex and numerous background microflora, such as milk and milk-based products. The major drawback in using DNA-based assays to detect pathogenic microorganisms is that DNA is detectable in both viable, injured and dead cells of a given microorganism also after a long period of time [86,146]. However, DNA amplification techniques may be combined with the use of molecules able to penetrate in dead or injured cells and bind to DNA making it insoluble so that it can be easily eliminated together with cell debris during genomic DNA extraction. Ethidium monoazide- (EMA) and propidium monoazide (PMA) conventional and real time PCRs have been applied to the detection and quantification of different food-borne pathogens [147-150]. PMA seems to have the important advantage over EMA of not penetrating living cells [149].

Stressed or injured pathogens cells may be quantitatively detected by PCR methods combined with a solid or a liquid based enumeration method in which stages of "revivification" have been introduced to restore and therefore bringing back in conditions of perfect viability and cultivability the greater number of stressed cells, not directly cultivable [79,123].

To address the need of detecting only living pathogens RNA may be detected rather than DNA [151-155].

Reverse Transcription (RT) PCR makes use of a reverse transcriptase, which, in presence of a complementary primer, can translate an RNA strand corresponding to a transcribed gene into complementary DNA (cDNA). The reaction is usually initiated by random oligonucleotide primers. Thereafter, the cDNA is used as template to amplify by PCR specific sequences using oligonucleotide primers and DNA polymerase under normal PCR or real time PCR conditions.

Another way to target mRNA as an indicator of cell viability is to employ the nucleic acid sequence-based amplification (NASBA). It is an isothermal nucleic acid amplification technology allowing the amplification of RNA or DNA targets (with a slight modification in the protocol) through a transcription process after the insertion of a T7 promoter, due to the concerted action of three enzymes: AMV Reverse Transcriptase for cDNA synthesis, RNase H to degrade the RNA in the heteroduplex RNA-DNA and T7 RNA polymerase to synthesize RNA from the T7 promotor [156].

Both the NASBA and the reverse transcription conventional and real time PCR techniques have been used for developing diagnostic tests to detect viable pathogenic microorganisms [157-159]. Since NASBA is performed in isothermal conditions, it does not require the use of a thermocycler. Therefore, it is less expensive than PCR and RT-real time PCR.

2.5. Biosensors, microarrays, micro and nano electro-mechanical-systems

Biosensors have been recognized as a means to provide a higher level of surveillance in a more automated and rapid manner. These analytical devices combine a biological sensing element (called receptor) with a chemical or physical transducer for selectively and quantitatively detecting a given compound. In complex matrices only the compound interacting specifically with the integrated biological component will generate the optical or electrical signal from the physical transducer, modulating the biosensors' selectivity [53]. Such specific interactions produce a physico-chemical change, detected and measured by the transducer that can output a signal proportional to the concentration of the target analyte, allowing for both real time quantitative and qualitative measurements. The chemico-physical change detected by the transducer may be: 1) absorption or evolution of heat (thermometric or calorimetric biosensors), 2) changes in the distribution of charges causing an electrical potential to be produced (potentiometric biosensors), 3) movement of electrons produced in a redox reaction (amperometric biosensors), 4) light radiation or difference in optical properties between the reactants and products (optical biosensors) and, 5) effects due to the mass or intermolecular interaction of the reactants or products (piezo-electric biosensors) [53,160]. Enzymes, antibodies, DNA, receptors organelles and microorganisms as well as plant cells or tissues are frequently used as biological sensing elements [161].

Many biosensors for food diagnostic application in the food and drink industry are currently being developed [161-166]. Among these, biosensors that can detect DNA are of particular interest. Intrinsic features of the DNA molecule, such as the high specificity of base pairing between homologous strands of single-stranded DNA, as well as its predisposition to electrical, fluorescent and mechanical measurements, make it a highly specific sensing element, and render it useful for signal transduction in a wide range of DNA based biosensors [167,168].

Hybridization biosensors rely on the immobilization of a species- or strain- specific single stranded DNA probe onto the transducer surface. Due to the characteristic negative charge of DNA, the duplex formation can be detected for example by following the association of an appropriate hybridization indicator. Hybridization events between analyte and probe DNA may be translated through electrochemical, optical, or mechanical output signals [169]. As for other types of biosensors, high selectivity is crucial for the success of DNA hybridization devices. The specificity of these sensors depends primarily on the selection of the probe, and secondarily upon the hybridization conditions (mainly the temperature).

Microarrays, based on the Watson-Crick base pairing principle [4], consist of genetic sensors, the so-called "spots," each containing single strands of species- or strain- specific DNA sequences termed probes immobilized at pre-determined position at high density. The DNA sequence of a target organism's genetic sample, previously labelled (through PCR amplification), will hybridize with its complementary sequence on the microarray to form a stable structure. After washing away non-specifically bound targets, the array is scanned using laser light of a wavelength designed to trigger fluorescence in the spots where binding has occurred. A specific pattern of array spots will fluoresce, which is then used to infer the genetic makeup in the test sample [170]. Microarray analysis is an emerging technology that has the potential to become a leading trend in bacterial identification in the dairy and overall in the in food and drink industry, allowing both the detection and/or genotyping of pathogenic microorganisms [171-179].

Due to the robust nature of PCR and the high sensitivity that can be achieved through amplification of target DNA, PCR-based biosensing has been widely used [168,180].

For PCR based-biosensing of bacteria optical methods of detection are widely used. In the case of the "real time PCR-based biosensor" the fluorescence emission is the measurable signal allowing the translation of the DNA amplification. DNA-based biosensor technologies are in constant evolution. In particular, there is a growing tendency toward miniaturization of these systems [167]. In recent years, micro- and nano- fabrication technologies, originally developed for producing silicon-based chips for the microelectronics industry, have spread out in a variety of applications as chemical and biochemical tools, commonly referred to as Biomedical or Biological Micro (and nano)-Electro-Mechanical Systems (BioMEMS or bioNEMS). BioMEMS and devices have been used as biosensor for the detection of bacteria, and the resulting biochips, also known as lab-on-a-chip devices, incorporate multiple laboratory processes in a semi-automated, miniaturized format, allowing rapid, sensitive and real-time measurements [163,164,180-183]. Obvious advantages of the miniaturized integrated detection technologies include higher sensitivity, as well as reduced reagent and sample volumes, reducing associated costs and time to result.

An example of such useful devices is given by the integrated microfluidic platform, known as the "microFLUIDICS DESKTOP" (Figure 4), developed by Cady and co-workers [185] for detecting *Listeria monocytogenes* by real time PCR. Monolithic DNA purification/real-time PCR silicon chips (Figure 5) were fabricated utilizing standard semiconductor processing technologies. These chips incorporated a microfabricated DNA purification chamber with a second PCR amplification chamber, connected by microfluidic channels.

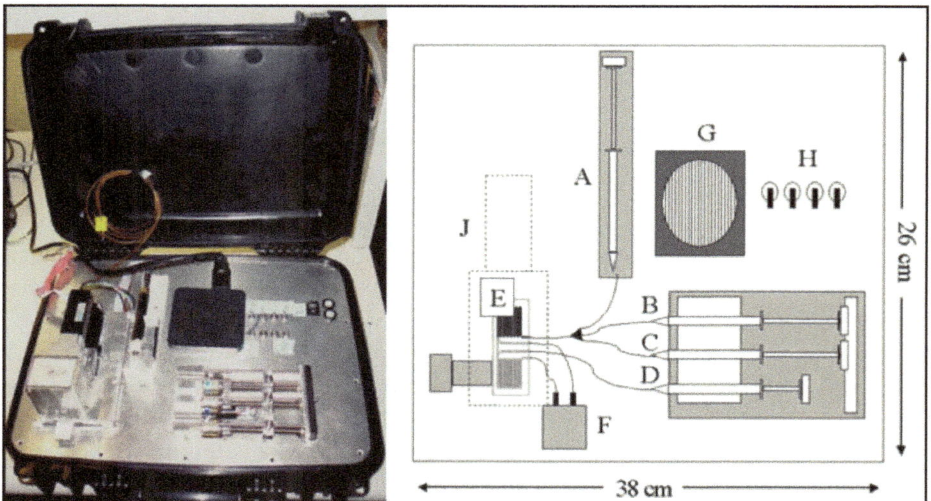

The "microFLUIDICS DESKTOP" includes integrated syringe pumps (A-D), Moog micro valve (F), cooling fan (G), LED-based fluorescence excitation/detection system (J and dotted outline), power toggle switches (H). The microfluidic purification/detection chip (E) is inserted into the unit directly above the thermoelectric heater cooler. The syringes are connected to the chip via TygonTM tubing (black lines) and contain the sample lysate (A), ethanol wash buffer (B), TE (C), and PCR master mix (D). The Moog micro valve (F) is also connected to the chip via tubing and controls pressurization and fluid flow through the chip outputs. The entire unit measures 36cm x 28cm x 15cm.

Figure 4. The "microFLUIDICS DESKTOP" [184,186].

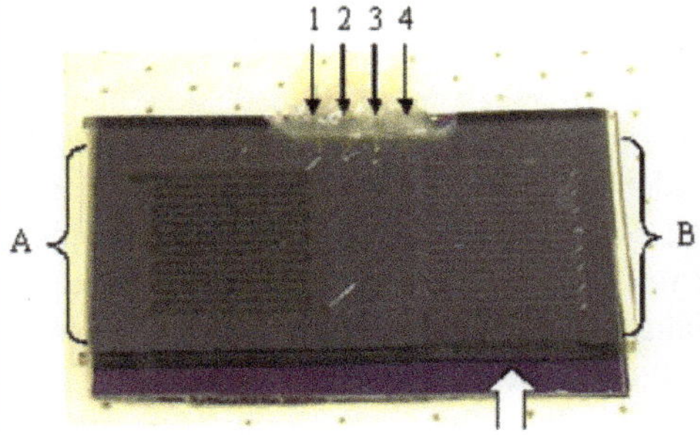

An optical micrograph of the DNA purification/real-time PCR microchip (2 cm x 4 cm). The nucleic acid purification region is shown in A while the real-time PCR region is shown in B. The fluid connections are 1) sample input, 2) waste outlet, 3) PCR reagent input, and 4) reaction outlet. The large white arrow denotes the lateral path for fluorescent excitation for real-time PCR.

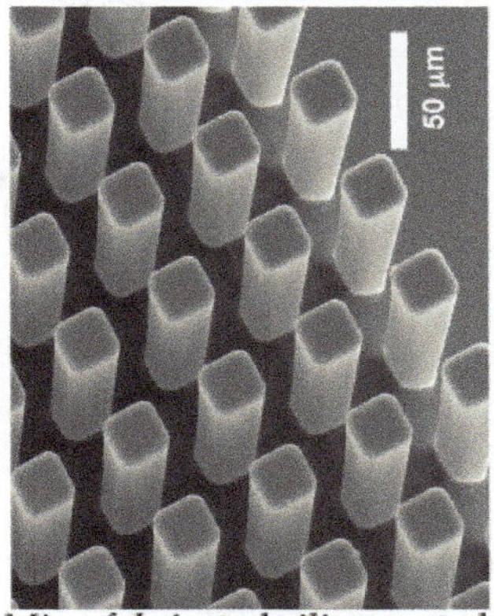

Microfabricated silica-coated pillars in the purification channels.
Each pillar is approximately 10 µm square by 50 µm tall and coated by a thin layer (100nm) of SiO2 allowing the DNA purification by chaotropic salts containing buffer.

Figure 5. The monolithic DNA purification/real time PCR microchip [184-186].

The DNA purification section contained an array of 10 µm square pillars that were etched 50 µm deep in silicon to form a microfluidic channel. These pillars were coated with a thin layer (100 nm) of SiO2 that could be used for DNA purification in chaotropic salt-containing buffers [185]. Using an automated detection platform with integrated microprocessor, pumps, valves, thermocycler and fluorescence detection modules, microchips were used to purify and detect bacterial DNA by real-time PCR amplification using SYBR Green fluorescent dye. This system was able to both purify and quantify DNA from 10^7 to 10^4 cells by SYBR Green real-time PCR-based detection, with an average turnaround time of 45 min. The "microFLUIDICS DESKTOP" has been successfully used for the more specific TaqMan real-time PCR detection of *S. aureus* [186].

In an improvement over other systems, which are time consuming and require multiple laboratory instruments, this device provides a fully automated method capable of purifying DNA from bacterial cells and preparing samples for PCR-based detection.

Obvious benefits of such device include: reduced time to result; reduced amount of expensive reagents used; reduced handling, avoiding sample contaminations; the possibility to perform a multiplex assay for revealing various pathogenic microorganisms, by incorporating on-board multiple detectors; the possibility to further miniaturize a multifunctional integrated system, which can be developed as a truly portable device to be used for on-line and on-site rapid control of the whole food/beverage chain.

3. Typing

In diagnostic microbiology, besides identifying a pathogenic microorganism, it is of outmost importance to type it. Typing, or subtyping (synonymous used in the American literature) has been defined as: "Phenotypic and/or genetic analysis of bacterial isolates, below the species/subspecies level, performed in order to generate strain/clone-specific fingerprints or datasets that can be used, for example, to detect or rule out cross-infections, elucidate bacterial transmission patterns and find reservoirs or sources of infection in humans" [187]. Several typing methods are to date available for discriminating microorganisms at strain level: the choice should be based on their appropriateness for each specific purpose. Whatever typing method is chosen, it has to be evaluated and validated in respect to several performance criteria (stability, typeability, discriminatory power, epidemiological concordance, reproducibility, test population) and convenience criteria (flexibility, rapidity, accessibility, easy of use, cost, amenability to computerised analysis and incorporation of typing results in electronic databases) prior to use it in a given study [187]. Conventional typing methods, based on the phenotypic features of microorganisms, including biotyping, serotyping, antibiogram-based typing (antimicrobial susceptibility testing), phage and bacteriocin typing, sodium dodecyl sulphate-polyacrilammide gel electrophoresis (SDS-PAGE) of cellular and extracellular components, multilocus enzyme electrophoresis (MLEE), mass spectrometry (MS), matrix-assisted laser desorption/ionization-time of flight (MALDI-TOF) MS, Infrared or Raman spectroscopy, Fourier transform Infrared spectroscopy, etc. are being gradually overtaken by nucleic acid-based methods. The advent of the PCR technology and the development of bioinformatics tools have allowed implementing protocols to investigate the inter- and intra- specific heterogeneity of pathogenic microorganisms based on genotypic characters, which, unlike the phenotypic ones (on which rely conventional typing methods) are certainly the most reliable, being little or not influenced at all by exogenous factors. As for the detection and identification methods, nucleic acid based methods may be grouped into three main clusters including nucleic acid hybridization-, amplification- and fragment based typing methods and a third cluster including methodologies that combine all the above mentioned techniques. Herein, an excursus of the most important molecular methods used to type the five most concerned pathogenic bacteria will be reported.

Acronyms such as RAPD- (Random Amplified Polymorphic DNA), DAF- (DNA Amplification Fingerprinting) and AP-PCR (arbitrarily Primed PCR), define a set of methodologies having, as lowest common denominator, the possibility to amplify anonymous DNA sequences by using a single oligonucleotide as a primer for the Taq Polymerase [188,189]. Under conditions allowing the annealing of the primer at several points, a complex banding pattern should be obtained that could be characteristic for each strain. Methods such as REP- (Repetitive Extragenic Elements Palyndromic), ERIC- (Enterobacterial repetitive Intergenic Consensus Sequence), BOX-, VNTRs- (Variable Number of Tandem Repeats) PCR, being based on the detection of repetitive regions of the genomic DNA, varying in length and number, have been individuated as more reliable and robust than RAPD-PCR to type pathogens [188,190-192].

Classical Restriction Fragment Length Polymorphism (RFLP) analysis combines the restriction endonuclease analysis with a Southern hybridization step. By this technique, genomic restriction fragments are separated by gel electrophoresis and then transferred to membranes, where they are hybridized with a labelled probe for specific DNA fragments that are present in the bacterial genome. Microbial strains are thereafter discriminated based on the size and number of restriction fragments that are homologous to the probe. A variation of RFLP, which is also one of the most used nucleic acid hybridization based method to type pathogenic bacteria, is ribotyping. This technique, originally used to establish phylogenetic relationships [193], is based on the restriction enzyme digestion of chromosomal DNA followed by Southern hybridization with a probe specific to conserved regions of the rRNA coding sequence. The resulting banding patterns are analysed to sort isolates into ribotypes also establishing the relatedness of isolates [193]. Using an automated ribotyping system, the RiboPrinter® (Qualicon Inc., Wilmington, Del., U.S.), this assay has been shortened from five days to eight hours. It has been demonstrated that the discriminatory power of this technique can considerably increase by using multiple ribopatterns to determine the overall ribotype of isolates [194]. Ribotyping has been widely used to characterise milk- and dairy- borne isolates of *L. monocytogenes* [195-200], *S. aureus* [201,202], *E. coli* O157:H7 [203,204], *Salmonella* spp. [205] and *Campylobacter* spp. [206,207]. rDNA based fingerprints can be obtained also by ARDRA (Amplified rDNA restriction analysis). In this case, bacterial rRNA gene(s) are amplified by PCR and digested with restriction endonuclease enzyme(s). The resulting restriction fragments are then resolved by gel or capillary electrophoresis to obtain a fingerprint [208]. However, although ARDRA is faster and easier to perform, it has a lower discriminatory power than ribotyping, because smaller areas of the rRNA operon are targeted. Other methods are available to obtain restriction maps providing strain-specific fingerprints of pathogenic bacteria. PCR-RFLP, based on the amplification of a given gene or operon, coupled with the digestion with appropriate endonuclease enzymes and electrophoresis of the resulting restriction fragments, has been used to characterise milk- and dairy- borne *S. aureus* [84,100,209], *L. monocytogenes* [2010-212], *E. coli* O157:H7 [213], *Salmonella* spp. [214,215] and *Campylobacter* spp. [216,217] isolates.

Restriction endonuclease analysis of genomic DNA fragments separated by pulsed-field gel electrophoresis (REA-PFGE) has become the "gold standard" for molecular typing [187,218]. By this technique, intact genomic DNA, obtained by performing the DNA extraction and purification of microbial cells imbibed in low melting agarose, is digested by rare-cutting endonuclease enzymes. The resulting restriction fragments, usually fewer than 30 ranging in size between 20 and 600 kilobasepairs, are then separated in agarose gels by the periodic alternation of the angle of the electric's field direction through PFGE, thus obtaining banding patterns that can be compared for each isolate to discriminate the different pulsotypes. Despite its lower convenience criteria, as it is more laborious, time consuming and expensive (it requires skilled labour, specialised equipment and expensive restriction endonucleases) than the other molecular typing techniques to date available, REA-PFGE has superior performance criteria (mainly discriminatory power and reproducibility) due to both the quality and quantity of banding patterns obtainable, which in turn are due to the intrinsic nature of the REA-PFGE. PulseNet (www.cdc.gov/pulsnet), which is a network of health and food regulatory laboratories created by the the Center for Disease Control and

Prevention [1], uses PFGE for typing foodborne pathogens, allowing a better tracking and earlier detection of possible common source outbreaks [219,220]. As several similar initiatives have been developed in other countries (129 laboratories from 70 countries trained on PulseNet methods in Latin America, Canada, Europe, the Middle East, Asia Pacific) the PulseNet international (www.pulsenetinternational.org) has been developed. Networks like these rely upon standardization of REA-PFGE and regular quality assessment through appropriate ring trials for all participating laboratories, in order to warrant consistently comparable data.

Informative polymorphic banding patterns can be obtained by amplified fragment length polymorphism (AFLP) allowing the differentiation with a high discriminatory power (as it involves, like REA-PFGE, the whole genome) even of phylogenetically closely related bacteria without any prior information on their genomes [110, 221]. This multi-locus fingerprinting technique combines the reliability of RFLP analysis with the flexibility and robustness of PCR using restriction site/adaptor-specific primers under stringent conditions [110, 222]. Semi-automated versions of this technique may be obtained by the fluorescent AFLP (fAFLP) using capillary array systems, fluorescently labelled primers and adequate analysis software [223], which provide digitised and complex DNA fingerprints, covering nearly the whole genome [110]. AFLP has been used to type *Salmonella* spp., *E. coli* O157, *Campylobacter* spp., *S. aureus* and *L. monocytogenes* [203,223-228].

Typing of pathogenic microorganisms can be achieved by PCR based methods. Conventional and, more proficiently, real time PCR, either in simplex or (better) in multiplex format of enterotoxins' and/or other virulence factors' encoding genes may provide valuable information on the potential pathogenicity of a given isolate (paragraph 2.2 and 2.3). The same achievement can be reached by using the microarray technology (described above).

PCR amplification and sequencing analysis of (usually) seven house-keeping genes is the basics of the Multilocus Sequence Typing (MLST), which, being based on sequence data, is an unambiguous procedure to characterize isolates of bacterial species [229,230]. If virulence and virulence-associated genes are used this technique is referred to as Multi-Virulence Locus Sequence Typing (MVLST) [231]. Since MLST relies upon specific nucleotide base changes to type microorganisms, it is easiest to perform (it can be automated by using an automated pipetting platform and an automated sequencer) and analyse unlike other typing procedures which compare complex fingerprints rather than sequence data and allelic profiles. Of course, high quality sequences results are essential in this method. Being a highly reproducible and reliable technique, by which results, i.e. strings of digits representing different alleles, are easily and unequivocally exchangeable, as for REA-PFGE, a multilocus sequence network has been developed (www.mlst.net).

4. Conclusion

As it emerges from the reading of the present chapter, a plethora of nucleic acid based methods are to date available for the detection, identification and typing of milk- and dairy-borne pathogenic microorganisms. However, none of these is perfect, harbouring wikenesses besides strenghts such as suboptimal reproducibility, sensitivity or discriminatory power. It can be inferred that the choice of the most appropriate method should rely upon the specific needs and, of course, on the available equipment for carrying out the task. Nucleic acid-based methods are gradually replacing traditional methods to identify, detect and type pathogenic microorganisms in milk and dairy products. Several efforts are being made to overcome limitations of these methods, mainly related to their sensitivity and accuracy besides the effectiveness in respect to the complex target food matrices, and achieve the validation and standardization of these approaches, which are basic requirements to become reference methods. As concerns identification and detection systems, the major perspective in the nearest future is the possibility to use portable miniaturized integrated devices allowing the rapid and reliable detection and quantification of pathogenic microorganisms directly from food. Such detection platform will decrease the risk of contaminating the food/feed/beverage supply, preventing a wide dissemination of contaminated foods and the possibility of a disease outbreak, as well as facilitate real-time preventive measures along the whole food/feed/beverage chains. On the other side, next generation sequencing and other user-friendly nucleic acid-based automated platform will provide promising way of achieving the genetic typing of milk- and dairy- borne pathogenic microorganisms.

Author details

Vincenzina Fusco and Grazia Marina Quero
National Research Council, Institute of Sciences of Food Production (CNR-ISPA), Bari, Italy

Acknowledgement

This review was prepared within the European Community's Seventh Framework Programme (FP7/ 2007- 2013) under the grant agreement no FP7-222 654-DREAM.

5. References

[1] Centre for Disease Control and Prevention (2011) CDC 2011 estimates of foodborne illness in the United States. Factsheet printable and downloadable from the following url: http://www.cdc.gov/foodborneburden/2011-foodborne-estimates.html

[2] European Food Safety Autority (EFSA), European Centre for Disease Control and Prevention (ECDC) (2011) Scientific report of EFSA and ECDC. The European Union summary report on trends and sources of zoonoses, zoonotic agents and food-borne outbreaks in 2009. EFSA Journal 2011 9:2090, 1-378.

[3] FDA (2007) Food Protection Plan. Department of health and human services. Maryland: US. Food and Drug Administration.

[4] Watson JD, Crick FHC (1953) Molecular structure of nucleic acids: a structure for deoxyribose nucleic acid. Nature 171:737-738.

[5] Strachan T, Read AP (1999) Nucleic acid hybridization assays. In: Human molecular genetics, 2nd edition. D: Strachan T., Read, A.P. New York: Wiley-Liss.
[6] Olsen JE, Aabo S, Hill W, Notermans S, Wernars K, Granum PE, Popvic T, Rasmussen HN, Olsvik O (1995) Probes and polymerase chain reaction for detection of foodborne bacterial pathogens. Int. J. Food Microbiol. 28: 1-78.
[7] Fang X, Liu X, Schuster S, Tan W (1999) Designing a novel molecular beacon for surface-immobilized DNA hybridization studies. J. Am. Chem. Soc. 121:2921-2922.
[8] Feinberg AP, Vogelstein B (1983) A technique for radiolabelling DNA restriction endonuclease fragments to high specific activity. Anal. Biochem. 132: 6–13.
[9] Jablonski E, Moomaw EW, Tullis RH (1986). Preparation of oligonucleotide-alkaline phosphatase conjugates and their use as hybridization probes. Nucl. Acid Res. 14: 6115-6128.
[10] Langer P, Waldrop AA, Ward DC (1981) Enzymatic synthesis of biotin-labeled polynucleotides: novel nucleic acid affinity probes. Proc. Natl. Acad. Sci. 78: 6633-6637.
[11] Kafatos FC, Jones CW, Efstratiadis A (1979) Determination of nucleic acid sequence homologies and relative concentrations by a dot hybridization procedure. Nucl. Acids Res. 7:1541-52.
[12] Southern EM (1975). Detection of specific sequences among DNA fragments separated by gel electrophoresis. J. Mol. Biol. 98: 503-517.
[13] Alwine JC, Kemp DJ, Stark GR (1977) Method for detection of specific RNAs in agarose gels by transfer to diazobenzyl-oxymethil-paper and hybridization with DNA probes. Proc. Natl. Acad. Sci. USA 74:5350-5354.
[14] Grunstein M, Hogness DS (1975) Colony hybridization: a method for the isolation of cloned DNA's that contain a specific gene. Proc. Natl. Acad. Sci. 72:3961-3965.
[15] Leary JJ, Brigati DJ, Ward DC (1983) Rapid and sensitive colorimetric method for visualizing biotinlabeled DNA probes hybridized to DNA or RNA immobilized on nitrocellulose: Bio-blots. Proc. Natl Acad. Sci. USA, 80:4045-4049.
[16] Achemchem F, Martínez-Bueno M, Abrini J, Valdivia E, Maqueda M (2005) *Enterococcus faecium* F58, a bacteriocinogenic strain naturally occurring in Jben, a soft, farmhouse goat's cheese made in Morocco. J. Appl. Microbiol. 99:141–150.
[17] Baylis CL (2009) Raw milk and raw milk cheeses as vehicles for infection by Verocytotoxin-producing *Escherichia coli*. Int. J. Dairy Technol. 62: 293–307.
[18] Bottari DA, Emmett CD, Nichols CE, Whippie KD, Rodriguez D, Durbin GW, Keough KM, Groody EP, Mozola MA, Reynolds GN (1995) Comparative study of a colorimetric DNA hybridization method and conventional culture procedures for the detection of *Listeria* spp. in foods. J. Food Protect. 58:1083-1090.
[19] Curiale MS, Klatt MJ (1990) Colorimetric deoxyribonucleic acid hybridization assay for rapid screening of *Salmonella* in food: collaborative study. J. Assoc. Off. Anal. Chem. 73: 248-256.
[20] Datta AR, Wentz BA, Hill WE (1987) Detection of hemolytic *Listeria monocytogenes* by using DNA colony hybridization. Appl. Environ. Microbiol. 53: 2256-2259.
[21] Flowers RS, Klatt MJ, Mozola MA, Curiale MS, Gabis DA, Silliker JH (1987) DNA hybridization assay for detection of *Salmonella* in foods: collaborative study. J. Assoc. Off. Anal. Chem. 70:521-9

[22] Gómez D, Miliwebsky E, Fernández Pascua C, Baschkier A, Manfredi E, Zotta M, Nario F, Piquín A, Sanz M, Etcheverría A, Padola N, Parma A, Rivas M (2002) Isolation and characterization of Shiga-toxin-producing *Escherichia coli* from frozen hamburgers and soft cheeses. Rev. Argent. Microbiol. 34:66-71.
[23] Mozola M, Halbert D, Chan S, Hsu H-Y, Johnson A, King W, Wilson S, Betts RP, Bankes P, Banks JG (2010) Detection of foodborne bacterial pathogens by a colorimetric DNA hybridization method. In Genetic Manipulation: Techniques and applications (eds J. M. Grange, A. Fox and N. L. Morgan), Blackwell Scientific Publishing, Oxford, UK.
[24] Grunstein M, Hogness DS (1975) Colony hybridization: a method for the isolation of cloned DNA's that contain a specific gene. Proc. Natl. Acad. Sci. 72:3961-3965.
[25] Herman LM, De Block JH, Moermans RJ (1995) Direct detection of *Listeria monocytogenes* in 25 milliliters of raw milk by a two-step PCR with nested primers. Appl. Environ. Microbiol. 61:817-819
[26] Hill WE, Madden JM, McCardell BA, Shah DB, Jagow JA, Payne WL, Boutin BK (1983) Foodborne enterotoxigenic *Escherichia coli*: detection and enumeration by DNA colony hybridization. Appl. Environ. Microbiol. 45:1324-1330.
[27] King, W., S. Raposa, J. Warshaw, A. Johnson, D. Halbert, J.D. Klinger (1989) A new colorimetric nucleic acid hybridization assay for *Listeria* in foods. Int. J. Food Microbiol. 8:225–232
[28] Kneifel W, Manafi M, Breit A (1992) Adaptation of two commercially available DNA probes for the detection of *E. coli* and *Staphylococcus aureus* to selected fields of dairy hygiene--an exemplary study. Zentralbl. Hyg. Umweltmed. 192:544-53.
[29] Lira WM, Macedo C, Marin JM (2004) The incidence of Shiga toxin-producing *Escherichia coli* in cattle with mastitis in Brazil. J. Appl. Microbiol. 97: 861–866.
[30] Madic J, Vingadassalon N, de Garam CP, Marault M, Scheutz F, Brugère H, Jamet E, Auvray F. (2011) Detection of Shiga toxin-producing *Escherichia coli* serotypes O26:H11, O103:H2, O111:H8, O145:H28, and O157:H7 in raw-milk cheeses by using multiplex real-time PCR. Appl. Environ. Microbiol. 77:2035-41.
[31] Ng LK, Kingombe CI, Yan W, Taylor DE, Hiratsuka K, Malik N, Garcia MM (1997) Specific detection and confirmation of *Campylobacter jejuni* by DNA hybridization and PCR. Appl. Environ. Microbiol. 63:4558-4563.
[32] O'Connor L, Joy J, Kane M, Smith T, Maher M (2000) Rapid Polymerase Chain Reaction/DNA Probe Membrane-Based Assay for the Detection of *Listeria* and *Listeria monocytogenes* in Food. J. Food Protect. 63:337-342.
[33] Padmapriya BP, Ramesh A, Chandrashekar A, Varadaraj MC (2003) Staphylococcal accessory gene regulator (*sar*) as a signature gene to detect enterotoxigenic staphylococci. J. Appl. Microbiol. 95:974–981.
[34] Pradel N, Bertin Y, Martin C, Livrelli V (2008) Molecular analysis of shiga toxin-producing *Escherichia coli* strains isolated from hemolytic-uremic syndrome patients and dairy samples in France. Appl. Environ. Microbiol. 74: 2118-2128.
[35] Samadpour M, Liston J, Ongerth JE, Tarr PI (1990) Evaluation of DNA probes for detection of Shiga-like-toxin-producing *Escherichia coli* in food and calf fecal samples. Appl. Environ. Microbiol. 56: 1212–1215.

[36] Stevens KA, Jaykus LA (2004) Direct detection of bacterial pathogens in representative dairy products using a combined bacterial concentration-PCR approach. J. Appl. Microbiol. 97:1115–1122.

[37] Vernozy-Rozand C, Mazuy C, Prevost G, Lapeyre C, Bes M, Brun Y, Fleurette J (1996) Enterotoxin production by coagulase-negative staphylococci isolated from goats' milk and cheese. Int. J. Food Microbiol. 30:271-80.

[38] Villard L, Kodjo A, Borges E, Maurin F, Richard Y (2000) Ribotyping and rapid identification of *Staphylococcus xylosus* by 16–23S spacer amplification. FEMS Microbiol. Lett. 185: 83–87.

[39] Wang G, Clark CG, Taylor TM, Pucknell C, Barton C, Price L, Woodward DL, Rodgers FG (2002) Colony multiplex PCR assay for identification and differentiation of *Campylobacter jejuni, C. coli, C. lari, C. upsaliensis,* and *C. fetus* subsp. *fetus.* J. Clin. Microbiol. 40:4744-4747.

[40] Wesley IV, Wesley RD, Heisick J, Harrell F, Wagner D (1990) Characterization of *Listeria monocytogenes* isolates by Southern blot hybridization. Vet. Microbiol. 24:341-53.

[41] Zweifel C, Giezendanner N, Corti S, Krause G, Beutin L, Danuser J, Stephan R (2010) Characteristics of Shiga toxin-producing *Escherichia coli* isolated from Swiss raw milk cheese within a 3-year monitoring program. J. Food Protect. 73: 88-91.

[42] Amann RI, Krumholz L, Stahl DA (1990a). Fluorescent oligonucleotide probing of whole cells for determinative, phylogenetic, and environmental studies in microbiology. J. Bacteriol. 172:762-770.

[43] Amann RI, Binder BJ, Olson RJ, Chisholm SW, Devereux R, Stahl DA (1990b) Combination of 16S rRNA-targeted oligonucleotide probes with flow cytometry for analyzing mixed microbial populations. Appl. Environ. Microbiol. 56:1919–1925.

[44] Fuchizawa I, Shimizu S, Kawai Y, Yamazaki K (2008) Specific detection and quantitative enumeration of *Listeria* spp. using fluorescent *in situ* hybridization in combination with filter cultivation (FISHFC). J. Appl. Microbiol. 105:502–509.

[45] Gunasekera TS, Veal DA, Attfield PV (2003) Potential for broad applications of flow cytometry and fluorescence techniques in microbiological and somatic cell analyses of milk. Int. J. Food Microbiol. 85:269-279.

[46] Oliveira M, Blasco L, Ferrer S, Bernardo F (2004) Rapid and simultaneous detection of *Salmonella* spp. and *Listeria monocytogenes* in milk by Fluorescent *In situ* Hybridisation. Revista Portuguesa de Ciencias Veterinárias. 99: 215-218.

[47] Schmid M, Walcher M, Bubert A, Wagner M, Wagner M, Schleifer K-H (2003) Nucleic acid-based, cultivation-independent detection of *Listeria* spp. and genotypes of *L. monocytogenes*. FEMS Immunol. Medical Microbiol. 35: 215–225.

[48] Stender H, Oliveira K, Rigby S, Bargoot F, Coull J (2001) Rapid detection, identification, and enumeration of *Escherichia coli* by fluorescence *in situ* hybridization using an array scanner. J. Microbiol. Methods 45:31-39.

[49] Mullis KB, Faloona FA (1987) Specific synthesis of DNA *in vitro* via a polymerase-catalyzed chain reaction. Meth. Enzimol. 155: 335-350.

[50] Saiki RK, Scharf SJ, Faloona F, Mullis KB, Horn GT, Erlich HA, Arnheim N (1985). Enzymatic amplification of β-globin genomic sequences and restriction site analysis for diagnosis of sickle cell anemia. Science 230: 1350.

[51] Mullis KB, Faloona FA, Scharf SJ, Saiki RK, Horn GT, Erlich HA (1986) Specific enzymatic amplification of DNA *in vitro*: The polymerase chain reaction. Cold Spring Harbor Symp. Quant. Biol. 51:263–273.

[52] Hong Y, Berrang ME, Liu T, Hofacre CL, Sanchez S, Wang L, Maurer JJ (2003) Rapid detection of *Campylobacter coli, C. jejuni*, and *Salmonella enterica* on poultry carcasses by using PCR-enzyme-linked immunosorbent assay. Appl. Environ. Microbiol. 69:3492–3499.

[53] Lazcka O, Del Campo FJ, Munoz FX (2007) Pathogen detection: a perspective of traditional methods and biosensors. Biosensors and bioelectronics 22: 1205-1217.

[54] Aslam M, Hogan J, Smith KL, (2003) Development of a PCR-based assay to detect Shiga toxin-producing *Escherichia coli, Listeria monocytogenes,* and *Salmonella* in milk. Food Microbiol. 20:345–350.

[55] Caro I, Garcìa-Armesto MR (2007) Occurrence of Shiga toxin-producing *Escherichia coli* in a Spanish raw ewe's milk cheese. Int. J. Food Microbiol. 116: 410-413.

[56] Caro I, Fernández-Barata VM, Alonso-Llamazares A, García-Armesto MR (2006) Detection, occurrence, and characterization of *Escherichia coli* O157:H7 from raw ewe's milk in Spain. J. Food Prot. 69:920-4.

[57] El-Sharoud WM, (2009) Prevalence and survival of *Campylobacter* in Egyptian dairy products. Food Res. Int. 42:622–626.

[58] Fooladi IAA, Naderi THR (2010) Detection of enterotoxigenic *Staphylococcus aureus* isolates in domestic dairy products. Iranian J. Microbiol. 2:137-142.

[59] Gilbert C, Winters D, O'Leary A, Slavik M (2003) Development of a triplex PCR assay for the specific detection of *Campylobacter jejuni, Salmonella* spp., and *Escherichia coli* O157:H7. Mol. Cell Probes 17:135-138.

[60] Herman LMF, De Ridder HFM, Vlaemynck GMM (1995) A multiplex PCR method for the identification of *Listeria* spp. and *Listeria monocytogenes* in dairy samples. J. Food Protect. 58:867-872.

[61] Kav K, Col R, Ardic M (2011) Characterization of *Staphylococcus aureus* isolates from white-brined Urfa cheese. J. Food Prot. 74:1788-96.

[62] Meng J, Zhao S, Doyle MP (1998) Virulence genes of Shiga toxin-producing *Escherichia coli* isolated from food, animals and humans. Int. J. Food Microbiol. 45: 229–235.

[63] Mora A, León SL, Blanco M, Blanco JE, López C, Dahbi G, Echeita A, González EA, Blanco J (2007) Phage types, virulence genes and PFGE profiles of Shigatoxin-producing *Escherichia coli* O157:H7 isolated from raw beef, soft cheese and vegetables in Lima (Peru). Int. J. Food Microbiol. 114:204–210.

[64] Peles F, Wagner M, Varga L, Hein I, Rieck P, Gutser K, Kereszturi P, Szabo A (2007) Characterization of *Staphylococcus aureus* strains isolated from bovine milk in Hungary. Int. J. Food Microbiol. 118:186-193.

[65] Rosengren A, Fabricius A, Guss B, Sylven S, Lindqvist R (2010) Occurrence of foodborne pathogens and characterization of *Staphylococcus aureus* in cheese produced on farm-dairies. Int. J. Food Microbiol. 144:263-269.

[66] Tamarapu S, McKillip JL, Drake M (2001) Development of a multiplex polymerase chain reaction assay for detection and differentiation of *Staphylococcus aureus* in dairy products. J. Food Protect. 64:664–668.

[67] Debruyne L, Samyn E, De Brandt E, Vandenberg O, Heyndrickx M, Vandamme P (2008) Comparative performance of different PCR assays for the identification of *Campylobacter jejuni* and *Campylobacter coli*. Res. Microbiol. 159:88-93.
[68] On SLW, Jordan PJ (2003) Evaluation of 11 PCR assays for species-level identification of *Campylobacter jejuni* and *Campylobacter coli*. J. Clin. Microbiol. 41:330-336.
[69] On SLW, Brandt SM, Cornelius AJ, Fusco V, Quero GM, Mackie E, Houf K, Bilbao A, Díaz AI , Benejat L, Megraud F, Collins-Emerson J, French NP, Gotcheva V, Angelov A, Alakomi H-L, Saarela M, Paulin SM (2012) PCR revisited: a case for revalidation of pcr assays for microorganisms using identification of *Campylobacter* species as an exemplar. Qual. Assur. Saf. Crop. (Submitted).
[70] Wang RF, Cao WW, Johnson MG (1992) 16S rRNA-based probes and polymerase chain reaction method to detect *Listeria monocytogenes* cells added to foods. Appl. Environ. Microbiol. 58:2827-2831.
[71] Furrer U, Candrian C, Luethy HJ (1991) Detection and identification of *Listeria monocytogenes* in cooked sausage products and in milk by *in vitro* amplification of haemolysin gene fragments J. Appl. Bacteriol. 70:372–379.
[72] Amagliani G, Brandi G, Omiccioli E, Casiere A, Bruce IJ, Magnani M (2004) Direct detection of *Listeria monocytogenes* from milk by magnetic based DNA isolation and PCR. Food Microbiol. 21: 597–603.
[73] Aznar R, Alarcón B (2003) PCR detection of *Listeria monocytogenes*: a study of multiple factors affecting sensitivity J. Appl. Microbiol. 95:958–966.
[74] Ercolini D, Fusco V, Blaiotta G, Sarghini F, Coppola S (2005) Response of *Escherichia coli* O157:H7, *Listeria monocytogenes, Salmonella typhimurium,* and *Staphylococcus aureus* to the thermal stress occurring in model manufactures Grana Padano cheese. J. Dairy Sci. 88:3818–3825.
[75] Longhi CA, Maffeo M, Petrone PG, Seganti L, Conte MP (2003) Detection of *Listeria monocytogenes* in Italian-style soft cheeses. J. Appl. Microbiol. 94: 879–885.
[76] Vanniasinkam T, Lanser JA, Barton MD (1999) PCR for the detection of *Campylobacter* spp. in clinical specimens Lett. Appl. Microbiol. 28:52–56.
[77] Meng J, Zhao S, Doyle MP, Mitchell SE, Kresovich S (1996) Polymerase chain reaction for detecting *Escherichia coli* O157: H7. Int. J. Food Microbiol. 32:103-13.
[78] Gooding CM, Choudary PV (1997) Rapid and sensitive immunomagnetic separation-polymerase chain reaction method for the detection of *Escherichia coli* O157:H7 in raw milk and ice-cream. J. Dairy Res. 64:87-93.
[79] Fusco V, Riccardi M, Quero GM (2012b) TAL method versus MPN-PCR to enumerate viable and stressed *E. coli* O157:H7 and application in a traditional raw milk pasta filata cheese. Int. J. Food Microbiol.
[80] McKillip JL, Jaykus LA, Drake MA (2000) A comparison of methods for the detection of *Escherichia coli* O157:H7 from artificially-contaminated dairy products using PCR. J. Appl. Microbiol. 89: 49–55.
[81] Ferretti R, Mannazzu I, Cocolin L, Comi G, Clementi F (2001) Twelve-hour PCR-based method for detection of *Salmonella* spp. in food. Appl. Environ. Microbiol. 67:977-978.

[82] Houser BA, Donaldson SC, Kehoe SI, Heinrichs AJ, Jayarao BM (2008) A survey of bacteriological quality and the occurrence of *Salmonella* in raw bovine colostrum. Foodborne Pathog. Dis. 5(6): 853-858.

[83] Khan MA, Kim CH, Kakoma I, Morin E, Hansen RD, Hurley WL, Tripathy DN, Baek BK (1998) Detection of *Staphylococcus aureus* in milk by use of polymerase chain reaction analysis. Am. J. Vet. Res. 59:807-13.

[84] Blaiotta G, Ercolini D, Pennacchia C, Fusco V, Casaburi A, Pepe O, Villani F (2004) PCR detection of staphylococcal enterotoxin genes in *Staphylococcus* spp. strains isolated from meat and dairy products. Evidence for new variants of seG and seI in *S. aureus* AB-8802. J. Appl. Microbiol. 97:719–730.

[85] Boerema JA, Clemens R, Brightwell G (2006) Evaluation of molecular methods to determine enterotoxigenic status and molecular genotype of bovine, ovine, human and food isolates of *Staphylococcus aureus*. Int. J. Food Microbiol. 107:192 – 201.

[86] Ercolini D, Blaiotta G, Fusco V, Coppola S (2004) PCR-based detection of enterotoxigenic *Staphylococcus aureus* in the early stages of raw milk cheese making. J. Appl. Microbiol. 96:1090–1096.

[87] Kim CH, Khan M, Morin DE, Hurley WL, Tripathy DN, Kehrli M Jr, Oluoch AO, Kakoma I (2001) Optimization of the PCR for detection of *Staphylococcus aureus nuc* gene in bovine milk. J. Dairy Sci. 84(1):74-83.

[88] Pinto B, Chenoll E, Aznar R (2005) Identification and typing of food-borne *Staphylococcus aureus* by PCR-based techniques. Sys. Appl. Microbiol. 28:340-352.

[89] Rosec JP, Gigaud O (2002) Staphylococcal enterotoxin genes of classical and new types detected by PCR in France. Int. J. Food Microbiol. 77: 61-70.

[90] Adwan G, Abu-Shanab B, Adwan K (2005) Enterotoxigenic *Staphylococcus aureus* in raw milk in the north of Palestine. Turk. J. Biol. 29:229-232.

[91] Arcuri EF, Angelo FF, Guimiraes MFM, Talon R, de Fatima Borges M, Leroy S, Loiseau G, Lange CC, de Andrade NJ, Montet D (2010) Toxigenic status of *Staphylococcus aureus* isolated from bovine raw milk and minas frescal cheese in Brazil. J. Food Protect. 73:2225-2231.

[92] Argudín MA, Mendoza MC, González-Hevia MA, Bances M, Guerra B,M. R. Rodicio (2012) Genotypes, exotoxin gene content, and antimicrobial resistance of *Staphylococcus aureus* strains recovered from foods and food handlers. Appl. Environ. Microbiol. 78:2930-2935.

[93] Bartolomeoli I, Maifreni M, Frigo F, Urli G, Marino M (2009) Occurrence and characterization of *Staphylococcus aureus* isolated from raw milk for cheesemaking. Int. J. Dairy Technol. 62:366-373.

[94] Bendahou A, Abid M, Bouteldoun N, Catelejine D, Lebbadi M (2009) Enterotoxigenic coagulase positive *Staphylococcus* in milk and milk products, lben and Jben, in northern Morocco. J. Infect. Dev. Ctries. 3:169-176.

[95] Intrakamhaeng M., Komutarin T., Pimpukdee K., Aengwanich W (2012) Incidence of enterotoxin-producing MRSA in bovine mastitis cases, bulk milk tanks and processing plants in Thailand. J. Anim. Vet. Adv. 11:655-661.

[96] Rall VLM, Vieira FP, Rall R, Vieitis RL, Fernandes Jr A, Candeias JMG, Cardoso KFG, Araujo Jr. JP (2008) PCR detection of staphylococcal enterotoxin genes in *Staphylococcus aureus* strains isolated from raw and pasteurized milk. Vet. Microbiol. 132:408-413.
[97] Wilson IG, Cooper JE, Gilmour A (1991) Detection of enterotoxigenic *Staphylococcus aureus* in dried skimmed milk: use of the polymerase chain reaction for amplification and detection of staphylococcal enterotoxin genes *entB* and *entC1* and the thermonuclease gene *nuc*. Appl. Environ. Microbiol. 57: 1793-1798.
[98] Chiang Y-C, Chang L-T, Lin C-W, Yang C-Y, Tsen H-Y (2006) PCR Primers for the detection of staphylococcal enterotoxins K, L, and M and survey of staphylococcal enterotoxin types in *Staphylococcus aureus* isolates from food poisoning cases in Taiwan. J. Food Protect. 69:1072-1079.
[99] Mantynen V, Niemela S, Kaijalainen S, Pirhonen T, Lindstrom K (1997) MPN-PCR-quantification method for staphylococcal enterotoxin c1 gene from fresh cheese. Int. J. Food Microbiol. 36:135-143.
[100] Blaiotta G, Fusco V, von Eiff C, Villani F, Becker K (2006) Biotyping of enterotoxigenic *Staphylococcus aureus* by enterotoxin gene cluster (*egc*) polymorphism and *spa* typing analyses. Appl. Environ. Microbiol. 72:6117-23.
[101] Fusco V, Quero GM, Morea M, Blaiotta G, Visconti A (2011a) Rapid and reliable identification of *Staphylococcus aureus* harbouring the enterotoxin gene cluster (*egc*) and quantitative detection in raw milk by real time PCR. Int. J. Food Microbiol. 144:528-37.
[102] Akineden Ö, Hassan AA, Schneider E, Usleber E (2008) Enterotoxigenic properties of *Staphylococcus aureus* isolated from goats' milk cheese. Int. J. Food Microbiol.124:211-216.
[103] Aires-de-Sousa M, Parente CE, Vieira-da-Motta O, Bonna IC, Silva DA, de Lencastre H (2007) Characterization of *Staphylococcus aureus* isolates from buffalo, bovine, ovine, and caprine milk samples collected in Rio de Janeiro State, Brazil. Appl. Environ. Microbiol. 73:3845-9.
[104] Hwang SY, Park YK, Koo HC, Park YH (2010) *spa* typing and enterotoxin gene profile of *Staphylococcus aureus* isolated from bovine raw milk in Korea. J. Vet. Sci. 11:125–131.
[105] Johler S, Layer F, Stephan R (2001) Comparison of virulence and antibiotic resistance genes of food poisoning outbreak isolates of *Staphylococcus aureus* with isolates obtained from bovine mastitis milk and pig carcasses. J Food Prot.74:1852-9.
[106] Stephan R., Annemuller C., Hassan A.A., Lammler CH (2001) Characterization of enterotoxigenic *Staphylococcus aureus* strains isolated from bovine mastitis in north-east Switzerland. Vet. Microbiol. 78:373-382.
[107] Mackay IM (2004) Real-time PCR in the microbiology laboratory. Clin. Microbiol. Infect. 10:190-212.
[108] Wilhelm J, Pingoud A (2003) Real –time Polymerase chain reaction. Chem. Bio Chem. 4:1120-1128.
[109] Förster T (1948) Zwischenmolekulare energiewanderung und fluoreszenz. Ann. Phys. 6: 55-75.
[110] Fusco V, Quero GM, Stea G, Morea M, Visconti A (2011b) Novel PCR-based identification of *Weissella confusa* using an AFLP-derived marker. Int. J. Food Microbiol. 145:437-443.

[111] Powell HA, Gooding CM, Garrett SD, Lund BM, McKee RA (1994). Proteinase inhibition of the detection of *Listeria monocytogenes* in milk using the polymerase chain reaction. Lett. Appl. Microbiol. 18:59–61.

[112] Ramesh A, Padmapriya BP, Chandrashekar A, Varadaraj MC (2002) Application of a convenient DNA extraction method and multiplex PCR for the direct detection of *Staphylococcus aureus* and *Yersinia enterocolitica* in milk samples. Mol. Cell. Prob. 16:307–314.

[113] Wilson IG (1997) Inhibition and facilitation of nucleic acid amplification. Appl. Environ. Microbiol. 63:3741–3751.

[114] Barkema HW, Schukken YH, Zadoks RN (2006) Invited review: The role of cow, pathogen, and treatment regimen in the therapeutic success of bovine *Staphylococcus aureus* mastitis. J. Dairy Sci. 89:1877–1895.

[115] Celik HA, Aydin I, Kav K (2009) Identification and antimicrobial susceptibility of subclinical mastitis pathogens isolated from hair goats' milk. J. Anim. Vet. Adv. 8:1086–1090.

[116] Vanderhaeghen W, Cerpentier T, Adriaensen C, Vicca J, Hermans K, Butaye P (2010) Methicillin-resistant *Staphylococcus aureus* (MRSA) ST398 associated with clinical and subclinical mastitis in Belgian cows. Vet. Microbiol. 144:166–171.

[117] Asao T, Kumeda Y, Kawai T, Shibata T, Oda H, Haruki K, Nakazawa H, Kozaki S (2003) An extensive outbreak of staphylococcal food poisoning due to low-fat milk in Japan: estimation of enterotoxin A in the incriminated milk and powdered skim milk. Epidemiol Infect. 130:33-40.

[118] De Buyser M-L, Dufour B, Maire M, Lafarge V (2001) Implication of milk and milk products in food-borne diseases in France and in different industrialised countries. Int. J. Food Microbiol. 67: 1–17.

[119] Jørgensen HJ, Mørk T, Rørvik LM (2005) The occurrence of *Staphylococcus aureus* on a farm with small-scale production of raw milk cheese. J. Dairy Sci. 88:3810–3817.

[120] Lindqvist R, Sylvén S, Vagsholm I (2002) Quantitative microbial risk assessment exemplified by *Staphylococcus aureus* in unripened cheese made from raw milk. Int. J. Food Microbiol. 78:155–170.

[121] Ostyn A, De Buyser ML, Guillier F, Groult J, Felix B, Salah S, Delmas G, Hennekinne JA (2010) First evidence of a food poisoning outbreak due to staphylococcal enterotoxin type E, France, 2009. Euro Surveill. 15(13) pii: 19528.

[122] Soejima T, Nagao E, Yano Y, Yamagata H, Kagi H, Shinagawa K (2007) Risk evaluation for staphylococcal food poisoning in processed milk produced with skim milk powder. Int. J. Food Microbiol. 115:29–34.

[123] Fusco V, Quero GM, Tempesta B (2012) Culture-dependent and independent TaqMan real time PCR-based identification and quantification of *Listeria monocytogenes* and its application in raw milk and soft cheese. (Manuscript in preparation).

[124] Nogva HK, Rudi K, Naterstad K, Holck A, Lillehaug D (2000) Application of 5′-nuclease PCR for quantitative detection of *Listeria monocytogenes* in pure cultures, water, skim milk, and unpasteurized whole milk. Appl. Environ. Microbiol. 66:4266.

[125] O'Brien M, Hunt K, McSweeney S, Jordan K (2009) Occurrence of foodborne pathogens in Irish farmhouse cheese. Food Microbiol. 26:910–914.

[126] O'Grady J, Sedano-Balbas S, Maher M, Smith T, Barry T (2008) Rapid real-time PCR detection of *Listeria monocytogenes* in enriched food samples based on the *ssr*A gene, a novel diagnostic target. Food Microbiol. 25:75-84.

[127] O'Grady J, Ruttledge M, Sedano-Balba´ s S, Smith TJ, Barry T, Maher M (2009) Rapid detection of *Listeria monocytogenes* in food using culture enrichment combined with real-time PCR. Food Microbiol. 26:4–7.

[128] D'Agostino M, Wagner M, Vazquez-Boland JA, Kuchta T, Kapriskova R, Hoorfar J, Novella S, Scortti M, Ellison J, Murray A, Fernandes I, Kuhn M, Pazlarova J, Heuvelink A, Cook N (2004) A validated PCR-based method to detect *Listeria monocytogenes* using raw milk as a food model-towards an international standard. J. Food Prot. 67:1646–1655.

[129] Rossmanith P, Krassnig M, Wagner M, Hein I (2006) Detection of *Listeria monocytogenes* in food using a combined enrichment/real-time PCR method targeting the *prf*A gene. Res. Microbiol. 157:763-771.

[130] De Martinis EC, Duvall RE, Hitchins AD. Real-time PCR detection of 16S rRNA genes speeds most-probable-number enumeration of foodborne *Listeria monocytogenes*. J. Food Prot. 2007 70:1650-5.

[131] Rantsiou K, Alessandria V, Urso R, Dolci P, Cocolin L (2008) Detection, quantification and vitality of *Listeria monocytogenes* in food as determined by quantitative PCR. Int. J. Food Microbiol. 121:99–105.

[132] Alessandria V, Rantsiou K, Dolci P, Cocolin L (2010) Molecular methods to assess *Listeria monocytogenes* route of contamination in a dairy processing plant. Int. J. Food Microbiol. 141:S156-S162.

[133] Chiang Y-C, Fan CM, Liao WW, Lin CK, Tsen HY (2007) Real-Time PCR detection of *Staphylococcus aureus* in milk and meat using new primers designed from the heat shock protein gene *htrA* sequence. J. Food Protect 70:2855-2859.

[134] Poli A, Guglielmini E, Sembeni S, Spiazzi M, Dellaglio F, Rossi F, Torriani S (2007) Detection of *Staphylococcus aureus* and enterotoxin genotype diversity in Monte Veronese, a protected designation of origin Italian cheese. Lett. Appl. Microbiol. 45: 529-534.

[135] Ibekwe AM, Watt PM, Grieve CM, Sharma VK, Lyons SR (2002) Multiplex fluorogenic real-time PCR for detection and quantification of *Escherichia coli* O157:H7 in dairy wastewater Wetlands. Appl. Environ. Microbiol. 68: 4853–4862.

[136] McKillip JL, Drake M (2000) Molecular beacon polymerase chain reaction detection of *Escherichia coli* O157:H7 in milk. J. Food Protect. 63: 855–859.

[137] Auvray F, Lecureuil C, Dilasser F, Tache´ J, Derzelle S (2009) Development of a real-time PCR assay with an internal amplification control for the screening of shiga toxin-producing *Escherichia coli* in foods. Lett. Appl. Microbiol. 48:554–559.

[138] Yang C, Jiang Y, Huang K, Zhu C, Yin Y (2003) Application of real-time PCR for quantitative detection of *Campylobacter jejuni* in poultry, milk and environmental water. FEMS Immunol. Med. Microbiol. 38:265-271.

[139] Malorny B, Paccassoni E, Fach P, Bunge C, Martin A, Helmuth R (2004) Diagnostic real-time PCR for detection of *Salmonella* in food. Appl. Environ. Microbiol. 70: 7046-7052.

[140] Bohaychuk VM, Gensler GE, McFall ME, King RK Renter DG (2007) A Real-Time PCR assay for the detection of *Salmonella* in a wide variety of food and food-animal matrices. J. Food Protect. 70:1080-1087.

[141] Paiva de Sousa C (2008) The impact of food manufacturing practices on food borne diseases. Braz. Arch. Biol. Technol. 51: 815-823.

[142] Wu VCH (2008) A review of microbial injury and recovery methods. Food Microbiol. 25:735-744.

[143] Dinu L-D, Bach S (2011) Induction of viable but nonculturable *Escherichia coli* O157:H7 in the Phyllosphere of lettuce: a food safety risk factor. Appl. Environ. Microbiol. 77: 8295-8302.

[144] House B, Kus JV, Prayitno N, Mair R, Que L, Chingcuanco F, Gannon V, Cvitkovitch DG, Barnett Foster D (2009) Acid-stress-induced changes in enterohaemorrhagic *Escherichia coli* O157:H7 virulence. Microbiology 155: 2907–2918.

[145] Robinson AL, McKillip J (2010) Biology of *Escherichia coli* O157:H7 in human health and food safety with enphasis on sublethal injury and detection. 1096-1105. In: Current research, technology and education topics in applied microbiology and microbial biotechnology. Mendez-Vilas A. (Ed.) Formatex.

[146] McKillip JL, Jaykus L-A, Drake M (1999) Nucleic acid persistence in heat-killed *Escherichia coli* O157:H7 from contaminated skim milk. J. Food Protection. 62: 839-844.

[147] Lu Y, Yang W, Shi L, Li L, Alam MJ, Guo S, Miyoshi S (2009) Specific detection of viable *Salmonella* cells by an ethidium monoazide-loop mediated isothermal amplification (EMA-LAMP) method. J. Health Sci. 55:820-824.

[148] Nogva HK, Drømtorp SM, Nissen H, Rudi K (2003) Ethidium monoazide for DNA-based differentiation of viable and dead bacteria by 5′-nuclease PCR. BioTechniques. 34:804-813.

[149] Pan Y, Breidt F (2007) Enumeration of viable *Listeria monocytogenes* cells by real-time PCR with propidium monoazide and ethidium monoazide in the presence of dead cells. Appl. Environ. Microbiol. 73:8028-8031.

[150] Rudi, K., Naterstad, K., Drømtorp, S.M. and Holo, H. (2005), Detection of viable and dead *Listeria monocytogenes* on gouda-like cheeses by real-time PCR. Lett. Appl. Microbiol. 40: 301–306.

[151] Critzer FJ, Dsouza DH, Golden DA (2008) Transcription analysis of *stx1*, *marA*, and *eaeA* genes in *Escherichia coli* O157:H7 treated with sodium benzoate. J Food Prot. 71:1469-74.

[152] Gonzalez-Escalona N, Hammack TS, Russell M, Jacobson AP, De Jesus AJ, Brown EW, Lampel KA (2009) Detection of live *Salmonella* sp. cells in produce by a TaqMan-based quantitative reverse transcriptase real-time PCR targeting invA mRNA. Appl. Environ. Microbiol. 75:3714–3720.

[153] Hanna SE, Wang HH (2006) Assessment of environmental factors on *Listeria monocytogenes* Scott A *inlA* gene expression by relative quantitative Taqman real-time reverse transcriptase PCR. J. Food Prot. 69:2754–2757.

[154] Lin S, Gilpin B, Scholes P, Podivinsky E, Klena J, Savill M (2009) Preliminary development and validation of a real-time reverse transcription PCR assay for the semi-quantification of viable *Campylobacter jejuni* in water samples. Water Sci. Technol. 60:3151-8.

[155] Wada M, Lkhagvadorj E, Bian L, Wang C, Chiba Y, Nagata S, Shimizu T, Yamashiro Y, Asahara T, Nomoto K (2010) Quantitative reverse transcription-PCR assay for the rapid detection of methicillin-resistant *Staphylococcus aureus*. J. Appl. Microbiol. 108:779-88.

[156] Compton J (1991) Nucleic acid sequence-based amplification. Nature 350: 91–2.
[157] Blais BW, Turner G, Sooknanan R, Malek LT (1997) A nucleic acid sequence-based amplification system for detection of *Listeria monocytogenes hlyA* sequences. Appl. Environ. Microbiol. 63: 310–313.
[158] Cools I, Uyttendaele M, D'Haese E, Nelis HJ, Debevere J (2006) Development of a real-time NASBA assay for the detection of *Campylobacter jejuni* cells. J. Microbiol. Methods. 66: 313–320.
[159] Simpkins SA, Chan AB, Hays J, Pöpping B, Cook N (2000) An RNA transcription-based amplification technique (NASBA) for the detection of viable *Salmonella enterica.* Lett. Appl. Microbiol. 30:75-9.
[160] Buck RP (2000) Bioanalytic Sensors. In: The biomedical engineering handbook: Second Edition. Ed. Joseph D. Bronzino Boca Raton: CRC Press LLC.
[161] Rasooly A, Herold KE. (2006) Biosensors for the analysis of food- and waterborne pathogens and their toxins. J. AOAC Int. 89:873-883.
[162] Anderson GP, Taitt CR(2008) Amplification of microsphere-based microarrays using catalyzed reporter deposition. Biosen. Bioelectron. 24:324-328.
[163] Deisingh AK, Thompson M (2004) Biosensors for the detection of bacteria. Can. J. Microbiol. 50:69-77(9).
[164] Lui C, Cady NC, Batt CA (2009) Nucleic acid-based detection of bacterial pathogens using integrated microfluidic platform systems. Sensors 9: 3713-3.
[165] Mello LD, Kubota LT (2002) Review of the use of biosensors as analytical tools in the food and drink industries Food Chem. 77:237-256.
[166] Rodriguez-Mozaz S, Marco MP, Lopez de Alda MJ, Barceló D (2004) Biosensors for environmental applications: future development trends. Pure Appl. Chem. 76:723–752.
[167] Hahn S, Mergenthaler S, Zimmermann B, Holzgreve W (2005) Nucleic acid based biosensors: the desires of the user. Bioelectrochemistry. 67:151-154.
[168] Junhui Z, Hong C, Ruifu Y (1997) DNA based biosensors. Biotechnology advances.15:43-58.
[169] Cady NC (2005) Nucleic acid-based biosensing. Ph.D. dissertation. Cornell University, Ithaca, NY, USA.
[170] Liu-Stratton Y, Roy S, Sen CK (2004) DNA microarray technology in nutraceutical and food safety. Toxicol. Lett. 150:29-42.
[171] Cao B, Li R, Xiong S, Yao F, Liu X, Wang M, Feng L, Wang L (2011) Use of a DNA microarray for detection and identification of bacterial pathogens associated with fishery products. Appl. Environ. Microbiol. 77:8219-25.
[172] Donhauser SC, Niessner R, Seidel M (2011) Sensitive quantification of *Escherichia coli* O157:H7, *Salmonella enterica,* and *Campylobacter jejuni* by combining stopped polymerase chain reaction with chemiluminescence flow-through DNA microarray analysis. Anal. Chem. Apr 15;83(8):3153-60. Epub 2011 Mar 21.
[173] Fox EM, Leonard N, Jordan K (2011) Physiological and transcriptional characterization of persistent and nonpersistent *Listeria monocytogenes* isolates. Appl. Environ. Microbiol. 77:6559-69.
[174] Garneau P, Labrecque O, Maynard C, Messier S, Masson L, Archambault M, Harel J (2010) Use of a bacterial antimicrobial resistance gene microarray for the identification of resistant *Staphylococcus aureus*. Zoonoses Public Health. 57:94-99.

[175] Grønlund H, Riber L, Vigre H, Löfström C, Folling L, Huehn S, Malorny B, Rådström P, Rudi K, Hoorfar J (2011) Microarray-based genotyping of *Salmonella*: inter-laboratory evaluation of reproducibility and standardization potential. Int. J. Food Microbiol. 145 Suppl 1:S79-85.

[176] Koyuncu S, Andersson G, Vos P, Häggblom P (2011) DNA microarray for tracing *Salmonella* in the feed chain. Int. J. Food Microbiol. 145 Suppl 1:S18-22.

[177] Rodin S, Andersson AF, Wirta V, Eriksson L, Ljungström M, Björkholm B, Lindmark H, Engstrand L (2008) Performance of a 70-mer oligonucleotide microarray for genotyping of *Campylobacter jejuni*. BMC Microbiol. 8:73.

[178] Salehi TZ, Tonelli A, Mazza A, Staji H, Badagliacca P, Tamai IA, Jamshidi R, Harel J, Lelli R, Masson L (2011) Genetic characterization of *Escherichia coli* O157:H7 strains isolated from the one-humped camel (*Camelus dromedarius*) by using microarray DNA technology. Mol Biotechnol. [Epub ahead of print].

[179] Spence RP, Wright V, Ala-Aldeen DA, Turner DP, Wooldridge KG, James R (2008) Validation of virulence and epidemiology DNA microarray for identification and characterization of *Staphylococcus aureus* isolates. J. Clin. Microbiol. 46:1620-7.

[180] Khandurina J, Guttman A (2002) Bioanalysis in microfluidic devices. J. Chromatog. A. 943:159-183.

[181] Auroux P-A, Koc Y, deMello A, Manz A, Day PJR (2004) Miniaturised nucleic acid analysis. Lab Chip. 4:534-546.

[182] Bashir R (2004) BioMEMS: state-of-the-art in detection, opportunities and prospects. Adv. Drug Deliver. Rev. 56:1565-1596.

[183] Bhattacharya S, Salamat S, Morisette D, Banada P, Akin D, Liu YS, Bhunia AK, Ladisch M, Bashir R (2008) PCR-based detection in a micro-fabricated platform. Lab. Chip. 8:1130-6.

[184] Cady NC, Stelick S, Kunnavakkam MV, Batt CA (2005) Real-time PCR detection of *Listeria monocytogenes* using an integrated microfluidic platform. Sensor. Actuat. B-Chem. 107:332-341.

[185] Cady NC, Stelick S, Batt CA (2003) Nucleic acid purification using microfabricated silicon structures. Biosens. Bioelectron. 19:59-66.

[186] Fusco V, Cady NC, Stelick S, Batt CA (2006) Real Time PCR detection of *Staphylococcus aureus* using a microfluidics biosensor module. FoodMicro 2006 food safety and food biotechnology: diversity and global impact. 29/8–2/9 2006, Bologna, Italy. Page 397.

[187] Van Belkum A, Tassios PT, Dijkshoorn L, Haeggman S, Cookson B, Fry NK, Fussing V, Green J, Feil E, Gerner-Smidt P, Brisse S, Struelens M. and for the European Society of Clinical Microbiology and Infectious Diseases (ESCMID) Study group on epidemiological markers (ESGEM) (2007) Guidelines for the validation and application of typing methods for use in bacterial epidemiology. Clin. Microbiol. Infect. 13: 1–46.

[188] Van Belkum A, Scherer S, van Alphen L, Verbrugh H (1998a) Short-Sequence DNA Repeats in Prokaryotic genomes. Microbiol. Mol. Biol. Rev. 62: 275-293.

[189] Welsh J, McClelland M (1990) Fingerprinting genomes using PCR with arbitrary primers. Nucleic Acid Res. 25: 7212-7218.

[190] van Belkum A, Sluijuter M, de Groot R, Verbrugh H, Hermans PWJ (1996) Novel BOX repeat PCR assay for high-resolution typing of *Streptococcus pneumoniae* strains. Clin. Microbiol. May 1996 34: 1176-1179.

[191] Versalovic J, Koeuth T, Lupski JR (1991) Distribution of repetitive DNA sequences in eubacteria and application to fingerprinting of bacterial genomes. Nucleic Acids Res. 19:6823–6831.

[192] Versalovic J, Schneider M, De Bruijn FJ, Lupski JR (1994) Genomic fingerprinting of bacteria using repetitive sequence-based polymerase chain reaction Methods in Molecular and Cellular Biology. Methods Mol. Cell. Biol. 5: 25-40.

[193] Grimont F, Grimont PAD (1986). Ribosomal ribonucleic acid gene restriction patterns as potential taxonomic tools. Ann. Inst. Pasteur Microbiol. 137B: 165-175.

[194] Grif K, Dierich MP, Much P, Hofer E, Allerberger F (2003) Identifying and subtyping species of dangerous pathogens by automated ribotyping. Diagn. Microbiol. Infect. Dis. 47:313-20.

[195] Arimi SM, Ryser E, Pritchard T, Donnelly CW (1997) Diversity of *Listeria* ribotypes recovered from dairy cattle, silage, and dairy processing environments J. Food Protect. 60:811-816.

[196] Dalton CB, Austin CC, Sobel J, Hayes PS, Bibb WF, Graves LM, Swaminathan B, Proctor ME, Griffin PM (1997) An outbreak of gastroenteritis and fever due to *Listeria monocytogenes* in milk. N. Engl J. Med. 336:100-106.

[197] Dogan B, Boor KJ (2003) Genetic diversity and spoilage potentials among *Pseudomonas* spp. isolated from fluid milk products and dairy processing plants. Appl. Environ. Microbiol. 69:130-138.

[198] Kabuki D.Y., Kuaye A.Y., Wiedmann M., Boor K.J. (2004) Molecular subtyping and tracking of *Listeria monocytogenes* in Latin-style fresh-cheese processing plants. J. Dairy Sci. 87 (9) , pp. 2803-2812.

[199] MacDonald PDM, Whitwam RE, Boggs JD, MacCormack JN, Anderson KL, Reardon JW, Saah JR, Graves LM, Hunter SB, Sobel J (2005) Outbreak of listeriosis among mexican immigrants as a result of consumption of illicitly produced mexican-style cheese. Clin. Infect. Dis. 40: 677-682.

[200] Manfreda G, De Cesare A, Stella S, Cozzi M, Cantoni C (2005) Occurrence and ribotypes of *Listeria monocytogenes* in Gorgonzola cheeses. Int. J. Food Microbiol. 102:287-293.

[201] Buzzola FRL, Quelle MI, Gomez M, Catalano L, Steele-Moore D, Berg E, Gentilini G, Denamiel G, Sordelli Do (2001) Genotypic analysis of *Staphylococcus aureus* from milk of dairy cows with mastitis in Argentina. Epidemiol. Infect. 126:445-452.

[202] Hennekinne JA, Kerouanton A, Brisabois A, De Buyser ML (2003) Discrimination of *Staphylococcus aureus* biotypes by pulsed-field gel electrophoresis of DNA macro-restriction fragments. J. Appl. Microbiol. 94: 321–329.

[203] Hahm B-K, Maldonado Y, Schreiber E, Bhunia AK, Nakatsu CH (2003) Subtyping of foodborne and environmental isolates of *Escherichia coli* by multiplex-PCR, rep-PCR, PFGE, ribotyping and AFLP. J. Microbiol. Methods 53:387-399.

[204] Pradel N, Bertin Y, Martin C, Livrelli V (2008) Molecular analysis of shiga toxin-producing *Escherichia coli* strains isolated from hemolytic-uremic syndrome patients and dairy samples in France. Appl. Environ. Microbiol. 74:2118-28.

[205] Nastasi A, Mammina C, Sacco C (2000) Epidemiology of *Salmonella bongori* 48:z35:- in Southern Italy, 1984-2000. J. Prev. Med. Hyg. 41: 31-33.

[206] Gibson JR, Fitzgerald C, Owen RJ (1995) Comparison of PFGE, ribotyping and phage-typing in the epidemiological analysis of *Campylobacter jejuni* serotype HS2 infections. Epidemiol. Infect. 115:215-225.
[207] Patton CM, Wachsmuth IK, Evins GM, Kiehlbauch JA, Plikaytis BD, Troup N, Tompkins L, Lior H (1991) Evaluation of 10 methods to distinguish epidemic-associated *Campylobacter* strains. J. Clin. Microbiol. 29:680-688.
[208] Vaneechoutte M, Rossau R, De Vos P, Gillis M, Janssens D, Paepe N, De Rouck A, Fiers T, Claeys G, Kersters K (1992) Rapid identification of bacteria of the *Comamonadaceae* with amplified ribosomal DNA-restriction analysis (ARDRA). FEMS Microbiol Lett. 72:227-33.
[209] Wichelhaus TA, Hunfeld KP, Boddinghaus B, Kraiczy P, Schafer V, Brade V (2001) Rapid molecular typing of methicillin resistant *Staphylococcus aureus* by PCR-RFLP. Infect. Control Hosp. Epidemiol. 22:294-8.
[210] Harvey J, Gilmour A (1992) Occurrence of *Listeria* species in raw milk and dairy products produced in Northern Ireland. J Appl Bacteriol. 72:119-25.
[211] Harvey J, Gilmour A (1994) Application of multilocus enzyme electrophoresis and restriction fragment length polymorphism analysis to the typing of *Listeria monocytogenes* strains isolated from raw milk, nondairy foods, and clinical and veterinary sources. Appl. Environ. Microbiol. 60:1547-53.
[212] Waak E, Tham W, Danielsson-Tham M-L (2002) Prevalence and fingerprinting of *Listeria monocytogenes* strains isolated from raw whole milk in farm bulk tanks and in dairy plant receiving tanks. Appl. Environ. Microbiol. 68: 3366–3370.
[213] Beutin L, Miko A, Krause G, Pries K, Haby S, Steege K, Albrecht N (2007) Identification of human-pathogenic strains of shiga toxin-producing *Escherichia coli* from food by a combination of serotyping and molecular typing of Shiga toxin genes. Appl. Environ. Microbiol. 73:4769-4775.
[214] Dauga C, Zabrovskaia A, Grimont PA (1998) Restriction fragment lenght polymorphism analysis of some flagellin genes of *Salmonella enterica*. J. Clin. Microbiol. 36: 2835-2843.
[215] Kwon HJ, Park KY, Yoo HS, Park JY, Park YH, Kim SJ (2000) Differentiation of *Salmonella enterica* serotype *gallinarum* biotype *pullorum* from biotype *gallinarum* by analysis of phase 1 flagellin C gene *fliC*. J. Microbiol. Methods 40:33–38.
[216] Nielsen EM, Engberg J, Fussing V, Petersen L, Brogren CH, On SL (2000) Evaluation of phenotypic and genotypic methods for subtyping *Campylobacter jejuni* isolates from humans, poultry, and cattle. J. Clin. Microbiol. 38: 3800– 3810.
[217] Kawasaki S, Fratamico PM, Wesley IV, Kawamoto S (2008) Species-specific identification of *Campylobacter*s by PCR-Restriction Fragment Length Polymorphism and PCR targeting of the gyrase B gene. Appl. Environ. Microbiol. 2529–2533.
[218] Van Belkum A, van Leeuwen W, Kaufmann ME, Cookson B, Forey F, Etienne J, Goering R, Tenover F, Steward C, O'Brein F, Grubb W, Tassios P, Legakis N, Morvan A, El Solh N, de Ryck R, Struelens M, Salmenlinna S, Vuopio-Varkila J, Kooistra M, Talens A, Witte W, Verbrugh H (1998b) Assessment of resolution and intercenter reproducibility of results of genotyping *Staphylococcus aureus* by pulsed-field gel electrophoresis of SmaI macrorestriction fragments: a multicenter study. J. Clin. Microbiol. 36:1653–1659.

[219] Kathariou S (2003) Foodborne outbreaks of listeriosis and epidemic-associated lineages of *Listeria monocytogenes*. In: Microbial food safety in animal agriculture: current topics. 243-256.
[220] Swaminthan B, Barrett TJ, Hunter SB, Tauxe RV, Force CPT (2001) PulseNet: the molecular subtyping network for bacterial disease surveillance, United States. Emerg. Infect. Dis. 7: 382-389.
[221] Janssen P, Coopman R, Huys G, Swings J, Bleeker M., Vos P, Zabeau M, Kersters K (1996). Evaluation of the DNA fingerprinting method AFLP as a new tool in bacterial taxonomy. Microbiology 142:1881–1893.
[222] Vos P, Hogers R, Bleeker M, Reijans M, Van de Lee T, Hornes M, Frijters A, Pot J, Peleman J, Kuiper M, Zabeau M (1995) AFLP: a new technique for DNA fingerprinting. Nucl. Ac. Res. 21:4407–4414.
[223] Zhao S, Mitchell SE, Meng J, Kresovich S, Doyle MP, Dean RE, Casa AM, Weller JW (2000) Genomic typing of *Escherichia coli* O157:H7 by semi-automated fluorescent AFLP analysis. Microbes Infect. 2:107-113.
[224] Aarts HJM, Hakemulder LE, Van Hoef AMA (1999) Genomic typing of *Listeria monocytogenes* strains by automated laser fluorescence analysis of amplified fragment length polymorphism fingerprint patterns. Int. J. Food Microbiol. 49: 95-102.
[225] Lindstedt BA, Heir E, Vardund T, Kapperud G (2000) Fluorescent amplified-fragment length polymorphism genotyping of *Salmonella enterica* subsp. *enterica* serovars and comparison with pulsed-field gel electrophoresis typing. J. Clin. Microbiol. 4:1623-1627.
[226] Lomonaco S, Nucera D, Parisi A, Normanno G, Bottero MT (2011) Comparison of two AFLP methods and PFGE using strains of *Listeria monocytogenes* isolated from environmental and food samples obtained from Piedmont, Italy. Int. J. Food Microbiol. 149:177-182.
[227] Tamada Y, Nakaoka Y, Nishimori K, Doi A, Kumaki T, Uemura N, Tanaka K, Makino SI, Sameshima T, Akiba M, Nakazawa M, Uchida I (2001) Molecular typing and epidemiological study of *Salmonella enterica* serotype Typhimurium isolates from cattle by fluorescent amplified-fragment length polymorphism fingerprinting and pulsed-field gel electrophoresis. J. Clin. Microbiol. 39:1057–1066.
[228] Tsai TY, Luo WC, Wu FT, Pan TM (2005) Molecular subtyping for *Escherichia coli* O157: H7 isolated in Taiwan. Microbiol. Immunol. 49:579-88.
[229] Maiden MCJ, Bygraves JA, Feil E, Morelli G, Russell JE, Urwin R, Zhang Q, Zhou J, Zurth K, Caugant DA, Feavers IM, Achtman M, Spratt BG (1998) Multilocus sequence typing: a portable approach to the identification of clones within populations of pathogenic microorganisms. Proc. Natl. Acad. Sci. USA 95:3140–3145.
[230] Maiden MC (2006) Multilocus sequence typing of bacteria. Annu. Rev. Microbiol. 60:561-88.
[231] Zhang W, Jayarao BM, Knabel SJ (2004) Multi-virulence-locus sequence typing of *Listeria monocytogenes*. Appl. Environ. Microbiol. 70: 913-920.

Chapter 14

The Emerging Role of the Yeast *Torulaspora delbrueckii* in Bread and Wine Production: Using Genetic Manipulation to Study Molecular Basis of Physiological Responses

Andreia Pacheco, Júlia Santos, Susana Chaves, Judite Almeida, Cecília Leão and Maria João Sousa

Additional information is available at the end of the chapter

1. Introduction

Saccharomyces cerevisiae is the yeast primarily responsible for both grape must and bread fermentation and has been used for centuries in wine and bread making. Commercial yeasts utilized by these industries are essentially strains of *S. cerevisiae* that have been selected and optimized for these applications. Nowadays, the desire to produce consumer-directed wines and bread of differentiated styles has led to the emergence of new standard selection criteria for desirable yeast strains and has expanded the selection of yeasts to other environments and to non-*Saccharomyces* species (1, 2). In this context, strains of *Torulaspora delbrueckii* display particular traits that have caught the attention of the bread and wine industries. However, commercial exploitation of *T. delbrueckii* in wine production has only recently begun, and this yeast is not yet used consistently in the bread industry, apart from Japan (3). This is mainly due to the lack of knowledge of its genetic and physiological background, particularly when compared to *S. cerevisiae,* the yeast traditionally used and one of the best-studied microorganisms. In latest years, there has been a substantial effort to characterize *T. delbrueckii* strains potentially interesting for industrial use, especially at the physiological and biochemical levels. However, there have been few studies regarding their molecular characterization, due to the lack of efficient molecular biology tools and of a sequenced genome. Two strains of interest for the bread industry have been isolated from traditional corn and rye bread dough from the North of Portugal. Both strains are potentially interesting to use in frozen dough technology due to their particularly high resistance to osmotic and freeze stress (4-6), and have been the subject of detailed studies. These studies focused on the characterization of traits that are relevant for a potential industrial application, such as growth rate and biomass yield on sucrose, maltose fermentative capacity, and patterns of sugar utilization and regulation when mixtures of different sugars are present (7).

Traceability is crucial in modern food technology, and it is thus essential to develop a method to discriminate between different *T. delbrueckii* strains. This would enable correct identification of the inoculated strain among the yeast flora present in bread dough or in wines. Recently, several methodologies of typing based on DNA polymorphisms that allow discriminating closely related yeast strains have been developed. Two different genetic fingerprinting techniques (karyotype analysis and mtDNA restriction analysis) have been used for detailed genotyping of *T. delbrueckii* strains. Mitochondrial DNA restriction analysis was not a good technique to differentiate among *T. delbrueckii* strains isolated from the same ecosystem that are genetically very closely related. In contrast, chromosome separation by pulsed-field gel electrophoresis revealed considerable variability in the chromosomal constitution of the strains studied, and turned out to be a useful method to discriminate among *T. delbrueckii* strains.

In the present chapter, we first review the current knowledge regarding the application of *T. delbrueckii* strains in the wine and bread industries, and discuss the physiological traits that make them valuable over *S. cerevisiae*. In the second part, we detail the biochemical and genetic characterization and disruption techniques used in *T. delbrueckii*, and conclude with future perspectives.

2. *Torulaspora delbrueckii* in wine production

2.1. *Saccharomyces cerevisiae* and non-*Saccharomyces* species in wine making: A brief overview

Saccharomyces cerevisiae is the yeast primarily responsible for grape must fermentation and has been used for centuries in wine making. Strains of this species are commonly found in nature on the surface of grapes and within the bioflora of wine cellars. Previous studies suggest that strains present in vats or in other winery equipment, rather than the strains found in the vineyards, are responsible for the fermentation of musts, although there is still some controversy on this subject (8, 9). Pure cultures of *S. cerevisiae* have been isolated from these environments and developed as starter cultures for conducting wine fermentations all over the world. Usually, start-up cultures of *S. cerevisiae* are employed to better control the fermentation process. In fact, in large scale fermentations and in the newer wine-producing countries, where a desirable natural flora may not be established in the vineyard and in the winery, there is a reluctance to rely on natural fermentation, and selected yeast cultures are inoculated into the grape must to induce fermentation (10, 11). Selection of strains for wine production is based on the identification of strains that can ferment grape must efficiently and produce wines of good quality. This selection is usually conducted within the genus *Saccharomyces* and using yeasts isolated in wine environments, particularly from cellars, which allow for growth of the better adapted strains (1, 12). Isolation of strains is followed by characterization of their traditional oenological properties (13), which are divided into technological and qualitative properties. Technological characteristics influence the efficiency of the fermentative process, while qualitative characteristics relate to the chemical composition and influence on the sensory properties of wines (1). Some of these characteristics can be evaluated by monitoring fermentation progress and by chemical analysis of the levels of compounds present at the end of fermentation (2). As these strains are marketed as dried yeast, they must also be capable of maintaining viability during the dehydration/ rehydration process (14). The desire to produce consumer-directed wines of differentiated styles has led to the emergence of new criteria when selecting yeast, such as: ability to enhance wine colour via metabolic formation of stable pigments; absence of β-glucosidase activity, ability to prevent colour degradation; facilitation of colloidal stabilization in red wines by allowing over-lees aging (to help stabilize colour); appropriate enhancement of aroma via the production of volatile compounds such as esters and higher alcohols, along with limited production of off-flavours; and the bestowing structure and body via the production of polyalcohols such as glycerol and 2,3-butanodiols, and the release of mannoproteins and yeast polysaccharides. Because it is unlikely to find *S. cerevisiae* strains with an ideal combination of oenological characteristics, it became necessary to expand the selection of strains to other environments and to non-*Saccharomyces* strains (1, 2). The latter strains cannot compete with *S. cerevisiae* under oenological conditions, but have distinct characteristics that positively influence the sensory profile of the wine. *S. cerevisiae* strains isolated from Parmesan cheese serum degrade malic acid present in wine musts, until 50% of its initial concentration (15). Likewise, *Schizosaccharomyces pombe* proved to be effective in the deacidification of must through consumption of malic acid (16-18). *Candida stellata* and *Kloeckera apiculata* produce large amounts of glycerol (19), whereas *Candida colliculosa* produces acetaldehyde and n-propanol (1). These positive characteristics can be used in wine production and therefore contribute to wine sensory composition. The future of this industry points to individualization of consumers based on their genetic differences and on their olfactory profiles, which will be determinant in decisions of production and marketing of wine (20).

Flavours and aromas of wines are due to the grape itself and to biological activities carried out by the microorganisms. Several species of yeasts and bacteria and sometimes filamentous fungi may be present during fermentation of the must, and are responsible for the final characteristics of the wine (1). Besides *Saccharomyces,* other genus such as *Hanseniaspora, Kluveromyces, Candida, Metschnikowia,* and *Pichia* are usually found. Presence and persistence of these non-*Saccharomyces* genera are conditioned by several factors such as temperature of fermentation, addition of nutrients, aeration, contact with the peel of the grape, nature of the *Saccharomyces* strain used, and inoculation practices. Other important yeasts in wine production are those responsible for wine deterioration. This category includes the genera *Brettanomyces* and some species of *Zygosaccharomyces, Candida,* and *Pichia* (14). The non-*Saccharomyces* yeasts, which also include species of *Hansenula, Kloeckera, Schizosaccharomyces,* and *Torulaspora,* are present in the early stages of fermentation, growing for several days until the fermentation is dominated by one or more of the *S. cerevisiae* strains (13, 21). During the initial stage of fermentation these yeasts, with low fermentation yield, produce high concentrations of long-chain alcohols, esters, aldehydes, and glycerol, which are important for the organoleptic characteristics of the wine (13). However, in the modern wine industry and particularly in large-scale production, as referred above, spontaneous fermentation is unlikely to be used due to lack of reproducibility of the wine quality. The main advantages of inoculated wine fermentations are a more rapid and even rate of fermentation and wine of more consistent quality (11, 22, 23). On the other hand, this practice has resulted in more uniform wines without the typical contribution of aromas and flavours from indigenous yeast flora. In an attempt to change this scenario, non-*Saccharomyces* yeasts have recently been used in commercial wine production, particularly in countries such as the United States and Australia. This practice represents a good alternative to problems that can result from spontaneous fermentation, without compromising the sensory profile of the wine. Due to the low ethanol tolerance demonstrated by some non-*Saccharomyces* yeasts, inoculation of these yeasts must be performed together with a more tolerant strain that assures fermentation is completed, usually mixed starter cultures or sequential inoculation (1). Mixed starter cultures of *Torulaspora delbrueckii* or *Kluyveromyces thermotolerans* together with *S. cerevisiae* have already been tested, and revealed promising results regarding the aroma and flavour obtained (24).

2.2. The use of *Torulaspora delbrueckii* in the wine industry

The use of non-*Saccharomyces* wine yeasts in pure cultures as fermentation starters has shown that these have both advantages and undesirable fermentation characteristics. Among the latter, there is the production of acetic acid, ethyl acetate, acetaldehyde and acetoin at high concentrations that usually impairs the use of such strains as starter cultures. A number of authors have reported on the impact of non-*Saccharomyces* yeast species on wine quality under usual winemaking conditions (25-29). *Torulaspora delbrueckii* is reported to have a positive influence on the taste and aroma of alcoholic beverages (19, 25, 26), at the same time exhibiting low production of acetaldehyde, acetoin, acetate, and ethyl acetate (25, 27, 30), even in high-sugar must (31). Due to its high fermentation purity, its usage under standard conditions, in mixed or sequential culture with *Saccharomyces cerevisiae,* has been suggested as a strategy to reduce the acetic acid content of wine (25, 26, 32). Even though it is a low ethanol producer, *T. delbrueckii* still has useful potential in sweet wine fermentation, as it does not seem to respond to osmotic stress in the same way as *S. cerevisiae*. For instance, mixed cultures of *T. delbrueckii* and *S. cerevisiae* have been used in pineapple wine production as a strategy for the production of a distinct flavour complexity (33). In these fermentations, volatile acidity production remained constant throughout the entire process, in contrast with fermentation carried out by *S. cerevisiae,* where over 35% of the total production occurs in the initial stage of fermentation. The mechanisms of osmotic resistance in *T. delbrueckii* have been investigated, but are still not completely understood and are significantly different from those of *S. cerevisiae* (34). Survival of wine strains during the fermentation process is also imperative for their application. Recent studies on wine ecology showed that non-*Saccharomyces* species survive during fermentation at significant levels for longer periods than previously thought (29). In high ethanol and moderate acetic acid concentrations, typical of stuck must fermentations, *T. delbrueckii* preserved its cell viability longer than *S. cerevisiae* (35), but on the other hand was unable to consume fructose under the same condition (35). On the contrary, *S. cerevisiae* presents a high fructose-consumption capacity but is much less resistant to ethanol and acetic acid. Attempting to combine the advantageous traits of these two yeasts, *T. delbrueckii*'s high ethanol and acetic acid tolerance and the high fructose consumption capacity of *S. cerevisiae,* Santos et al (35) created a hybrid strain of the two species by protoplast fusion. This hybrid (F1-11) displayed a fructose consumption capacity comparable to that of the *S. cerevisiae* parent strain and exhibited increased resistance to ethanol and acetic acid, displaying a lower cell death rate under the harsh conditions present in stuck fermentations. In addition to its potential to restart stuck fermentations, this hybrid could also be useful to conduct an entire fermentation that would benefit from its flavour properties.

3. *Torulaspora delbrueckii* in bread production

3.1. Baker's yeast and its important traits for baking applications

In the history of human nutrition, a diversity of bakery products has been created which continues today. Bread is mostly made from flour dough that is allowed to rise (leaven) before baking in the oven. Making bread requires three main ingredients: flour, water, and yeast. The yeast's main role in the bread making process is to promptly ferment the sugars available in the flour of the dough or that have been added to it. As a result of an efficient fermentation, the yeast produces carbon dioxide (CO_2) and ethanol; the CO_2 is trapped within the gluten matrix of the dough, causing the leavening or rising, while the ethanol contributes to flavour development, along with other volatile compounds and flavour precursors that are formed during the fermentation process. Technically, the most important properties of baker's yeast comprise (1) the leavening ability in the dough, (2) the ability to adapt to different carbon sources, by expressing invertase and maltase activities, and (3) stress resistance, particularly osmo- and cryo-tolerance. Obviously yeast should also contribute to the flavour of the baked products, as well as grow rapidly in molasses, which are used in the culture media in their industrial production. Commercial baker's yeasts are domesticated strains, essentially of *Saccharomyces cerevisiae,* that have been selected and optimized for baking applications. These particular features are the result of natural adaptation from the continuous selective pressure generated by yeast manufacturers for many years. However, some parameters are still far from optimal. Fermentative capacity is one of the most important biotechnological challenges in the baking industry. Yeast's gassing rate is crucial in baking technology and mostly depends on characteristics of baker's yeast. Tolerance to different stresses, like osmotic or freezing, is also clearly insufficient. Baker's yeast cells subjected to osmotic stress dehydrate rapidly, which limits growth and gas-production capacity (36, 37). Consequently, proofing time (the period where the yeast is allowed to leaven or raise the dough) increases and the bread volume is reduced. In frozen-sweet dough, freezing and thawing further reduce the water activity, aggravating this situation. Furthermore, freezing and frozen storage of dough has a negative impact on the baking performance due to cell damage (36, 37). Therefore, developing yeast strains with better gassing power in frozen and frozen-sweet dough is of great economic interest.

The physiological requirements of baker's yeast for optimal production and application represent an apparent contradiction (Figure 1). In fact, sugar-limited respiro-fermentative fed batch cultivation (yeast production phase) must render a yeast product that has developed a high fermentative capacity, although this requirement is not important during this phase. Subsequently, the gassing capacity (fermentation) is used in the application phase in the dough, under anaerobic, excess sugar conditions. Therefore the physiological flexibility of baker's yeast must be exceptional.

In addition to good fermentative capacities and high stress resistance, another major trait must be considered when selecting a yeast strain for the baking industry (38): effective biomass production in molasses. Because molasses are cheap and easily available and contain some nitrogen and several vitamins and minerals necessary for yeast growth, they are the main substrate used for large-scale baker's yeast production. However, molasses are considered a major factor of variation in the quality of baker's yeast (39). These substrates are highly variable and contain different proportions of sugars. Though sucrose is the major sugar present, there is also a quite high amount of glucose and fructose. Sucrose is cleaved outside the cell by invertase into glucose and fructose. Invertase is also capable of cleaving raffinose, a trisaccharide also present in molasses, into fructose and melibiose (glucose-galactose), but melibiose is generally not assimilated (40). The main fermentable sugar in plain bread dough is maltose, liberated from starch by amylase activity (α-glucosidase) in flour. This disaccharide is transported through a maltose permease and is subsequently hydrolyzed into glucose by maltase (figure 2). The order by which these different carbohydrates are fermented by *S. cerevisiae* is not random; rather, it is based on a specific hierarchy, with glucose being the preferred sugar. Consequently, in dough containing glucose, sucrose, and maltose, the disaccharides are fermented only when all the glucose is consumed. The monosaccharides glucose, fructose, and mannose are transported into cells across the plasma membrane by hexose transporter (Hxt) proteins, and are subsequently further metabolized in glycolysis. The endpoint of glycolysis is pyruvate, and whether it is used for respiration or fermentation depends on the growth conditions. When cells are grow in an anaerobic environment, sugars are fermented into CO_2 and ethanol. However, cells are also able to use glucose for fermentation when grown aerobically if glucose is present at high concentrations. This phenomenon is generally referred to as the Crabtree effect (42, 43). In an aerobic environment, glucose is catabolized exclusively through respiration only when cells are grown in low sugar concentrations. In addition to being the favourite carbon source for yeast, glucose controls different mechanisms that ensure its primary utilization, as well as general regulation of metabolism, cell growth, and development.

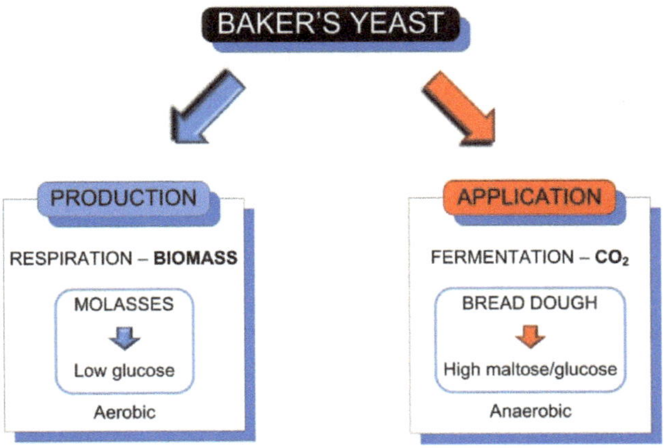

Figure 1. Baker's yeast production and application paradox. Baker's yeast must be able to readily ferment sugar to CO_2 (and ethanol) in doughs. At the same time it must grow on molasses in sugar-limited respiro-fermentative fed batch cultivations for yeast production. Therefore, when the metabolic flux is directed toward cell growth and biomass production, yeast is expected to display a good fermentative capacity. Adapted from (41).

For the most part, regulation is mediated by catabolite repression, acting at early steps in various catabolic pathways. The aim of the regulation is to induce utilization of most favoured carbon source (glucose), and to exclude utilization of other carbon sources if a sufficient amount of glucose is available. Therefore, in most strains of *S. cerevisiae*, glucose represses genes responsible for maltose transport and hydrolysis, as well as invertase that hydrolyzes sucrose to glucose and fructose (42). Although high levels of invertase activity are required for growth in molasses, there is evidence that the capacity of *S. cerevisiae* to ferment high sucrose concentrations, like those present in sweet bread doughs, is inversely related to the activity of this enzyme (43).

3.2. *Torulaspora delbrueckii* as an emergent yeast in the baking industry

Nowadays, the baker's yeast strains used have been developed as a result of centuries of experience and selection, resulting in a high degree of domestication best suited for bread making. Nevertheless, research to improve yeast strains continues. Although methods of classical genetics (selection, mutation, and hybridization) are still very useful, novel methods such as protoplast fusion and genetic engineering have resulted in baker's yeast strains with even better technological properties (35, 37, 44). *Saccharomyces cerevisiae* strains are generally used as baker's yeast, nevertheless the use of alternative species in bread making may allow coping with the new and more demanding challenges in the baking industry. Dough leavening ability has been reported for yeast other than *S. cerevisiae*. *Issatchenkia orientalis*, *Pichia membranaefaciens*, and *Torulaspora delbrueckii* were the most abundant non-*Saccharomyces* species present in homemade corn and rye bread dough (6, 45). Among them, the biotechnological interest in *T. delbrueckii* has increased in recent years due to its particularly high freezing and osmotic tolerance

(4, 5, 46). Thus, some *T. delbrueckii* strains are commercialized in Japan for regular (5% sucrose) frozen dough applications (47). In rural areas of Portugal, bread is sometimes prepared with dough carried over from a previous making. Following an isolation program of yeasts from homemade corn and rye bread doughs, two strains of *T. delbrueckii*, PYCC 5321 and PYCC 5323, were selected on the basis of simultaneously combining high growth rates and leavening ability, characteristics desirable in baker's yeast, as well as an exceptional freeze tolerance (6). The leavening activity of *T. delbrueckii* PYCC 5321 and PYCC 5323 upon freeze-thaw appeared to be unaltered, in contrast with the decrease of fermentative capacity of *S. cerevisiae* baker's yeast (4). In addition, these report from Almeida and Pais (1996b) (4) showed that *T. delbrueckii* strains are not affected by a period of fermentation before freezing. This yeast also displayed higher leavening ability than *S. cerevisiae*, under conditions of hyperosmotic stress in bread dough containing 20% sucrose and 2% salt (46). This feature is in agreement with a low invertase activity, a slow rate of trehalose mobilisation and the ability to respond rapidly to osmotic stress. In summary, *T. delbruecckii* strains studied gather the most important traits needed in a baker's yeast as rapid growth, high biomass, a high leavening activity in lean and frozen dough and extend these properties to sweet and frozen sweet dough. Thus, application of these strains in bread making would have additional advantages, since a single strain can be used for most or all baking applications with the highest performance.

Although this yeast is widely commercialized in Japan, the regular utilization of this species in the bread-making industry has not been established due to several drawbacks. *T. delbrueckii* strains are usually resistant to osmotic stress, but they show considerable variation in their abilities to ferment and to assimilate carbon compounds, as galactose or maltose (48), as well as variable maltase activity and fermentative capacity (46). Moreover, its small cell size is a major disadvantage in the industrial dehydration process, as filtration of cells for dehydration requires a long time and, even worse, cannot be performed continuously because the filter used for dehydration becomes clogged and must be changed frequently (49). Additionally, there is a lack of knowledge on the physiology and molecular biology of this yeast. Despite its phylogenetic closeness to *S. cerevisiae*, the differences observed between the two species demonstrate that the behaviour of *T. delbrueckii* cannot be directly inferred from that of *S. cerevisiae*.

4. Sugar metabolism in *Torulaspora delbrueckii*

As referred above, *Torulaspora delbrueckii* is an important case study among the non-*Saccharomyces* yeast species, of particular relevance to the baking and wine industries. In order to better evaluate the potential offered by *T. delbrueckii*, several physiological and biochemical studies have been carried out with this yeast. In particular, our group studied sugar utilization patterns, maltase and invertase activities, respiration/fermentation rates, and sugar uptake, thus contributing to a better understanding of the mechanisms underlying regulation of sugar metabolism in this yeast.

4.1. Sugar utilization patterns and respiro-fermentative metabolism

One of the leading characteristics of a valuable baker's yeast is its dough-leavening ability, which implies there is efficient fermentation of both the maltose and glucose that are present in the dough. Another important feature is its ability to generate high biomass yields on sucrose, the major sugar present in molasses. On the other hand, glucose and fructose are the main sugars present in grape musts, and thus efficient fermentation of these sugars is also of great importance for utilization of a yeast strain in wine fermentation.

The behavior of *T. delbrueckii* PYCC 5321 in Yeast Peptone (YP) medium with glucose, sucrose, and maltose, either as single carbon and energy source or in mixtures, has been compared with that of a commercial baker's yeast, used as reference (7). When single-sugar media were used, the growth rate of both yeasts was similar in both glucose and sucrose-containing media and slightly lower in maltose-containing medium. For *Saccharomyces cerevisiae,* these values were similar to those obtained in mixed-sugar media, but for *T. delbrueckii* the values obtained in mixed-sugar media with maltose (Glucose-Maltose (G-M) and Sucrose-Maltose (S-M) mixtures) were again slightly lower. Since analysis of the specific sugar consumption rates and transport capacities suggested that sugar transfer rates into the cell limit fermentation efficiency (50), the lower growth rates obtained in the presence of maltose were probably a result of inhibition of glucose uptake by maltose, which is known to occur in both *S. cerevisiae* (51) and *T. delbrueckii* (7). In G-M mixtures, maltose was consumed only after glucose was exhausted from the medium, and the increase in maltose transport and maltase activities clearly correlated with the beginning of maltose consumption. Therefore, it seems that the maltose transporter and maltase are co-regulated, and are subject both to induction by maltose and to repression by glucose. On the contrary, commercial baker's yeast began to consume maltose when glucose was still available, suggesting glucose control over maltose metabolism is higher in *T. delbrueckii* than in commercial baker's yeast. Under laboratory conditions, this distinct performance of the two species could imply an unwanted delay in CO_2 production in maltose-containing bread dough by *T. delbrueckii*. However, this apparent advantage of *S. cerevisiae* is counteracted by the higher osmotolerance of *T. delbrueckii* under the conditions prevailing in bread dough. In S-M mixtures, maltose consumption is also inhibited until sucrose is exhausted, although this inhibition is not particularly efficient and maltose levels slightly decrease concomitantly with sucrose consumption. These results suggest that glucose released from sucrose through invertase activity inhibits maltose metabolism. When cells were grown in media with either only one sugar or with mixtures of sugars as carbon sources, the invertase activities obtained were also in accordance with a regulation pattern similar to that observed in *S. cerevisiae*. In fact, while invertase activity was subject to glucose control, it was not dependent on induction by sucrose.

T. delbrueckii displayed biomass and ethanol yields typical of fermentative metabolism in all media. Despite this clear fermentative metabolism exhibited by *T. delbrueckii* in batch cultures with each of the sugars tested (glucose, maltose and sucrose), the specific rates of CO_2 production and O_2 consumption, estimated with the Warburg manometric technique, showed there was a higher contribution of respiration to the overall sugar metabolism than that observed in *S. cerevisiae*. It is worth noting that during batch cultivation the available oxygen rapidly reaches limiting concentrations, thereby favoring fermentative metabolism. In fact, when biomass yields were determined in YP medium containing glucose, sucrose, or maltose, using higher aeration rates resulted in a very significant increase in biomass yields (from 20 % in glucose or sucrose medium to 80 % in maltose medium). The more efficient modulation of the respiratory metabolism in *T. delbrueckii* under aerobic conditions thus represents an asset for the large-scale production of this yeast.

Aiming to provide new insights into the molecular mechanisms underlying energy source signaling in *T. delbrueckii*, a gene coding for a putative protein with high similarity to the *S. cerevisiae* carbon catabolic-derepressing Ser/Thr protein kinase Snf1 was recently identified, and named Tdsnf1p (52). It seems that, like Snf1p, Tdsnf1p is required for growth in ethanol, low glucose (0.05%) and raffinose-containing medium, although in the last case the phenotype was not as pronounced as that of the *S. cerevisiae snf1*Δ mutant. In contrast, the *tdsnf1*Δ mutant displayed increased Li^+ tolerance, a phenotype not observed in the *S. cerevisiae snf1*Δ mutant, further highlighting the differences between these two yeasts.

In summary, the overall patterns of sugar utilization and regulation in mixed sugar media by *T. delbrueckii* are equivalent to those described for *S. cerevisiae* (42, 53, 54), though some critical differences were identified. In addition, given that specific growth and fermentation rates in maltose-containing media were lower than those obtained in glucose- and sucrose-containing media, and that growth rate seems to be limited by transport capacity, these studies also indicated that maltose uptake is a good target for metabolic engineering and improvement of *T. delbrueckii*'s performance in bread doughs.

4.2. Hexose transport in *Torulaspora delbrueckii*

Yeast sugar transporters: a brief introduction

The sugar porter family is the largest within the major facilitator superfamily (MFS), which includes proteins from Bacteria, Achaea, and Eukarya, with very diverse sequences and function (55-57). Proteins belonging to the MFS exhibit highly structural conservation, though they share little sequence similarity (58). Generally, these permeases have 12 putative transmembrane segments, and consist of a single integral membrane protein with two sets of six hydrophobic transmembrane-spanning (TMS) α-helices connected by a hydrophilic loop, whose amino- and carboxy-terminal regions localize to the cytoplasm (59-61). The strong similarity between the two sets of hydrophobic TMS of MFS proteins and their overall structure supports the theory that they result from a gene duplication event that probably took place before de divergence of MFS families (57, 59). Sugar transport across the plasma membrane is the first and obligatory step of its utilization. Yeasts can use different carbon sources for growth, but evolution has selected mechanisms for the preferential utilization of glucose. As permeability of biological membranes is quite restricted, most of the cellular nutrients must enter the cell via specific transport systems and both facilitated diffusion and proton-symport transport systems for sugars have been described in yeasts. In facilitated diffusion, solutes are transported down a concentration gradient by a uniport mechanism. Secondary active transport uses accumulated energy from an electrochemical gradient to transport molecules against their concentration gradient, coupled with the simultaneous movement of another molecule (normally H^+ or Na^+) in the same (symport) or opposite (antiport) direction (62). Such a mechanism becomes fundamental during growth in very low extracellular sugar concentrations, when an intracellular accumulation of hexoses may be necessary to allow hexose kinases to function optimally. Evidently, yeast species possessing proton-hexose symport systems are better adapted to grow in low hexose concentrations (63, 64). Since a facilitated diffusion transport system is most efficient only under reasonably constant levels of the carrier substrate, this system might not be appropriate for yeasts like *Saccharomyces cerevisiae*. However, this yeast uptakes hexoses only by facilitated diffusion, and has clearly overcome this setback by developing an unusual diversity of hexose transporter proteins (Hxtp) with specific individual properties and kinetics (65).

Multiple hexose carriers have been characterized genetically in *S. cerevisiae*. Among about 5600 protein-coding genes in *S. cerevisiae*, at least 271 encode predicted or established permeases (66). Twenty encode hexose transporters and related proteins, the so-called HXT gene family (60, 65). Of these, only *HXT1–HXT7* encode transporters that are important for growth and metabolism of glucose (51, 67). The galactose transporter, encoded by *GAL2*, is also a member of the *HXT* gene family (68, 69). Two members of the family, encoded by *SNF3* and *RGT2*, have lost the ability to transport hexoses; instead they function as sensors of the extracellular glucose concentration. This glucose signal is involved in the transcriptional regulation of various *HXT* genes (70). The remaining members of the family (*HXT8–HXT17*) are phenotypically silent, and may not be expressed under normal physiological conditions (51). As discussed by Wieczorke et al. (71), the large number of hexose transporter proteins in *S. cerevisiae* seems to reflect its adaptation to the variety of environmental conditions to which yeast cells are exposed. Two kinetically distinct glucose uptake systems were initially described in *S. cerevisiae*: a constitutive low-affinity system (K_m=15–20 mM) and a glucose-repressible high-affinity system (K_m=1–2 mM) (reviewed in 70). It is now considered these two components consist of several different transporters contributing to the overall kinetic properties of the systems. The affinity of the major Hxt proteins for glucose was determined by individual expression of these transporters in a *hxt* null strain (72), and differs significantly; for example, Hxt1p and Hxt3p have a low-affinity for glucose, whereas Hxt2p, Hxt6p, and Hxt7p have a high-affinity (72). The low-affinity transporters are expressed when cells are grown in media containing high concentrations of glucose, whereas the high-affinity transporters are expressed when cells are grown in media containing low concentrations of glucose (70). It should however be noted that most of the data pertaining to the kinetics of glucose transport was obtained from individual expression of *HXT* genes in a *hxt* null mutant, and thus the results may not reflect the *in vivo* functions of Hxt proteins. When expressed in the absence of other Hxt proteins, an individual Hxt protein might display a different affinity for glucose, as it may be modulated by means of interactions between different Hxt proteins. Furthermore, any one *HXT* gene may be important for regulation of the expression of the other *HXT* genes (70, 73).

The number of hexose transporters is very variable among yeasts, ranging from 20 hexose transporters in *S. cerevisiae* and *Candida albicans* to six in *Schizosaccharomyces pombe*. *In silico* analysis of the *Kluyveromyces lactis* genome (74) showed that this yeast has 20 sugar transporter genes, but only seven of them have been characterized. Based on protein sequence homology (TBLAST search) with CaHgt1, the first gene encoding a glucose transporter in *C. albicans* (75), 19 other putative glucose transporters were uncovered and designated Hgt2-Hgt20 (76, 77). In the fission yeast *S. pombe*, six hexose transporter genes (*GHT1-GHT6*) have been identified (78), which are highly similar at both the nucleotide and amino acid level. In *Pichia stipitis*, three genes encoding glucose transporters *SUT1*, *SUT2* and *SUT3* (sugar transporters) have been identified, which probably constitute only a subfamily of glucose transporters (79). With the complete sequencing of the *P. stipitis* genome, several additional putative sugar transporter genes were also uncovered (80). Stasyk et al. (81) described Hxt1p, the first functional hexose transporter identified in the methylotrophic yeast *Hansenula polymorpha*. Wei et al. (82) found at least 17 putative hexose transporters in the genome of *Aspergillus nidulans*. So far, only one hexose (particularly fructose) transporter has been described in *S. pastorianus* (83) and in *Zygosaccharomyces bailii (84)*. Fsy1p (fructose symport) is a specific fructose/H+ symporter which mediates high-affinity fructose uptake (it does not transport glucose) in *S. pastorianus* (83). In *Z. bailii*, Ffz1p (fructose facilitator of *Zygosaccharomyces*) does not accept glucose as a substrate and displays low affinity for fructose (84).

Hexose transport in *Torulaspora delbrueckii* and cloning of *LGT1*

Two natural habitats of *Torulaspora delbrueckii* are bread doughs and fruit juices, such as grape juice, environments that are rich in sugars. As a consequence of growth and fermentation of these sugars, this yeast experiences dramatic changes in its physicochemical environment, and thus must adapt to these varying conditions in order to sustain its growth. The sugar concentration may decline from 1 M to 10^{-5} M during fermentation, and the overall composition of the medium will be altered by yeast metabolism. The sugar transport activity of the cell and the proteins that mediate sugar transport must be responsive to these changing conditions, and thus the capacity and kinetic complexity of hexose transport in the yeast may be a reflection of the existence of a large number of sugar transporter genes in its genome. Based on this assumption, multiple hexose transporters with different affinities for glucose probably exist in *T. delbrueckii*. This yeast displays a mediated glucose transport activity best fitted assuming a biphasic Michaelis–Menten kinetics with a low- (apparent $K_m = 8.32 \pm 0.55$ mM) and a high-affinity component (apparent $K_m = 1.30 \pm 0.34$ mM) (50). A kinetic compatible with the presence of these two components was observed in either glucose-, frutose- or maltose-grown cells. Aiming to identify glucose transporters in *T. delbrueckii*, a complementation screen of a *S. cerevisiae* hexose transport-null mutant strain (71) yielded a genomic DNA fragment containing a gene encoding Lgt1p, a low-affinity glucose transporter (50). When expressed in the *hxt* null strain, Lgt1p exhibited an apparent Km value for glucose of 36.5 ± 3.1 mM, in the range of the low-affinity component, and a V_{max} of 1.1 ± 0.04 nmol/s/mg dry weight. This transporter is also able to mediate significant fructose uptake in the *hxt* mutant, although with a lower affinity than that for glucose, with an apparent K_m value of 51.4 ± 3.0 mM. Glucose transport in this mutant (expressing Lgt1p) was inhibited by the presence of frutose, manose and maltose. Expression studies of the *T. delbrueckii LGT1* gene in *S. cerevisiae* strains, including wild-type, using a fusion of the *LGT1* promoter to the reporter gene l*acZ*, revealed that it was induced by high glucose concentrations, and its expression was elevated in media containing 4% glucose and almost undetectable in medium containing galactose as the sole carbon source. The transcription factor Rgt1p was necessary for repression of *LGT1* in the absence of glucose; however, and in contrast with the activity of Rgt1p as a bifunctional regulator in *S. cerevisiae* strains, full induction of *LGT1* by high glucose concentrations does not require functional Rgt1p. Even though Mig1p-binding sequences were identified in the promoter region of the *LGT1* gene, the general repressor of *S. cerevisiae* had no effect in the regulation of *LGT1* gene expression. However, disruption of *MIG2* in a *mig1* background led to high levels of *LGT1* expression in high glucose concentrations, indicating that either Mig2p or both Mig1p and Mig2p acting redundantly, function as repressors of *LGT1* expression under these conditions, consistently with their function in *S. cerevisiae*.

Even though just one glucose transporter has been identified in *T. delbrueckii* until now (the low-affinity glucose transporter *LGT1* (50)), it is likely that *T. delbrueckii* possesses high-affinity transporters, which is supported by the biphasic Michaelis–Menten kinetics of glucose transport. These results suggest there are additional physiological relevant glucose transporters. Identification of novel transporters will provide new clues regarding the mechanisms underlying regulation of sugar transport, and as a consequence the fermentative capacity of this biotechnologically relevant yeast.

4.3. Cloning and characterization of the *Torulaspora delbrueckii MAL11*, encoding a high-affinity maltose transporter

The genes involved in the utilization of maltose have been characterized in detail in laboratory strains of *Saccharomyces cerevisiae* (85). Genetic analysis revealed there are five MAL loci, *MAL1-MAL4* and *MAL6*, located on different chromosomes (86), but with a high degree of similarity (87). Each locus contains a set of three different genes that encode a maltose transporter (MALT - *MALx1*, where x represents the number of each locus), α-glucosidase (MALS - *MALx2*), and a regulatory protein (MALR - *MALx3*) (53). *MALx1* genes code for high affinity maltose-H^+ simporters with a K_m of approximately 5 mM (88), but with different specificities for various substrates. While Mal11p/Agt1p transports a wide range of substrates, including several α-glucosides (89), Mal31p and Mal61p seem to primarily use maltose, maltotriose, and turanose (89, 90). Nevertheless, only one fully functional locus seems to be found in standard laboratory strains, MAL1, which is heavily regulated through repression by glucose and induction by maltose (91). Comparatively, industrial yeasts contain multiple fully or partially functional MAL loci (92, 93). Additional analysis showed there are considerable variations in the *MALR* gene (94), leading to non-sensitivity to glucose and lack of control by maltose. These special features were the result of applying successive programs directed to a rapid adaptation to the fermentation of maltose.

Characterization of maltose transport rates in *Torulaspora delbrueckii* indicated it contains an inducible active transport system that co-transports protons with maltose, with the following kinetic parameters: V_{max} 1.03 ± of 0.05 nmol s^{-1} (mg dry weight) $^{-1}$ and K_m 2.26 ± 0.27 mM maltose (95). This transport system was subject to glucose repression and was competitively inhibited by the presence of sucrose, melizitose, and melibiose, suggesting these sugars likely share the same transporter(s) with maltose.

A DNA fragment containing the *MAL11* gene from *T. delbrueckii* (*TdMAL11*) was isolated by complementation cloning in *S. cerevisiae* with a *T. delbrueckii* genomic library (95, 96). DNA sequence analysis revealed the presence of an ORF of 1884 bp encoding a putative 627-amino acid membrane protein with 10 transmembrane domains, highly similar to other yeast maltose transporters. Upstream of *TdMAL11*, the DNA insert included a partial ORF (*TdMAL12*) on the opposite strand and direction, highly similar to the *S. cerevisiae MAL12* gene. Two consensus binding sites of the repressor Mig1p (positions -362 to -378 and -459 to -475) are evident in the divergent promoter. Interestingly, the first overlaps with a binding site of a MAL activator (-451 to -461). A similar situation was described in the intergenic region *MAL61-MAL62* in *S. cerevisiae*, where one of the two Mig1p binding sites is very close to a UAS_{MAL} site (97). This overlap seems to have a functional role in the transcriptional regulation of *MAL61* and *MAL62* genes, as occupation of the UAS_{MAL} site will result from direct competition between the two regulators for the binding region (98). Sequence analysis, Northern blot, and transport measurements, indicated that *TdMAL11* expression is regulated by carbon source, and is subjected to repression by glucose and induction by maltose. Attempts to disrupt *TdMAL11* and Southern blot analyses revealed the presence of two functional *MAL loci*. Disruption of a single copy decreased the V_{max} of maltose transport, but not the K_m, whereas the double disruption abolished the uptake of this sugar in *T. delbrueckii* (95).

As referred above, the activity and regulation of the maltose uptake system in *T. delbrueckii* cells could be a good target for improvement of its leavening ability in bread dough, and the identification of maltose transporter genes in this yeast can now open the door to these studies.

5. Freezing tolerance of *Torulaspora delbrueckii*: Cellular and biochemical basis

Baker's yeast and stress resistance

In the baking industry, yeasts encounter numerous stresses. During production, they must adapt to low sugar and high aeration, repressing fermentation to produce large amounts of biomass. Cells are then preserved in a cold, frozen or dry state until use, when rehydration or thawing and inoculation cause osmotic shock in a new environment that requires the induction of enzymes for maltose utilization under semianaerobic conditions. The low stress resistance of yeast during active fermentation is disadvantageous for its use in industrial applications, and it would be highly advantageous to have yeast strains available that do not lose their stress resistance during fermentation (36). Furthermore, human food habits have changed in the past few years and there has been an increasing usage of frozen dough for bread production. Yeasts under such stress conditions reduce fermentation performance, compromising product quality (36, 99). Namely, reduced yeast vitality after freezing and thawing the dough causes loss of fermentation capacity and makes it necessary to use higher yeast amounts and longer proofing times (i.e., the resting period after mixing during which fermentation takes place), consequently decreasing product volume (100). As a result, these effects have a great technological and economic impact in the baking industry. Undoubtedly, the ability of baker's yeast to cope with stress conditions is an essential physiological requirement in this industry, which evidently would greatly benefit from the availability of yeast strains with improved freeze resistance.

One of the first stresses encountered by baker's yeast cells during the preparation of frozen dough is the cold-shock, i.e. the decrease in the environment temperature after mixing. This change impairs the correct functioning of both the membrane and the translational apparatus as a result of reduced membrane fluidity and stabilization of the secondary structures of DNA and RNA (101, 102). While positive cold temperatures lead to the synthesis of specific proteins associated with the development of transient phenotypic adaptation (103, 104), freezing is frequently a lethal stress to cells. At sub-zero temperatures, the damaging effects on yeast cells depend on the freezing rate. In the case of rapid freezing, cells are injured by the formation of intracellular ice crystals, which leads to membrane disruption (105). Structural examination of these cells shows discontinuous nuclear membranes, disappearance of vacuoles, and spreading of DNA all over the cells (106). On the other hand, in cells exposed to low freezing rates, osmotic shrinkage of the cells and frozen extracellular water is observed. Therein, cells become exposed to hyperosmotic solutions and try to compensate by moving water across the membranes (107). In this case, cells suffer cellular damage similar to that caused by dehydration. During frozen storage, the growth of ice crystals can further deteriorate the plasma membrane and damage the activity of different cellular systems. Taken together, these findings indicate that freezing is a very complex stress, in which different stress responses appear to play important roles. Therefore, freezing tolerance likely involves different mechanisms working in concert.

Torulaspora delbrueckii and freezing resistance

The high freezing resistance of *T. delbrueckii* PYCC 5321 and 5323 strains was the main characteristic that set them apart as potential candidates for the baking industry. In fact, even after freezing these *T. delbrueckii* cells at –20 ºC for at least 120 days, they retained nearly 100% cell viability, estimated as colony forming units (CFU). In contrast, viability of a *Saccharomyces cerevisiae* commercial baker's yeast was less than 20% viability by the 15th

day (5). The resistance of *T. delbrueckii* was due to its ability to maintain the integrity of the plasma membrane, and it diminished in the presence of cycloheximide in the freezing medium. Membrane integrity, evaluated by flow cytometry with propidium iodide staining, correlated directly with the CFU counts for both yeasts, validating the utilization of flow cytometry to measure viability of yeast cells subject to freezing stress. The ability of *T. delbrueckii* to preserve plasma membrane integrity during freezing does not correlate with the concentration of intracellular trehalose (T_{in}) at the time of freezing, since the values of T_{in} were high and in the same order of magnitude in both *T. delbrueckii* and *S. cerevisiae* strains. In addition, *T. delbrueckii* was able to retain much higher cell viability when subject to a period of fermentation before freezing. Under those circumstances, there was also no correlation between membrane integrity and T_{in}. During the period of prefermentation, the concentration of T_{in} fell at a high rate and to similar values in both strains, consistent with a similar pattern of activation of trehalase (s) by glucose, determined in cellular extracts of *T. delbrueckii* and *S. cerevisiae*. In contrast, the higher capacity to preserve the plasma membrane observed in *T. delbrueckii* seems to be related to a smaller increase of lipid peroxidation during the freeze storage period. These results suggest that the ability of *T. delbrueckii* to avoid damage from oxidative stress in the plasma membrane during freezing can contribute to its freeze-resistant phenotype. In agreement with these results, which seem to imply oxidative damage is involved in the loss of plasma membrane integrity during freezing, pre-treatment of *S. cerevisiae* cells with the radical scavenger N-tert-Butyl-α-phenylnitrone (PBN) led to a reduction in the loss of membrane integrity.

6. Gene disruption in *Torulaspora delbrueckii*

The use of gene deletion mutants is an important tool to decipher the role and physiological relevance of the proteins encoded by different genes. Results from these studies can increase the knowledge of the physiology, biochemistry and molecular biology of an organism. To this purpose, construction and analyses of *Torulaspora delbrueckii* mutant strains are of highest importance; however, the genetic tools available for this yeast are very scarce. Disruption of genes followed by phenotypic analyses is a vital tool for understanding yeast gene function. However, the efficiency of gene disruption is highly variable among species, and is often quite low for non-conventional yeasts. In yeast, gene disruption is usually accomplished by transforming cells with a gene-targeting section (cassette) containing a selectable marker, conferring drug resistance or nutrient autotrophy, flanked by upstream and downstream sequences of the gene of interest (108). These cassettes are frequently generated by PCR, using primers containing both bordering regions of the target gene and part of the selectable marker gene, and subsequently used to transform yeast cells through various transformation protocols, but usually by the lithium acetate TRAFO method (109). There has been widespread use of gene disruption cassettes generated by PCR (110, 111), since very short sequences of yeast DNA flanking the marker gene are sufficient for efficient integration into the *Saccharomyces cerevisiae* genome by homologous recombination (112).

Therefore, it is possible to screen a relatively small number of the transformants growing on the selection media (usually by PCR) and confirm correct integration of the cassette. When recombination efficiency is very low but cells are under selection pressure, cassettes are often integrated in the wrong locus, giving rise to a large number of false positives (cells that are able to grow on the selection media, but not actual disruptants), and the number of true positives may be lower than 1 in a 100. It then becomes necessary to perform a secondary screen, which is only possible when deletion of the gene of interest results in a readily identifiable phenotype. Evidence of the low efficiency of *T. delbrueckii* homologous recombination emerged when disruption of the *TdMAL11* gene was attempted (95); however, in that case it was possible to screen the transformants through phenotypic analysis, since *TdMAL11* null mutants exhibit deficient growth on medium containing maltose as the sole carbon source. However, it was not possible to use the same strategy to disrupt *LGT1*, the first gene identified as coding for a hexose transporter in *T. delbrueckii*, since there was evidence that other hexose transporters exist in this yeast (50). Therefore, we could not screen for potential *T. delbrueckii LGT1* disruptants by searching for a clear-cut phenotype, because loss of *LGT1* might be compensated by the activities of other genes and was thus not expected to impair glucose growth capacity. Therefore, using the conventional method of transforming a PCR-amplified disruption cassette with a short flanking homology (SFH-PCR) (113), we were unable to generate *Δlgt1* mutants. Our strategy was thus first to obtain a TdLGT1-targeting cassette harboring longer homology arms, and then to attempt further optimization of the yield of *LGT1* disruption, by testing how different parameters inherent to the lithium acetate method described by (109) contributed to the transformation efficiency (114). We constructed a cassette with longer flanking regions by inserting a marker-resistance module into the core of the *LGT1* gene, and then used this construction as a template for PCR amplification of a TdLGT1 long-flanking homology disruption cassette (figure 2). Then, to further improve the yield of *LGT1* disruption, we reformulated some parameters of the transformation method. Mainly, after heat shock, cells were pelleted and resuspended in rich medium supplemented with low concentrations of geneticin (100 mg mL^{-1}) and incubated overnight (instead of the usual 4-h recovery time). Finally, cells were plated onto selective YPD plates supplemented with higher concentrations of geneticin (300 mg mL^{-1}), to select against false positives, and incubated for up to 4 days.

Integration efficiency using this strategy was extremely high when compared with the conventional method (no disruptants from using the conventional method and 12/16 using this improved method). We thus concluded that two important modifications were the most relevant for our global strategy: the size of the disruption cassette and the new recovery period of cells during the transformation protocol. As a result, this method demonstrated to be a valuable alternative to the conventional PCR-based gene disruption for the yeast *T. delbrueckii*. This methodology could also be advantageously applied to other non-conventional yeasts, where correct gene disruption with the commonly used short flanking homology cassettes is frequently very low.

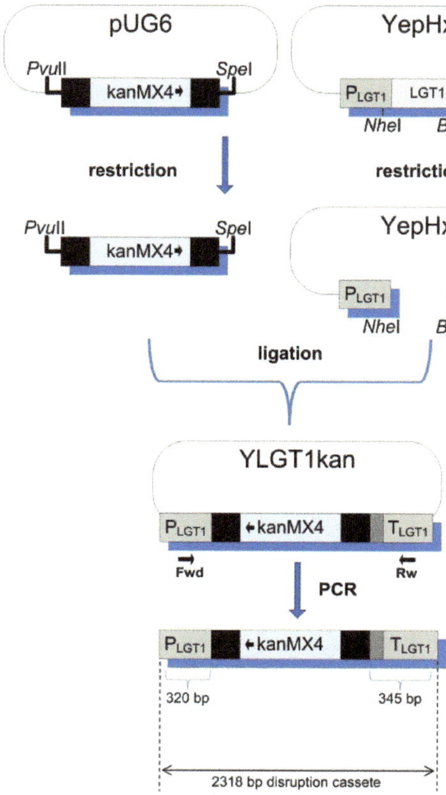

Figure 2. **Schematic illustration of the construction of *Torulaspora delbrueckii* LGT1 disruption** cassete. pUG6 plasmid was digested with *Pvu*II and *Spe*I to release de KanMX4 module, which confers resistance to geneticin (left side of the scheme). In parallel YepHxt6 plasmid (which contains *LGT*1 ORF and part of the gene promoter and terminator regions) was restricted with *Nhe*I and *Bal*I, removing nearly the entire ORF (right side of the scheme). Afterward the PvuII-loxP-KanMX-loxP-SpeI released from pUG6 was cloned into *Bal*I/*Nhe*I restricted YepHxt6, creating YLGT1kan plasmid, the template to generate the *LGT*1 disruption cassette. Using specific primers to *LGT*1 promoter and terminator regions, the disruption cassette (2318 bp) containing the marker module flanked by 320 and 345 bp (5′ and 3′ sides, respectively) *LGT*1 homologous regions was generated by PCR. This cassette was used to transform *T. delbrueckii* with a modified LiAc transformation protocol (described in the text). P_{LGT1} and T_{LGT1} are the promoter and terminator regions, respectively, of *T. delbrueckii LGT1* gene. Restriction enzyme sites and the sizes of the DNA fragments are shown. Arrows at either end of the module represent the oligonucleotides used for PCR.

7. Molecular characterization of *Torulaspora delbrueckii* strains

Before we can exploit the potential of *Torulaspora delbruecckii* in industrial processess, we must be able to identify it and distinguish between strains, using reliable techniques. The accessibility of typing techniques that enable a rapid and accurate differentiation at the strain level is imperative for both wine and baker's yeast users and producers, to assure that the commercialized yeast corresponds to the strain selected originally. Thus, developing practical typing techniques that enable discrimination between *T. delbrueckii* strains is an essential tool for its implementation in the baking and wine industries.

In order to determine the suitability of mitochondrial DNA restriction analysis for *T. delbrueckii* strain differentiation, we selected additional autochthonous yeast strains from the yeast flora present on the home-made corn and rye bread doughs in the northern area of Portugal (unpublished results from our laboratory). We first screened 134 isolates by restriction pattern analysis of both PCR-amplified 5.8S rRNA gene and internally transcribed spacers ITS1 and ITS2, as previously described (115), selecting only *T. delbrueckii* species. The total length of the ITS1-5.8S-ITS2 regions of the 5.8S rRNA gene is identical for all *T. delbrueckii* strains, and for that reason this method cannot discriminate at the strain level (45, 115-117). Three isolates (45A, 45D and 62C) were selected and placed in the CBMA yeast culture collection, Department of Biology, University of Minho, Braga-Portugal. To discriminate between *T. delbrueckii* strains, both mitochondrial DNA restriction fragment length polymorphism and pulsed-field gel electrophoresis (PFGE) were applied to the three selected isolates, to *T. delbrueckii* PYCC 5321 and PYCC 5323, and to type strain ISA1082 (Portuguese Yeast Culture Collection, Institute Gulbenkian de Ciência, Oeiras –Portugal) for a comparative pattern.

Mitochondrial DNA restriction fragment length polymorphism (mtRFLP) analysis has been widely applied to the characterization of reference and commercial *Saccharomyces cerevisiae* wine yeast strains (118-123), as well as strains belonging to other species (116, 124). Not all the enzymes used in this method detect the same degree of polymorphisms, which depend greatly on the species. mtRFLP using *HinfI* is associated with the detection of a high polymorphism and is a widely used genetic marker to distinguish *S. cerevisiae* wine strains (118, 122, 125, 126). On the other hand, GC clusters of the mitochondrial genome are the main source of the polymorphisms, and a large portion of these contains restriction sites for *Hae*lll.

For mtRFLP, DNA was isolated from yeast cells grown in YPD and digested with *HinfI* or *Hae*III restriction enzymes. Restriction fragments were separated in horizontal agarose gels (figure 3). mtRFLP's of the six strains using *HinfI* or *Hae*lll resulted in two distinct profiles, with slight variability. The major difference was found in the upper bands, where the resolution is better (figure 3 arrows). Apart from these bands, the pattern of the profiles provided by each enzyme is identical, indicating that these strains are genetically very closely related. Restriction with *HinfI* resulted in one profile including *T. delbrueckii* ISA1082 (type strain), *T. delbrueckii* PYCC 5321, 45A and 45D, and in another profile including *T. delbrueckii* PYCC 5323 and 62C strains (Fig.3A). Restriction analysis with *Hae*III resulted in a profile including *T. delbrueckii* PYCC 5321 and 45A, and a second profile that includes *T. delbrueckii* ISA1082, *T. delbrueckii* PYCC 5323, 45D, and 62C strains (figure 3B). This method is therefore not suitable to discriminate between *T. delbrueckii* strains.

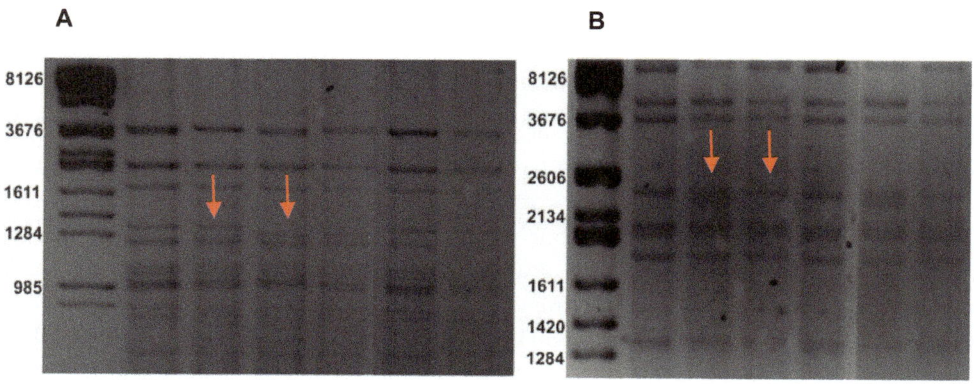

Figure 3. Mitochondrial DNA restriction profiles of *T. delbrueckii* strains obtained with the (A) *Hinf*I and (B) *Hae*III restriction endonucleases in 1.5% agarose gel (unpublished results from our laboratory). Lane 1 – Molecular Marker Lambda DNA/Eco47I (*Ava*II) from Fermentas; Lane 2 – *T. delbrueckii* ISA1082 (reference strain); Lane 3 – *T. delbrueckii* PYCC 5321; Lane 4 – *T. delbrueckii* PYCC 5323; Lane 5 – 62C; Lane 6 – 45A; Lane 7 – 45D. Arrows indicate main differences between profiles I and II. Yeast cells were cultivated in 1 ml YPD (1% yeast extract, 2% peptone, 2% glucose) for 24 h at 30 ºC and 160 r.p.m and DNA isolation was performed as previously described (126). Digestion was carried out with *Hinf*I or *Hae*III restriction enzymes overnight at 37 ºC, in a final volume of 20 μl as previously described (123). Restriction fragments were separated in horizontal 1.5% agarose gels run in 0.5X TBE buffer at 60 V for six hours and visualized in a UV transilluminator (Eagleeye II Image Acquisition System, Stratagene, La Jolla, CA) after ethidium bromide staining.

Karyotype analysis is a highly efficient technique to differentiate strains of *S. cerevisiae*, and was applied by numerous authors to characterize reference and commercial yeasts belonging to different species (118, 120, 122, 123, 127). The electrophoretic karyotypes of the strains under study were therefore also compared. Intact DNA for pulsed field gel electrophoresis (PFGE) was prepared in plugs as previously described (128) and PFGE was run in a CHEF-DRII Chiller System (Bio-Rad, Hercules, CA). Under the conditions used, six chromosome bands were detected in all the strains, which is in agreement with previous studies indicating *T. delbrueckii* has six chromosomes (129). PFGE gel electrophoresis revealed that the chromosomal DNA banding profiles of the strains differ substantially (figure 4), and six different karyotypes could be defined on the basis of the size of putative chromosomes, thereby allowing the discrimination of 4 strains that were not indistinguishable by mtRFLP.

The different karyotypes of the six strains are consistent with their different phenotypes. Indeed, *T. delbrueckii* PYCC 5321, PYCC 5323, and *T. delbrueckii* ISA1082 (type strain) have already been established as different strains (4) and several physiological and biochemical studies of the other three isolates indicated they also correspond to different strains (our unpublished results). Contrary to reports of molecular typing of other yeasts (123), where both methods allowed discriminating strains in a similar manner, our results show that karyotyping analysis displayed a much higher discriminative power than mtRFLP for *T. delbrueckii* strains. A reasonable explanation for this difference may be the need for higher stability of an intact mitochondrial genome in this species than in *S. cerevisiae*. For instance, we have already shown that the relative contribution of respiration to sugar catabolism is higher in *T. delbrueckii* than in *S. cerevisiae* (7). Although mitochondrial genomes contain a very similar set of genes common to all organisms, mtDNA molecules among species are extremely variable in size and organization (130). Furthermore, the stability of the mitochondrial genome can be evaluated by the ability to form petite mutants. *S. cerevisiae* spontaneously produces these mutants, which are deficient in the capacity to respire aerobically. The petite phenotype is correlated with gross alterations and extensive deletions or loss of mtDNA (131, 132). On the contrary, *T. delbrueckii* is a petite-negative species, as it doesn't have the ability to form these respiratory mutants even after prolonged treatment with ethidium bromide (133-135). The high resolutive capability of the CHEF technique allowed us to differentiate between strains isolated from the same environment and that could not be distinguished by mtRFLP. These results underline this technique as a powerful tool for *T. delbrueckii* strain differentiation, although there are some factors that limit its applicability, since it is complex and time-consuming and not suitable as a routine technique for strain identification.

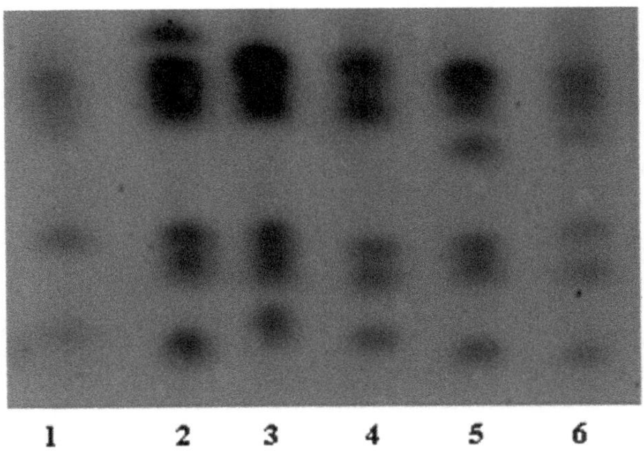

Figure 4. Electrophoretic karyotype comparison of *T. delbrueckii* strains (unpublished results from our laboratory). Lane 1- *T. delbrueckii* ISA1082 (reference strain): Lane 2 - *T. delbrueckii* PYCC 5321; Lane 3 - *T. delbrueckii* PYCC 5323; Lane 4 - 62C; Lane 5 - 45A. Lane 6 - 45D. DNA for pulsed field gel electrophoresis (PFGE) was prepared in plugs as previously described (128). PFGE was run in a CHEF-DRII Chiller System (Bio-Rad, Hercules, CA). PFGE gels were run in 0.5% Tris borate-EDTA buffer at 12 °C with an angle of 120° with the following voltage and switch times: 480s → 900s, 3 v/cm for 10 hours; 240s → 480s, 3 v/cm for 15 hours; 120s → 240 s, 3 v/cm for 15 hours; 90 s, 6 v/cm for 10 hours and 60 s, 6 v/cm for 5 hours. Thereafter, gels were stained with 0.8% ethidium bromide for 45 min and de-stained for 20 min. Gels were visualized under UV light and analyzed using the EagleEye II Image Acquisition System (Stratagene, La Jolla, CA).

In summary, the karyotyping profiles and RFLP's of mitochondrial DNA of the *T. delbrueckii* strains PYCC 5321 and PYCC 5323 were clearly different. These data corroborate and complement the results obtained in the past by the classical biochemical methodology (6), representing an update to the understanding of *T. delbrueckii* populations present in bread doughs. Furthermore, the availability of functional typing tools that enable differentiation at the strain level is extremely important to the bread and wine industries, to assure traceability of the selected strains.

8. Conclusion

The biotechnological interest in *Torulaspora delbrueckii* has increased in recent years due to its particularly high freezing and osmotic tolerance (4, 5, 46). These features made this yeast species a candidate of potential value for the baking industry. However, the existing knowledge on this yeast is still far from the vast knowledge on the traditional baker's yeast *Saccharomyces cerevisiae*. Therefore, studies have been developed to gain insight into the physiology, biochemistry, and molecular genetics of *T. delbrueckii*.

While two of the most important traits for large-scale baker's yeast production are its growth rate and biomass yield on sucrose, its leavening ability depends mainly on its capacity to ferment maltose. The pattern of sugar utilization and regulation also determines the yeast capacity to rapidly adapt when changing from sucrose-rich growth medium to the dough. Physiological and biochemical studies of *T. delbrueckii* in batch cultures with the sugars present in molasses and in bread dough, both alone and in mixtures, showed that *T. delbrueckii* behaves very similarly to *S. cerevisiae* with respect to sugar utilization and regulation patterns. However, this yeast modulates respiratory metabolism under aerobic conditions more efficiently, an asset for large-scale production of the yeast. Furthermore, comparative analysis of specific sugar consumption rates and transport capacities suggested that it is the transport step that limits both glucose and maltose metabolism.

So far, only one glucose transporter has been identified in *T. delbrueckii*, the low-affinity glucose transporter *LGT1* (50). Southern blot analysis of the *T. delbrueckii* genome revealed the existence of several genes with high similarity to *LGT1*, suggesting there are several hexose transporters in this yeast, which hampered disruption of the *LGT1* gene. The existence of several hexose transporters had first been suggested by the isolation of several plasmids from a genomic library of this strain that could complement the glucose growth defect of the *S. cerevisiae* hexose transport-null mutant (50). Despite the phylogenetic closeness of *T. delbrueckii* and *S. cerevisiae*, the differences observed between the two species show that the behavior or even the methods that can be applied to the former yeast cannot always be inferred from those of *S. cerevisiae*. For instance, when we attempted to disrupt the *T. delbrueckii LGT1* gene, the current methods used for *S. cerevisiae* were not suitable, and an optimized disruption method had to be developed.

In modern food technology, traceability is a crucial requirement, and thus establishing a rapid method to discriminate between *T. delbrueckii* strains is of upmost relevance. This technique would enable correct identification of the inoculated strain from the remaining yeast flora present in the bread dough. In the last years, several methodologies of typing based on DNA patterns have been developed which allowed discriminating closely related yeast strains. In this chapter, two different genetic fingerprinting techniques (karyotype analysis and mtDNA restriction analysis) were presented for detailed genotyping of *T. delbrueckii* strains. Mitochondrial DNA restriction analysis was not a good technique to differentiate among *T. delbrueckii* strains isolated from the same ecosystem and genetically very closely related. Chromosome separation by pulsed-field electrophoresis revealed considerable variability in the chromosomal constitution of the strains studied, and turned out to be a useful method to discriminate among *T. delbrueckii* strains. However, this method of chromosome karyotyping may be too complex, laborious, and time-consuming for the analysis of numerous yeast isolates, in contrast with mtDNA restriction analysis.

Nowadays, yeast strains used in bread industry are involved in large-scale processes and hence are exposed to more extreme stress conditions. On the other hand, development of new products and more versatile processes also require yeast strains with new traits. This chapter aimed to highlight some of these emergent problems/needs in the wine and bread-making industries, including selecting, characterizing, and constructing resistant yeast strains, and strains with important qualities for application in the baking and wine industries, as is the case of some studied strains of *T. delbruekii.*

Despite the accomplishments reported in this chapter, many important questions remain to be answered regarding sugar transporters and freezing resistance in *T. delbrueckii.* How many hexose transporters are present in *T. delbrueckii*? What are their affinities and regulation? Is *T. delbrueckii* similar to *K. lactis,* as speculated by Alves-Araújo (50), based on comparison of sequencing data and regulatory studies of *LGT1* expression? Or is this yeast more comparable to *S. cerevisiae* as is suggested by their similar sugar utilizations patterns (7)? Evidently, it would be important to continue the characterization of *T. delbrueckii* strains, as their biotechnological potential has already been established (5, 7, 46). It is clear that answers to these questions may only arise from future studies. Characterization of *T. delbrueckii* at the different levels will narrow the gap towards its industrial exploitation and increase knowledge on the so-called non-conventional yeast species.

Author details

Andreia Pacheco, Susana Chaves and Judite Almeida
Molecular and Environmental Research Centre (CBMA)/Department of Biology, University of Minho, Braga, Portugal

Júlia Santos and Cecília Leão
Life and Health Sciences Research Institute (ICVS), School of Health Sciences, University of Minho, Braga, Portugal
ICVS/3B's - PT Government Associate Laboratory, Braga/Guimarães, Portugal

Maria João Sousa
Corresponding Author
Molecular and Environmental Research Centre (CBMA)/Department of Biology, University of Minho, Braga, Portugal

9. References

[1] Rainieri S, Pretorius IS. Selection and improvement of wine yeasts. Ann Microbiol. 2000;50(1):15-31.

[2] Schuller D, Casal M. The use of genetically modified *Saccharomyces cerevisiae* strains in the wine industry. Applied microbiology and biotechnology. 2005;68(3):292-304.

[3] Randez-Gil F AJ, Codón A, Rincón AM, Estruch F & Prieto JA. Baker's yeast: challenges and future aspects. Functional Genetics of Industrial Yeasts. de Winde JH ed2003.

[4] Almeida MJ, Pais C. Leavening ability and freeze tolerance of yeasts isolated from traditional corn and rye bread doughs. Applied and environmental microbiology. 1996;62(12):4401-4.

[5] Alves-Araujo C, Almeida MJ, Sousa MJ, Leão C. Freeze tolerance of the yeast Torulaspora delbrueckii: cellular and biochemical basis. FEMS microbiology letters. 2004;240(1):7-14.

[6] Almeida MJ, Pais C. Characterization of yeast population from traditional corn and rye bread doughs. Letters in applied microbiology. 1996;23:154-8.

[7] Alves-Araujo C, Pacheco A, Almeida MJ, Spencer-Martins I, Leao C, Sousa MJ. Sugar utilization patterns and respiro-fermentative metabolism in the baker's yeast Torulaspora delbrueckii. Microbiology. 2007;153(Pt 3):898-904.

[8] Mercado L, Dalcero A, Masuelli R, Combina M. Diversity of *Saccharomyces* strains on grapes and winery surfaces: analysis of their contribution to fermentative flora of Malbec wine from Mendoza (Argentina) during two consecutive years. Food microbiology. 2007;24(4):403-12.

[9] Mortimer R, Polsinelli M. On the origins of wine yeast. Research in microbiology. 1999;150(3):199-204.

[10] Lafon-Lafourcade S, Geneix C, Ribereau-Gayon P. Inhibition of alcoholic fermentation of grape must by Fatty acids produced by yeasts and their elimination by yeast ghosts. Applied and environmental microbiology. 1984;47(6):1246-9.

[11] Kunkee RE, Goswell RW. Table wines. Rose AH, editor: Acad. Press. Inc. (London); 1977.

[12] Pretorius IS. Tailoring wine yeast for the new millennium: novel approaches to the ancient art of winemaking. Yeast. 2000;16(8):675-729.

[13] Esteve-Zarzoso B, Manzanares P, Ramon D, Querol A. The role of non-Saccharomyces yeasts in industrial winemaking. International microbiology : the official journal of the Spanish Society for Microbiology. 1998;1(2):143-8.

[14] Bisson LF. Biotechnology of wine yeast. Food Biotechnol. 2004;18(1):63-96.

[15] Rainieri S, Zambonelli C, Tini V, Castellari L, Giudici P. The enological traits of thermotolerant *Saccharomyces* strains. Am J Enol Viticult. 1998;49(3):319-24.

[16] Sousa MJ, Teixeira JA, Mota M. Malo-alcoholic fermentation: the influence of operating conditions on the kinetics of deacidification. Journal of Wine Reseach. 1991;2:115-24.

[17] Sousa MJ, Teixeira JA, M. M. Must deacidification with na induced flocculant yeast strain of *S. pombe*. Appl Microbiol Biotechnol 1993;39:189-93.

[18] Taillandier P, Riba JP, Strehaiano P. Malate utilization by *S. pombe* Biotechnology Letters. 1988;10:469-72.
[19] Ciani M, Maccarelli F. Oenological properties of non-Saccharomyces yeasts associated with wine-making. World J Microb Biot. 1998;14(2):199-203.
[20] Bisson LF, Waterhouse AL, Ebeler SE, Walker MA, Lapsley JT. The present and future of the international wine industry. Nature. 2002;418(6898):696-9.
[21] Mills DA, Johannsen EA, Cocolin L. Yeast diversity and persistence in botrytis-affected wine fermentations. Applied and environmental microbiology. 2002;68(10):4884-93.
[22] Rankine BC, Lloyd B. Quantitative assessment of 9 dominance of added yeast in wine fermentations. J Sci Food Agric. 1963;14:793-8.
[23] Kunkee RE, Amerine MA. Yeasts in winemaking: Acad. Press. Inc. (London),; 1970. 5-72 p.
[24] Sommer P, Stolpe E, Kramp B, Heinemeyer J. Dried Starter Cultures of Non-*Saccharomyces* Yeasts for Alcoholic Fermentation: Impact of *Kluyveromyces thermotolerans* or *Torulaspora delbrueckii* with *Saccharomyces cerevisiae* on Aroma and Flavor Development in Wine. ASEV 54th Annual Meeting. 2003:83.
[25] Herraiz T, Reglero, G., Herraiz, M., Martin-Alvarez, P.J., Cabezudo, M. The influence of the yeast and type of culture on the volatile composition of wine fermented without sulphur dioxide. Am J Enol Viticult. 1990;41:313-8.
[26] Ciani M, Picciotti, G. The growth kinetics and fermentation behaviour of some non-Saccharomyces yeasts associated with wine-making. Biotechnology Letters 1995;17:1247-50.
[27] Ciani M, Ferraro L. Combined use of immobilized *Candida stellata* cells and Saccharomyces cerevisiae to improve the quality of wines. Journal of applied microbiology. 1998;85(2):247-54.
[28] Romano P, Fiore C, Paraggio M, Caruso M, Capece A. Function of yeast species and strains in wine flavour. International journal of food microbiology. 2003;86(1-2):169-80.
[29] Fleet GH. Yeast interactions and wine flavour. International journal of food microbiology. 2003;86(1-2):11-22.
[30] Martinez J, Toledano, F., Millan, C. Development of alcoholic fermentation in non-sterile musts from Pedro Ximenez grapes inoculated with pure cultures of selected yeasts. Food microbiology. 1990;7:217-25.
[31] Bely M, Stoeckle P, Masneuf-Pomarede I, Dubourdieu D. Impact of mixed Torulaspora delbrueckii-Saccharomyces cerevisiae culture on high-sugar fermentation. International journal of food microbiology. 2008;122(3):312-20.
[32] Ciani M, Beco L, Comitini F. Fermentation behaviour and metabolic interactions of multistarter wine yeast fermentations. International journal of food microbiology. 2006;108(2):239-45.
[33] Panjai L, Ongthip K, Chomsri N. Complex fruit wine produced from dual culture fermentation of pineapple juice with *torulaspora delbrueckii* and *saccharomyces cerevisiae.* International Conference on the Role of Universities in Hands-On Education; Rajamangala University of Technology Lanna, Chiang-Mai, Thailand 2009.

[34] Hernandez-Lopez MJ, Randez-Gil F, Prieto JA. Hog1 mitogen-activated protein kinase plays conserved and distinct roles in the osmotolerant yeast *Torulaspora delbrueckii*. Eukaryotic cell. 2006;5(8):1410-9.

[35] Santos J, Sousa MJ, Cardoso H, Inacio J, Silva S, Spencer-Martins I, et al. Ethanol tolerance of sugar transport, and the rectification of stuck wine fermentations. Microbiology. 2008;154(Pt 2):422-30.

[36] Attfield PV. Stress tolerance: the key to effective strains of industrial baker's yeast. Nature biotechnology. 1997;15(13):1351-7.

[37] Randez-Gil F, Sanz P, Prieto JA. Engineering baker's yeast: room for improvement. Trends Biotechnol. 1999;17(6):237-44.

[38] Benitez B, Gasent-Ramirez, J. M., Castrejon, F. & Codon, A. C. Development of new strains for the food industry. Biotechnol Prog. 1996;12:149–63.

[39] Sinda E., Parkkinen E, editors. Problems with Molasses in the Yeast Industry.: Helsinki: Kauppakirjapainooy.; 1979.

[40] Vaughan-Martini A, Martini A. *Saccharomyces* Meyen ex Reess In: Kurtzman CP, Fell JW, editors. The Yeasts, a Taxonomic Study. 4 ed. Amsterdam: Elsevier; 1998.

[41] de Winde JH. Functional Genetics of Industrial Yeasts.2003.

[42] Mormeneo S, Sentandreu R. Regulation of invertase synthesis by glucose in *Saccharomyces cerevisiae*. Journal of bacteriology. 1982;152(1):14-8.

[43] Attfield PV, Kletsas S. Hyperosmotic stress response by strains of bakers' yeasts in high sugar concentration medium. Lett Appl Microbiol. 2000;31(4):323-7.

[44] Hernandez-Lopez MJ, Pallotti C, Andreu P, Aguilera J, Prieto JA, Randez-Gil F. Characterization of a *Torulaspora delbrueckii* diploid strain with optimized performance in sweet and frozen sweet dough. International journal of food microbiology. 2007;116(1):103-10.

[45] Valmorri S, Tofalo R, Settanni L, Corsetti A, Suzzi G. Yeast microbiota associated with spontaneous sourdough fermentations in the production of traditional wheat sourdough breads of the Abruzzo region (Italy). Antonie van Leeuwenhoek. 2010;97(2):119-29.

[46] Hernandez-Lopez MJ, Prieto JA, Randez-Gil F. Osmotolerance and leavening ability in sweet and frozen sweet dough. Comparative analysis between *Torulaspora delbrueckii* and *Saccharomyces cerevisiae* baker's yeast strains. Antonie van Leeuwenhoek. 2003;84(2):125-34.

[47] Winde J. Functional Genetics of Industrial Yeasts. Hohman S, editor: Springer; 2003. 367 p.

[48] Kurtzman CP. Torulaspora Lindner. The Yeasts: A Taxonomic Study. 4th ed: Elsevier, Amsterdam, the Netherlands; 1998.

[49] Sasaki T, Ohshima Y. Induction and Characterization of Artificial Diploids from the Haploid Yeast Torulaspora delbrueckii. Appl Environ Microbiol. 1987;53(7):1504-11.

[50] Alves-Araújo C, Hernandez-Lopez MJ, Prieto JA, Randez-Gil F, Sousa MJ. Isolation and characterization of the LGT1 gene encoding a low-affinity glucose transporter from Torulaspora delbrueckii. Yeast. 2005;22(3):165-75.

[51] Diderich JA, Schepper M, van Hoek P, Luttik MA, van Dijken JP, Pronk JT, et al. Glucose uptake kinetics and transcription of *HXT* genes in chemostat cultures of Saccharomyces cerevisiae. The Journal of biological chemistry. 1999;274(22):15350-9.

[52] Hernandez-Lopez MJ, Prieto JA, Randez-Gil F. Isolation and characterization of the carbon catabolite-derepressing protein kinase Snf1 from the stress tolerant yeast *Torulaspora delbrueckii*. Yeast. 2010;27(12):1061-9.

[53] Needleman R. Control of maltase synthesis in yeast. Molecular microbiology. 1991;5(9):2079-84.

[54] Gancedo JM. Yeast carbon catabolite repression. Microbiology and molecular biology reviews : MMBR. 1998;62(2):334-61.

[55] Baldwin SA, Henderson PJ. Homologies between sugar transporters from eukaryotes and prokaryotes. Annu Rev Physiol. 1989;51:459-71.

[56] Henderson PJ, Maiden MC. Homologous sugar transport proteins in *Escherichia coli* and their relatives in both prokaryotes and eukaryotes. Philos Trans R Soc Lond B Biol Sci. 1990;326(1236):391-410.

[57] Maiden MC, Davis EO, Baldwin SA, Moore DC, Henderson PJ. Mammalian and bacterial sugar transport proteins are homologous. Nature. 1987;325(6105):641-3.

[58] Vardy E, Arkin IT, Gottschalk KE, Kaback HR, Schuldiner S. Structural conservation in the major facilitator superfamily as revealed by comparative modeling. Protein Sci. 2004;13(7):1832-40.

[59] Pao SS, Paulsen IT, Saier MH, Jr. Major facilitator superfamily. Microbiol Mol Biol Rev. 1998;62(1):1-34.

[60] Kruckeberg AL. The hexose transporter family of Saccharomyces cerevisiae. Arch Microbiol. 1996;166(5):283-92.

[61] Saier MH, Jr. Families of transmembrane sugar transport proteins. Mol Microbiol. 2000;35(4):699-710.

[62] Lodish H, Baltimore, D., Berk, A. Molecular Cell Biology. 3 ed. books SA, editor. New York: W.H. Freeman & Co.; 1995.

[63] Postma E, Kuiper A, Tomasouw WF, Scheffers WA, van Dijken JP. Competition for glucose between the yeasts *Saccharomyces cerevisiae* and Candida utilis. Appl Environ Microbiol. 1989;55(12):3214-20.

[64] van Urk H, Postma E, Scheffers WA, van Dijken JP. Glucose transport in crabtree-positive and crabtree-negative yeasts. J Gen Microbiol. 1989;135(9):2399-406.

[65] Boles E, Hollenberg CP. The molecular genetics of hexose transport in yeasts. FEMS Microbiol Rev. 1997;21(1):85-111.

[66] Van Belle D, Andre B. A genomic view of yeast membrane transporters. Curr Opin Cell Biol. 2001;13(4):389-98.

[67] Reifenberger E, Freidel K, Ciriacy M. Identification of novel HXT genes in Saccharomyces cerevisiae reveals the impact of individual hexose transporters on glycolytic flux. Mol Microbiol. 1995;16(1):157-67.

[68] Nehlin JO, Carlberg M, Ronne H. Yeast galactose permease is related to yeast and mammalian glucose transporters. Gene. 1989;85(2):313-9.

[69] Szkutnicka K, Tschopp JF, Andrews L, Cirillo VP. Sequence and structure of the yeast galactose transporter. J Bacteriol. 1989;171(8):4486-93.
[70] Ozcan S, Johnston M. Function and regulation of yeast hexose transporters. Microbiology and molecular biology reviews : MMBR. 1999;63(3):554-69.
[71] Wieczorke R, Krampe S, Weierstall T, Freidel K, Hollenberg CP, Boles E. Concurrent knock-out of at least 20 transporter genes is required to block uptake of hexoses in *Saccharomyces cerevisiae*. FEBS letters. 1999;464(3):123-8.
[72] Reifenberger E, Boles E, Ciriacy M. Kinetic characterization of individual hexose transporters of Saccharomyces cerevisiae and their relation to the triggering mechanisms of glucose repression. Eur J Biochem. 1997;245(2):324-33.
[73] Meijer MM, Boonstra J, Verkleij AJ, Verrips CT. Glucose repression in Saccharomyces cerevisiae is related to the glucose concentration rather than the glucose flux. J Biol Chem. 1998;273(37):24102-7.
[74] De Hertogh B, Hancy F, Goffeau A, Baret PV. Emergence of species-specific transporters during evolution of the hemiascomycete phylum. Genetics. 2006;172(2):771-81.
[75] Varma A, Singh BB, Karnani N, Lichtenberg-Frate H, Hofer M, Magee BB, et al. Molecular cloning and functional characterisation of a glucose transporter, CaHGT1, of *Candida albicans*. FEMS Microbiol Lett. 2000;182(1):15-21.
[76] Fan J, Chaturvedi V, Shen SH. Identification and phylogenetic analysis of a glucose transporter gene family from the human pathogenic yeast Candida albicans. J Mol Evol. 2002;55(3):336-46.
[77] Arnaud MB, Costanzo MC, Skrzypek MS, Shah P, Binkley G, Lane C, et al. Sequence resources at the Candida Genome Database. Nucleic Acids Res. 2007;35(Database issue):D452-6.
[78] Heiland S, Radovanovic N, Hofer M, Winderickx J, Lichtenberg H. Multiple hexose transporters of *Schizosaccharomyces pombe*. J Bacteriol. 2000;182(8):2153-62.
[79] Weierstall T, Hollenberg CP, Boles E. Cloning and characterization of three genes (SUT1-3) encoding glucose transporters of the yeast *Pichia stipitis*. Mol Microbiol. 1999;31(3):871-83.
[80] Jeffries TW, Grigoriev IV, Grimwood J, Laplaza JM, Aerts A, Salamov A, et al. Genome sequence of the lignocellulose-bioconverting and xylose-fermenting yeast *Pichia stipitis*. Nat Biotechnol. 2007;25(3):319-26.
[81] Stasyk OG, Maidan MM, Stasyk OV, Van Dijck P, Thevelein JM, Sibirny AA. Identification of hexose transporter-like sensor *HXS1* and functional hexose transporter *HXT1* in the methylotrophic yeast *Hansenula polymorpha*. Eukaryotic cell. 2008;7(4):735-46.
[82] Wei H, Vienken K, Weber R, Bunting S, Requena N, Fischer R. A putative high affinity hexose transporter, hxtA, of *Aspergillus nidulans* is induced in vegetative hyphae upon starvation and in ascogenous hyphae during cleistothecium formation. Fungal genetics and biology : FG & B. 2004;41(2):148-56.

[83] Gonçalves P, Rodrigues de Sousa H, Spencer-Martins I. FSY1, a novel gene encoding a specific fructose/H(+) symporter in the type strain of *Saccharomyces carlsbergensis.* J Bacteriol. 2000;182(19):5628-30.

[84] Pina C, Goncalves P, Prista C, Loureiro-Dias MC. Ffz1, a new transporter specific for fructose from *Zygosaccharomyces bailii.* Microbiology. 2004;150(Pt 7):2429-33.

[85] Attfield AV, Bell PJ. Genetics and classical genetic manipulations of industrial yeasts2003. 17–55 p.

[86] Mortimer RK, Contopoulou CR, King JS. Genetic and physical maps of Saccharomyces cerevisiae, Edition 11. Yeast. 1992;8(10):817-902.

[87] Volckaert G, Voet M, Robben J. Sequence analysis of a near-subtelomeric 35.4 kb DNA segment on the right arm of chromosome VII from *Saccharomyces cerevisiae* carrying the *MAL1* locus reveals 15 complete open reading frames, including *ZUO1, BGL2* and *BIO2* genes and an ABC transporter gene. Yeast. 1997;13(3):251-9.

[88] Day RE, Higgins VJ, Rogers PJ, Dawes IW. Characterization of the putative maltose transporters encoded by YDL247w and YJR160c. Yeast. 2002;19(12):1015-27.

[89] Han EK, Cotty F, Sottas C, Jiang H, Michels CA. Characterization of *AGT*1 encoding a general alpha-glucoside transporter from *Saccharomyces.* Molecular microbiology. 1995;17(6):1093-107.

[90] Day RE, Rogers PJ, Dawes IW, Higgins VJ. Molecular analysis of maltotriose transport and utilization by Saccharomyces cerevisiae. Applied and environmental microbiology. 2002;68(11):5326-35.

[91] Michels CA, Needleman RB. The dispersed, repeated family of MAL loci in *Saccharomyces* spp. Journal of bacteriology. 1984;157(3):949-52.

[92] Oda Y, Tonomura K. Detection of maltose fermentation genes in the baking yeast strains of *Saccharomyces cerevisiae.* Letters in applied microbiology. 1996;23(4):266-8.

[93] Hayford AE, Jespersen L. Characterization of *Saccharomyces cerevisiae* strains from spontaneously fermented maize dough by profiles of assimilation, chromosome polymorphism, PCR and MAL genotyping. Journal of applied microbiology. 1999;86(2):284-94.

[94] Bell PJ, Higgins VJ, Attfield PV. Comparison of fermentative capacities of industrial baking and wild-type yeasts of the species Saccharomyces cerevisiae in different sugar media. Letters in applied microbiology. 2001;32(4):224-9.

[95] Alves-Araújo C, Hernandez-Lopez MJ, Sousa MJ, Prieto JA, Randez-Gil F. Cloning and characterization of the MAL11 gene encoding a high-affinity maltose transporter from Torulaspora delbrueckii. FEMS yeast research. 2004;4(4-5):467-76.

[96] Hernandez-Lopez MJ, Prieto JA, Randez-Gil F. Isolation and characterization of the gene *URA3* encoding the orotidine-5'-phosphate decarboxylase from *Torulaspora delbrueckii.* Yeast. 2002;19(16):1431-5.

[97] Sirenko OI, Ni B, Needleman RB. Purification and binding properties of the Mal63p activator of *Saccharomyces cerevisiae.* Current genetics. 1995;27(6):509-16.

[98] Wang J, Sirenko O, Needleman R. Genomic footprinting of Mig1p in the *MAL62* promoter. Binding is dependent upon carbon source and competitive with the Mal63p activator. The Journal of biological chemistry. 1997;272(7):4613-22.

[99] Ivorra C, Perez-Ortin JE, del Olmo M. An inverse correlation between stress resistance and stuck fermentations in wine yeasts. A molecular study. Biotechnology and bioengineering. 1999;64(6):698-708.

[100] Teunissen A, Dumortier F, Gorwa MF, Bauer J, Tanghe A, Loiez A, et al. Isolation and characterization of a freeze-tolerant diploid derivative of an industrial baker's yeast strain and its use in frozen doughs. Appl Environ Microbiol. 2002;68(10):4780-7.

[101] Thieringer HA, Jones PG, Inouye M. Cold shock and adaptation. Bioessays. 1998;20(1):49-57.

[102] Inouye M. Cold-shock response and adaptation. J Mol Microbiol Biotechnol. 1999;1(2):191.

[103] Sahara T, Goda T, Ohgiya S. Comprehensive expression analysis of time-dependent genetic responses in yeast cells to low temperature. J Biol Chem. 2002;277(51):50015-21.

[104] Rodriguez-Vargas S, Estruch F, Randez-Gil F. Gene expression analysis of cold and freeze stress in Baker's yeast. Appl Environ Microbiol. 2002;68(6):3024-30.

[105] Morris GJ, Coulson GE, Clarke KJ. Freezing injury in S. cerevisiae. The effects of growth conditions. Cryobiology. 1988;25:471-2.

[106] Kaul SC, Obuchi K, Komatsu Y. Cold shock response of yeast cells: induction of a 33 kDa protein and protection against freezing injury. Cell Mol Biol (Noisy-le-grand). 1992;38(5-6):553-9.

[107] Wolfe J, Bryant G. Freezing, drying, and/or vitrification of membrane- solute-water systems. Cryobiology. 1999;39(2):103-29.

[108] Rothstein R. Targeting, disruption, replacement, and allele rescue: integrative DNA transformation in yeast. Methods Enzymol. 1991;194:281-301.

[109] Schiestl RH, Gietz RD. High efficiency transformation of intact yeast cells using single stranded nucleic acids as a carrier. Curr Genet. 1989;16(5-6):339-46.

[110] Winzeler EA, Shoemaker DD, Astromoff A, Liang H, Anderson K, Andre B, et al. Functional characterization of the S. cerevisiae genome by gene deletion and parallel analysis. Science. 1999;285(5429):901-6.

[111] Dujon B. European Functional Analysis Network (EUROFAN) and the functional analysis of the Saccharomyces cerevisiae genome. Electrophoresis. 1998;19(4):617-24.

[112] Manivasakam P, Weber SC, McElver J, Schiestl RH. Micro-homology mediated PCR targeting in Saccharomyces cerevisiae. Nucleic Acids Res. 1995;23(14):2799-800.

[113] Wach A, Brachat A, Pohlmann R, Philippsen P. New heterologous modules for classical or PCR-based gene disruptions in Saccharomyces cerevisiae. Yeast. 1994;10(13):1793-808.

[114] Pacheco A, Almeida MJ, Sousa MJ. Improved gene disruption method for *Torulaspora delbrueckii*. FEMS yeast research. 2009;9(1):158-60.

[115] Esteve-Zarzoso B, Belloch C, Uruburu F, Querol A. Identification of yeasts by RFLP analysis of the 5.8S rRNA gene and the two ribosomal internal transcribed spacers. Int J Syst Bacteriol. 1999;49 Pt 1:329-37.

[116] Petersen KM, Moller PL, Jespersen L. DNA typing methods for differentiation of *Debaryomyces hansenii* strains and other yeasts related to surface ripened cheeses. International journal of food microbiology. 2001;69(1-2):11-24.

[117] Guillamon JM, Sabate J, Barrio E, Cano J, Querol A. Rapid identification of wine yeast species based on RFLP analysis of the ribosomal internal transcribed spacer (ITS) region. Arch Microbiol. 1998;169(5):387-92.
[118] Querol A, Barrio E, Huerta T, Ramon D. Molecular monitoring of wine fermentations conducted by active dry yeast strains. Applied and environmental microbiology. 1992;58(9):2948-53.
[119] Guillamon JM, Cano J, Ramon D, Guarro J. Molecular differentiation of *Keratinomyces* (Trichophyton) species. Antonie van Leeuwenhoek. 1996;69(3):223-7.
[120] Fernandez-Espinar MT, Lopez V, Ramon D, Bartra E, Querol A. Study of the authenticity of commercial wine yeast strains by molecular techniques. International journal of food microbiology. 2001;70(1-2):1-10.
[121] Esteve-Zarzoso B, Fernandez-Espinar MT, Querol A. Authentication and identification of *Saccharomyces cerevisiae* 'flor' yeast races involved in sherry ageing. Antonie van Leeuwenhoek. 2004;85(2):151-8.
[122] Schuller D, Pereira L, Alves H, Cambon B, Dequin S, Casal M. Genetic characterization of commercial *Saccharomyces cerevisiae* isolates recovered from vineyard environments. Yeast. 2007;24(8):625-36.
[123] Schuller D, Valero E, Dequin S, Casal M. Survey of molecular methods for the typing of wine yeast strains. FEMS microbiology letters. 2004;231(1):19-26.
[124] Guillamon JM, Sanchez I, Huerta T. Rapid characterization of wild and collection strains of the genus *Zygosaccharomyces* according to mitochondrial DNA patterns. FEMS microbiology letters. 1997;147(2):267-72.
[125] Fernadez-Espinar MT, Esteve-Zarzoso B, Querol A, Barrio E. RFLP analysis of the ribosomal internal transcribed spacers and the 5.8S rRNA gene region of the genus Saccharomyces: a fast method for species identification and the differentiation of flor yeasts. Antonie van Leeuwenhoek. 2000;78(1):87-97.
[126] Lopez V, Querol A, Ramon D, Fernandez-Espinar MT. A simplified procedure to analyse mitochondrial DNA from industrial yeasts. International journal of food microbiology. 2001;68(1-2):75-81.
[127] Petersen KM, Jespersen L. Genetic diversity of the species *Debaryomyces hansenii* and the use of chromosome polymorphism for typing of strains isolated from surface-ripened cheeses. Journal of applied microbiology. 2004;97(1):205-13.
[128] Ribeiro GF, Corte-Real M, Johansson B. Characterization of DNA damage in yeast apoptosis induced by hydrogen peroxide, acetic acid, and hyperosmotic shock. Molecular biology of the cell. 2006;17(10):4584-91.
[129] Oda Y, Tonomura K. Electrophoretic karyotyping of the yeast genus Torulaspora. Lett Appl Microbiol. 1995;21(3):190-3.
[130] Wolf K, Del Giudice L. The variable mitochondrial genome of ascomycetes: organization, mutational alterations, and expression. Adv Genet. 1988;25:185-308.
[131] Ferguson LR, von Borstel RC. Induction of the cytoplasmic 'petite' mutation by chemical and physical agents in Saccharomyces cerevisiae. Mutat Res. 1992;265(1):103-48.
[132] Bernardi G. The petite mutation in yeast. Trends Biochem Sci 1979;4:197-201.

[133] Fekete V, Cierna M, Polakova S, Piskur J, Sulo P. Transition of the ability to generate petites in the *Saccharomyces/Kluyveromyces* complex. FEMS yeast research. 2007;7(8):1237-47.

[134] Spirek M, Soltesova A, Horvath A, Slavikova E, Sulo P. GC clusters and the stability of mitochondrial genomes of *Saccharomyces cerevisiae* and related yeats. Folia microbiologica. 2002;47(3):263-70.

[135] Soltesova A, Spirek M, Horvath A, Sulo P. Mitochondria--tool for taxonomic identification of yeasts from *Saccharomyces* sensu stricto complex. Folia microbiologica. 2000;45(2):99-106.

Permissions

The contributors of this book come from diverse backgrounds, making this book a truly international effort. This book will bring forth new frontiers with its revolutionizing research information and detailed analysis of the nascent developments around the world.

We would like to thank Prof. Dr. Ayman Hafiz Amer Eissa, for lending his expertise to make the book truly unique. He has played a crucial role in the development of this book. Without his invaluable contribution this book wouldn't have been possible. He has made vital efforts to compile up to date information on the varied aspects of this subject to make this book a valuable addition to the collection of many professionals and students.

This book was conceptualized with the vision of imparting up-to-date information and advanced data in this field. To ensure the same, a matchless editorial board was set up. Every individual on the board went through rigorous rounds of assessment to prove their worth. After which they invested a large part of their time researching and compiling the most relevant data for our readers. Conferences and sessions were held from time to time between the editorial board and the contributing authors to present the data in the most comprehensible form. The editorial team has worked tirelessly to provide valuable and valid information to help people across the globe.

Every chapter published in this book has been scrutinized by our experts. Their significance has been extensively debated. The topics covered herein carry significant findings which will fuel the growth of the discipline. They may even be implemented as practical applications or may be referred to as a beginning point for another development. Chapters in this book were first published by InTech; hereby published with permission under the Creative Commons Attribution License or equivalent.

The editorial board has been involved in producing this book since its inception. They have spent rigorous hours researching and exploring the diverse topics which have resulted in the successful publishing of this book. They have passed on their knowledge of decades through this book. To expedite this challenging task, the publisher supported the team at every step. A small team of assistant editors was also appointed to further simplify the editing procedure and attain best results for the readers.

Our editorial team has been hand-picked from every corner of the world. Their multi-ethnicity adds dynamic inputs to the discussions which result in innovative outcomes. These outcomes are then further discussed with the researchers and contributors who give their valuable feedback and opinion regarding the same. The feedback is then collaborated with the researches and they are edited in a comprehensive manner to aid the understanding of the subject.

Apart from the editorial board, the designing team has also invested a significant amount of their time in understanding the subject and creating the most relevant covers. They scrutinized every image to scout for the most suitable representation of the subject and create an appropriate cover for the book.

The publishing team has been involved in this book since its early stages. They were actively engaged in every process, be it collecting the data, connecting with the contributors or procuring relevant information. The team has been an ardent support to the editorial, designing and production team. Their endless efforts to recruit the best for this project, has resulted in the accomplishment of this book. They are a veteran in the field of academics and their pool of knowledge is as vast as their experience in printing. Their expertise and guidance has proved useful at every step. Their uncompromising quality standards have made this book an exceptional effort. Their encouragement from time to time has been an inspiration for everyone.

The publisher and the editorial board hope that this book will prove to be a valuable piece of knowledge for researchers, students, practitioners and scholars across the globe.

List of Contributors

Renato Souza Cruz
Technology Department, State University of de Feira de Santana, Feira de Santana, BA, Brazil

Geany Peruch Camilloto and Ana Clarissa dos Santos Pires
Food Tecnhology Department, Federal University of Viçosa, Viçosa, MG, Brazil

Thawien Wittaya
Prince of Songkla University, Department of Material Product Technology, Thailand

Gary M. Booth
Brigham Young University, Department of Plant and Wildlife Sciences, Provo, Utah, USA

Tory L. Parker
Brigham Young University, Department of Nutrition, Dietetics, and Food Science, Provo, Utah, USA

Christopher M. Lee
Research for Cancer Care Northwest and Gamma Knife of Spokane, Department of Radiation Oncology at the University of Washington, Spokane, WA, USA

Alžbeta Medveďová
Department of Nutrition and Food Quality Assessment, Faculty of Chemical and Food Technology, Slovak University of Technology, Bratislava, Slovak Republic

Ľubomír Valík
Department of Nutrition and Food Quality Assessment, Faculty of Chemical and Food Technology, Slovak University of Technology, Bratislava, Slovak Republic

Yoshimasa Sagane, Ken Inui, Shin-Ichiro Miyashita, Keita Miyata, Koichi Niwa and Toshihiro Watanabe
Department of Food and Cosmetic Science, Faculty of Bioindustry, Tokyo University of Agriculture, Abashiri, Japan

Tomonori Suzuki
Department of Bacteriology, Okayama University Graduate School of Medicine, Dentistry and Pharmaceutical Sciences, Okayama, Japan

Grazina Juodeikiene and Loreta Basinskiene
Department of Food Technology, Kaunas University of Technology, Kaunas, Lithuania

Elena Bartkiene and Paulius Matusevicius
Department of Food Safety and Quality, Lithuanian University of Health Sciences, Kaunas, Lithuania

Vladimir Kendrovski and Dragan Gjorgjev
Medical Faculty, University"St. Cyrile and Methodius", Department for Hygiene and Environmental Health, Skopje, R. Macedonia

Maria Graça Campos and Maria Luísa Costa
Observatory of Herb-Drug Interactions & Faculty of Pharmacy, University of Coimbra, Polo III, Azinhaga de Santa Comba, Coimbra, Portugal
Drug Discovery Group, Center of Pharmaceutical Studies, Laboratory of Pharmacognosy, Faculty of Pharmacy, University of Coimbra, Polo III, Azinhaga de Santa Comba, Coimbra, Portugal

Maged E.A. Mohamed
Department of Agriculture Engineering, Faculty of Agricultural, Minoufiya University, Egypt
Date Palm Research Center, King Faisal University, Saudi Arabia

Ayman H. Amer Eissa
Department of Agriculture Engineering, Faculty of Agricultural, Minoufiya University, Egypt
Department of Agriculture Systems Engineering, College of Agricultural and Food Sciences, King Faisal University, Saudi Arabia

Ayman H. Amer Eissa
Department of Agriculture Engineering, Faculty of Agricultural, Minoufiya University, Egypt
Department of Agriculture Systems Engineering, College of Agricultural and Food Sciences, King Faisal University, Saudi Arabia

Ayman A. Abdel Khalik
Agriculture Engineering Research Institute (AEnRI), Giza, Egypt

Paulo César Stringueta
Department of Food Science and Technology - Federal University of Viçosa, Brazil

Maria da Penha Henriques do Amaral and Miriam Aparecida de Oliveira Pinto
Department of Pharmaceutical Sciences - Federal University of Juiz de Fora, Brazil

Larissa Pereira Brumano
Federal University of Juiz de Fora, Brazil

Mônica Cecília Santana Pereira
Sanitary Surveillance Department of the Municipality of Juiz de Fora, Brazil

Vincenzina Fusco and Grazia Marina Quero
National Research Council, Institute of Sciences of Food Production (CNR-ISPA), Bari, Italy

Andreia Pacheco, Susana Chaves and Judite Almeida
Molecular and Environmental Research Centre (CBMA)/Department of Biology, University of Minho, Braga, Portugal

Júlia Santos and Cecília Leão
Life and Health Sciences Research Institute (ICVS), School of Health Sciences, University of Minho, Braga, Portugal
ICVS/3B's - PT Government Associate Laboratory, Braga/Guimarães, Portugal

Maria João Sousa
Molecular and Environmental Research Centre (CBMA)/Department of Biology, University of Minho, Braga, Portugal